WATER QUALITY INDICES

WATER QUALITY INDICES

TASNEEM ABBASI

S. A. ABBASI

AMSTERDAM • BOSTON • HEIDELBERG • LONDON
NEW YORK • OXFORD • PARIS • SAN DIEGO
SAN FRANCISCO • SINGAPORE • SYDNEY • TOKYO

Elsevier
The Boulevard, Langford Lane, Kidlington, Oxford OX5 1GB, UK
Radarweg 29, PO Box 211, 1000 AE Amsterdam, The Netherlands

British Library Cataloguing in Publication Data
A catalogue record for this book is available from the British Library

Library of Congress Cataloging-in-Publication Data
A catalog record for this book is available from the Library of Congress

ISBN: 978-0-444-63836-6

12 13 14 10 9 8 7 6 5 4 3 2 1

Dedicated to

Neelu Aunty, Sophia Aunty, Rubi Aunty, and the memory of Daada - Daadi

-TA

Aapa, Didi, Rubi Behen, and the memory of Daddy and Mummy

-SAA

Contents

13. The Biotic Indices 219

14. Indices of Biological Integrity or the Multi-metric Indices 249

15. Multivariate Approaches for Bioassessment of Water Quality 337

III

LOOKING BACK, LOOKING AHEAD

16. Water-Quality Indices: Looking Back, Looking Ahead 353

Foreword

Till as late as the beginning of the current millennium, environmental education was largely confined to postgraduate and doctoral programmes, with only a few undergraduate programmes offering specialisation in environmental studies. But, in recent years, environmental education is increasingly featuring not only in collegiate education but preparatory school level as well. The exceedingly desirable consequence of this happening is that lay persons all over the world are developing familiarity with terms such as air quality, water quality and ecorestoration. Weather reports are no longer confined to temperature, humidity and air speeds but are beginning to speak of levels of particulate matter, NO_x and SO_x.

Given the pivotal role of water in supporting and shaping our existence, it is expected that sooner than later everyone would like to know, in quantifiable terms, how good or bad is the water one uses. One would like to know whether the piped water one gets is turning for better or worse and, all aspects considered, whether Brand A of bottled drinking water is *quantifiably* superior to Brand B. Awareness towards the overall water-quality situation of a town, a city, a region or a country will also increase. This emerging scenario would necessitate widespread use of water-quality indices (WQIs) because WQIs are the only medium by which the highly multi-attribute and multivariate concept of water quality can be conveyed to lay persons in the form of a single score.

As has been brought out by the authors in their introductory chapter, WQIs indicate water quality much in the same manner as the Sensex reflects the level of the Mumbai Stock Market and the Dow Jones Index reflects the status of the New York Stock Market. Even a person with no acquaintance with the intricacies of stock prices and economics can get an idea of the state of economy by these index values. Likewise, the simple figure of the per cent by which a share price index has risen or fallen on a day gives an idea of how thousands of different companies have fared *vis a vis* market capitalisation on that day. WQIs will perform the task of 'measuring' water quality and communicating it to the water users in a similar fashion. We will hear sentences like, "the water quality of our river at the water supply station had fallen to 35 last year; now it has improved to 45 but we are trying to improve it beyond 60 so it goes into the 'very good' class". Or a bottled drinking water supplier announcing, "our water always has a score well above 75 on the National Drinking Water Index".

Given the importance of water in the life of every single living being, not to speak of the veritable 'life-and-death' impact it has on human beings, the importance of this book cannot be overemphasised. That it is the first-ever book on the subject makes its publication an important global event. I congratulate *Elsevier* for their initiative in publishing this book and wish it great success.

Prof J. A. K. Tareen
Vice Chancellor,
Pondicherry University

CHAPTER

1

Why Water-Quality Indices

OUTLINE

1.1. INTRODUCTION

1.1.1. Water Quantity and Water Quality

Of all natural resources, water is unarguably the most essential and precious. Life began in water, and life is nurtured with water. There are organisms, such as anaerobes, which can survive without oxygen. But no organism can survive for any length of time without water.

The crucial role of water as the trigger and sustainer of civilizations has been witnessed throughout human history. But, until as late as the 1960s, the overriding interest in water has been *vis a vis* its quantity. Except in manifestly undersirable situations, the available water was automatically deemed utilisable water. Only during the last three decades of the twentieth century the concern for water *quality* has been exceedingly felt so that, by now, water quality has acquired as much importance as water quantity.

What is water quality? This question is immensely more complex than the question: *What is water quantity?*

We can say: this reservoir contains 2 million m^3 of water or the present flow in this river is $15\ m^3\ sec^{-1}$. Expressing water *quantity* is as simple as this.

But how do we express water quality of the same stream? The quality may be good enough for drinking but not suitable for use as a coolant in an industry. It may be good for irrigating some crops but not good for irrigating some other crops. It may be suitable for livestock but not for fish culture. Whereas water *quantity* is determined by a single parameter – the water mass – water quality is a function of anything

Water Quality Indices DOI: 10.1016/B978-0-444-54304-2.00001-4

and everything the water might have picked up during its journey from the clouds to the earth to the water body: in dissolved, colloidal, or suspended form. Given the fact that water is a 'universal solvent', it picks up a lot!

One way to describe the quality of a given water sample is to list out the concentrations of everything that the sample contained. Such a list would be as long as the number of constituents analyzed and that may be anything from the 20-odd common constituents to hundreds! Moreover, such a list will make little sense to anyone except well-trained water-quality experts.

How to compare the quality of different water sources? It can't be done easily by comparing the list of constituents each sample contains. For example, a water sample which contains six components in 5% higher-than-permissible (hence objectionable) levels: pH, hardness, chloride, sulphate, iron and sodium may not be as bad for drinking as another sample with just one constituent – mercury – at 5% higher-than-permissible.

Water-quality indices seek to address this vexing problem.

1.2. WATER-QUALITY INDICES (WQIs)

Water-quality indices aim at giving a single value to the water quality of a source on the basis of one or the other system which translates the list of constituents and their concentrations present in a sample into a single value. One can then compare different samples for quality on the basis of the index value of each sample.

1.2.1. A Novel Idea?

The concept of using an index to represent in a single value the status of several variables is not a novel idea; it has been well-entrenched in economics and commerce (Fisher, 1922; Diewert and Nakamura, 1993). Most countries have their 'consumer price index' in which, on the basis of an integration of the prices of certain commodities, a single value is obtained to determine whether the market is, overall, cheaper or costlier at any given instant compared to any other past instant.

The commodities for such indices are selected on the basis of their 'driver power' – in other words, the 'power' or the 'reach' of the commodity vis a vis influencing the prices of several other commodities. If a shampoo becomes costlier or cheaper, it will not affect the prices of other commodities significantly, whereas any change in the price of cement or petroleum would.

Then we have share-market indices such as the Dow Jones Index of the New York stock exchange and the Sensex of Mumbai's stock exchange. These indices are also composed of the prices of certain shares of high driver power (such as cement). In time these indices have become measures not merely of the stock traded at the respective exchanges but also of the economies of the respective countries. When indices of economically advanced countries such as the Dow Jones Index suffer a slump, it impacts the stock exchanges of most other countries as well, often taking their indices down with it.

Indices have also been used in ecology to represent species richness, evenness, diversity etc. Accordingly, we have the Shannon Index, the Simpson Index, and so on. In numerous other fields – such as of medicine, sociology, process safety, etc – indices are extensively used.

It can be said that indices (the singular of *indices* is *index*) are composite representations of a condition or situation derived from a combination, done in certain ways, of several relevant but noncommensurate observed facts/measurements. The combination leads to a single ordinal number that facilitates understanding and interpretation of the overall import of the facts that have contributed to that number.

Environmental indices – of which water-quality indices form a major component – are used as communication tools by regulatory agencies to describe the 'quality' or 'health' of a specific environmental system (e.g., air, water, soil and sediments) and to evaluate the impact of regulatory policies on various environmental management practices (Song and Kim, 2009; Pusatli et al., 2009; Sadiq et al., 2010). Environmental indices have also been used in life-cycle assessment (Weiss et al., 2007; Khan et al., 2004) and to characterise different types of environmental damages, including global warming potential (Goedkoop and Spriensma, 2000).

1.3. BACK TO WATER-QUALITY INDICES (WQIs)

WQIs may have gained currency during the last 3 decades but the concept in its rudimentary form was first introduced more than 150 years ago – in 1848 – in Germany where the presence or absence of certain organisms in water was used as indicator of the fitness or otherwise of a water source.

Since then various European countries have developed and applied different systems to classify the quality of the waters within their regions. These water classification systems usually are of two types:

1. those concerned with the amount of pollution present and
2. those concerned with living communities of macroscopic or microscopic organisms.

Rather than assigning a numerical value to represent water quality, these classification systems categorised water bodies into one of several pollution classes or levels. By contrast, indices that use a numerical scale to represent gradations in water-quality levels are a recent phenomenon, beginning with Horton's index in 1965, detailed below.

1.4. THE FIRST MODERN WQI: HORTON'S INDEX

Horton (1965) set for himself the following criteria when developing the first-ever modern WQI:

1. The number of variables to be handled by the index should be limited to avoid making the index unwidely.
2. The variables should be of significance in most areas.
3. Only such variables of which reliable data are available, or obtainable, should be included.

Horton selected 10 most commonly measured water-quality variables for his index, including dissolved oxygen (DO), pH, coliforms, specific conductance, alkalinity, and chloride. Specific conductance was intended to serve as an approximate measure of total dissolved solids (TDSs), and carbon chloroform extract (CCE) was included to reflect the influence of organic matter. One of the variables, sewage treatment (percentage of population served), was designed to reflect the effectiveness of abatement activities on the premise that chemical and biological measures of quality are of little significance until substantial progress has been made in eliminating discharges of raw sewage. The index weight ranges from 1 to 4. Notably, Horton's index did not include any toxic chemicals.

The index score is obtained with a linear sum aggregation function. The function consists of the weighted sum of the subindices I_i divided by the sum of the weights W_i and multiplied by two coefficients M_1 and M_2, which reflect temperature and obvious pollution, respectively:

$$QI = \frac{\sum_{i=1}^{n} W_i I_i}{\sum_{i=1} W_i} M_1 M_2$$

Horton's index is easy to compute, even though the coefficients M_1 and M_2 require some tailoring to fit individual situations. The index structure, its weights, and rating scale are highly subjective as they are based on the judgement of the author and a few of his associates.

Horton's pioneering effort has been followed up by several workers who have striven to develop less and less subjective but more and more sensitive and useful water-quality indices.

1.5. MORE ON THE BENEFITS OF WQI

The formulation and use of indices have been strongly advocated by agencies responsible for water supply and control of water pollution. Once the water-quality data have been collected through sampling and analysis, a need arises to translate it into a form that is easily understood. Once the WQIs are developed and applied, they serve as a convenient tool to examine trends, to highlight specific environmental conditions, and to help governmental decision-makers in evaluating the effectiveness of regulatory programmes.

WQIs, of course, are not the only source of information that is brought to bear on water-related decisions. Many other factors are considered besides indices and the monitoring data on which the indices are based.

Indeed, nearly all the purposes for which one monitors water quality – assessment, utilisation, treatment, resource allocation, public information, R&D and environmental planning – are all served by indices as well. In addition, indices make the transfer and utilisation of water-quality data enormously easier and lucid. To wit, water-quality indices help in:

1. Resource allocation
   Indices may be applied in water-related decisions to assist managers in allocating funds and determining priorities.
2. Ranking of allocations
   Indices may be applied to assist in comparing water quality at different locations or geographical areas.
3. Enforcement of standards
   Indices may be applied to specific locations to determine the extent to which legislative standards and existing criteria are being met or exceeded.
4. Trend analysis
   Indices may be applied to water-quality data at different points in time to determine the changes in the quality (degradation or improvement) which have occurred over the period.
5. Public information
   Index score being an easy-to-understand measure of water-quality level, indices can be used to keep the public informed of the overall water quality of any source, or of different alternative sources, on a day-to-day basis just as Sensex score tells in one word whether the stocks, by and large, went up or down.
6. Scientific research
   The inherent quality of an index – which translates a large quantity of data to a single score – is immensely valuable in scientific research, for example, in determining the efficacy of different ecorestoration measures or water-treatment strategies with reference to a water body, the impact of developmental activities on water quality, etc.

1.6. WQIS BASED ON BIOASSESSMENT

On the basis of parameters that are incorporated in an index to judge water quality, the WQIs can be loosely classified into 'indices predominantly based on physcio-chemical characteristics' and 'indices based on bioassessment'. The first modern WQI, the Horton's index, described in Section 1.4, belonged to the first category, so are all the indices described in *Part I* of this book. Just as Horton's index has one parameter (out of 10 chosen by him) which requires bioassessment – coliforms – several other indices described in *Part I* have one or more 'biological' parameters but they are all predominantly based on physico-chemical parameters. In contrast, the WQIs based on bio-assessment, described in *Part II*, are predominantly based on sampling, identification, and enumeration of biological organisms.

References

Diewert, W.E., Nakamura, A.O., 1993. Indices. In: Essays in Index Number Theory, vol. 1. North Holland Press, Amsterdam. 71–104.

Fisher, I., 1922. The Making of Index Numbers. Houghton Mifflin, Boston, MA.

Goedkoop, M., Spriensma, R., 2000. The Eco-Indicator 99-a Damage Oriented Method for Life Cycle Impact Assessment, Methodology Report. http://www.pre.nl

Horton, R.K., 1965. An index number system for rating water quality. Journal of Water Pollution Control Federation. 37 (3), 300–306.

Khan, A.A., Paterson, R., Khan, H., 2004. Modification and application of the Canadian Council of Ministers of the Environment Water Quality Index (CCME WQI) for the communication of drinking water quality data in Newfoundland and Labrador. Water Quality Research Journal of Canada 39 (3), 285–293.

Pusatli, O.T., Camur, M.Z., Yazicigil, Z.H., 2009. Susceptibility indexing method for irrigation water management planning: applications to K. Menderes river basin Turkey. Journal of Environmental Management 90 (1), 341–347.

Sadiq, R., Haji, S.A., Cool, G., Rodriguez, M.J., 2010. Using penalty functions to evaluate aggregation models for environmental indices. Journal of Environmental Management 91 (3), 706–716.

Song, T., Kim, K., 2009. Development of a water quality loading index based on water quality modeling. Journal of Environmental Management 90 (3), 1534–1543.

Weiss, M., Patel, M., Heilmeier, H., Bringezu, S., 2007. Applying distance-to-target weighing methodology to evaluate the environmental performance of bio-based energy, fuels, and materials. Resources, Conservation and Recycling 50, 260–281.

CHAPTER

2

Approaches to WQI Formulation

OUTLINE

2.1. INTRODUCTION

2.1.1. Indices for 'Water Quality' and 'Water Pollution'

Water-quality indices can be formulated in two ways: one in which the index numbers increase with the degree of pollution (increasing scale indices) and the other in which the index numbers decrease with the degree of pollution (decreasing scale indices). One may classify the former as 'water pollution indices' and the latter as 'water-quality indices'. But this difference is essentially cosmetic; 'water quality' is a general term of which 'water pollution' – which indicates 'undesirable water quality' – is a special case.

 DOI: 10.1016/B978-0-444-54304-2.00002-6

2.2. THE COMMON STEPS

The following four steps are most often associated with the development of any WQI; depending on the sophistication being aimed at, additional steps may also be taken:

1. Parameter selection.
2. Transformation of the parameters of different units and dimensions to a common scale.
3. Assignment of weightages to all the parameters.
4. Aggregation of subindices to produce a final index score.

Of these, steps 1, 2 and 4 are essential for all indices. Step 3 is also commonly taken through some indices may be formed without this step as well.

Water-quality indices make it very easy for a lay person to judge whether a water source is usable or not and how one source compares to another, but the development of WQI is by no means an easy task. It, in fact, is fraught with several complications and uncertainties.

As we may see from the following discussion, and from the numerous examples given in the following chapters on how different WQIs have been developed for different needs, it would be clear that a great deal of subjective opinion and judgement is associated with each step, particularly steps 1 and 3. There is no technique or device by which 100% objectivity or accuracy can be achieved in these steps. Even parameter selection through statistical analysis of past data (detailed in Chapter 4), though apparently objective, is fraught with inherent uncertainties and incompleteness.

One can only try to *reduce* subjectivity and inaccuracy by involving large number of experts in collecting opinion, and doing it by well-developed opinion-gathering techniques such as Delphi (Abbasi 1995, Abbasi and Arya 2000).

2.3. PARAMETER SELECTION

As we have elaborated in Chapter 1, a water sample may have hundreds of constituents, including elements in neutral or ionic form (metals, non-metals, metalloids); organics (pesticides, detergents, other organics of industrial or natural origin); anions such as carbonate, bicarbonate, sulphate, nitrate, nitrite etc. It may also have suspended solids which, in turn, constitute a bewildering range of chemicals. It may also have radioactivity, and may have colour and odour. Then it may have pathogenic bacteria, fungi, helminthic cysts etc.

A WQI would become unwieldy if each and every possible constituent is included in the index. Instead, one needs to choose a set of parameters which, together, *reflect* the overall water quality for the given end use. It is an exercise similar to the one involved in the selection of a few dozen or a few hundred shares out of thousands to construct a share market index. To reduce the number of shares to be incorporated in the index, yet keep the index representative of the overall stock market situation, those shares are picked up which have a high 'driver power' — i.e shares of whose movement influences a large number of other shares. In this manner, the share price index is made 'sensitive' to the stock market as a whole even as it is computed on the basis of the prices of only a fraction of the listed shares.

Water-quality indices are also made on the basis of a few parameters chosen for their 'forcing' or 'driver power'. But it is here that subjectivity creeps in. Different experts and end users may have different perceptions of the importance of a parameter *vis a vis* a given end use. For example, a medical expert may perceive water carrying a faint odour but otherwise free from harmful constituents as good. In his/her opinion, odour may be a parameter of very little significance. But to others even the faintest odour in their drinking water may be totally unacceptable. They will advocate that odour must be included as a key parameter in any WQI dealing with drinking water.

Even the water-quality standards, on which much of our decisions on fitness or otherwise of a water source depend, are not common to all countries. Further, as new research brings to light new facts on the beneficial or harmful effects of a constituent, or gives new information on the concentration beyond which a constituent becomes harmful or the concentration below which a constituent *ceases to be helpful*, the standards are continuously revised.

The criteria of 'acceptability' also vary from region to region. In regions well-endowed with water resources – for example the State of Kerala, India – drinking water containing more than 500 mg/L of total dissolved solids (which is the BIS (Bureau of Indian Standards) limit for ideal drinking water) may be considered unfit for drinking because the state may find ample drinking water in an alternative source meeting with the BIS standard. On the other hand, people in arid or semi-arid regions, for example in the Indian states of Rajasthan, Tamil Nadu and elsewhere, routinely drink water with TDS well above 500 mg/L.

Therefore, parameter selection is as fraught with uncertainty and subjectivity as it is crucial to the usefulness of any index. Enormous care, attention, experience, and consensus-gathering skills are required to ensure the most representative parameters are included in a WQI.

In an attempt to reduce the subjectivity in parameter selection, statistical approaches have been attempted, as described in Chapter 4. While in theory such approaches are objective – because they select parameters on the basis of considerations such as frequency of occurrence of different parameters, the number of other parameters to which they seem to correlate, etc – they can lead to erroneous results because correlations can occur by sheer chance and need not have a cause-effect link. For example, data on chloride may strongly correlate with data on abundance of a fish species but the two may, in reality, have no link whatsoever.

2.4. TRANSFORMATION OF THE PARAMETERS OF DIFFERENT UNITS AND DIMENSIONS TO A COMMON SCALE: MAKING SUBINDICES

Different water-quality parameters are expressed in different units. For example, temperature is expressed in degrees celsius or fahrenheit, coliforms in numbers, electrical conductivity in micro-mhos, and most chemicals in milligramme per litre (or microgram per ml). Further, the *ranges* of levels to which different parameters can occur vary greatly from parameter to parameter. For example, dissolved oxygen would rarely be beyond the range 0–12 mg/L but sodium can be in the range 0–1000 mg/L or beyond. Toxic elements such as mercury rarely occur above 1 mg/L level, whereas acidity/alkalinity, hardness, chloride and sulphate nearly always occur at levels above 1 mg/L. Yet again, water containing 10 mg/L of chloride is as fit for drinking as water containing fifteen times higher chloride. But a water sample containing 0.001 mg/L of mercury is acceptable while a water with even twice this concentration of mercury is not.

In other words, different parameters occur in different ranges, are expressed in different units, and have different *behaviour* in terms of concentration–impact relationship. Before an index can be formulated, all this has to be transformed into a single scale – usually beginning with zero and ending at 1. Some index scales have the range 0–100. But this, again, makes only a cosmetic difference.

2.4.1. Developing Subindices

Subindices, one for each parameter selected for the index, are developed so that different parameters, their units and the range of concentrations (from highly acceptable to highly unacceptable) are all transformed onto a single scale.

If we consider a set of n pollutant variables denoted as (x_1, x_2, x_3 ... x_i, x_n), then for each pollutant variable x_i, a subindex I_i is computed using subindex function f_i, (x_i):

$$I_i = F_i(x_i) \tag{2.1}$$

In most indices, different mathematical functions are used to compute different pollutant variables, yielding the subindex functions $f_1(x_1), f_2(x_2) \ldots f_n(x_n)$. Such functions may consist of simple multiplier, or the pollutant variable raised to a power, or some other functional relationship.

Once the subindices are calculated, they usually are aggregated together in a second mathematical step to form the final index:

$$I = g(I_1, I_2 \ldots I_n) \tag{2.2}$$

The aggregation function, (2.2), usually consists either of a summation operation, in which individual subindices are added together, or a multiplication operation, in which a product is formed of some or all the subindices, or some other operation; some of the common aggregation methods are given in Section 6.

The overall process: calculation of subindices and aggregation of subindices to form the index can be illustrated as in Figure 2.1.

2.4.2. Different Types of Subindices

Subindices can be classified as one of four general types:

1. Linear
2. Nonlinear
3. Segmented linear
4. Segmented nonlinear

2.4.2.1. *Linear Function Subindices*

The simplest subindex function is the linear equation:

$$I = \alpha x + \beta \tag{2.3}$$

where I is the subindex, x the pollutant variable and α, β the constants.

With this function, a direct proportion exists between the subindex and the pollutant variable. The linear indices are simple to compute and easy to understand but have limited flexibility.

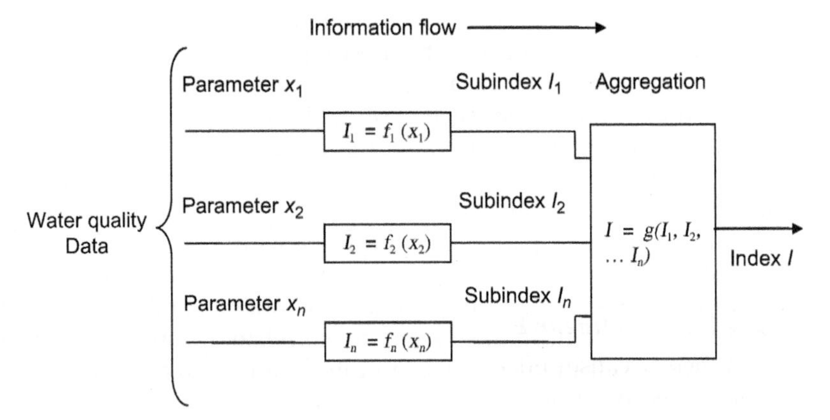

FIGURE 2.1 The index development process.

2.4.2.2. *Segmented Linear Function Subindices*

A segmented linear function consists of two or more straight line segments joined at break points (threshold level). It offers more flexibility than linear function and is especially useful for incorporating administratively recommended limits, such as Bureau of Indian Standards (BIS) limits, WHO limits etc. One of the important types of segmented linear functions is the step function, which exhibits just two states and therefore is called a dichotomous function. The segmented linear function subindices may also consist of a staircase of steps, giving a multiple-state function. For example, Horton (1965) has used subindex functions containing three, four, and five steps. In Horton's dissolved oxygen subindex, one finds $I = 0$ for x less than 10% saturation, $I = 30$ for x between 10% and 30% saturation, and $I = 100$ for x above 70% saturation.

Mathematically, the general form of segmented linear function can be formulated as follows.

Given that x and I coordinates of the break points are represented by (a_1, b_2), (a_2, b_2),..., (a_j, b_j), any segmented linear function with m segments can be presented by the following general equation:

$$I_i = \frac{b_{i+1} - b_i}{a_{i+1} - a_i}(x - a_i) + b_i, \quad a_i \leq x \leq a_{i+1} \quad (2.4)$$

where, $i = 1,2,3,...,m$.

Although segmented linear functions are flexible, they are not ideally suited to some situations, particularly those in which the slope changes very gradually with increasing levels of pollution. In these instances, a nonlinear function is usually more appropriate.

2.4.2.3. *Nonlinear Function*

When a cause–effect relationship does not vary linearly, it leads to a curvature when plotted on a graph sheet. Such nonlinear functions are of two basic types:

1. an implicit function, which can be plotted on a graph but for which no equation is given
2. an explicit function, for which a mathematical equation is given.

Implicit functions usually arise when some empirical curve has been obtained from a process under study. For example, Brown et al. (1970) proposed an implicit nonlinear subindex function for pH.

Explicit nonlinear functions automatically lead to nonlinear curves. An important general nonlinear function is one in which the pollutant variable is raised to a power other than one, the power subindex function:

$$I_i = x^c \quad (2.5)$$

where $c \neq 1$

Walski and Parker (1974) used the following general parabolic form in evolving the subindices for temperature and pH:

$$I_i = -\frac{b}{a^2}(x - a)^2 + b, \; 0 \leq x \leq 2a \quad (2.6)$$

Another common nonlinear function is the exponential function, in which pollutant variable x is the exponent of a constant:

$$I_i = C^x \quad (2.7)$$

The constant usually selected is either 10 or e, the base of the natural logarithm. If a and b are constants, the general form of an exponential function is written as follows:

$$I_i = ae^{bx} \quad (2.8)$$

2.4.2.4. *Segmented Nonlinear Function*

Segmented nonlinear function consists of line segments similar to the segmented linear function; however, at least one segment is nonlinear. Usually, each segment is represented

TABLE 2.1 The Segments, Ranges and Functions Used by Prati et al. (1971) for pH

Segment	Range	Function
1	$0 \leq x \leq 5$	$I_i = -0.4x^2 + 14$
2	$5 \leq x \leq 7$	$I_i = -2x + 14$
3	$7 \leq x \leq 9$	$I_i = x^2 - 14x + 49$
4	$9 \leq x \leq 14$	$I_i = -0.4x^2 + 11.2x - 64.4$

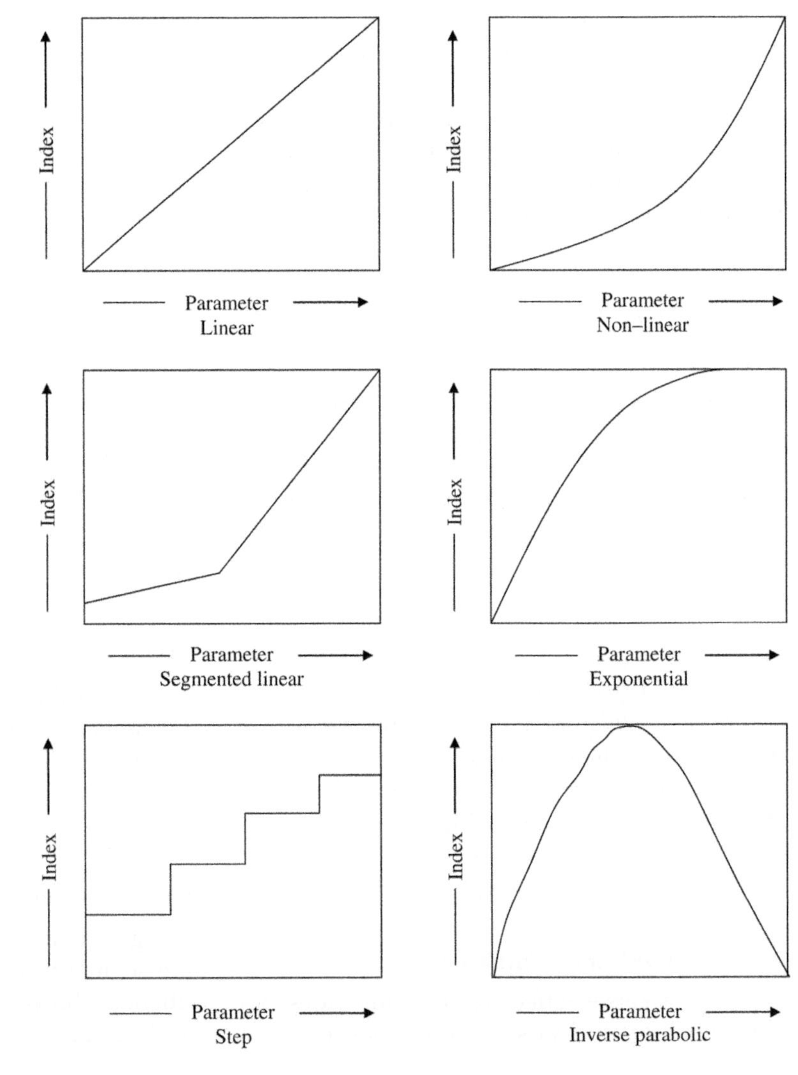

FIGURE 2.2 Typical subindex functions.

by a different equation which applies over a specific range of the pollutant variable. Segmented nonlinear function being more flexible than the segmented linear function has been used for a number of water-quality subindices. For example, Prati et al. (1971) used a segmented nonlinear function for the pH subindex in their water-quality index. The pH subindex function contained four segments (Table 2.1).

Typical subindex functions are presented in Figure 2.2.

2.5. ASSIGNMENT OF WEIGHTAGES

We have explained at some length in Chapter 1 and in Section 2.3 of this chapter that it is a necessary as well as a very challenging task to shortlist a few from among hundreds of water-quality parameters so that a balance is achieved between size of an index, genuineness of the water-quality data and effectiveness of the index.

But even after a shortlist of 10–20 parameters has been made, it still remains a major task to assign weightage to each parameter. Because even as all short-listed parameters are deemed to be important as water-quality indicators, they would still not be equally important. Within the selected parameters, some would be of greater importance than some others.

Some of the indices assume equal weightage of all the parameters. But in great many, different weightage is given to the different parameters. The assignment of weightage is, like selection of parameters, a matter of opinion, hence subjective. For this, too, well-formulated techniques of opinion gathering such as Delphi (Abbasi & Arya 2000) are utilised to minimise subjectivity and enhance credibility. It must be brought to the attention of the readers that Delphi is a rather cumbursome and time-consuming exercise. Perhaps this is one reason why we do not find as many new indices based on Delphi being proposed as we otherwise might have.

2.6. AGGREGATION OF SUBINDICES TO PRODUCE A FINAL INDEX

In the final step, the subindices are aggregated to obtain the final index. Several methods of aggregation are possible; the three most basic ones are:

Additive: In additive aggregation, the subindices (referring to transformed values of parameters) are combined through summation (e.g., arithmetic mean). This has been the most oft-used aggregation method. Indices of Horton (1965), Brown et al., (1970), Prati et al., (1971), Dinius (1972), Otto (1978) – among others – have been based on this model.

Multiplicative: In multiplicative aggregation, the subindices are combined through product operation (e.g., geometric mean). Indices of Landwehr et al. (1974), Walski and Parker (1974), Bhargava (1985), Dinius (1987), etc, have been based on this model.

Logical: In logical aggregation, the subindices are combined through logical operation (e.g., minimum or maximum). The index of Smith (1990) is an example.

The combination of operations is also used, for example the WQI of Inhaber (1975) is based on the weighted root sum square function and the WQI of Dojlido et al. (1994) employs square root harmonic mean.

In recent years, attempts have been made to relate the indices to some sort of 'acceptability' measure that can be interpreted as the membership of a fuzzy set (Sadiq et al., 2007, Sadiq and Rodriguez, 2004, Lu and Lo, 2002, Chang et al., 2001, Lu et al., 1999, and Tao and Xinmiao, 1998). There is increasing use of factor analysis, principal-component analysis, and other concepts such as entropy and genetic algorithms in making 'hybrid' indices or in enhancing the applicability of conventional indices (Peng, 2004; Nasiri et al., 2007; Bonnet et al., 2008; Meng et al., 2009; Taheriyoun et al., 2010; Thi Minh Hanh et al., 2011). In addition, there is a class of generalised mean operators developed by Yager (1988), known as Ordered Weighted Averaging (OWA) operators, which have been used to develop a WQI (Sadiq and Tesfamariam, 2007). These aspects are covered in more detail in the following chapters. However, the three types of aggregation models described above continue to be used very extensively.

The manner in which these aggregation models are used is illustrated in the following examples.

2.6.1. Linear Sum Index

A linear sum index is computed by the addition of unweighted subindices, in which no subindex is raised to a power other than 1:

$$I = \sum_{i=1}^{n} I_i \tag{2.9}$$

where I_i is the subindex for pollutant variable i and n is the number of pollutant variables

The simplicity of linear summation is outweighed by the disadvantage that the resulting index can project poor water quality even when no individual parameter is below acceptable level as explained below.

A linear sum water pollution index is formed consisting of just two subindices, I_1 and I_2:

$$I = I_1 + I_2 \tag{2.10}$$

Assuming that $I_1 = 0$ and $I_2 = 0$ represent zero concentration and $I_1 \geq 100$ or $I_2 \geq 100$ represent concentration at the cut-off or below the permissible level. If the summation leads to $I > 100$, the users would infer that permissible level is violated by at least one subindex, whereas, in reality, one can get $I > 100$ even when both the individual parameters constituting I are within permissible limits. This phenomenon is called 'ambiguity problem'.

2.6.2. Weighted Sum Index

A weighted sum index is given by

$$I = \sum_{i=1}^{n} W_i I_i \tag{2.11}$$

where I_i is the subindex for i^{th} variable and W_i the weight for i^{th} variable:

$$\sum_{i=1}^{n} W_i = 1 \tag{2.12}$$

This index steers clear of the kind of ambiguity which dogs the linear sum index but suffers an equally serious problem called 'eclipsing'. Eclipsing occurs when at least one subindex reflects poor water quality as explained below:

For the two variable case,

$$I = W_1 I_1 + W_2 I_2 \tag{2.13}$$

$$W_1 + W_2 = 1 \tag{2.14}$$

Equations (2.13) and (2.14) can be written in a single equation as

$$I = W_1 I_1 + (1 - W_1) I_2 \tag{2.15}$$

From Equation (2.15), it is clear that $I = 0$ when both I_1 and $I_2 = 0$, i.e. the zero pollution is indicated properly. Further, I will not be 100 until and unless one of the subindices is more or equal to 100. Hence, the problem of ambiguity is also removed.

But in situations such as the ones arising when $I_1 = 50$ and $I_2 = 110$ with W_1 and W_2 both equal to 0.5, gives $I = 80$. In other words, the overall score indicates acceptable water quality even though one of the constituents as reflected in I_2 was above the permissible limit of 100. This type of situation when the index score 'hides' the unacceptable level of one or more constituent parameters is called 'eclipsing'.

2.6.3. Root Sum Power Index

The root sum power index is formed by a nonlinear aggregation function:

$$I = \left[\sum_{i=1}^{n} Ii^{p} \right]^{1/p} \tag{2.16}$$

where p is a positive real number, greater than 1. As p becomes larger, the ambiguous region becomes smaller. For large values of p, the ambiguous region is almost entirely eliminated. The root sum power function is a good means for aggregating subindices, because it neither yields an eclipsing region nor an ambiguous region. However, because it is a limiting function, it is somewhat unwieldy.

2.6.4. Multiplicative Form Indices

The most common multiplicative aggregation function in such indices is the weighted product, which has the following general form:

$$I = \left[\prod_{i=1}^{n} I_i^{Wi} \right] \tag{2.17}$$

where

$$\sum_{i=1}^{n} W_i = 1 \tag{2.18}$$

In this aggregation function, as with all multiplicative forms, an index is zero if any one subindex is zero. This characteristic eliminates the eclipsing problem, because if any one subindex exhibits poor water quality, the overall index will exhibit poor water quality. Conversely, $I = 0$ if and only if at least one subindex is zero; this characteristic eliminates the ambiguity problem.

If the weights in Equation (2.18) are set equal, $W_i = w$ for all I, then Equation (2.18) can be written as follows:

$$\sum_{i=1}^{n} W_i = nw = 1 \tag{2.19}$$

For this situation, $w = 1/n$, Equation (2.17) becomes the geometric mean of subindices:

$$I = \left[\prod_{i=1}^{n} Ii \right]^{w} = \left[\prod_{i=1}^{n} Ii \right]^{1/n} \tag{2.20}$$

Thus, the geometric mean is a special case of the weighted product aggregation function. A common version of the weighted product is the geometric aggregation function:

$$I = \left[\prod_{i=1}^{n} I_i^{gi} \right]^{1/\gamma} \tag{2.21}$$

where

$$\gamma = \sum_{i=1}^{n} g_i \tag{2.22}$$

2.6.5. Maximum Operator Index

The maximum operator index can be viewed as the limiting case of the root sum power index as p approaches infinity. The general form of the maximum operator is as follows:

$$I = \max\{I_1, I_2, \ldots I_n\} \tag{2.23}$$

In the maximum operator, I takes on the largest of any of the subindices, and $I = 0$ if and only if $I_i = 0$ for all i. It is ideally suited to determine if a permissible value is violated and by how much.

The limitation of the maximum operator becomes apparent when fine gradations of water quality, rather than discrete events, are to be reported and a number of subindices are to be aggregated.

The maximum operator is ideally suited to applications in which an index must report if at least one recommended limit is violated and by how much. Of course, if several subindices violate a recommended limit, the maximum operator will accordingly yield increasingly desirable subindex. The suitability of the maximum operator for use in water pollution indices has not been explored as it ought to have been, however, and none of the published water-quality indices has employed this aggregation function.

2.6.6. Minimum Operator Index

The minimum operator index, when summing decreasing scale subindices, performs in a fashion similar to the increasing scale maximum operator index. The general form of the minimum operator is

$$I = \min\{I_1, I_2, \ldots I_n\} \tag{2.24}$$

As g is within the maximum operator functions, eclipsing does not occur with this aggregation method, nor does an ambiguous region exist. Consequently, the minimum operator appears to be a good candidate for aggregating decreasing scale subindices. However, none of the published environmental indices employ the minimum operator, and its potential, too, remains unexplored.

An overview of parameters, the type of subindices, weightages and aggregation methods used in some of the commonly used indices is presented in Table 2.2.

2.7. CHARACTERISTICS OF AGGREGATION MODELS

2.7.1. Ambiguity and Eclipsing

In the preceding section, examples were given of how certain aggregation methods cause *ambiguity* or *eclipsing*. These are two of the characteristics of aggregation methods. To wit, ambiguity is caused in an aggregation method when index I exceeds the critical level (unacceptable value) without any of the subindices exceeding the critical level, and eclipsing is caused when index I does *not* exceed the critical level (unacceptable value) despite one or more of the subindices exceeding the critical level.

The other characteristics of aggregation methods are *compensation* and *rigidity.*

2.7.2. Compensation

An aggregation method model with good compensation is one that is not biased towards extremes (i.e., highest or lowest subindex value). But this attribute comes in the way when ambiguity-free and eclipsing-free models are desired. For example, maximum (or minimum) operators which are free from ambiguity and eclipsing have poor compensation as they are biased toward the highest (or the lowest) subindex values. Hence, the virtues of compensation have to be balanced with the disadvantages of ambiguity (and eclipsing). Generally, aggregation methods are regarded as having good compensation when they satisfy the following constraint:

$$\min_{i=1}^{N}(s_i) \leq Agg(s_1, s_2, \ldots, s_N) = I \leq \max_{i=1}^{N}(s_i) \tag{2.25}$$

However, this constraint does not necessarily apply to some methods where the compensatory property is undefined. It has been reported that operators resulting in aggregate values less than minimum and greater than maximum, respectively, lack compensation properties (Zimmermann and Zysno, 1980).

2.7.3. Rigidity

Rigidity is manifested when necessity arises for additional variables to be included in an index to address specific water quality concerns, but the aggregation model does not allow this. For example, it is common for a regulatory agency to have an existing overall index, but the agency would like to add one or more additional parameters. This situation may arise when at a particular site the index may show the water quality to be good and yet the water may be adversely impacted by constituents not

TABLE 2.2 Formulations of Some Oft – used Water Quality Indices

Index	Parameter	Subindex, SI_i	W_i	Aggregation formulation	Range of WQI
NSF–WQI (Brown et al. 1970)	DO (%)	142 experts drew curves for raw data and assigned a value ranging from 0 (*worst*) to 100 (*best*) and final curves were obtained with the weighting curves for each parameter	0.17	$\sum_{i=1}^{N} SI_i W_i$	0-25 = very bad
	FC, MPN/100 mL		0.16		26-50 = bad
	pH		0.11		51-70 = regular
	BOD_5 (ppm)		0.11		71-90 = good
	Nitrates (ppm)		0.10		91-100 = excellent
	Total phosphates (ppm)		0.10		
	Temp. (°C)		0.10		
	Turbidity, NTU/JTU		0.08		
	Total solids (ppm)		0.07		
O-WQI (Dunnette 1979; Cude 2001)	Temp (°C)	1,*a*		$\sqrt{\frac{N}{\sum_{i=2}^{N} \frac{1}{SI_i^2}}}$	10-59 = very poor
	DO (%)	1,2			60-79 = poor
	BOD_5 (mg/L)	2			80-84 = fair
	pH	2			85-89 = good
	Ammonia+Nitrate nitrogen (mg/L)	2			90-100 = excellent
	Total phosphorus (mg/L)	1,*b*			
	Total solids (mg/L)	2			
	FC (#/100 mL)	2			
PW-WQI (Pesce and Wunderlin 2000)	DO (mg/L)	4		$\frac{\sum_{i=1}^{3} SI_i}{3}$	0 = minimum quality

(Continued)

TABLE 2.2 Formulations of Some Oft – used Water Quality Indices (*cont'd*)

Index	Parameter	Subindex, SI_i	W_i	Aggregation formulation	Range of WQI
	Conductivity (μS/cm)				100 = maximum quality
	Turbidity, NTU				
CPCB – WQI (Sarkar and Abbasi 2006)	DO (%)	3	0.31	$\sum_{i=1}^{N} SI_i W_i$	<38 = bad to very bad 38-50 = bad
	BOD_5 (mg/L)	3	0.19		50-63 = medium to good
	pH	3	0.22		63-100 = good to excellent
	FC (MPN/100 mL)	5	0.28		
River pollution index (RPI) (Liou et al. 2004)	DO (mg/L) BOD_5 (mg/L)	4		$SI_{temp} SI_{pH} SI_{tox} \left[\sum_{i=1}^{3} SI_i W_i \times \sum_{j=1}^{2} SI_j W_j \times \sum_{k=1}^{1} SI_k \right]^{1/3}$	Value varies from 0-64.8 and are divided into nonpolluted, lightly polluted, moderately polluted, and grossly polluted
	Ammonia nitrogen (mg/L)			SI_j = subindex for two particulate parameters	
	Suspended solids (mg/L)			SI_k = sub-index for FC	
	Turbidity (NTU)			SI_i = sub -index for last three parameter	
	Temp. (°C)				
	FC (MPN/100 m/L)				
	pH				
	Toxicity				

U-WQI (Boyacioglu, 2007)	Cadmium, cyanide, mercury, selenium, arsenic, fluoride, nitrate-nitrogen, DO, BOD_5, total phosphorus, pH and total coliform	N/A	N/A	$\frac{\sum_{i=1}^{N} SI_i}{N}$	N/A
S-WQI (Said et al. (2004)	DO (%)			$\log\left[\frac{DO^{1.5}}{(3.8)^{TP}(Turb)^{TP}15^{FC/1000}+0.14(Con)^{0.5}}\right]$	<1 = poor
	Con (μS/cm)				<2 = marginal and remediation
	Turbidity, Turb (NTU)				3-2 = acceptable
	FC, MPN/100 mL				3 = very good
	Total phosphorus, TP (mg/L)				0 = minimum quality
ISQA*	Temp. (°C)	3		$SI_{TEMP}(SI_{TOC}+SI_{SS}+SI_{DO}+SI_{Con})$	100 = maximum quality
	TOC (mg/L)	3			
	SS (mg/L)	3			
	DO (mg/L)	3			
	Con (μS/cm)	5			
	Not fixed	F_1: scope (% of variables that do not meet their objectives at least once); F_2: frequency (% of individual tests that do not meet their objectives); F_3: amplitude (amount by which failed tests do not meet their objectives)		$100-\left(\frac{\sqrt{F_1^2+F_2^2+F_3^2}}{1.732}\right)$	0-44 = poor 45-64 = marginal 65-79 = fair 80-94 = good 95-100 = excellent

(Continued)

TABLE 2.2 Formulations of Some Oft − used Water Quality Indices (*cont'd*)

Index	Parameter	Subindex, SI_i	W_i	Aggregation formulation	Range of WQI
S-T WQI	Not fixed	Monotonically decreasing		$\left[1 - N + \sum_{i=1}^{N} SI_i^{-\log_2^{(N-1)}}\right]^{-1/\log_2 N-1}$	0-0.25 = poor
(Swamee and Tyagi 2007)		Subindices,			0.26-0.50 = fair
		$SI = \left(1 + \frac{P}{P_C}\right)^{-m}$			0.51-0.70 = medium/ average
		Nonuniformly decreasing subindices,			0.71-0.90 = good
		$SI = \frac{1 + \left(\frac{P}{P_T}\right)^4}{1 + 3\left(\frac{P}{P_T}\right)^4 + 3\left(\frac{P}{P_T}\right)^8}$			0.91-1.0 = excellent
		Unimodal subindices,			
		$SI = \frac{qr + (n+q)(1-r) + \left(\frac{P}{P_C}\right)^r}{q + n(1-r)\left(\frac{P}{P_C}\right)^{n+q}}$			

Adopted with Permission from Islam et al., 2011.

included in the index. Or an agency may like to use an index, developed for one region, in another region where weather and other environmental conditions may be significantly different. To do this some changes may be needed in the number of water-quality variables but the index may not be able to accommodate it.

Hence, rigidity is related to the number (N) of subindices. When new subindices are added in an aggregation model, it may, due to rigidity, artificially reduce the index value irrespective of the magnitude of subindices (Swamee and Tyagi, 2000; 2007). Product-type operators and nonlinear summation-type operators generally exhibit this behaviour. For example, in two subindices $s_1 = 0.2$ and $s_2 = 0.3$, on using a root sum power addition (for $p = -2$), index (I) becomes 0.166. If an additional subindex $s_3 = 0.35$ is included, the final result becomes $I = 0.15$, which is less than the value obtained for two subindices though the subindex s_3 was greater in magnitude than index (I) earlier obtained.

On the other hand, index (I) increases with the increase in number of subindices when nonlinear summation-type operators are used (e.g., root sum power addition with $p > 1$ for pollution index).

Most of the aggregation methods do not have any provision to add an additional parameter into its preidentified set of water-quality constituents. If those methods are used for aggregation, then the value of the overall index decreases as the number of subindices increases irrespective of their magnitude. This decrease in the value of the overall index exacerbates the issue of ambiguity in indices that are already suffering from this problem and reintroduces the issue of ambiguity in indices that were free from this problem.

References

Abbasi, S.A., Arya, D.S., 2000. Environmental Impact Assessment. Discovery Publishing House, New Delhi.

Bhargava, D.S., 1985. Water quality variations and control technology of Yamuna River. Environmental Pollution Series A: Ecological and Biological 37 (4), 355–376.

Bonnet, B.R.P., Ferreira, L.G., Lobo, F.C., 2008. Water quality and land use relations in Goias: a watershed scale analysis. Revista Arvore 32 (2), 311–322.

Brown, R.M., McClelland, N.I., Deininger, R.A., Tozer, R.G., 1970. A water quality index – do we dare? Water Sewage Works 117, 339–343.

Chang, N.-B., Chen, H.W., Ning, S.K., 2001. Identification of river water quality using the fuzzy synthetic evaluation approach. Journal of Environmental Management 63 (3), 293–305.

Dinius, S.H., 1972. Social accounting system for evaluating water. Water Resources Research 8 (5), 1159–1177.

Dinius, S.H., 1987. Design of an index of water quality. Water Resources Bulletin 23 (5), 833–843.

Dojlido, J., Raniszewsk, I.J., Woyciechowska, J., 1994. Water quality index – application for rivers in Vistula river basin in Poland. Water Science and Technology 30, 57–64.

Horton, R.K., 1965. An index number system for rating water quality. Journal of Water Pollution Control Federation 37 (3), 300–306.

Inhaber, H., 1975. An approach to a water quality index for Canada. Water Research 9 (9), 821–833.

Islam, N., Sadiq, R., Rodriguez, M.J., Francisque, A., 2011. Reviewing source water protection strategies: A conceptual model for water quality assessment. Environmental Reviews 19, 68–105.

Landwehr, J.M., Deininger, R.A., Mcclelland, N.L., Brown, R.M., 1974. An objective water quality index. Journal of the Water Pollution Control Federation 46 (7), 1804–1807.

Lu, R.S., Lo, S.L., 2002. Diagnosing reservoir water quality using self-organizing maps and fuzzy theory. Water Research 36 (9), 2265–2274.

Lu, R.S., Lo, S.L., Hu, J.Y., 1999. Analysis of reservoir water quality using fuzzy synthetic evaluation. Stochastic Environmental Research and Risk Assessment 13 (5), 327–336.

Meng, W., Zhang, N., Zhang, Y., Zheng, B., 2009. Integrated assessment of river health based on water quality, aquatic life and physical habitat. Journal of Environmental Sciences 21 (8), 1017–1027.

Nasiri, F., Maqsood, I., Huang, G., Fuller, N., 2007. Water quality index: a fuzzy river-pollution decision support expert system. Journal of Water Resources Planning and Management 133 (2), 95–105.

Ott, W.R., 1978. Environmental Indices: Theory and Practice. Ann Arbor Science Publishers Inc, Ann Arbor, MI.

Peng, L., 2004. A Universal Index Formula Suitable to Multiparameter Water Quality Evaluation. Numerical Methods for Partial Differential Equations 20 (3), 368–373.

Prati, L., Pavanello, R., Pesarin, F., 1971. Assessment of surface water quality by a single index of pollution. Water Research 5, 741–751.

Sadiq, R., Rodriguez, M.J., 2004. Fuzzy synthetic evaluation of disinfection by-products – a risk-based indexing system. Journal of Environmental Management 73 (1), 1–13.

Sadiq, R., Rodriguez, M.J., Imran, S.A., Najjaran, H., 2007. Communicating human health risks associated with disinfection by-products in drinking water supplies: a fuzzy-based approach. Stochastic Environmental Research and Risk Assessment 21 (4), 341–353.

Sadiq, R., Tesfamariam, S., 2007. Probability density functions based weights for ordered weighted averaging (OWA) operators: an example of water quality indices. European Journal of Operational Research 182 (3), 1350–1368.

Smith, D.G., 1990. A better water quality indexing system for rivers and streams. Water Research 24 (10), 1237–1244.

Swamee, P.K., Tyagi, A., 2000. Describing water quality with aggregate index. ASCE Journal of Environmental Engineering 126 (5), 451–455.

Swamee, P.K., Tyagi, A., 2007. Improved method for aggregation of water quality subindices. Journal of Environmental Engineering 133 (2), 220–225.

Taheriyoun, M., Karamouz, M., Baghvand, A., 2010. Development of an entropy-based Fuzzy eutrophication index for reservoir water quality evaluation. Iranian Journal of Environmental Health Science and Engineering 7 (1), 1–14.

Tao, Y., Xinmiao, Y., 1998. Fuzzy comprehensive assessment, fuzzy clustering analysis and its application for urban traffic environment quality evaluation. Transportation Research Part D: Transport and Environment 3 (1), 51–57.

Thi Minh Hanh, P., Sthiannopkao, S., The Ba, D., Kim, K.-W., 2011. Development of water quality indexes to identify pollutants in vietnam's surface water. Journal of Environmental Engineering 137 (4), 273–283.

Walski, T.M., Parker, F.L., 1974. Consumers water quality index. ASCE Journal of Environmental Engineering Division 100 (EE3), 593–611.

Yager, R.R., 1988. On ordered weighted averaging aggregation in multicriteria decision making. IEEE Transactions on Systems, Man and Cybernetics 18, 183–190.

Zimmermann, H.J., Zysno, P., 1980. Latent connectives in human decision making. Fuzzy Sets and Systems 4, 37–51.

CHAPTER

3

'Conventional' Indices for Determining Fitness of Waters for Different Uses

OUTLINE

Water Quality Indices DOI: 10.1016/B978-0-444-54304-2.00003-8

3.1. GENERAL

In Chapter 1, Section 4 we have described Horton's water-quality index (Horton, 1965) which can be considered as the forerunner of the modern WQIs.

In this chapter, we present a wide cross-section of 'conventional' indices which have followed the Horton's WQI. These indices a) are based on parameter selection, assignment of weightage and aggregation by methods other than advanced statistical techniques, artificial intelligence or probability theory (which have been detailed in Chapters 4–6), and b) deal predominantly with physico-chemical characteristics of water. To distinguish these indices from the ones described in subsequent chapters we have referred these indices as 'conventional' WQIs.

3.2. BROWN'S OR THE NATIONAL SANITATION FOUNDATION'S WATER-QUALITY INDEX (NSF-WQI)

Brown et al. (1970) developed a water-quality index similar in structure to Horton's index but with much greater rigour in selecting parameters, developing a common scale and assigning weights for which elaborate Delphic exercises were performed. This effort was supported by the National Sanitation Foundation (NSF). For this reason, Brown's index is also referred to as NSF-WQI.

A panel of 142 persons with expertise in water-quality management was formed for the study. The panelists were asked to consider 35 parameters for possible inclusion in the index. They were free to add to the list any parameter of their choice. Each parameter was to be assigned one of the following choices: 'do not include', 'undecided' or 'include'. The panelists were asked to rank the parameters marked as 'include' according to their significance as contributor to the overall quality. The rating was done on a scale of 1 (highest) to 5 (lowest). The responses of the panel were brought to the knowledge of every member of the panel and the members were allowed to review their individual judgement in the light of the full panel's response.

Finally, the panelists were asked to select not more than 15 parameters which they considered to be the most important. The complete list of parameters arranged in decreasing order of significance, as determined by average rating of the panel, was presented to each member. Continuing in this fashion, a list of eleven parameters was finalised (Table 3.1).

The panelists were asked to assign values for the variation in the level of water quality produced by different concentrations of the parameters selected as above. The concentration-value relationship of each parameter was obtained in the form of a graph. These graphs were produced by the panelists to denote curves which, in their judgement, best represented the variation in level of water quality produced by possible measurements of a each respective parameter. The judgement of all the respondents was averaged to produce a set of curves, one for each parameter. Figure 3.1 shows the rating curve for DO. For pesticides

TABLE 3.1 List of Parameters Chosen as Most Significant by a Delphi Conducted by Brown et al. (1970) for the NSF-WQI

Parameter	Rank of Importance
Dissolved oxygen	1
Biochemical oxygen demand	2
Turbidity	3
Total solids	4
Nitrate	5
Phosphate	6
pH	7
Temperature	8
Faecal coliforms	9
Pesticides	10
Toxic elements	11

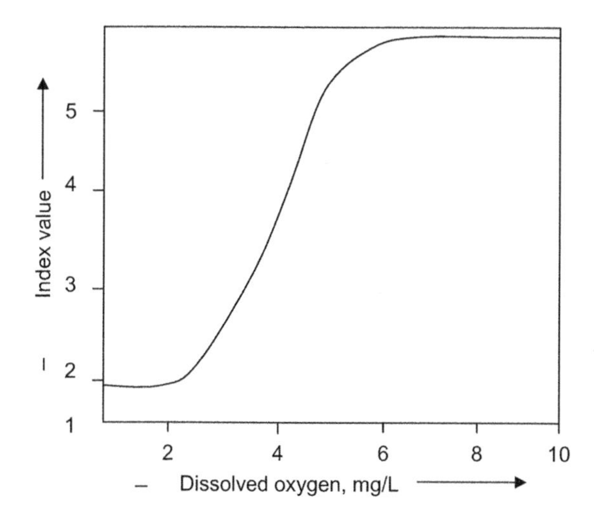

FIGURE 3.1 A subindex for dissolved oxygen (DO). For lower values of DO the subindex score is close to the lower end of the rating and at DO levels of 7 mg/I and above the index score is close to the highest.

and toxic elements, it was proposed that, if the total contents of detected pesticides or toxic elements (of all types) exceed 0.1 mg/1, the water-quality index be automatically registered zero.

The panelists were asked to compare relative overall water quality using a scale of 1 (highest) to 5 (lowest) for the finally selected parameters. Arithmetic mean was calculated for the ratings of experts.

To convert the rating into weights, a temporary weight of 1.0 was assigned to the parameter which received the highest significance rating. All other temporary weights were obtained by dividing the highest rating by the individual mean rating. Each temporary weight was then divided by the sum of all the temporary weights to arrive at the final weight. Table 3.2 gives the mean rating, temporary weights and final weights of the selected parameters.

TABLE 3.2 Significance Ratings and Weights for Parameters Included in Brown's (NSF) WQI

Parameters	Mean of all Significance Ratings Returned by Respondents	Temporary Weights	Final Weights
Dissolved oxygen	1.4	1.0	0.17
Faecal coliform density	1.5	0.9	0.16
pH	2.1	0.7	0.11
BOD (5-day)	2.3	0.6	0.11
Nitrates	2.4	0.6	0.10
Phosphates	2.4	0.6	0.10
Temperature	2.4	0.6	0.10
Turbidity	2.9	0.5	0.08
Total solids	3.2	0.4	0.07
Total			1.00

The index originally proposed by Brown et al. (1970) has the form

$$WQI = \sum_{i=1}^{9} w_i T_i(p_i) = \sum_{i=1}^{9} w_i q_i \quad (3.1)$$

where

p is the measured value of the ith parameter and

T_i is the quality rating transformation (curve) of the ith parameter value, p_i, into a quality rating q_i, such that

$T_i p_i = q_i$; and

w_i is relative weight of the ith parameter such that

$$\sum_{i=1}^{9} w_i = 1 \quad (3.2)$$

The method of calculation of the index value is illustrated in Table 3.3.

Brown's index represents general water quality. It does not recognise and incorporate specific water functions such as drinking water supply, agriculture, industry, etc. Related to this difficulty has been an apparent tendency for some respondents to be heavily influenced in their judgement of parameter suitability for inclusion in a WQI by factors such as data availability and existing analytical methodologies for measuring the various parameters.

In the course of using the index, it was found that arithmetic or additive formulation, although easy to understand and calculate, lacked sensitivity in terms of the effect a single bad parameter value would have on the WQI. This led Brown et al. (1973) to propose a variation of NSF-WQI in the following multiplicative form:

$$WQI = \prod_{i=1}^{n} S_i^{wi} \quad (3.3)$$

The subsequent investigations tended to show that multiplicative formulation agreed better with expert opinion than did the additive one. However, both of them have continued to be in use (Lumb et al., 2011).

TABLE 3.3 Illustrative Example of a Calculation of Brown's (NSF) WQI

Parameters	Measured Values	Individual Quality Rating (q_I)	Weights (w_i)	Overall Quality Rating (q_i x w_I)
DO, Percent sat	100	98	0.17	16.7
Faecal coliform density. 1b/100 ml	0	100	0.16	16.0
pH	7	92	0.11	10.1
BOD_5 mg/L	0	100	0.11	11.0
Nitrate, mg/L	0	98	0.10	9.8
Phosphate, mg/L	0	98	0.10	9.8
Temperature C departure from equal	0	94	0.10	9.4
Turbidity, units	0	98	0.08	7.8
Total solids, mg/L	25	84	0.07	5.9
			$WQI = \Sigma w_i q_i = 96.5$	

3.3. NEMEROW AND SUMITOMO'S POLLUTION INDEX

This increasing scale water-supply index has been proposed by Nemerow and Sumitomo (1970) on behalf of the US Environmental Protection Agency. It consists of indices for three types of use:

1. human contact use ($j = 1$)
2. indirect contact use ($j = 2$)
3. remote contact use ($j = 3$)

The first category includes drinking (including water used for beverage manufacturing) and swimming. The second category includes fishing, food processing and agriculture. The last category includes uses in which human contact is very indirect, such as in navigation, industrial cooling and some recreational activities (aesthetics, picnicking, hiking and visits to the area).

Each specific use index PI_j is represented by a function of the relative values, where:

$$\text{Relative value} = \frac{C_i}{L_{ij}} \tag{3.4}$$

Here C_i is the level of the water quality parameter i and L_{ij} is the permissible quality level of i at a location of a water use j. Thus C_i/L_{ij} is the relative pollution contributed by the water quality parameter i.

For cases where the contaminant level decreases in value as pollution increases, such as DO, the relative value is computed as follows:

$$\frac{C_i}{L_{ij}} = \frac{C_{im} - C_i}{C_{im} - L_{ij}} \tag{3.5}$$

Where C_{im} is the maximum value that C_i can attain. For the parameter DO, C_{im} is the DO level at saturation.

For the cases where the contaminant has permissible levels ranging from $L_{ij\ min}$ to $L_{ij\ max}$, such as pH,

$$\frac{C_i}{L_{ij}} = \frac{C_i - \left[\frac{L_{ij\ \mathrm{min}} + L_{ij\ \mathrm{max}}}{2}\right]}{L_{ij\ \mathrm{min}} \text{ or } L_{ij\ \mathrm{max}}; \textit{whichever is closer to } C_i - \left[\frac{L_{ij\ \mathrm{min}} + L_{ij\ \mathrm{max}}}{2}\right]} \tag{3.6}$$

To reduce the eclipsing problem, the relative values were aggregated with a root mean square operation. For each specific use j, the maximum C_i/L_{ij} value for all i was combined with the arithmetic mean of C_i/L_{ij} in a root mean square operation to obtain the value of PI_j:

$$PI_j = \sqrt{\frac{\left(\frac{C_i}{L_{ij}}\right)^2_{maximum} + \left(\frac{C_i}{L_{ij}}\right)^2_{mean}}{2}} \tag{3.7}$$

Using this approach, each specific use index reflects both the highest relative value (a measure of the extreme) and the average of all relative values (a measure of central tendency). The investigators recommended the use of 14 pollutant variables in the index.

The general water-quality index is computed as the weighted sum of the three specific use indices:

$$PI = \sum_{j=1}^{3} w_j\, PI_j \tag{3.8}$$

Here, w_j, the weight coefficient, is determined by the relative importance of the water use j in the region or society.

3.4. PRATI'S IMPLICIT INDEX OF POLLUTION

This index was developed by Prati et al. (1971) on the basis of water-quality standards. The concentration values of all the pollutants were transformed into levels of pollution expressed in new units through mathematical expressions. These mathematical expressions were constructed in such a way that the new units were proportional to the polluting effect relative to other factors. In this way, even if a pollutant is to be present in smaller concentrations than other pollutants, it still will exert a large impact on the index score if its polluting effect is greater.

In the first step, water quality was classified vis a vis all the parameters based on water-quality standards (Table 3.4).

In the second step, one pollutant was taken as reference and its actual value was considered directly as reference index.

In the third step, mathematical expressions were formed to transform each of the values of the other pollutants into subindices. This transformation took into account the polluting capacity of the parameters related to a selected reference parameter. In the construction of these functions, the analytical properties of various curves were used to

TABLE 3.4 Classification of Water Quality for the Development of Prati's Index

Parameter	Excellent	Acceptable	Slightly Polluted	Polluted	Heavily Polluted
pH	6.5–8.0	6.0–8.4	5.0–9.0	3.9–10.1	<3.9–>10.1
DO (% Sat)	88–112	75–125	50–150	20–200	<20–>200
BOD_5 (ppm)	1.5	3.0	6.0	12.0	>12.0
COD (ppm)	10	20	40	80	>80
Permanganate (mg $l^{-1}O_2$) (Kubel test)	2.5	5.0	10.0	20.0	>20.0
Suspended solids (ppm)	20	40	100	278	>278
NH_3 (ppm)	0.1	0.3	0.9	2.7	>2.7
NO_3 (ppm)	4	12	36	108	>108
Cl (ppm)	50	150	300	620	>620
Iron (ppm)	0.1	0.3	0.9	2.7	>2.7
Manganese (ppm)	0.05	0.17	0.5	1.0	>1.0
ABS (ppm)	0.09	1.0	3.5	8.5	>8.5
CCE (ppm)	1.0	2.0	4.0	8.0	>8.0

TABLE 3.5 Subindex Functions of Parti's Index

S.No	Parameter	Subindex
1	Dissolved Oxygen (%)	$I_i = -0.08x + 8, \quad 50 \leq x < 100,$ $I_i = 0.08x - 8, \quad 100 \leq x.$
2	pH (units)	$I_i = -0.4x^2 + 14, \quad 0 \leq, x < 5,$ $I_i = -2x + 14, \quad 5 \leq x < 7,$ $I_i = x^2 - 14x + 49, \quad 7 \leq x < 9,$ $I_i = -0.4x^2 + 11.2x - 64.4, \quad 9 \leq x < 14$
3	5-Day BOD (mg/L)	$I_i = 0.66666x$
4	COD (mg/L)	$I_i = 0.10x$
5	Permanganate (mg/l)	$I_i = 0.04x$
6	Suspended Solids (mg/L)	$I_i = 2^{[2.1 \log (0.1x-1)]}$
7	Ammonia (mg/L)	$I_i = 2^{[2.1 \log (10x)]}$
8	Nitrates (mg/L)	$I_i = 2^{[2.1 \log (0.25)]}$
9	Chlorides (mg/L)	$I_i = 0.000228x^2 + 0.0314x, \quad 0 \leq x < 50,$ $I_i = 0.0000132x^2 + .0074x + 0.6, \quad 50 \leq x < 300,$ $I_i = 3.75\ (0.02x - 5.2)^{0.5}, \quad 300 \leq x$
10	Iron	$I_i = 2^{[2.1 \log(10x)]}$
11	Manganese (mg/L)	$I_i = 2.5x + 3.9\sqrt{x}, \quad 0 \leq x < 0.5,$ $I_i = 5.25x^2 + 2.75, \quad 0.5 \leq x$
12	Alkyl Benzene sulphonates (mg/L)	$I_i = -1.2x + 3.2\ \sqrt{x}, \quad 0 \leq x < 1,$ $I_i = 0.8x + 1.2, \quad 1 \leq x$
13	Carbon Chloroform Extract (mg/L)	$I_i = x$

ensure that the resulting transformation would be applicable not only to small values of pollutant concentrations but also to those exceeding class V.

The resulting functions (subindices) are given in Table 3.5

The index was computed as the arithmetic mean of the 13 subindices:

$$I = 1/13 \sum_{i=1}^{13} I_i \tag{3.9}$$

The index ranges from 0 to 14 (and above) and was applied by Prati et al. to data on surface waters in Ferrara, Italy.

3.5. DEININGER AND LANDWEHR'S PWS INDEX

Deininger and Landwehr (1971) presented an index pertaining to water used for public water supply (PWS). It employed 11 parameters for

TABLE 3.6 Comparison of Weights in the NSF-WQI and the Two (Additive) Water-Supply Indices

		Deininger and Landwehr	
Pollutant Variable	**NSF-WQI**	**PWS_{11}**	**PWS_{13}**
Dissolved Oxygen	0.17	0.06	0.05
Faecal Coliforms	0.15	0.14	0.12
pH	0.12	0.08	0.07
5-Day BOD	0.10	0.09	0.08
Nitrates	0.10	0.10	0.09
Phosphates	0.10		
Temperature	0.10	0.07	0.06
Turbidity	0.08	0.09	0.08
Total Solids	0.08		
Dissolved Solids		0.10	0.08
Phenols		0.10	0.08
Colour		0.10	0.08
Hardness		0.08	0.07
Fluorides			0.07
Iron			0.07
Total	1.00	1.01	1.00

surface water sources and 13 for groundwater sources.

Two aggregation functions were considered: an additive form and a geometric mean. The 11-variable and the 13-variable versions of the indices were computed for each aggregation function:

Additive

$$PWS = \sum_{i=1}^{n} W_i I_i \quad (3.10)$$

Geometric mean

$$PWS = \left(\prod_{i=1}^{n} I_i^{w_i} \right)^{1/n} \quad (3.11)$$

where $n = 11$, for 11-variable version and $n = 13$, for 13-variable version.

The variables along with their associated weights for the two versions are compared with NSF-WQI in Table 3.6.

3.6. MCDUFFIE AND HANEY'S RIVER POLLUTION INDEX (RPI)

It is a relatively simple water-quality index in which eight pollutant variables are included. Most subindices are of the general linear form:

$$I \text{ A Proposed River Pollution Index} = \frac{X}{X_N} \quad (3.12)$$

where

I_i is the subindex of the ith pollutant variable,
X is the observed value of the pollutant variable and
X_N is the natural level of the pollutant variable

Six of the eight subindices described by McDuffie and Haney were explicit linear functions, and two (coliform count and temperature) were explicit nonlinear functions (Table 3.7). The index did not include pH or toxic substances.

The overall index is computed as the sum of n subindices times a scaling factor $10/(n+1)$:

$$RPI = \frac{10}{n+1}\left(TF + \sum_{i=1}^{n} 10 \times I_i \right) \quad (3.13)$$

Here TF is the temperature factor and n is number of parameters other than temperature. The RPI was applied on a test basis using data from New York State's water-quality surveillance network and from other sources.

TABLE 3.7 Subindex Functions of McDuffie's Index

S. No	Parameter	Subindex
1	Percent Oxygen Deficit	$I_1 = 100 - x, x = DO\%$
2	Biodegradable Organic Matter	$I_2 = 10x$, $x = BOD_5(ppm)$
3	Refractory Organic Matter	$I_3 = 5\ (x - y)$, $x = COD$, $y = BOD_5$
4	Coliform Count (no./100 ml)	$I_4 = 10\ (\log x/\log 3)$
5	Nonvolatile Suspended Solids	$I_5 = x,$
6	Average Nutrient Excess	$I_6 = 25x + 50yx$ = Total N(ppm) y = Total PO_4(ppm)
7	Dissolved Salts	$I_7 = 0.25x$, x = Specific conductance micro mho/cm
8	Temperature	$I_8 = \frac{x^2}{6} - 65$

3.7. DINIUS' WATER-QUALITY INDEX (1972)

This index broke new ground in the sense that through it an attempt was made to design a rudimentary social accounting system which would measure the costs and impact of pollution control efforts. In this sense, Dinius' WQI is a forerunner of the 'planning' or 'decision-making' indices described in Chapter 5. Eleven parameters were selected. Like Horton's index and the NSF-WQI, it had decreasing scale, with values expressed as a percentage of perfect water quality which corresponds to 100%.

Like Prati's, and McDuffie-Haney's indices, the subindices in Dinius' index were developed from a review of the published scientific literature. Dinius examined the water quality described by various authorities to different levels of pollutant variables, and from this information generated 11 subindex equations (Table 3.8).

The index was calculated as the weighted sum of the subindices, like Horton's index, and the additive version of the NSF-WQI:

TABLE 3.8 Subindex Functions of Dinius' Index (Ott 1978)

S.No	Parameter	Subindex
1	Dissolved Oxygen (%)	$I_1 = x$
2	5-Day BOD (mg/L)	$I_2 = 107x^{-0.642}$
3	Total Coliforms (MPN/100 ml)	$I_3 = 100\ (x)^{-0.3}$
4	Faecal Coliforms (MPN/100 ml)	$I_4 = 100\ (5x)^{-0.3}$
5	Specific Conductance (μmho/cm)	$I_5 = 535x^{-0.3565}$
6	Chlorides (mg/L)	$I_6 = 125.8x^{-0.207}$
7	Hardness ($CaCO_3$, ppm)	$I_7 = 10^{1.974-0.00132x}$
8	Alkalinity ($CaCO_3$, ppm)	$I_8 = 108x^{-0.178}$
9	pH	$I_9 = 10^{0.2335+0.44}, x < 6.7$ $I_{10} = 100, 6.7 \le x \le 7.58$ $I_{11} = 10^{4.22-0.293x}, x > 7.58$
10	Temperature (°C)	$I_{12} = -4\ (x_a - x_a) + 112$, x_a = actual temp, x_a = std. Temp
11	Colour (C units)	$I_{13} = 128x^{-0.288}$

$$WQI = \frac{1}{21}\sum_{i=1}^{11} w_i I_i \quad (3.14)$$

The weights ranged from 0.5 to 5 on a basic scale of importance. On this scale, 1,2,3,4 and 5 denote, respectively, very little, little, average, great and very great importance. The sum of the weights was 21, which is the denominator in the index equation.

The index was applied by Dinius on an illustrative basis to data on several streams in Alabama, USA.

3.8. O'CONNOR'S INDICES

O'Connor (1972) developed two water-quality indices: for fish and wild life (FAWL), and for public water supply (PWS). Both indices were developed using Delphi technique for reducing subjectivity of the judgements on parameter selection. The parameters and their weights for the two indices are compared with the Brown's or the National Science Foundation's Water-Quality Index (NSF-WQI) (Brown et al. 1970) in Table 3.9.

TABLE 3.9 Comparison of Weights Used in Three Water-Quality Indices

		O'Connor's Indices	
Pollutant Variable	**NSF-WQI**	**FAWL**	**PWS**
Dissolved Oxygen	0.17	0.206	0.056
Faecal Coliforms	0.15		0.171
pH	0.12	0.142	0.079
5-Day BOD	0.10		
Nitrates	0.10	0.074	0.070
Phosphates	0.10	0.064	
Temperature	0.10	0.169	
Turbidity	0.08	0.088	0.058
Total Solids	0.08		
Dissolved Solids		0.074	0.084
Phenols		0.099	0.104
Ammonia		0.084	
Fluorides			0.079
Hardness			0.077
Chlorides			0.060
Alkalinity			0.058
Colour			0.054
Sulphates			0.050
Total	1.00	1.00	1.00

The FAWL and PWS indices were computed as the weighted sum of the subindices times, a factor which takes into account pesticides and toxic substances:

$$I_{\mathrm{FAWL}} = \delta\sum_{i=1}^{9} W_i I_i \quad (3.15)$$

$$I_{\mathrm{PWS}} = \delta\sum_{i=1}^{13} W_i I_i \quad (3.16)$$

where $\delta = 0$, if pesticides or toxic substances exceed recommended limits, and is otherwise 1 (Ott 1978).

3.9. WALSKI AND PARKER'S INDEX

This index (Walski and Parker, 1974) is based on empirical information on the suitability of water for a particular use, and was developed specifically for the recreational water (such as used for swimming and fishing). The authors introduced four general categories of variables:

1. those which affect aquatic life (e.g., DO, pH and temperature),
2. those which affect health (e.g., coliforms),
3. those which affect taste and odour (e.g., threshold odour number) and
4. those which affect the appearance of the water (e.g., turbidity, grease and colour).

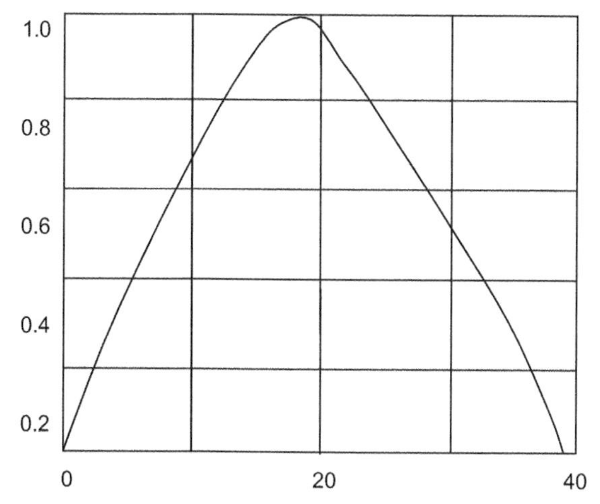

FIGURE 3.2 Inverted parabola as the subindex function for temperature used in Walski–Parker index.

In the second step, the sensitivity functions were determined to assign each parameter a value between one and zero, representing ideal conditions and completely unacceptable conditions, respectively. The nature of the sensitivity function was determined by the impact of a change in the value of the parameter on water quality. For substances that are inversely related to water quality, a negative exponential curve was thought to best represent the sensitivity function.

The authors determined values for the parameters which would be considered perfect, good, poor and intolerable, and assigned to each of these values the numbers 1, 0.9, 0.1 and 0.01, respectively. With these sets of values, the sensitivity functions could be found easily.

For example, to determine a sensitivity function for temperature, a listing of lethal temperatures was consulted. As the maintenance of aquatic life is difficult at low as well as high temperatures, it was felt that the sensitivity function for temperature should consist of an inverted parabola (Figure 3.2) described as follows:

$$F(T) = \frac{a^2 - (T-a)^2}{a^2} \tag{3.17}$$

where a is the ideal temperature such that $f(a) = 1$.

A total of 12 different pollutant variables were used in the index. The subindices consist of nonlinear and segmented nonlinear explicit functions (Table 3.10). Except for the two unimodel variables, pH and temperature, all subindices are represented by negative exponential equations. The pH was represented by parabolic equation as was temperature (noted above). Two subindices were used for temperature: one for actual temperature and another for departure from equilibrium temperature. To aggregate the subindices, a geometric mean was employed over an arithmetic mean to avoid the problem of eclipsing. Their aggregation function is as follows:

$$I = \left(\prod_{i=1}^{12} I_i^{w_i} \right)^{1/12} \tag{3.18}$$

3.10. STONER'S INDEX

This index, aimed for use in public water supply and irrigation, employed a single aggregation function which selects from two sets of recommended limits and subindex equations. Although Stoner (1978) applied the index to just two water uses, it could be adapted to additional water uses as well.

Two types of water-quality parameters are used in the Stoner's index:

Type I parameters normally considered toxic at low concentrations (for example, lead, chlordane, radium-226)
Type II parameters which affect health or aesthetic characteristics (for example, chlorides, sulphur, colour, taste and odour)

The *type I* pollutant variables were treated in a dichotomous manner, giving subindex step functions. Each *type I* subindex is assigned the value of zero if the concentration is less than or equal to the recommended limit and the value 100 if the recommended limit is exceeded. The *type II* pollutant variables are represented by explicit mathematical functions.

TABLE 3.10 Subindex Functions of Walski–Parker Index (Ott 1978)

Pollutant Variable	Equation	Range
Dissolved oxygen (mg/L)	$I = e^{[0.3\ (x-8)]}$	$0 < x \leq 8$
	$I = 0$	$8 < x$
pH (Std. Units)	$I = 0$	$x < 2$
	$I = 0.04\ [25 - (x - 7)^2]$	$2 \leq x \leq 12$
	$I = 0$	$12 < x$
Total Coliforms (no./100 ml)	$I = e^{-0.0002x}$	
Temperature (°C)	$I = 0.0025\ [1 - (x - 20)^2]$	$0 \leq x \leq 40$
	$I = 0$	$\Delta x < -10$
	$I = 0.01\ (100 - \Delta x)^2$	$-10 \leq \Delta x \leq 10$
	$I = 0$	$10 < \Delta x$
Phosphates (mg/L)	$I = e^{-2.5x}$	
Nitrates (mg/L)	$I = e^{-0.16x}$	
Suspended Solids (mg/L)	$I = e^{-0.02x}$	
Turbidity (JTU)	$I = e^{-0.001x}$	
Colour (c units)	$I = e^{-0.002x}$	
Grease (Concentration (mg/L))	$I = e^{-0.016x}$	
Grease (Thickness, μ)	$I = e^{-0.35x}$	
Odour	$I = e^{-0.1x}$	
Secchi Disk Transparency (m)	$I = \log\ (x + 1)$	$X \leq 9$
	$I = 1$	$9 < x$

A total of 26 *type I* pollutant variables were used in the public water-supply version of the index, and 5 *type I* variables in the irrigation version (Tables 3.11 and 3.12).

The overall index was computed by combining the unweighted *type I* subindices with the weighted *type II* subindices:

$$I = \sum_{i=1}^{m} I_i + \sum_{j=1}^{n} W_j I_j \qquad (3.19)$$

where

I_i is the subindex for the ith *type I* pollutant variable,

W_j is the weight for the jth *type II* pollutant variable and

I_j is the subindex for the jth *type II* pollutant variable.

3.11. BHARGAVA'S INDEX (1983, 1985)

This is one of the first reported indices by an Asian author, and addresses the issue of drinking water supply.

To develop the index, Bhargava (1983, 1985) identified 4 groups of parameters. Each group

TABLE 3.11 Subindex Functions of Stoner's Index for Public Water Supply

Variable	Subindex Function
Group – A ($w = 0.134$)	
Ammonia–Nitrogen (mg/L)	$100 - 200x$
Nitrate–Nitrogen (mg/L)	$100 - 100x^2$
Faecal–Coliforms (no./100 ml)	$100 - 0.000025x^2$
Group – B ($w = 0.089$)	
pH (Standard Units)	$-1125 + 350x - 25x^2$
Fluorides	$98.8 + 24.7x - 123x^2$
Group – C ($w = 0.067$)	
Chlorides (mg/L)	$100 - 0.4x$
Sulphates (mg/L)	$100 - 0.4x$
Group – D ($w = 0.053$)	
Phenols (μg/L)	$100 - 100x$
Methylene Blue Active Sub. (mg/L)	$100 - 200x$
Group – E ($w = 0.045$)	
Copper (mg/L)	$100 - 100x^2$
Iron (mg/L)	$100 - 33.3x$
Zinc (mg/L)	$100 - 20x$
Colour (Pt–Co units)	$100 - 0.0178x^2$

TABLE 3.12 Subindex Functions for Stoner's Index for Irrigation Water

Variable	Subindex Function
Group – A ($w = 0.111$)	
Sodium Absorption Ratio	$100 - x^2$
Specific Conductance (μmho)	$100 - 0.0002x^2$
Faecal Coliforms (no./100 ml)	$100 - 0.0001x^2$
Group – B ($w = 0.074$)	
Arsenic (mg/L)	$100 - 1000x$
Boron (mg/L)	$100 - 100x^2$
Cadmium (mg/L)	$100 - 10^6x^2$
Group – C ($w = 0.0555$)	
Aluminium (mg/L)	$100 - 4x^2$
Beryllium (mg/L)	$100 - 10^4 x^2$
Chromium (mg/L)	$100 - 10^4 x^2$
Cobalt (mg/L)	$100 - 2000x$
Manganese (mg/L)	$100 - 500x$
Vanadium (mg/L)	$100 - 1000x$
Group – D ($w = 0.028$)	
Copper (mg/L)	$100 - 2500x^2$
Fluorides (mg/L)	$100 - 100x^2$
Nickel (mg/L)	$100 - 2500x^2$
Zinc (mg/L)	$100 - 25x^2$

contained sets of one type of parameters. The first group included the concentrations of coliform organisms to represent the bacterial quality of drinking water. The second group included toxicants, heavy metals, etc., some or all of which have a cumulative toxic effect on the consumer. The third group included parameters that cause physical effects, such as odour, colour and turbidity. The fourth group included the inorganic and organic nontoxic substances such as chloride, sulphate, foaming agents, iron, manganese, zinc, copper, total dissolved solids (TDS) etc. The variables, with their maximum allowable contaminant level, C_{MCL} (as per the US Environmental Protection Agency), and the subindices worked out by Bhargava, which include the effects of concentrations of different parameters and their weightage, are given in Table 3.13.

The subindices were aggregated as follows:

$$WQI = \left[\prod_{i=1}^{n} f_i\right]^{1/n} \quad (3.20)$$

in which, $f_i(P_i)$ is the sensitivity function of the ith variable, and n is the number of variables considered.

TABLE 3.13 Subindex Functions of Bhargava's Drinking Water-supply Index

Variables	Subindex Function	C_{MCL}
Group I Coliform organisms, e.g., coliform bacteria	$f_1 = \exp[-16 (C-1)]$	Coliform bacteria/100 ml
Group II Heavy metals, other toxicants, etc., e.g., Cr, Pb Ag etc.	$f_1 = \exp[-4 (C-1)]$	0.05 mg/L each
Group III Physical variables, e.g., turbidity, colour.	$f_1 = \exp[-2 (C-1)]$	1 TU 15 Colour units
Group IV Organic & Inorganic nontoxic substances, e.g., chlorides, sulphates, TDS.	$f_1 = \exp[-2 (C-1)]$	250 mg/L each 500 mg/L

The index was applied to the raw water-quality data at the upstream and downstream of river Yamuna at Delhi. The author suggested that the public drinking water supplies should have a WQI larger than 90.

3.12. DINIUS' SECOND INDEX

A multiplicative water-quality index was developed by Dinius (1987) with liberal use of Delphi in decision making (Helmer & Rescher 1959, Dalkey & Helmer 1963, Abbasi & Arya 2000). The index included 12 pollutants – dissolved oxygen, 5-day BOD, coliform count, *E. coli*, pH, alkalinity, hardness, chloride, specific conductivity, temperature, colour and nitrate – for six water uses – public water supply, recreation, fish, shellfish, agriculture and industry. The subindex functions were worked out as summarised in Table 3.14.

TABLE 3.14 Subindex Functions of the Second Dinius' Index of Water Quality

Parameter	Dimension	Weight	Function
DO	%Saturation	0.109	$0.82DO + 10.56$
5-Day BOD	mg/L, at 20 °C	0.097	$108(BOD)^{-0.3494}$
Coli	MPN–Coli/100 ml	0.090	$136(COLI)^{-0.1311}$
E. coli	Faecal–Coli/100 ml	0.116	$106(E-COLI)^{-0.1286}$
Alkalinity	ppm $CaCO_3$	0.063	$110(ALK)^{-0.1342}$
Hardness	ppm $CaCO_3$	0.065	$552\ (HA)^{-0.4488}$
Chloride	mg/L, fresh water	0.074	$391(CL)^{-0.3480}$
Sp. Conductance	μmhos/cm 25 °C	0.079	$506(SPC)^{-0.3315}$
pH	pH < 6.9		$10^{0.6803+0.1856\ (pH)}$
	pH – units (6.9 – 7.1)	0.077	1
	pH > 7.1		$10^{3.65-0.2216\ (pH)}$
Nitrate	as NO_3, mg/l	0.090	$125(N)^{-0.2718}$
Temperature	°C	0.077	$10^{2.004-0.0382(T_a-T_s)}$
Colour	Colour units – Pt std	0.063	$127(C)^{-0.2394}$

The individual subindex functions were combined using a multiplicative aggregation function in which the weight of each subindex equation was based on evaluation of the importance by the Delphi panel members of each parameter to overall quality. The final multiplicative aggregation function had the general form:

$$\mathrm{IWQ} = \prod_{i=1}^{n} I_i^{w_i} \tag{3.21}$$

where

IWQ is the the index of water quality, a number between 0 and 100;
I_i is the subindex of pollutant variable, a number between 0 and 100;
W_i is the unit weight of pollutant variable, a number between 0 and 1; and
n is the number of pollutant variables.

The weighted function $I_i^{W_i}$ for each pollutant was calculated by substituting the corresponding value of subindex function and its weightage. For example, weighted function for BOD is

$$W_{I_{BOD}} = [108(\mathrm{BOD})^{-0.3494}]^{0.097}$$

3.13. VIET AND BHARGAVA'S INDEX (1989)

This index was developed for the evaluation of the water-quality status of the Saigon River for its various desired uses. It is based on the Welski–Parker index with slight modifications:

$$WQI = \left[\prod_{i=1}^{n} f_i(P_i)\right]^{1/n} \times 100 \tag{3.22}$$

in which n is the number of variables considered more relevant to the use than the rest of the variable and $fi\ (P_i)$ is the sensitivity function of the ith variable.

Typical subindex function values are shown in Figure 3.3.

For example for fish culture and wild life, the following parameters were considered relevant:

1. **1.** Temperature: value of the sensitivity function – 0.58
2. **2.** Chloride: value of the sensitivity function – 1.0
3. **3.** DO: value of the sensitivity function – 0.70
4. **4.** BOD: value of the sensitivity function – 0.73

The overall index is calculated as follows:

$$\mathrm{WQI} = (0.58 \times 1.0 \times 0.70 \times 0.73)^{1/4} \times 100$$

The classification of water resources based on this index is given in Table 3.15.

3.14. THE RIVER GANGA INDEX OF VED PRAKASH ET AL.

The index was developed to evaluate the water-quality profile of river Ganga in its entire stretch and to identify the reaches where the gap between the desired and the existing water quality is significant enough to warrant urgent pollution control measures.

The index had the weighted multiplication form:

$$\mathrm{WQI} = \sum_{i=1}^{p} W_i I_i \tag{3.23}$$

where I_i denotes subindex for i^{th} water-quality parameter,

W_i is weight associated with i^{th} water-quality parameter and
p is the number of water-quality parameters.

This index was based on the NSF-WQI (Brown et al., 1970), with slight modifications in terms of weightages to confirm to the water-quality criteria for different categories of uses as set by

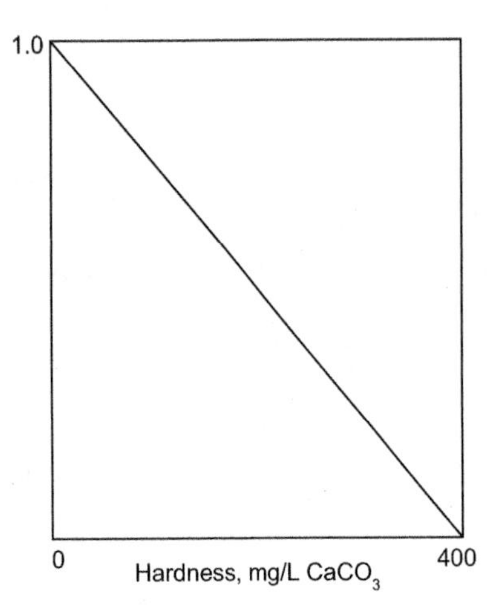

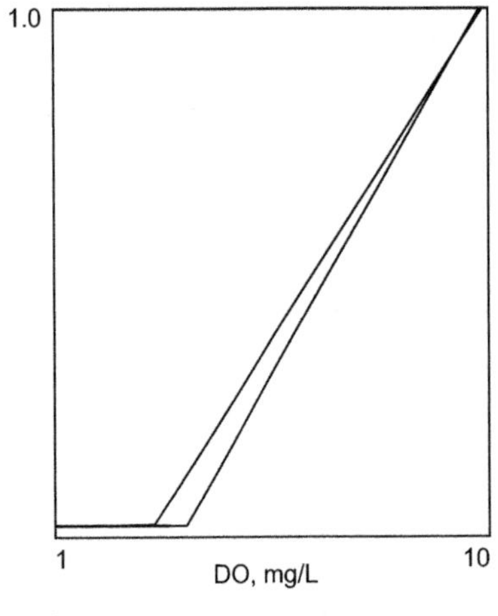

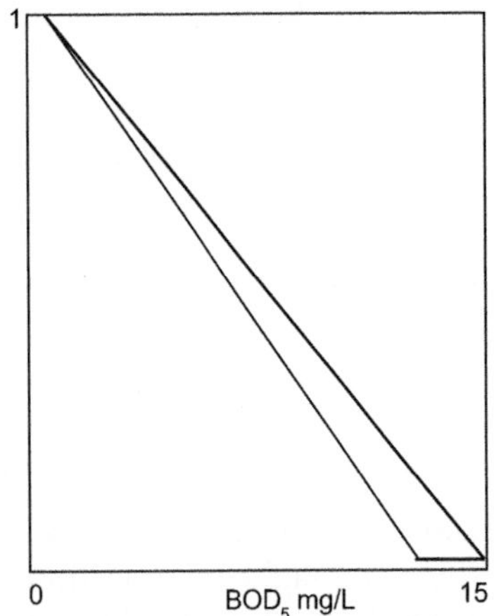

FIGURE 3.3 Representative sub-indices employed by Viet and Bhargava, 1989.

TABLE 3.15 Water Resource Classification as Per Viet and Bhargava's Index

WQI Value	Classification
90>	I Excellent
65–89	II Permissible
35–64	III Marginally suitable
11–34	IV Inadequate for use
10<	V Totally unsuitable

the Central Water Pollution Board, India (Sarkar and Abbasi 2006).

A list of parameters was selected through Delphi. Subindex values were obtained by using subindex equations as shown in Table 3.16.

To assign weightages, significance ratings were given to all the selected parameters. A temporary weight of 1 was assigned to the parameter which received highest significance rating. All other temporary weights were obtained by dividing each individual mean rating with the highest. Each temporary weight was then divided by the sum of all weights to arrive at the final weights. These weights were modified to suit the water-quality criteria for different categories of uses.

The method of obtaining weights and modified weights is illustrated in Table 3.17. The classification of water *vis a vis* the final index values is given in Table 3.18.

3.15. SMITH'S INDEX (1990)

The distinguishing feature of Smith's index is that it is a hybrid of the two common index types and is based on expert opinion as well as water-quality standards. Moreover, the index addresses four types of water use, which include contact as well as noncontact use (Table 3.19):

1. General
2. Regular public bathing
3. Water supply
4. Fish spawning

TABLE 3.16 Subindex Equations of the Index Reported by Ved Prakash et al.

Parameter	Range Applicable	Equation	Correlation
DO (percent Staturation)	0–40% saturation	IDO = 0.18 + 0.66x (% sat)	0.99
	40–100% saturation	IDO = −13.5 + 1.17x (% sat)	0.99
	100–140% saturation	IDO = 263.34 − 0.62x (% sat)	−0.99
BOD (mg/L)	0–10	IDO = 96.67 − 7 (BO)	−0.99
	10–30	IBOD = 38.9 − 1 (BOD)	−0.95
	>30	IBOD = 2	
pH	2–5	IpH = 16.1 + 7.35 x (pH)	0.925
	5–7.3	IpH = −142.67 + 33.5 x (pH)	0.99
	7.3–10	IpH = 316.96 − 29.85 x (pH)	−0.98
	10–12	IpH = 96.17 − 8.0 x (pH)	−0.93
	<2, >12	IpH = 0	
Faecal coliform (counts/100 ml)	$1–10^3$	Icoli = 97.2 − 26.60 x log(FC)	−0.99
	$10^3–10^5$	Icoli = 42.33 − 7.75 x log(FC)	−0.98
	10^5	Icoli = 2	

The selection of parameters for each water class, developing subindices and assigning weightages were all done using Delphi. But Smith (1990) employed additional rounds of questionnaire besides the usual Delphic steps, to arrive at greater convergence of opinion. The panel members were allowed to telephone the co-ordinator to seek clarifications. This was a departure from the standard Delphic procedure wherein direct contact and discussions among panel members are discouraged on the premise that such discussions may lead to some members unduly influencing some others. However, in the procedure adopted by Smith, it was considered prudent to let the experts interact and thrash out points of doubt.

Due to these unique attributes, the procedure employed by Smith (1990) is presented in some detail below.

As the starting point, Smith drew up a panel of 18 water-quality experts of differing backgrounds

TABLE 3.17 Method of Obtaining Weights and Modified Weights, *cf* Table 3.16

Parameters	Mean of all Significance Ratings	Temporary Weights	Final Weights	Modified Weights
DO	1.4	1.0	0.17	0.31
Faecal coliforms	1.5	0.9	0.15	0.28
pH	2.1	0.7	0.12	0.22
BOD	2.3	0.6	0.1	0.19
Total			0.54	1.00

TABLE 3.18 Water Class as Per Index Score; *cf* Table 3.16

S. N	WQI	Description	Class
1	63–100	Good to excellent	A
2	50–63	Medium to good	B
3	38–50	Bad	C
4	38 & less	Bad to very bad	D,E

TABLE 3.19 The Four Water Uses Which are Covered by Smith's Index

Category	Defining Characteristics
General	These waters have no specifically designated use, but can have competing uses. They are meant to be protected for the following uses: **(a)** the maintenance of a substantially unaltered aquatic community; **(b)** the general aesthetic amenity; **(c)** fishing; **(d)** stock watering; **(e)** irrigation; **(f)** public water supply after extensive treatment; **(g)** occasional contact use such as swimming; **(h)** waste assimilation The proposed standards for this water classification would generally be lower than for the specific uses.
Bathing	This is for water in rivers for regular public bathing. Such water will have other uses; in particular, aquatic life would require protection.
Water supply	This is source water for potable supply or for the preparation and processing of food for sale for human consumption, where treatment at least equivalent to flocculation, filtration and disinfection could be reasonably expected. Aquatic life also needs to be protected, but at a lower level than in the bathing and fish-spawning uses.
Fish spawning	This is water specially protected for fish-spawning purposes. For simplicity, it was assumed that these waters are salmonid waters.

to provide him with a range of expert opinion. The development process had several phases as outlined in Figure 3.4. Delphi method was employed to obtain and moderate the expert opinion at each stage. As mentioned in an earlier chapter, a series of questionnaires are used to elicit a group response from the panelists in Delphi. At each stage subsequent to the first, panel members are sent the pooled group response for the preceding stage and the individuals are asked to reassess their previous response in the light of the pooled group response. Attempt is thus made to obtain convergence of opinion. Anonymity between the panel members is preserved and communication is by mail.

Smith (1990) used five questionnaires and after the first questionnaire (which sought panel members' preliminary opinions on which determinands should be included for each water use) he employed an additional two rounds of questionnaires and sent supplementary material and questions to panel members to assist their deliberations. This helped in the resolution of difficult or contentious issues (Smith, 1987). The index development proceeded as follows:

Phase 1: Selection of determinands

Determinands were finally agreed upon as being the most desirable and appropriate for each of the four water uses. They are given in Table 3.20. They comprise some of the most commonly measured attributes of natural water. Initially, there was wide divergence of opinion among the panel members. The index does not include toxic substances because a wide range of different chemicals would have to be

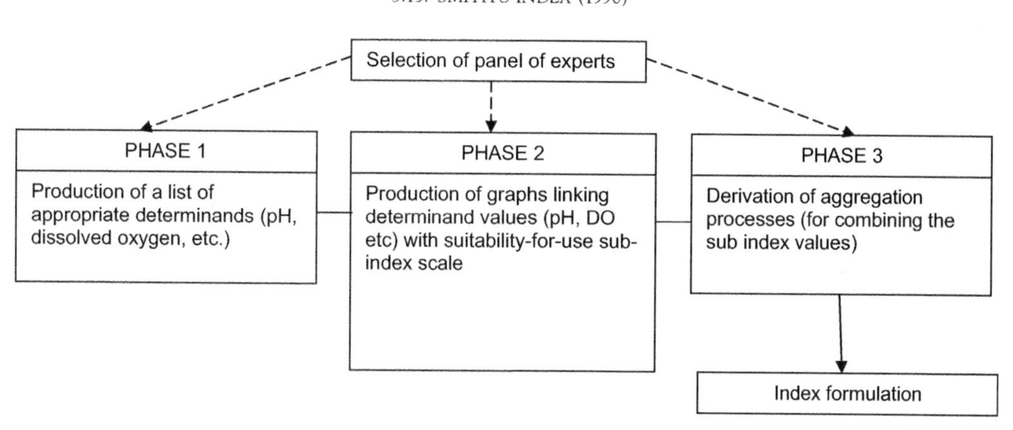

FIGURE 3.4 The three main phases of the development of Smith's Index. The information sought in the five rounds of questionnaire is briefly presented in the Appendix.

TABLE 3.20 Determinands used to produce index scores for Smith's Index

	Water Use			
Characteristic	**General**	**Bathing**	**Water Supply**	**Fish Spawning (Salmonids)**
Dissolved oxygen	*RD,NS*	*RD,NS*	*RD,NS*	*RD,NS*
pH	*RD,NS*	*RD,NS*	*RD,NS*	*RD,NS*
Suspended solids	*RD*	*RD*	*RD*	*RD*
Turbidity	*RD*	*RD*	*RD*	*RD*
Temperature (actual)	*RD,NS*	*RD,NS*	*RD,NS*	*RD,NS*
Temperature (elevation)	*RD,NS*	*RD,NS*	*RD,NS*	*RD,NS*
BOD_5 (unfiltered)	*RD*	*RD*	O	*RD*
Ammonia	–	–	*RD,NS*	–
Faecal coliforms	*RD*	*RD,NS*	*RD,NS*	–

RD means a required determinand.
RD, NS *means a required determinand but for which there is also a proposed numerical standard.*
O *Although BOD_5 is not a stipulated water supply use determinand, it may be included if local conditions warrant it. In this instance for the general water use BOD_5 sub-index curve was agreed to by the panel.*

measured, and toxicity was not a widespread problem in the country for which Smith's index was being developed, viz., New Zealand. Ammonia was included in the water-supply index but not as a consequence of its toxicity.

Phase 2: Subindex curve development

To obtain subindex curves, blank graph formats were supplied to the panel members. The *x*-axis represented the expected range of determinand values likely to be encountered in New Zealand during nonflood conditions. The *y*-axis ranged from 0 to 100. It represented the suitability-for-use axis (subindex ratings, /sub). Panel members were supplied with a mutually agreed set of descriptors for the range of subindex values (Table 3.20) and asked to draw a curve to indicate graphically the subindex rating (and hence suitability-for-use) as a function of determinand magnitude. Thus, a water pH considered eminently suitable for all designated uses would score 100, whereas a water totally unsuitable for the main and/or many uses could score as low as 0. In all, 32 curves were requested, one for each determinand listed under the water use headings in Table 3.20, except temperature for which two curves were required (actual temperature and elevation above natural temperature). In addition, for fish-spawning waters two actual temperature curves were required, one for the spawning season and one for the rest of the year. This was in accordance with the proposed standards in which maximum temperatures for both periods of the year are stipulated.

In some cases (i.e. where there was a proposed numerical standard), the curves were to be forced through a fixed point (the proposed standard's value and Isub = 60), corresponding to the lowest value in the 'suitable-for-all-uses' category (Table 3.21). For instance, in the proposed schedule for general use waters, the pH must not fall below 6.0, nor must it rise above 9.0. Thus, there are fixed points in the subindex graph at (6.0, 60) and (9.0, 60). The use of such fixed points was agreed to in advance by the panel members.

The panel members were informed that if they felt they had insufficient expertise to draw a particular curve, they were under no compulsion to do so. For most curves, there were responses from 10 to 12 panel members.

The curves drawn by the panel members were then averaged to produce the final graphs. This was accomplished by taking about 20 points along the x-axis of each individual curve, obtaining subindex values and arithmetically averaging them over the whole panel.

The new curves, drawn up using these averaged points together with 95% confidence limit curves (produced by linear interpolation of the 95% confidence limit points), were returned to panel members for comment. All were approved (i.e. no modifications were requested) except one and in this instance (fish-spawning waters, suspended solids subindex curve) a fishery expert suggested that the lower 95% confidence limit curve was more appropriate than the averaged curve. This was put to the panel by letter and met with agreement. The eight curves for general water use are illustrated in Figure 3.5; the curves for other water uses are similar in appearance.

TABLE 3.21 Descriptors for the Range of Subindex Values (Isub) in Smith's Index

Range	Descriptor
$100 \geq$ Isub ≥ 80	Eminently suitable for all uses
$80 >$ Isub ≥ 60	Suitable for all uses
$60 >$ Isub ≥ 40	Main use and/or some uses may be compromised
$40 >$ Isub ≥ 20	Unsuitable for main and/or several uses
$20 >$ Isub ≥ 0	Totally unsuitable for main and/or many uses

Phase 3: The aggregation process

The minimum operator function represented by the equation

$$I = \min \sum (Isub_1, Isub_2 \ldots Isub_n)$$

was used for aggregation of the subindex scores. It uses the lowest subindex rating to produce the find index score.

The logic behind the use of this function is that a water's suitability-for-use is largely governed by its poorest characteristic; this concept is similar to the limiting nutrient idea in eutrophication studies where one component can define the state of the water.

Other advantages of the minimum operator function as seen by Smith (1990) are:

1. There need be no restriction on the number of determinands employed (numbers have to be limited with commonly employed

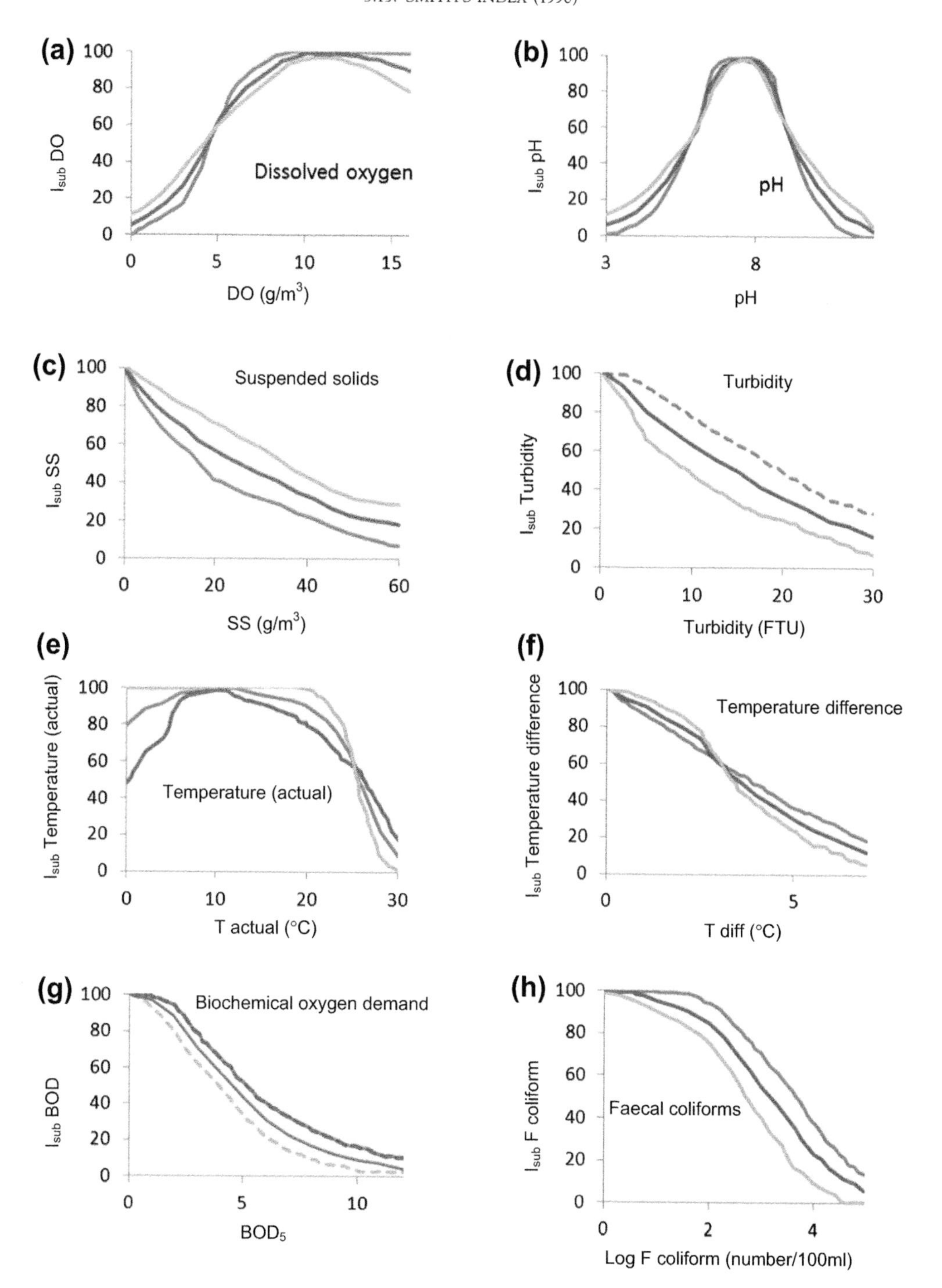

FIGURE 3.5 Typical subindex curves of Smith's WQI. The curves on either side of each middle curve indicate range within 95% confidence interval.

aggregation methods to reduce the eclipsing effect of the many over the few).
2. New determinands can easily be introduced (or current ones omitted) at a later stage without affecting index computation (this is difficult with methods which apply a relative weighting to the determinands).
3. Weightings are not required, thereby simplifying index development.

3.15.1. Testing of the Index

Prior to the dissemination of the index for national use, a preliminary test was carried out in which a totally different panel consisting of water managers, consultants and researchers was associated. Each panel member was sent a questionnaire consisting of four tables (one for each water use) of synthetic water-quality data (10 waters for each use category) and asked to derive their own index score for each of the 40 waters based purely on their own experience. They were also asked to identify the limiting determinand for a water's suitability-for-use. The panel members were informed of the use of a subindex value of 60 for determinands just meeting proposed numerical standards. They were also supplied with a list of specified water uses (as in Table 3.19), the relevant parts of the proposed legislation and the scale descriptors (Table 3.21).

It turned out that the panel members were able to identify the limiting determinand for a water's suitability-for-use except at theoretical

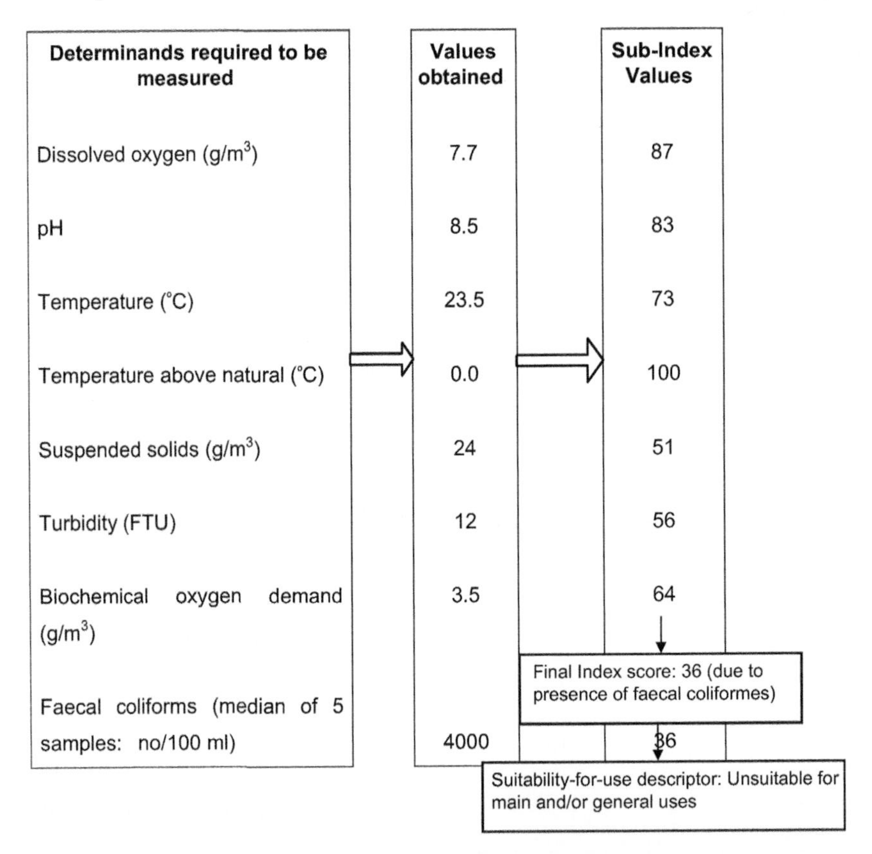

FIGURE 3.6 An example of how an index scores is obtained. The water use in this instance is general.

index scores greater than 90 when many subindex values are very similar. Turbidity and suspended solids index-determining values proved the most difficult to identify correctly.

An example of how index scores are obtained is presented in Figure 3.6.

3.15.2. Field Experience with the Index

The index occasionally produces interesting results. For example, if the temperature of a bathing water is 26.5 °C, it corresponds to a subindex value of 44 which is the lowest subindex value. This also happens to be the index score for the water which will fall into the category 'main use and/or some uses may be compromised' in spite of the fact that such a temperature may be conducive to bathing. Although this may seem strange at first glance, the fact is that bathing waters also contain aquatic life for which, in a temperate country such as New Zealand, 26.5 °C may represent stressful heat. At a temperature of 26.5 °C species such as trout are likely to be stressed. What this means is that if a water has multiple designated uses, it may never be 'perfect' for all its uses.

3.16. CHESAPEAKE BAY WATER-QUALITY INDICES (HAIRE ET AL. 1991)

The Maryland Department of Environment has developed a nutrient loading index and a eutrophication index for Chesapeake Bay and its major tributaries to provide easily understandable summary information concerning nutrient reduction and water-quality trends in the bay to legislators, administrators and the public. The nutrient loading index documents the average daily load of N and P to the system.

Some features of the eutrophication index include comparison of observed conditions to model projections of desired water quality, segmentation of each system into salinity zones and analysis of data collected through the comprehensive Chesapeake Bay monitoring programme.

For the Patuxent Estuary, the Potomac Estuary and the Chesapeake Bay mainstream, the results of the nutrient loading index show that significant progress has been made in the reduction of phosphorous point source loads. The eutrophication index for each of these estuaries indicates that the water quality is variable, largely because of flow conditions, but has generally improved over the last decade.

3.17. THE AQUATIC TOXICITY INDEX

The Aquatic Toxicity Index (ATI) was developed by Wepener et al. (1992) to assess the health of aquatic ecosystems. Since extensive toxicity database are available for fishes, the toxic effects of different water quality on fishes have been employed in ATI development as health indicators of the aquatic ecosystem. The physical water-quality parameters employed are pH, dissolved oxygen and turbidity, while the chemical determinants include ammonium, total dissolved salts, fluoride, potassium and orthophosphates. Among the potentially hazardous metals covered in the index are total zinc, manganese, chromium, copper, lead and nickel. An ATI scale, similar to the WQI scale proposed by Smith (1990) for salmonid spawning was used. The Solway modified unweighted additive aggregation function (House and Ellis, 1980) was initially employed to aggregate the values obtained from the rating curves

$$\mathrm{ATI} = \frac{1}{100} \cdot \left(\frac{1}{n} \sum_{i=1}^{n} q_i \right)^2 \tag{3.24}$$

where q_i is the quality of the ith parameter (a value between 0 and 100) and n is the number of determinants in the indexing system. Wepener et al. (1992) did not employ the weighted sum system, as too little information is available about the importance of one determinant compared to another under different local conditions and the inherent chemistry of the system as a whole. Moreover, it is impossible to compare the factors which have a direct and interactive effect upon one another, in order to avoid concealing the identity of the determinant which limits the water's suitability-for-use, the minimum operator function was also employed. A computer software package, named WATER, was developed to compute both the additive and the minimum operator final index values. Later, Wepener et al. (1999) reported the spatial and temporal trends of water quality of Olifants River and Selati River in Kruger National Park during the course of metal mining project over a two-year period (February 1990–April 1992) based on their index.

3.18. LI'S REGIONAL WATER RESOURCE QUALITY ASSESSMENT INDEX (1993)

A comprehensive quality index including function damage rate of water bodies combined with water quality, which has reasonable structure and strong synthetic ability, was proposed by Li (1993). It assesses the water quality for not only sections of rivers but also systems of regional water resources.

Application of the index, as illustrated by the author, can play an important role in water resource development as well as in water pollution control.

3.19. A TWO-TIER WQI

Dojlido et al. (1994) have used a two-tier index. At its 'basic parameters level' it relies on seven parameters which are frequently used in water-quality monitoring: BOD_5, suspended solids, phosphate, ammonia, dissolved solids, COD and dissolved oxygen.

At the higher 'additional parameters' level it considers several other parameters: COD, nitrate, lead, mercury, copper, chromium (VI), total chromium, zinc, cadmium, nickel and free cyanides.

The justification for the two-tier indexing as given by the authors is that the index based on basic parameters can be used to compare the water quality of different watercourses while the additional parameter index is utilisable to get deeper insights into a specific watercourse.

Subindices for individual parameters were developed on the basis of the Polish Ministry Council Standards, using four classes of water quality ranging from excessively polluted, to 'clean'. The index was calculated with the 'square root of the harmonic mean of squares' aggregation function:

$$WQI = \sqrt{\frac{n}{\sum_{i=1}^{n} \frac{1}{x_i^2}}}, \quad \text{if } x_i \neq 0 \quad \text{for each } i, \tag{3.25}$$

In this, if $x_i = 0$ for any i, WQI will be 0.

3.20. USE OF WQI TO ASSESS POND WATER QUALITY (SINHA, 1995)

The portability of the water of two ponds used by the villagers of Muzaffarpur District, Bihar, was assessed by Sinha (1995). He used an index similar to Brown's (NSF-WQI, 1973), which has been described earlier. Ten parameters – pH, hardness, DO, chloride, Na, K, Zn, Fe, turbidity and coliform – contributed to the sub-indices forming the WQI. The manner of computation of the index is illustrated in

Table 3.22. The water-quality characteristics of the two ponds and the monthly variation in the index scores are presented in Tables 3.23 and 3.24. The author concluded that the pond waters, though used for drinking, were actually not potable and needed proper treatment.

$$WQI = \sum_{i=1}^{10} w_i\, q_i \tag{3.26}$$

TABLE 3.22 Calculation of Water-Quality Index of Susta Pond for the Month of January, 1986

Water Quality Parameter	I.C.M.R. Standard	Unit Weight (W_i)	Value of Water Sample	Quality Rating (d_i)	Parameter Subindex ($d_i w_i$)
pH	7.0–8.5	0.229	7.8	53.33	1.2213
Hardness	300 mg/L	0.0006	52.0	17.33	0.0104
Dissolved oxygen	5 mg/L[@]	0.0352	6.2	87.5	3.08
Chloride	250 mg/L	0.0007	48.5	19.4	0.0136
Sodium	20 mg/L	0.0088	5.2	26.0	0.2288
Potassium	10 mg/L*	0.0176	5.8	58.0	1.0208
Zinc	5 mg/L	0.0352	0.05	1.0	0.0352
Iron	0.3 mg/L	0.5859	0.18	60.0	35.154
Turbidity	1.5 mg/L*	0.1172	10.0	666.66	78.1325
S.P.C. of coliform	1/100 ml	0.1758	1000	400.0	70.00

[@] *European Economic Community (E.E.C.) standard.*
* *Soviet State Standard (TOCT) No.2874–73.*

TABLE 3.23 Physico-Chemical Characteristics of the Water

		Susta Pond Range		Madhaul Pond Range	
Sl. No.	**Parameters**	**1st Year**	**2nd Year**	**1st Year**	**2nd Year**
1.	Turbidity (mg/L)	17.5 ± 7.5	20 ± 10	20 ± 9	20 ± 10
2.	Conductivity (ml mhos/cm)	0.34 ± 0.11	0.36 ± 0.12	0.435 ± 0.095	0.43 ± 0.08
3.	Dissolved Oxygen (mg/L)	5.8 ± 1.4	6.5 ± 1.1	6.2 ± 1.4	6.1 ± 1.3
4.	pH	7.85 ± 0.35	7.9 ± 0.5	7.85 ± 0.65	7.95 ± 0.45
5.	Sodium (mg/L)	5.2 ± 0.6	5.8 ± 0.9	5.35 ± 0.95	5.6 ± 0.8
6.	Potassium (mg/L)	6.25 ± 0.45	6.4 ± 1	5.75 ± 0.65	5.9 ± 1.4
7.	Zinc (mg/L)	0.095 ± 0.085	0.095 ± 0.065	2.6 ± 1	2.5 ± 1.1
8.	Iron (mg/L)	0.235 ± 0.075	0.155 ± 0.155	0.2 ± 0.08	0.14 ± 0.14
9.	Chloride (mg/L)	60.37 ± 11.88	54.75 ± 10.25	45.35 ± 6.85	52.11 ± 13.12
10.	Hardness (mg/L)	60 ± 12	63.5 ± 17.5	65.6 ± 14.4	64 ± 12.8

TABLE 3.24 Water-Quality Index of Two Ponds (January 1986 to September 1987)

Month	Susta Pond	Madhaul Pond
January (1986)	188.89	290.51
March	233.99	227.81
May	253.70	248.39
July	321.54	276.47
September	300.61	296.89
November	252.78	226.68
January (1987)	203.42	238.59
March	307.77	248.67
May	285.54	268.98
July	342.71	303.91
September	239.59	313.69

TABLE 3.25 Assignment of Weightage to Water-Quality Parameters

Parameters	Standards	Weights	Unit Weights
pH	7.0–8.5	4	0.16
Total Hardness (as $CaCO_3$) mg/L	100–500	2	0.08
Calcium mg/L	75–200	2	0.08
Magnesium mg/L	30–150	2	0.08
Total Alkalinity mg/L	<120	3	0.12
Dissolved Oxygen mg/L	>6	4	0.16
Total Solids (mg/L)	500–1500	4	0.16
Total Suspended Solids (mg/L)	<100	2	0.08
Chloride (mg/L)	200–500	2	0.08

3.21. USE OF WQI TO STUDY HANUMAN LAKE, JABALPUR (DHAMIJA AND JAIN 1995)

Hanumantal was studied on the basis of a WQI formed with 9 parameters which were assigned weights as summarised in Table 3.25. The unit weight (w_i) for each parameter was calculated as

$$W_i = \frac{W_i}{\sum_{i=1}^{9} W_i} \tag{3.27}$$

Each subindex was given by

$$(SI)_i = q_i w_i$$

where q_i is the quality rating of the ith parameter.

Then,

$$WQI = \sum_{i=1}^{i=9} q_i w_i \tag{3.28}$$

The rating scale was set up in the 0–100 range (Table 3.26). A typical calculation of WQI is illustrated in Table 3.27. The seasonal fluctuations in the WQI as a function of the fluctuations in the values of the various parameters are reflected in Table 3.28.

3.22. COASTAL WATER-QUALITY INDEX FOR TAIWAN (SHYUE ET AL. 1996)

A Coastal Water Quality Index (CWQI) was established to better understand the coastal water quality for the general public. Six coastal water-quality experts in Taiwan were surveyed by using Delphi to select several parameters from Marien Water Quality Standard.

The fourth-order polynomial regression was performed to process the surveyed data for each parameter as the scoring function. The minimum scoring method gave more diverse results for different water-quality monitored sites than the geometric weighted method. Therefore, the minimum scoring method was

TABLE 3.26 Rating Scale for Water-Quality Parameters

	Permissible	Slight	Moderate	Severe
Degree of Pollution Rating (q_i)	100	80	50	0
pH	7–8.5	8.6–8.8	8.9–9.2	>9.2
		6.8–7.0	6.5–6.7	<6.5
Total Hardness (mg/L)	<100	101–300	310–500	>500
Calcium Hardness (mg/L)	<75	76–137	138–200	>200
Magnesium Hardness (mg/L)	<30	31–90	91–150	150
Total Alkalinity (mg/L)	50	51–85	86–120	>120
Dissolved Oxygen (mg/L)	6	4.4–4.9	3–4.5	<3
Total Solids (mg/L)	500	500–1000	1000–1500	>1500
Total Suspended Solids (mg/L)	<30	30–65	65–100	>100
Chloride (mg/L)	<200	201–400	401–600	>600

favoured in order to distinguish the degree of the pollution.

The parameters identified for the CWQI are pH, DO, BOD, cyanide, coliform, Cu, Zn, Pb, Cd and Cr.

TABLE 3.27 Calculation of WQI of Hanumantal Lake for Site I in Summer

	Value in Summer		
Parameters	q_i	W_i	q_iW_i
pH	100	0.16	16.0
Total Hardness (mg/L)	100	0.08	8.0
Calcium Hardness (mg/L)	100	0.08	8.0
Magnesium Hardness (mg/L)	100	0.08	8.0
Total Alkalinity (mg/L)	0	0.12	0.0
Dissolved Oxygen (mg/L)	100	0.16	16.0
Total Solids (mg/L)	100	0.16	16.0
Total Suspended Solids (mg/L)	80	0.08	6.4
Chloride (mg/L)	100	0.08	80.0
WQI			86.4

3.23. THE MODIFIED OREGON WATER-QUALITY INDEX (CUDE, 2001)

3.23.1. Oregon Water-Quality Index (OWQI)

The OWQI was developed in the 1970s by the Oregon Department of Environmental Quality, USA, for the purpose of summarising and evaluating water-quality status and trends for the legislatively mandated water-quality status assessment reports. It was modelled after the National Sanitation Foundation's WQI (Brown et al., 1970, 1973) and employed the Delphi technique for the selection of water-quality variables. The water-quality variables were classified according to the impairment categories, i.e., oxygen depletion, eutrophication or potential for excess biological growth, dissolved substances and health hazards. However, the original OWQI was discontinued in 1983 on account of the enormous resources required for calculating and reporting the results. With the advancements in computer technology, enhanced tools of data

TABLE 3.28 Seasonal Fluctuations in Different Parameters and WQI at Hanumanutal, Jabalpur

	Summer		Monsoon		Winter	
Parameters	**Site I**	**Site II**	**Site I**	**Site II**	**Site I**	**Site II**
pH	7.51 (100)	7.41 (100)	7.45 (100)	7.37 (100)	7.68 (100)	7.61 (100)
Total hardness (mg/L)	94.0 (100)	92.5 (100)	153.0 (80)	167.5 (80)	130.0 (80)	111.87 (80)
Calcium hardness (mg/L)	73.7 (100)	73.7 (100)	118.0 (80)	127.5 (80)	83.73 (100)	73.75(100)
Magnesium hardness (mg/L)	20.25(100)	18.75 (100)	35.0 (80)	27.5 (100)	38.75 (80)	45.62 (80)
Total alkalinity (mg/L)	123.5 (0)	128.0 (0)	191.25 (0)	201.25 (0)	107.5 (50)	95.62 (50)
Dissolved oxygen (mg/L)	6.97 (100)	7.25 (100)	7.87 (100)	6.625 (100)	7.0 (100)	6.65 (100)
Total solids (mg/L)	300.25 (100)	305.25 (100)	340.0(100)	357.75 (100)	370.5 (100)	372.25 (100)
Total suspended solids (mg/L)	63.25 (80)	64.75 (80)	75.5 (50)	84.25 (50)	71.0 (50)	81.0 (50)
Chloride (mg/L)	103.96 (100)	106.46 (100)	58.73 (100)	58.72 (100)	69.97 (100)	64.93 (100)
WQI	86.4	86.4	79.2	80.8	86.8	86.8

display and visualisation and a better understanding of water quality, the OWQI was updated in 1995 by refining the original sub-indices, adding temperature and total phosphorus sub-indices and improving the aggregation calculation (Cude, 2001). The resulting index reflects the water quality of Oregon's streams with respect to general recreational use including fishing and swimming. The overall water quality is expressed as a single digit by integrating measurements of eight different water-quality variables, namely temperature, dissolved oxygen, biochemical oxygen demand, pH, ammonia + nitrate nitrogen, total phosphorus, total solids and faecal coliform. The subindex transformation formulae were derived using nonlinear regression, from the transform table developed from the originally hand-drawn OWQI subindex transformation curves (Dunnette, 1980). It was felt that the minimum operator aggregator had proved to be too sensitive to the most impacted variable and did not integrate the other variables (Cude, 2002). Consequently, the unweighted harmonic square mean formulae was employed for the purpose of aggregation of the subindex scores as an improvement over the weighted arithmetic mean formula used in the original version:

$$WQI = \sqrt{\frac{n}{\sum_{i=1}^{n} \frac{1}{SI_i^2}}} \qquad (3.29)$$

where n is the number of sub-indices and S_i is the subindex i.

The OWQI helps to evaluate the effectiveness of water-quality management activities. It may also be employed to develop environmental indicators, such as percentage of river monitoring sites with significantly improving water quality, or the percentage of sites with excellent water quality.

3.24. THE 'OVERALL INDEX OF POLLUTION'

A WQI called 'Overall Index of Pollution' (OIP) has been proposed by Sargoankar and

TABLE 3.29 Classification of Water Quality in the 'Overall Index of Pollution' (Sargoankar and Deshpande, 2003)

	Excellent	Acceptable	Slightly Polluted	Polluted	Heavily Polluted
Classification	C1	C2	C3	C4	C5
Class Index (score)	1	2	4	8	16
Parameters	**Concentration Limit/Ranges**				
Turbidity (NTU)	5	10	100	250	>250
pH	6.5–7.5	6.0–6.5 and 7.5–8.0	5.0–6.0 and 8.0–9.0	4.5–5 and 9–9.5	<4.5 and >9.5
Colour (Hazen Unit), max	10	150	300	600	1200
DO (%)	88–112	75–125	50–150	20–200	<20 and >200
BOD_5 (20 °C), (mg/L), max	1.5	3	6	12	24
TDS (mg/L), max	500	1500	2100	3000	>3000
Hardness $CaCO_3$ (mg/L), max	75	150	300	500	>500
Cl (mg/L), max	150	250	600	800	>800
NO_3 (mg/L), max	20	45	50	100	200
SO_4 (mg/L), max	150	250	400	1000	>1000
Total coliform (MPN), max	50	500	5000	10000	15000
As (mg/L), max	0.005	0.01	0.05	0.1	1.3
F (mg/L), max	1.2	1.5	2.5	6	>6.0

Deshpande (2003), when working at the National Environmental Engineering Research Institute (NEERI), Nagpur India. The OIP aims to assess the status of surface waters, specifically under Indian conditions. A general classification scheme has been formulated based on a concept similar to the one proposed by Prati et al. (1971) and giving due consideration to the classification scheme developed by the Central Pollution Control Board (CPCB), India, and the Indian Standards Institution (ISI). The scheme reflects the status of water quality in terms of pollution effects of parameters under consideration. Five classes, namely C1: Excellent/pristine, C2: Acceptable/requires disinfection, C3: Slightly polluted/requires filtration and disinfection, C4: Polluted/requires special treatment and disinfection and C5: Heavily polluted/cannot be used, have been considered. Different concentration levels of the parameters were put into these classes or categories on the basis of the standards/criteria employed by CPCB, ISI or other agencies (Table 3.29). In order to bring the different water-quality parameters into a commensurate unit, an integer value 1, 2, 4, 8 and 16 has been assigned to each of the five classes C1, C2, C3, C4 and C5, respectively in geometric progression.

These numbers are termed as class indices and they indicate the level of pollution in numeric terms. The parameter concentration is then assigned to the respective mathematical expression to obtain a numerical value called an index (P_i) which indicates the level of pollution for that parameter. The Overall Index of Pollution (OIP) is then evaluated as a mean of all the individual pollution indices (P_i) as follows:

$$\text{OIP} = \frac{1}{n}\sum_{i=1}^{n} P_i \tag{3.30}$$

where P_i is the pollution index for the ith parameter, $i = 1, 2, \ldots, n$ and n = number of parameters. The index was tested by the authors for the assessment of surface water status as well as the formulation of pollution control strategies in terms of treatment required at different levels. The index was employed to ascertain the suitability of water at a few sampling stations along the Yamuna River, India.

3.25. THE CANADIAN WATER-QUALITY INDEX (CCME, 2001) AND THE INDEX OF SAID ET AL. (2004)

Please see Chapter 10 for details.

3.26. A 'UNIVERSAL' WATER-QUALITY INDEX

Boyacioglu (2007) took into consideration the water-quality standards set by the Council of European Communities (EC 1991), the Turkish water pollution control regulations and other scientific information to select 12 water-quality parameters as the most representative for drinking water quality (Table 3.30). They set three

TABLE 3.30 Classification of Water Quality for the Development of the 'Universal' WQI of Boyacioglu (2007)

Parameter	Unit	Class I (Excellent)	Class II (Acceptable)	Class III (Polluted)	Remark
Total Coliform	CPU/100 mL	50	5000	50000	It is used to indicate whether other potentially harmful bacteria may be present
Cadmium	mg/L	0.003	0.005	0.01	Chemicals from industrial and domestic discharges
Cyanide	mg/L	0.01	0.05	0.1	
Mercury	mg/L	0.0001	0.0005	0.002	
Selenium	mg/L	0.01	0.01	0.02	Naturally occurring chemicals
Arsenic	mg/L	0.02	0.05	0.1	
Fluoride	mg/L	1	1.5	2	
Nitrate–nitrogen	mg/L	5	10	20	Chemicals from agricultural activities
DO	mg/L	8	6	3	Operational monitoring parameters
pH		6.5–8.5	5.5–6.4 8.6–9	<5.5 >9	
BOD	mg/L	<3	<5	<7	Indicator of organic pollution
Total phosphorus-PO_4-P	mg/L	0.02	0.16	0.65	It is included to satisfy the ecological requirements of certain types of environment

classes of water – representing 'excellent', 'acceptable' and 'polluted' categories (Table 3.30).

To assign weights to the water-quality variables the following factors taken into account:

- Chemical parameters had a lower weight than microbiological parameters, because microbial contaminants belong to the greatest health impact category
- Higher weight was given to those parameters which are of known health concern

The temporary weights ranged from 1 to 4 on a basic scale of importance. On this scale 1, 2, 3 and 4 denote, respectively, average, great and very great importance. Each weight was then divided by the sum of all weights to arrive at the final weight factor (Table 3.31).

The index is given by

$$\mathrm{UWQI} = \sum_{i=1}^{n} w_i I_i \tag{3.31}$$

TABLE 3.31 Significance Ratings and Weights Assigned to Different Parameters in the UWQI of Boyacioglu (2007)

Category	Variable	Rating	Weight Factor
Health hazard	Total coliform	4	0.114
	Cadmium	3	0.086
	Cyanide	3	0.086
	Mercury	3	0.086
	Selenium	3	0.086
	Arsenic	4	0.113
	Fluoride	3	0.086
	Nitrate–nitrogen	3	0.086
Operational	DO	4	0.114
Monitoring	pH	1	0.029
Oxygen	BOD	2	0.057
Depletion	Total phosphorus	2	0.057

where w_i is the weight for i^{th} parameter

I_i is the subindex for i^{th} parameter

The index value in the range of 0 to less than 25 represents poor quality, 25 to less than 50 marginal quality, 50 to less than 75 fair quality, 75 to less than 95 good quality and, above it, excellent quality.

3.27. IMPROVED METHODS OF AGGREGATION

Swamee and Tyagi (2000, 2007) took a close look at the methods of aggregation (described in the preceding chapter) that have been used in developing all the previous WQIs, and proposed newer methods to overcome one or the other problems of ambiguity, eclipsing and rigidity that have been associated with the previous indices.

As noted in Chapter 2, ambiguity is caused in an aggregation method when an index exceeds the critical level (unacceptable value) without any of its constituent sub-indices exceeding the critical level, and eclipsing is caused when an index does *not* exceed the critical level (unacceptable value) despite one or more of its constituent sub-indices exceeding the critical level.

The other major problem is rigidity. It occurs when it is necessary to include additional variables in the index to address specific water-quality concerns, but the aggregation function does not allow it without upsetting the index. For instance, it is common for a regulatory agency to have an existing overall index, but the agency would like to add one or more additional parameters. A situation may arise when a particular site receives a good water-quality index, and yet has water quality impaired by constituents not included in the index. It is also common for a similar regulatory agency in another region or area to require a different number of water-quality variables in its aggregated index. However, most of the pre-existing aggregation forms do not have any provision to add an additional parameter into its pre-identified set of water-quality constituents. Attempts to do it often leads to a decrease in the value of the overall index as the number of sub-indices increases, irrespective of their magnitude. When this happens, the ambiguity, if already present in that index, gets magnified. On the other hand, some indices which were free from ambiguity, none acquire it.

To overcome these problems, the authors have proposed new methods of aggregation taking the well-known NSF-WQI as the base. In the NSF-WQI, curves are used to relate concentrations or measurements of various constituents to sub-indices and then individual scores are aggregated into a single number (Brown et al., 1970). These curves are considered as the consensus of national criteria, state standards, information developed in the technical literature and professional judgement (Peterson and Bogue, 1989). Initially, using these curves Swamee and Tyagi (2000) developed three generic subindex equations to relate concentrations of various water-quality variables included in the NSF-WQI to their respective index scores. For monotonically decreasing sub-indices with their water-quality concentrations, they gave the subindex equation as

$$s = \left(1 + \frac{q}{q_c}\right)^{-m} \tag{3.32}$$

where q is the concentration of a water-quality variable, q_c is the characteristic value of q and m is the exponent.

The generic equation for nonuniformly decreasing sub-indices was given as

$$s = \frac{1 + \left(\frac{q}{q_T}\right)^4}{1 + 3\left(\frac{q}{q_T}\right)^4 + 3\left(\frac{q}{q_T}\right)^8} \tag{3.33}$$

where q_T is the threshold concentration for various water-quality variables.

For unimodal sub-indices having maxima at $s = 1$ at $q = q^*$, the generic subindex equation was described as

$$s = \frac{pr + (n+p)(1-r)\left(\frac{q}{q_*}\right)^n}{p + n(1-r)\left(\frac{q}{q_*}\right)^{n+p}} \tag{3.34}$$

where r is the subindex for $q = 0$, q_* is the characteristic value of q and n and p are exponents.

After converting concentrations of various water-quality variables included in an index to their respective sub-indices the individual scores were aggregated into a single number to get an overall water-quality index. In order to be free from ambiguity and eclipsing problems, the aggregation form that aggregated N sub-indices s_1, s_2, s_3, ..., s_N into a single number I was expected to exhibit the following properties (Swamee and Tyagi, 2000):

$$I(1, 1, \ldots, 1, s_i, 1, \ldots, 1) = s_i \tag{3.35}$$

$$I(s_1, s_2, \ldots, s_{i-1}, 0, s_{i+1}, \ldots, s_N) = 0 \tag{3.36}$$

The aggregation form free from ambiguity and eclipsing problems that satisfied the requirements (3.35) and (3.36) was proposed by Swamee and Tyagi (2000) as

$$I = \left(1 - N + \sum_{i=1}^{N} s_i^{-1/k}\right)^{-k} \tag{3.37}$$

where k is a positive constant. But this mathematical structure was found to introduce rigidity in the analysis. If the number of sub-indices N increased, the aggregated index I became smaller, producing an ambiguity problem as a result of rigidity. Hence, in order to solve rigidity problems and make the aggregated form flexible, it was deemed necessary to vary k with N to expand the WQI to include more water-quality variables.

In order to determine the exponent k an additional condition was imposed by Swamee and Tyagi (2007): For $s_i = 0.5$, $i = 1, 2, 3, \ldots, N$; $I = 0.25$. Imposing this condition, Equation (3.37) reduced to

$$2^{2/k} - N2^{1/k} + N - 1 = 0 \qquad (3.38)$$

Solving Equation (3.38) as a quadratic equation led to

$$k = \frac{1}{\log_2(N-1)} \qquad (3.39)$$

Thus, the aggregation process was modified to

$$I = \left[1 - N + \sum_{i=1}^{N} s_i^{-\log_2(N-1)}\right]^{-1/\log_2(N-1)} \qquad (3.40)$$

In the special case of all the sub-indices being equal, that is, $s_i = s$, Equation (3.37) changes to

$$I = 1 + N\left(s^{-1/k} - 1\right)^{-k} \qquad (3.41)$$

Accordingly, Equation (3.40) reduces to

$$I = [I - N(s^{-\log_2(N-1)} - 1)]^{-1/\log_2(N-1)} \qquad (3.42)$$

The authors have showed, with illustrative examples, that this aggregation form is free from ambiguity, eclipsing and rigidity problems. It allows the inclusion of additional water-quality parameters, and provides consistent results for the overall index irrespective of the number of parameters selected to define the overall index. Thus, using this aggregation form, one is not limited to a pre-identified set of water-quality parameters as have been used in the past. Further, a user can use either a smaller or a larger set of water-quality parameters based on data availability and the objectives of the index application.

3.28. A FIRST-EVER WQI FOR VIETNAM

Hanh et al. (2011) have proposed the first-ever water-quality index which has been aimed at monitoring and managing the quality of the surface waters in Vietnam.

The index encompasses twenty-seven water-quality parameters covering a wide range of physico-chemical variables, oil and grease, coliform and pesticides.

The authors have employed the rating curves method to (Figure 3.7) transform the concentrations of water-quality variables into quality scores. A hybrid aggregation function of additive and multiplicative forms initially proposed by Liou et al. (2004) was used to aggregate sub-indices to produce a final index score. Principal component analysis (PCA) was applied to divide the selected parameters into groups – the original variables were transformed into new uncorrelated variables, or the principal components (PC):

$$Z_{ij} = a_{i1}x_{ij} + a_{i2}x_{2j} + a_{i3}x_{3j} + \cdots + a_{im}x_{mj} \qquad (3.43)$$

where z is component score, a is the component loading, x is the measured value of variable, i is the component number, j is the sample number and m is the total number of variables. The number of PC to remain and their component loadings were characterised by eigenvalues, percent of total variance and cumulative percentage.

The range of water-quality parameters and their incremental levels defined for rating curves are as presented in Table 3.32. On the basis of these rating curves, parameter concentrations were given final scores between 1 (the

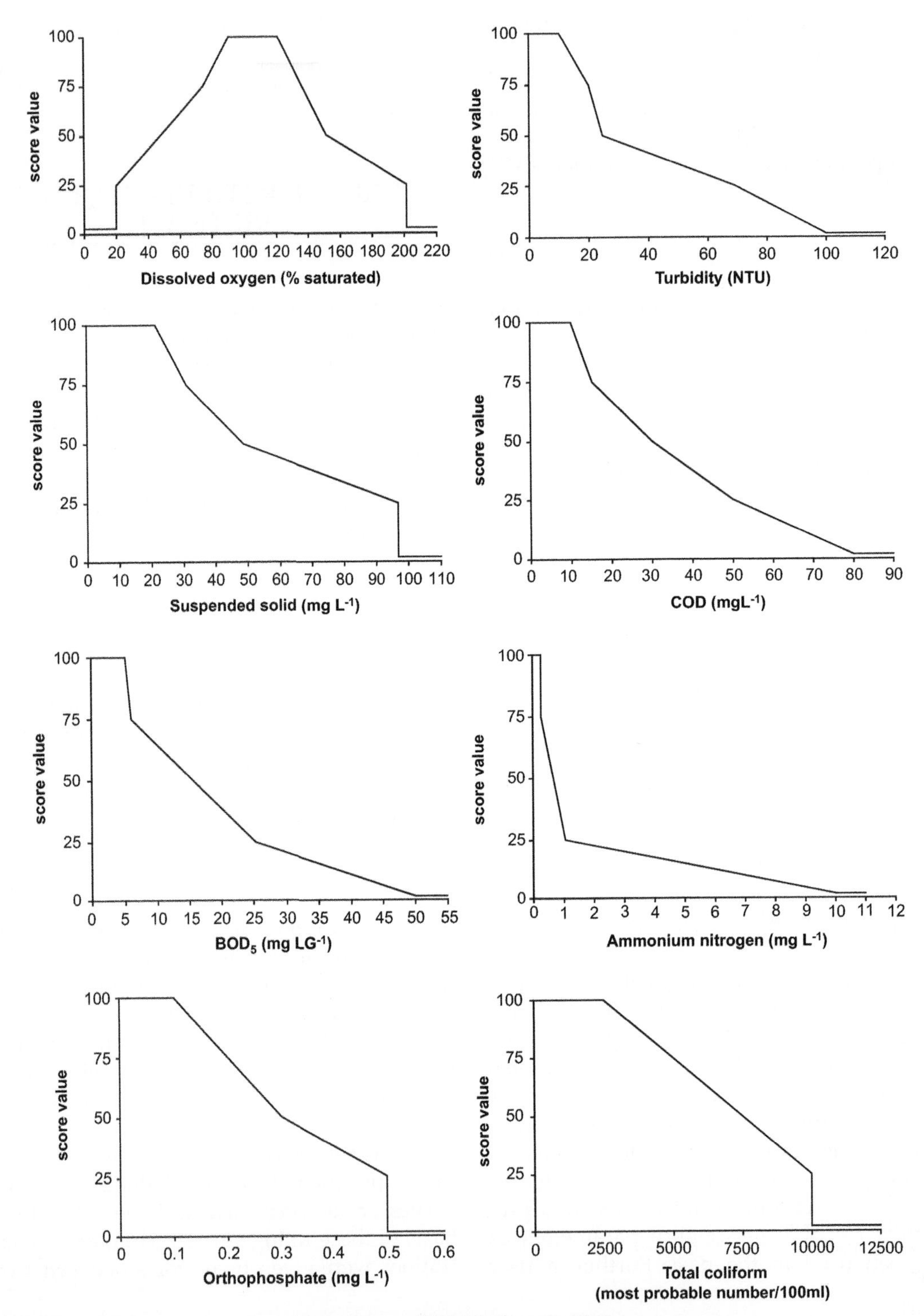

FIGURE 3.7 Rating curves used in the Vietnamese WQI (Hanh et al., 2011).

TABLE 3.32 Weightage Assigned to Water-Quality Parameters in the Vietnamase WQI

	Score Value				
	100	75	50	25	1
Parameter	**Level 1**	**Level 2**	**Level 3**	**Level 4**	**Level 5**
pH	6–8.5	6–8.5	5.5–9	5.5–9	5.5–9.0
Temperature, °C	–	–	–	–	40
DO saturated, %	88–112	75–88 and 112–125	50–75 and 125–150	20–50 and 150–200	<20 and >200
Turbidity, NTU	5	20	30	70	100
SS, mg/L	20	30	50	100	100
COD, mg/L	10	15	30	50	80
BOD, mg/L	4	6	15	25	50
Ammonium, mg/L	0.1	0.2	0.5	1	10
Nitrite, mg/L	0.01	0.02	0.04	0.05	–
Nitrate, mg/L	2	5	10	15	–
Orthophosphate, mg/L	0.1	0.2	0.3	0.5	–
Chlorine, mg/L	250	400	600	–	600
Fluorine, mg/L	1	1.5	1.5	2	10
Cyanide, mg/L	0.005	0.01	0.02	0.02	0.1
Arsenic, mg/L	0.01	0.02	0.05	0.1	0.1
Cadmium, mg/L	0.005	0.005	0.01	0.01	0.01
Lead, mg/L	0.02	0.02	0.05	0.05	0.5
Chrome (3), mg/L	0.05	0.1	0.5	1	–
Chrome (6), mg/L	0.01	0.02	0.04	0.05	–
Copper, mg/L	0.1	0.2	0.5	1	2
Zinc, mg/L	0.5	1	1.5	2	3
Ni, mg/L	0.1	0.1	0.1	0.1	0.5
Total iron mg/L	0.5	1	1.5	2	5
Mercury, mg/L	0.001	0.001	0.001	0.002	0.01
Manganese, mg/L	0.1	–	0.8	–	1
Oils and grease, mg/L	0.01	0.02	0.1	0.3	–
Phenol, mg/L	0.005	0.005	0.01	0.02	0.5
E. coli or thermotolerant coliform bacteria, MPN per 100 ml	20	50	100	200	–
Total coliform, MPN per 100 ml	2,500	5,000	7,500	10,000	–

TABLE 3.33 An Overview of Types of Subindices, Aggregation Functions and Flaws

Index	Subindices	Aggregation Function	Flaws
Horton's	Segmented linear (step functions)	Weighted sum multiplied by 2 dichotomous terms	Eclipsing region
Brown et al. (NSF WQI_a)	Implicit nonlinear	Weighted sum	Eclipsing region
Landwehr (NSF WQI_m)	Implicit nonlinear	Weighted product	Non-linear
Parti et al.	Segmented non-linear	Weighted sum (arithmetic mean)	Eclipsing region
McDuffie & Haney	Linear	Weighted Sum	Eclipsing region
Dinius	Non-linear	Weighted sum	Eclipsing region
Dee et al.	Implicit nonlinear	Weighted sum	Eclipsing region
O'Connor's (FAWL, PWS)	Implicit nonlinear	Weighted sum	Eclipsing region
Deininger & Landwehr (PWS)	Implicit nonlinear	Weighted sum	Eclipsing region
		Weighted product	Non-linear
Walski & Parker	Non-linear	Weighted product Geometric mean	Non-linear
Stoner	Non-linear	Weighted sum	
Nemerow & Sumitomo	Segmented linear	Root mean square of max. & arithmetic mean	Can give negative value
Smith	Multiple types	Minimum operator	–
Viet & Bhargava	Multiple types	Weighted product	–

worst case) and 100 (the best case). The curves are in the piecewise-linear-membership-functions form.

PCA led to three components of the basic parameter group; of these, the first component accounted for 46.56% of total variance, indicating strong positive loadings on BOD_5, COD, NH_4^+ and PO_4^{3-}, and moderate negative loading on DO. As high levels of organic matter and nutrients consume large amounts of dissolved oxygen, that component was taken to represent organic and nutrient pollution. The second component, assigned for particulate pollution, correlated strongly with suspended solids and turbidity and explained 24.02% of total variance. The third component, accounting for 12.54% of total variance, was contributed by *E. coli*.

Based on this, the aggregation function for the basic water-quality indicator (WQI_B) came to be

$$WQI_B = \left[\frac{1}{5}\sum_{i=1}^{5} q_i \times \frac{1}{2}\sum_{j=1}^{2} q_j \times q_k\right]^{1/3} \quad (3.44)$$

where q_i is subindex value of the organic and nutrients group containing DO, BOD_5, COD, NH_4^+, –N PO_4^{3-} and –P; q_j is the subindex value of the particulate group containing SS and turbidity; and q_k is the subindex value of the bacteria group represented by *E. coli*.

The sub-indices for additional water-quality parameters were then calculated. Each subindex was compared with the WQI_B and taken into account only if it was lower. The Tw and pH coefficients were calculated directly from their respective sub-indices. The toxic coefficient was calculated by averaging all scores of toxic substances. Since the overall index, WQI_O, was scaled from 1 to 100, the Tw, pH and toxic coefficients were scaled from 0.01 to 1. With this the WQI_O aggregation function read as

$$WQI_O = \left[\prod_i^n C_i\right]^{1/n} \left[\frac{1}{5}\sum_{i=1}^{5} q_i \times \frac{1}{2}\sum_{j=1}^{2} q_j \times q_k\right]^{1/3} \quad (3.45)$$

where C_i represents the coefficients addressing the sub-indices of Tw, pH and toxic substances; and n is number of coefficients.

Water quality was classified on the basis of the WQI_B or WQI_O scores as

91–100: excellent water quality;
76–90: good water quality;
51–75: fair;
26–50: marginal; and
1–25: poor water quality.

3.29. A COMPARISON

A comparison of various indices is presented in Table 3.33.

References

Abbasi, S.A., Arya, D.S., 2000. Environmental Impact Assessment. Discovery Publishing House, New Delhi.

Bhargava, D.S., 1983. Use of a water quality index for river classification and zoning of Ganga River. Environmental Pollution Series B: Chemical and Physical 6 (1), 51–67.

Bhargava, D.S., 1985. Water quality variations and control technology of Yamuna River. Environmental Pollution Series A: Ecological and Biological 37 (4), 355–376.

Boyacioglu, H., 2007. Development of a water quality index based on a European classification scheme. Water SA 33 (1), 101–106.

Brown, R.M., McClelland, N.I., Deininger, R.A., Landwehr, J.M. (1973). Validating the WQI. The paper presented at national meeting of American society of civil engineers on water resources engineering, Washington, DC.

Brown, R.M., McClelland, N.I., Deininger, R.A., Tozer, R.G., 1970. A water quality index – do we dare? Water Sewage Works 117, 339–343.

Canadian Council of Ministers of the Environment (CCME), 2001. Canadian water quality index 1.0 technical report and user's manual. Canadian Environmental Quality Guidelines Water Quality Index Technical Subcommittee, Gatineau, QC, Canada.

Cude, C.G., 2001. Oregon water quality index: A tool for evaluating water quality management effectiveness. Journal of the American Water Resources Association 37 (1), 125–137.

Cude, C.G., 2002. Reply to discussion – Oregon water quality index: a tool for evaluating water quality management. Journal of the American Water Resources Association 38 (1), 315–318.

Dalkey, N., Helmer, O., 1963. An experimental application of the Delphi method to the use of experts. Management Science 9, 458–467.

Deininger, R.A., Landwehr, J.M., 1971. A water quality index for public water supplies. Unpublished report, School of Public Health, University of Michigan, Ann Arbor.

Dhamija, S.K., Jain, Y., 1995. Studies an the water quality index of a lentic water body at Jabalpur M.P. Poll. Res. 14 (3), 341–346.

Dinius, S.H., 1972. Social accounting system for evaluating water. Water Resources Research 8 (5), 1159–1177.

Dinius, S.H., 1987. Design of an index of water quality. Water Resources Bulletin 23 (5), 833–843.

Dojlido, J., Raniszewsk, I.J., Woyciechowska, J., 1994. Water quality index – application for rivers in Vistula river basin in Poland. Water Science and Technology 30, 57–64.

Dunnette, D.A., 1980. Oregon Water Quality Index Staff Manual. Oregon Department of Environmental Quality, Portland, Oregon.

Haire, M.S., Panday, N.N., Domotor, D.K., Flora, D.G., 1991. USEPA Report, No. EPA-600/9–91/039.

Hanh, P., Sthiannopkao, S., Ba, D., Kim, K.W., 2011. Development of water quality indexes to identify pollutants in vietnam's surface water. Journal of Environmental Engineering 137 (4), 273–283.

Helmer, O., Rescher, N., 1959. On the epistemology of the inexact sciences. Management Science 6 (1).

Horton, R.K., 1965. An index number system for rating water quality. Journal of Water Pollution Control Federation 37 (3), 300–306.

House, M., Ellis, J.B., 1980. Water quality indices (UK): an additional management tool? Progress in Water Technology 13, 413–423.

Li, C., 1993. Zhongguo Nuanjing Kexue (Chinese) 13, 63.

Liou, S.M., Lo, S.L., Wang, S.H., 2004. A generalized water quality index for Taiwan. Environmental Monitoring and Assessment 96 (40603), 35–52.

Lumb, A., Sharma, T.C., Bibeault, J.-F., 2011. A Review of genesis and evolution of water quality index (WQI) and some future directions. Water Quality, Exposure and Health, 1–14.

Nemerow, N.L., Sumitomo, H., 1970. Benefits of Water Quality Enhancement, Report No. 16110 DAJ, prepared for the U.S. Environmental Protection Agency. December 1970. Syracuse University, Syracuse, NY.

O'Connor, F.M., 1972. The application of multi-attribute scaling procedures to the development of indices of water quality. Ph.D. dissertation, Univ. of Michigan.

Otto, W.R., 1978. Environmental Indices: Theory and Practice. Ann Arbor Science Publishers Inc, Ann Arbor, MI.

Peterson, R., Bogue, B., 1989. Water quality index (used in environmental Assessments), EPA Region 10. Seattle.

Prati, L., Pavanello, R., Pesarin, F., 1971. Assessment of surface water quality by a single index of pollution. Water Research 5, 741–751.

Said, A., Stevens, D.K., Sehlke, G., 2004. An innovative index for evaluating water quality in streams. Environmental Management 34 (3), 406–414.

Sargoankar, A., Deshpande, V., 2003. Development of an overall index of pollution for surface water based on a general classification scheme in Indian context. Environmental Monitoring and Assessment 89, 43–67.

Sarkar, C., Abbasi, S.A., 2006. Qualidex - A new software for generating water quality indice. Environmental Monitoring and Assessment 119, 201–231.

Shyue, S.-W., Lee, C.-L., Chen, H.-C., 1996. Approach to a coastal water quality index for Taiwan 904–907.

Sinha, S.K., 1995. Potability of some rural ponds water at Muzaffarpur (Bihar)-A note on water quality index. Journal of Pollution Research 14 (1), 135–140.

Smith, D.G., 1987. Water Quality Indexes for Use in New Zealand's Rivers and Streams. Water Quality Centre Publication No.12, Water Quality Centre, Ministry of Works and Development, Hamilton, New Zealand.

Smith, D.G., 1990. A better water quality indexing system for rivers and streams. Water Research 24 (10), 1237–1244.

Stoner, J.D., 1978. Water-quality indices for specific water uses. Us Geol. Surv. Circ. (770).

Swamee, P.K., Tyagi, A., 2000. Describing water quality with aggregate index. ASCE Journal of Environmental Engineering 126 (5), 451–455.

Swamee, P.K., Tyagi, A., 2007. Improved method for aggregation of water quality subindices. Journal of Environmental Engineering 133 (2), 220–225.

Viet, N.T., Bhargava, D.S., 1989. Indian J. Environ. Health 31, 321.

Walski, T.M., Parker, F.L., 1974. Consumers water quality index. Asce j environ. eng. div. 100 (EE3), 593–611.

Wepener, V., Euler, N., van Vuren, J.H.J., Du Preez, H.H., Kohler, A., 1992. The development of an aquatic toxicity index as a tool in the operational management of water quality in the Olifants River (Kruger National Park). Koedoe 35 (2), 1–9.

Wepener, V., Van Vuren, J.H.J., Preez, H.H.D.U., 1999. The implementation of an aquatic toxicity index as a water quality monitoring tool in the Olifants River (Kruger National Park). Koedoe 42 (1), 85–96.

CHAPTER

4

Combating Uncertainties in Index-based Assessment of Water Quality: The Use of More Advanced Statistics, Probability Theory and Artificial Intelligence

All the approaches to water-quality index formulation described in the preceding chapters are 'crisp' and 'deterministic' in the sense that they rely on accurate determination of water-quality characteristics and then on crisp manners of aggregation using deterministic tools that can resolve sets of characteristics into overall water quality.

But all experimental methods to assess water quality are also strongly 'reductionist' in the sense that they try to assess the nature of the whole on the basis of a few parts of the whole. For example, any natural or anthropogenically perturbed source of water consists of hundreds of chemicals. It may also have radioactivity that may differ in nature from source to source. If we wish to have an accurate and precise assessment of the water quality of a source, we need to analyse the water for each and every chemical, physical and biological characteristic it may possess. But such an assessment will be prohibitively costly. Hence, a 'reductionist' approach is followed by choosing a few among the likely constituents for analysis by assuming that the chosen components provide a fair representation of all the rest of the components. This introduces elements of subjectivity and uncertainty in the endeavour at the outset.

Moreover, the characteristics of any natural water course differ strongly from time to time and across its space (Abbasi, 1998). For example, the water quality of a lake may differ strongly at any two points across its depth or breadth at any instant of time. And the characteristics may also change with time; the physico-chemical characteristics and the biological assemblage at any point in a lake may change significantly with the time of the day. This may be understood from illustrative examples presented in Figures 4.1 and 4.2.

In Figure 4.1 the patterns of dissolved oxygen (DO) concentration and temperature as functions of depth, found in tropical lakes, are presented. If a sample taken from the surface would show a DO of ~10 mg l^{-1} and temperature ~30°C, a sample taken from a depth of 5 meters may have a DO of merely ~2 mg l^{-1} and a temperature as much as 10°C lesser.

Water Quality Indices DOI: 10.1016/B978-0-444-54304-2.00004-X

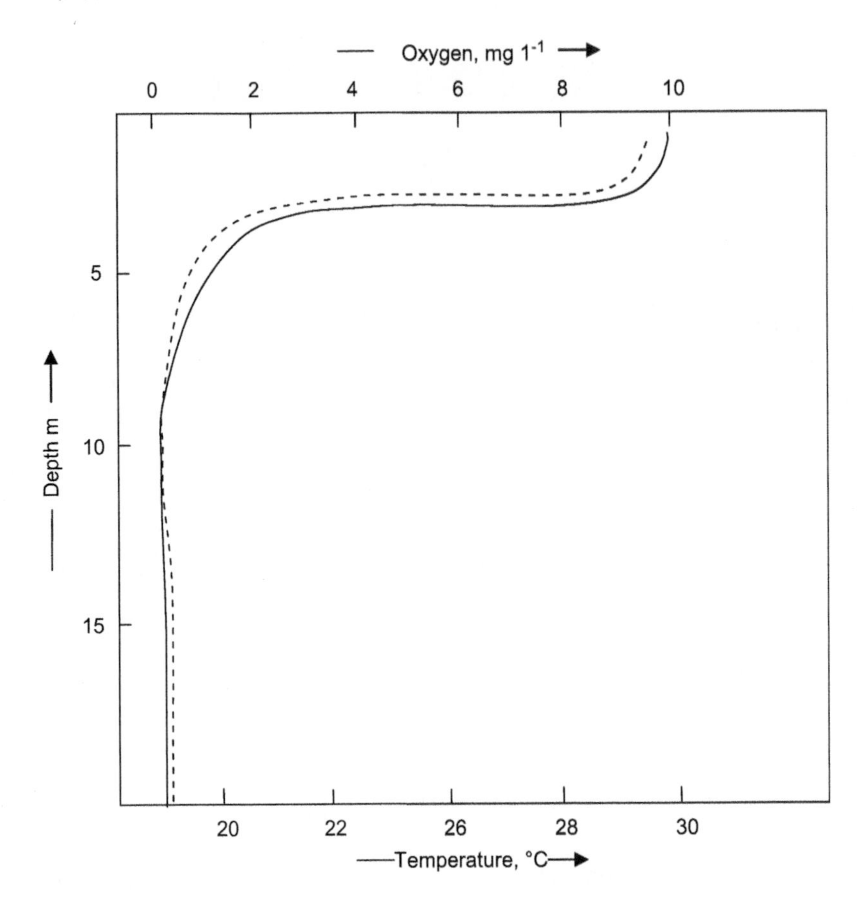

FIGURE 4.1 Dissolved oxygen (—) and temperature (---) across the depth of a typical lake.

Even these patterns are not rigid or constant and would change substantially over the course of even a day. Figure 4.2 illustrates how the two key water quality parameters – pH and CO_2 – can vary from dawn to dusk in a river. Such changes occur because water bodies are not just inanimate spans of water as they may appear from above the surface but are dynamic and 'live' systems. As the day breaks and sunlight begins to fall on a water body, the process of photosynthesis starts. This gives rise to utilisation of dissolved CO_2 from water and release of oxygen. At the high noon, when the solar influx is maximum, these processes attain their peaks. As the day wears on and the solar radiation begins to decline in intensity, respiration begins to dominate again so that CO_2 level begins to rise and DO level begins to fall. There is, therefore, a wide fluctuation in pH and CO_2 over the course of a day as illustrated in Figure 4.2. These fluctuations induce similar diurnal fluctuations in all other water-quality variables such as hardness, alkalinity, trace metals, phosphorus, and nitrogen.

On a day of cloud cover, or partial cloud cover, these patterns would be different. So, they differ across seasons, as the ambient temperature and solar flux would vary. The patterns will also be greatly influenced by factors such as withdrawal of appreciable quantities of

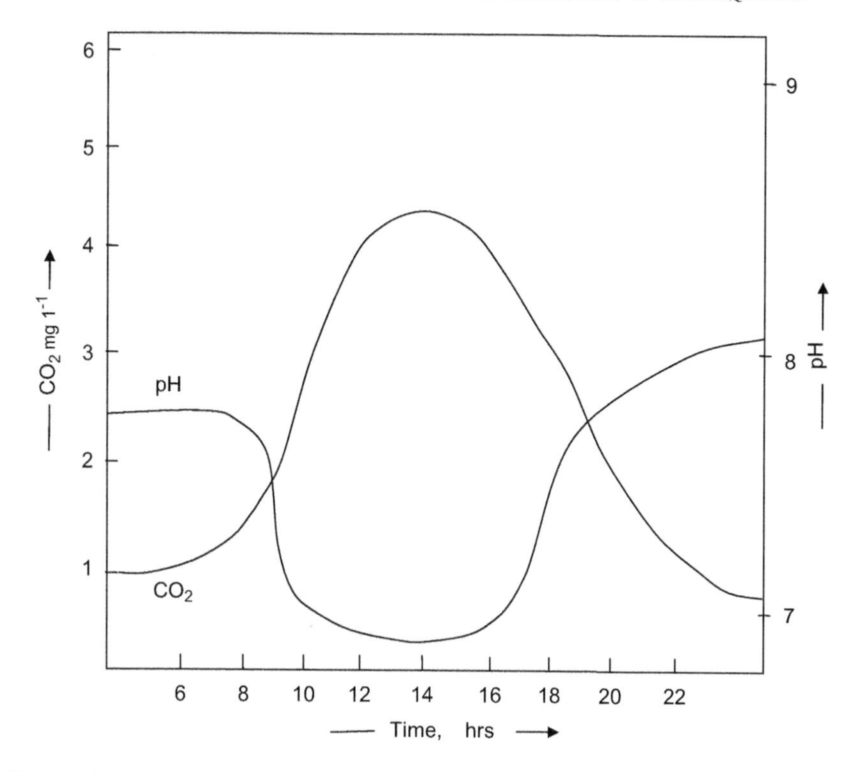

FIGURE 4.2 Typical changes in the pH and the concentration of CO_2 over a day in a lake.

water or inflow of municipal or industrial wastes. The biological assemblages at different levels in a water body also vary with the variation in water chemistry and diurnal rhythm.

Indeed, seasonal and diurnal variations have such a profound impact on the quality of natural water bodies that much of the historic information on the impact of acid rain on the pH of the lakes across the world has been considered inconclusive. This is because the pH was measured only once a day or sometimes even less frequently (Abbasi and Arya, 2000; Abbasi and Abbasi, 2011).

When drawing samples for analysis, a great deal of care is taken to ensure that a sample, or a set of samples, truly represent the lake as a whole. But, given the dynamic and stochastic nature of what happens in a lake, as illustrated above it is well neigh impossible to draw a perfectly representative sample (or a perfectly representative set of samples). Uncertainties of different types plague sampling of other lentic, lotic and underground water bodies in a similar manner.

Indeed, uncertainties are introduced into all water-quality-monitoring efforts at each and every stage – from drawing samples to their storage, transportation and physical–chemical–biological analysis and then in the analysis of data. When the data are fed into an index-generating system, newer uncertainties are introduced. What was supposed to be a crisp and deterministic exercise from the beginning eventually leads to a statistic that is far from being accurate or precise with reference to the purpose for which it was determined.

This is but one dimension of the problem. Another major dimension is related to the vagueness inherent in the objective itself: in the very definitions of 'acceptability' and the degrees of acceptability (i.e 'excellent', 'very good', 'good', 'fair', etc).

In fact, notwithstanding the various uncertainties associated with index development mentioned earlier, we still can *measure* water quality far more accurately (in relative terms) than we can evaluate its significance (Silvert, 2000).

This is because 'significance' is decided by measuring not instruments but the perception of stakeholder, and in any society these perceptions can differ very strongly across people of different socio-economic levels, cultural origins and attitudes. These differences come into sharper relief as we scan different regions and nations – which is one reason why no index is, or can be, applied as a global standard. These aspects have been dealt in greater detail in the following chapters.

Then, again, quite often it becomes necessary to use data acquired by agencies in the past which, for various reasons, may not be free from omissions or gross errors. If one is to totally disregard such data, one will have no way to know how the water quality at that time was. But such a knowledge is essential to determine how the water quality has changed with time, and to what extent. Hence, techniques of probability and/or fuzzy logic, in conjunction with other computational tools, are used to make best-possible sense of the uncertain past data.

As a major thrust towards addressing these complexities, concepts of statistics more advanced than the rudimentary ones used in conventional WQIs (described in the preceding chapters), probability theory and artificial intelligence are being increasingly applied to index development.

A number of indices have been developed using statistical techniques such as multivariate analysis, principal-component analysis and factor analysis. Fuzzy mathematics and probability theories have been separately applied to WQI development, and there are hybrid indices which use fuzzy logic as well as probability theory, with or without the assistance of other statistical or artificial intelligence methods. It is very difficult to slot indices into mutually exclusive categories; hence, the slotting in the following chapters is rather fuzzy; we have tried to put different indices in the chapters we have, based on our perception of what has been the *dominant* approach in the development of a particular index.

References

Abbasi, S.A., 1998. Water quality: sampling and analysis. 200–250. Discovery Publishing House, New Delhi.

Abbasi, S.A., Arya, D.S., 2000. Environmental Impact Assessment. Discovery Publishing House, New Delhi.

Abbasi, T., Abbasi, S.A., 2011. Water quality indices based on bioassessment: the biotic indices. Journal of Water and Health 9 (2), 330–348.

Silvert, W., 2000. Fuzzy indices of environmental conditions. Ecological Modelling 130 (1–3), 111–119.

CHAPTER

5

Indices Based on Relatively Advanced Statistical Analysis of Water-Quality Data

OUTLINE

5.1. INTRODUCTION

5.1.1. An Overview

We have seen in the preceding chapters that simple statistical methods such as averaging and summation have been used in all indices from Horton's index onwards. But the indices were always based on parameters, concentration–acceptability relationships of the parameters, weightages, etc., selected or defined by the index developer with or without the help of other experts. In contrast, for the type of indices described in this chapter, the parameters of importance and the extent of their importance are determined on the basis of analysis of water-quality and related data by statistical techniques such as factor analysis and principal-component analysis.

Water Quality Indices DOI: 10.1016/B978-0-444-54304-2.00005-1

This approach has the advantage that there are fewer subjective assumptions than in the traditional indices; however, the indices based on statistical analysis are more complex and more difficult to apply.

Of the statistical techniques on which some of these indices are based, the ones focussing on correlation explore associations among variables to determine the importance of each as a determinant of water quality. Shoji et al. (1966) applied factor analysis to the Yodo river system in Japan for interrelationships among 20 pollutant variables. By comparing the correlation of each variable with every other variable and selecting combinations with the highest correlations, they identified three major factors that affect the river water quality: pollution, temperature and rainfall.

In an attempt to examine the very basis of the concept of indices, Landwehr (1979) observed, "regardless of its construct, an index is a random variable in as much as the water-quality constituents upon which it depends are themselves random variables". He derived and compared the statistical properties of the most widely used functional structures of indices.

Joung et al. (1978) used factor analysis to develop water-quality indices by examining water-quality data from Carson Valley, Nevada. Ten pollutant variables were considered. By manipulating the matrix of correlation coefficients, the authors were able to identify linear combinations of the variables which best explain the variance but which have low correlations with each other. The approach retains the most important information in the raw data while eliminating redundant variables. The authors used the approach to identify the most significant variables and index weights for two water-quality indices containing five variables each: the Index of Partial Nutrients and the Index of Total Nutrients. These indices were then applied to the Snake and the Colorado River basins in Nevada, USA. Of the two, the Index of Total Nutrients (with the variables DO, BOD_5, total phosphates, temperature and conductivity) was selected, and its performance was compared with that of the Brown's NSFWQI (Brown et al. 1970) using water-quality data from 20 locations in the U.S.

In another effort at correlation, Coughlin et al. (1972) studied the relationship between the Brown's NSFWQI and the uses of a stream made by the nearby residents. They used principal-component analysis to examine the relationships among individual NSFWQI variables and such factors as distance of residence from the stream, land values and tendency for residents to walk along the stream or to wade in it or fish in it. They reported that water pollution was correlated with wading, fishing, picnicking, bird watching, walking and other activities.

5.2. HARKIN'S INDEX

Based on the premise elaborated in the preceding section that conventional indices, such as the NSF–WQI developed by Brown et al. (1970) and others described in Chapter 3, lack objectivity, Harkin (1974) presented a statistical approach for analysing water-quality data based on the rank order of observations.

Harkin's index was an application of Kendall's (1955) nonparametric classification procedure. It begins with ranking the observations for each pollutant variable, including a control value, which is usually a water-quality standard or recommended limit. For each observation j of pollutant variable i, the transform Z_{ij} was computed as the difference between the rank order of the observation and the rank order of the control value (R_{ic}), divided by the standard deviation of the ranks S:

$$Z_{ij} = (R_{ij} - R_{ic})/S_i \tag{5.1}$$

where R_{ij} is the rank of the jth observation of the ith variable,

R_{ic} is the rank of the control value for the ith variable and

S_i is the standard deviation of the ranks for the ith variable

The index was computed for each observation by adding the square of the transform for n pollutant variables:

$$I = \sum_{i=1}^{n} Z_{ij}^2 \quad (5.2)$$

The standard deviation is given by

$$S_i = m_i^2 - 1/12 \quad (5.3)$$

where m_i is the number of values (observation + control value) for pollutant variable i.

In Harkin's treatment, the same value often appears more than once; these repeated values reduce the variance and must be taken into account. When repeated values occur, the standard deviation S_i is calculated as follows:

$$S_i = \left[1/12m_i\left\{m_i^3 - m_i - \sum_{k=1}^{q_i}\left(t_k^3 - t_k\right)\right\}\right]^{1/2} \quad (5.4)$$

where m_i is the number of values for each variable i, t_k is the number of repeated values (ties) and q_i is the number of separate occurrences of ties.

Harkin's index is a relative index rather than an absolute one; values generated with one data set cannot be compared directly with those generated with a different data set.

5.3. BETA FUNCTION INDEX

The approach of Harkin (1974) was extrapolated by Shaefer and Janardan (1977) into a statistical index which has a fixed range: the Beta Function Index. It uses the same ranking procedure as employed in Harkin's index. Two additional values were computed from the ranks: the sum of the square of the z-transforms given by Equation (5.1) and the sum of all the ranks excluding the control values:

$$S = \sum_{i=1}^{a}\sum_{i=1}^{m_i} Z_{ij}^2 \quad (5.5)$$

$$T = \sum_{i=1}^{a}\sum_{i=1}^{m_i-1} R_{ij} \quad (5.6)$$

where m_i is the number of values for pollutant variable i.

The Beta Function Index was calculated using the transform of S and T:

$$I = 1/b\left[\frac{S}{S+T}\right]^{1/2} \quad (5.7)$$

$$b = \left[\frac{2\sum_{i=1}^{n} m_i^2}{3\sum_{i=1}^{n} m_i^2 + \sum_{i=1}^{n} m_i - 2n}\right]^{1/2} \quad (5.8)$$

where n is the number of pollutant variables. If the number of observations for each variable is the same (i.e., $m_i = m$, for all I), then Equation (5.8) can be simplified as

$$b = \left[2m^2/3m^2 + m - 2\right]^{1/2} \quad (5.9)$$

Since it was assumed to have a chi-square distribution and T was approximately constant, the authors concluded that the index follows a beta probability distribution. The index is thus nonparametric; its distribution is the same regardless of the underlying distribution of the data.

The number of variables and the scales used in the indices described above are summarised in Table 5.1.

5.4. AN INDEX WITH A MULTI-PRONGED ('MIXED') AGGREGATION FUNCTION

Given that larger the number of parameters handled by an index, more susceptible it becomes

TABLE 5.1 Number of Variables and Scales used in the Indices Described in Sections 1–4

Index	No. of Variables	Scale
Shoji et al. Composite Pollution Index (CPI)	18	−2 to 2
Joung Index of Partial Nutrients	5	0 to 100
Joung Index of Total Nutrients	5	0 to 100
Coughlin et al. Principal Component Index	*	N.A.
Harkin's Index (Kendall ranking)	*	0 to 100
Schaeffer & Janardans Beta Function Index	*	0 to 1
Kung et al., Fuzzy clustering	*	@

* Variables not fixed; worked out by the index.
@ *Matches the results with the standards.*

to the problems of ambiguity and eclipsing (explained in Chapter 2), Liou et al. (2004) employed a multi-pronged aggregation method to develop a 'generalised' WQI for Taiwan. It consisted of arithmetic and geometric aggregations as moderated by scaling coefficients.

Thirteen variables – DO, BOD_5, $NH_3{-}N$, faecal coliforms, turbidity, suspended solids, temperature, pH, Cd, Pb, Cr, Cu and Zn – were converted to values on a 0–100 scale; higher the score better the water quality. The criteria for conversion were based on the classification of national water sources; water criteria adopted by other countries, the background data of water quality, the legislated standards of Taiwan, etc. As Cd, Pb, Cr, Cu and Zn are regarded as toxic substances, the rating curves for each of them were calculated by the equation

$$r_i = \frac{C_i}{S_i} \tag{5.10}$$

where r_i is concentration ratio of ith substance, C_i is substance with concentration in mg L^{-1} and S_i is maximum permissible concentration (mg L^{-1}).

The resulting index scores corresponding to levels of different water-quality parameters are given in Table 5.2. Principal-components analysis (PCA) was then applied to recognise the common features among the variables: it identifies k new principal components, which are obtained as weighted linear combinations of p original components (Johnson and Wichern, 1998).

TABLE 5.2 Index Scores for Different Levels of Water-Quality Parameters in the WQI of Liou et al. (2004)

Score Value	Faecal Coliform (MPN/ 100 mL)	Dissolved Oxygen (mg L^{-1})	Biochemical Oxygen (mg L^{-1})	Ammonia Nitrogen (mg L^{-1})	Suspended Solid (mg L^{-1})	Turbidity (NTU)	Temperature (°C)	pH	Toxicity (r^*)[a]
100	0	≥6.55	≤2.95	≤0.45	≤19.5	0	≤38	≥6; ≤9	0
90	6								
80	50					4			
70	5000	5.55	3.95	0.7	34.5	15			
50	10,000				30				
30		3.25	10	2	75	50			
0	1,000,000	≤2.04	≥15.05	≥3.05	≥100	120	>38	>9; <6	≥1

[a] *r^* denotes ratio of the particular toxic substance to its maximum permissible concentration.*

The PCs became inputs to the aggregation function for overall index, given by

$$\mathrm{WQI} = C_{\mathrm{tem}} C_{\mathrm{pH}} C_{\mathrm{tox}} \times \left[\left(\sum_{i=1}^{3} I_i W_i \right) \left(\sum_{j=1}^{Z} I_j W_j \right) \left(\sum_{k=1}^{1} I_k \right) \right]^{1/3} \quad (5.11)$$

where I_i denotes the subindex values for the 'organics' and contains three variables DO, BOD_5 and ammonia nitrogen (I_1 for the subindex of DO, I_2 for the subindex of BOD_5 and I_3 for the subindex of NH_3-N); I_j represents the subindex values for the 'particulates', which consist of suspended solids and turbidity; I_k is the measurement of faecal coliform, which represents the 'microorganisms'. Three scaling coefficients were prefixed to account for the influence of the subindices of temperature (C_{tem}), pH (C_{pH}) and toxic substances (C_{tox}), respectively. The multipliers are commended with regards to the characteristics of effects and the rating structures of the three variables.

5.5. WQI FOR MEDITERRANEAN COSTAL WATER OF EGYPT BASED ON PRINCIPAL-COMPONENT ANALYSIS

In an attempt to develop a framework with which to identify the most polluted areas (hot spots) and the cleanest ones along the Mediterranean coastal water of Egypt, principal-component models were formulated by El-Iskandarani et al. (2004). For the variance convariance matrices, the data, comprising levels of NO_2, NH_4, $NO_2 + NO_3$, total nitrogen, PO_4, total phosphorus and Si, in the coastal water at 42 stations and 11 different representing the annual mean levels of the eutrophication parameters for two years were used. A principal-component model was also developed for the data set. A water-quality index was then developed and applied to indicate the quality of the Mediterranean coastal water as impacted by the enforcement of the Egyptian environmental law.

5.6. WQI FOR RIO LERMA RIVER

Sedeno-Diaz and Lopez-Lopez (2007) assessed spatial and temporal variations in water quality of Rio Lerma River basin over the last 25 years with two approaches: the use of a multiplicative and weighted WQI, and a principal-component analysis (PCA).

The WQI was based on the earlier index of Dinius (1987) and had the form

$$\mathrm{WQI} = \prod_{i=1}^{n} I_i^{w_i}$$

where I_i is subindex of a parameter, a number from 0 to 100; w_i is unit weight of a parameter, a number from 0 to 1, with $\sum_{i=1}^{n} w_i = 1$, and n is number of parameters.

Twelve parameters were used in the formulation of the WQI: DO, BOD, total and faecal coliforms, alkalinity, hardness, chloride, conductivity, pH, nitrate, colour and temperature. The mean annual values of each parameter were used to calculate subindex functions (Table 5.3) from which the WQI was then derived. PCA was applied to assess the significance of parameters that explain the patterns of the monitoring stations. For this data sets of the mean annual values of the 12 water-quality parameters were employed. The index had a 0–100 scale.

5.7. A NEW WQI BASED ON A COMBINATION OF MULTIVARIATE TECHNIQUES

Qian et al. (2007) combined several multivariate techniques and a comprehensive

TABLE 5.3 Subindices and Weighted Values of Variables in the WQI of Sedeno-Diaz and Lopez-Lopez (2007)

Parameter	Subindex I_i	Weighted Value (w_i)
Dissolved oxygen (percent saturation)	$I_{OD} = 0.82\ (DO) + 10.56$	0.109
Biochemical oxygen demand 5-day (mg/L)	$I_{DBO} = 108\ (BOD)^{-0.3494}$	0.097
Nitrate (mg/L)	$I_{NO3} = 125\ (N)^{-0.2718}$	0.090
Total coliforms (MPN coliforms/ml)	$I_{ColTot} = 136\ (TotCol)^{-0.1311}$	0.090
Faecal coliforms (MPN coliforms/ml)	$I_{Col\ Fec} = 106\ (E.\ coli)^{-0.1286}$	0.116
Alkalinity (mg/L)	$I_{ALC} = 110\ (ALK)^{-0.1342}$	0.063
Hardness (mg/L)	$I_{DUR} = 552\ (Ha)^{-0.4488}$	0.065
Chloride (mg/L)	$I_{Cloruros} = 391\ (Cl)^{-0.4488}$	0.074
Temperature (°C) where T_a = air temperature T_s = water temperature	$I_{T°C} = 10^{2.004-0.0382\|T_a-T_s\|}$	0.077
Conductivity (μmhos/cm)	$I_{COND} = 506\ (SPC)^{-3315}$	0.079
pH (units)	$I_{pH} = 10^{0.6803+0.1856(pH)} pH < 6.9$	0.077
	$I_{pH} = 100;\ 6.9 \leq pH \leq 7.1$	
	$I_{pH} = 10^{3.65-0.2216\ (pH)};\ pH > 7.1$	
Colour (Pt—Co units)	$I_{COLOUR} = 127\ (C)^{-0.2394}$	0.063

water-quality index (WQI) (Table 5.4) to evaluate water quality in the south Indian River Lagon (IRL), Florida.

Thirteen water-quality parameters at six stations were selected for analysis because of their continuous length of record and location within the selected study area: dissolved oxygen, specific conductivity, pH, turbidity, colour, total suspended solids, nitrite nitrogen, nitrate nitrogen, ammonia nitrogen, total Kjeldhal nitrogen, orthophosphate as phosphorus, total phosphorus and total iron.

Constituent concentrations were log-transformed in order to account for the log normal distribution of water-quality data and to minimise the effect of outliers within the data. Mean values of log-transformed constituent concentrations were then used in clustering. Agglomerative hierarchical clustering method using Euclidean dissimilarity measure and Ward's linkage algorithm was adopted to group the six monitoring stations. To minimise the effects of the scale of units on the clustering, the log-transformed data were standardised prior to clustering, using Equation (5.12) so that each variable could be considered equally important:

$$z_{if} = \frac{x_{if} - m_f}{s_f} \tag{5.12}$$

where m_f and s_f are the mean and the mean absolute deviation of log-transformed variable f for all stations; x_{if} and z_{if} are the original and the standardised value of log-transformed variable f at station i, respectively. The mean absolute deviation was used instead of the usual standard deviation so that the effect of outliers could be reduced. The dissimilarity

TABLE 5.4 Summary of the Statistical Techniques Employed by Qian et al. (2007)

Technique	Brief Description	Purpose in this Study
Cluster analysis (or clustering)	Finding groups in data	To identify the extents of similarity among the six monitoring stations
Principal-component analysis (PCA)	Reducing the information contained in several measured variables into a smaller set of components without losing important information	To characterise water quality using a smaller data set (principal components) which were extracted from the larger original data set
Exploratory Factor analysis (EFA)	Revealing latent factors which cannot be measured directly	To provide more information on relative importance of water-quality constituents
		To identify the latent information within the original data set
		To further reduce the number of variables
		To serve as a check for the results from PCA
Trend analysis	Detecting changes in water quality over time using linear regression	To determine the monotonic trends using different time series data
Water-quality index (WQI)	Developing an index characterising the overall water quality at a given location	To give an overall picture of water quality in the study areas

(i.e., distance) between two objects, $d(i, j)$, was computed as follows (Euclidean distance):

$$d(i,j) = \sqrt{\sum_{f=1}^{p} (z_{i_f} - z_{j_f})^2} \quad (5.13)$$

where i and j are different stations; p is the number of variables.

For each monitoring station group (identified by clustering), PCA was applied to both annual and seasonal data sets (standardised log-transformed concentrations). For each clustering group, exploratory factor analysis (EFA) was applied to both annual and seasonal data sets (standardised log-transformed concentrations). Varimax rotation was used in EFA. Constituents with loadings >0.75 were considered as principal constituents.

For each clustering group, the time series of the scores of first five factors derived from the all-data EFA and seasonal EFA were analysed for annual and seasonal trends from 1979 to 2004 using simple linear regression (factor scores versus time) at a significance level of $p < 0.1$.

For each clustering group, the WQI was calculated according to Equation (5.14) based on both entire data sets as well as seasonal subsets. Simple linear regression was used to determine the annual and seasonal trends for each WQI at a significance level of $p < 0.1$:

$$WQI = \sum_{i=1}^{n} \left\{ \frac{I_i \times L_i}{\sum_{i}^{n} (I_i \times L_i)} \times \frac{[C_i]}{[S_i]} \right\} \quad (5.14)$$

where WQI is water-quality index; n is the number of principal constituents; I_i is the proportion variance explained by the factor in which the ith principal constituent is involved; L_i is the loading (absolute value) of the ith principal constituent; $[C_i]$ is the concentration of the ith principal constituent and $[S_i]$ is the surface water-quality criterion of the ith principal constituent.

In Equation (5.3), I_i and L_i characterise the relative importance of the ith factor involving the principal constituents among all factors and the ith principal constituent among all constituents in that factor, respectively. Therefore,

$$\frac{I_i \times L_i}{\sum_{i}^{n}(I_i \times L_i)}$$

This measures the relative contribution of the ith principal constituent to the overall WQI in a combinational way. $[C_i]/[S_i]$ describes the status of water quality characterised individually by the ith principal constituent and thus has a criticality value of 1.

If the measured concentration of the ith principal constituent meets the criterion, $[C_i]/[S_i]$ will be less than 1; otherwise, it will be greater than 1.

The criteria of DO (the higher DO, the better water quality) and pH (an appropriate range of pH, e.g., 6.0–8.5 in this case, indicates satisfying water quality) have been specified by means different from those used to specify the criteria for other contaminants (usually higher the concentration, more the degraded water quality). This difference has been taken into account during the aforementioned transformation for DO and pH. Theoretically and ideally, WQI will equal to 1 if measured concentrations of all principal constituents equal the criteria levels given for corresponding water-quality constituents. A less than 1 value of WQI could be indicative of satisfactory condition of given water body considering its designed uses, whereas a more than 1 value would indicate that designed uses are not being met due to the quality of the water body. It should be noted that this WQI provides an average evaluation of water quality. Even if some of the critical values are exceeded, it is still possible that the WQI could be lower than 1.

5.8. INDICES FOR LIAO RIVER STUDY

Meng et al. (2009) studied several water-quality parameters of Liao River with principal-component analysis (PCA). They employed PCA to identify the water-quality factors which strongly impact the river health. In order to interpret the principal components easily, the maximum variance rotation method was used to discriminate the factors which have larger load values. Each component value of the index was then calculated and summed to generate the river health assessment score (RH):

$$RH = \sum_{i=1}^{n}(EH_i \times W_i)$$

where EH_i is the value of the ith assessing index and W_i is the weight of the ith assessing index. The indices, of which value decreased with the increase in the human disturbance, were formulated according to Equation (5.15), and the indices which attained higher numerical values with increasing pollution were formulated as per Equation (5.16):

$$EH = \frac{EH_{max} - EH_{fact}}{EH_{max} - EH_{III}} \quad (5.15)$$

$$EH = \frac{EH_{fact}}{EH_{III}} \quad (5.16)$$

where EH_{max} is the maximum of the index; EH_{fact} is the actual value of the index; EH_{III} is the category III value of the index. The weight, W_i, was determined by the PCA. In addition, cluster analysis (CA) was used to obtain spatial distribution of water quality and physical habitat quality (Figure 5.1), where Ward's amalgamation method and squared Euclidean distances measure were used.

A comprehensive integrated assessing system of river ecological health was then

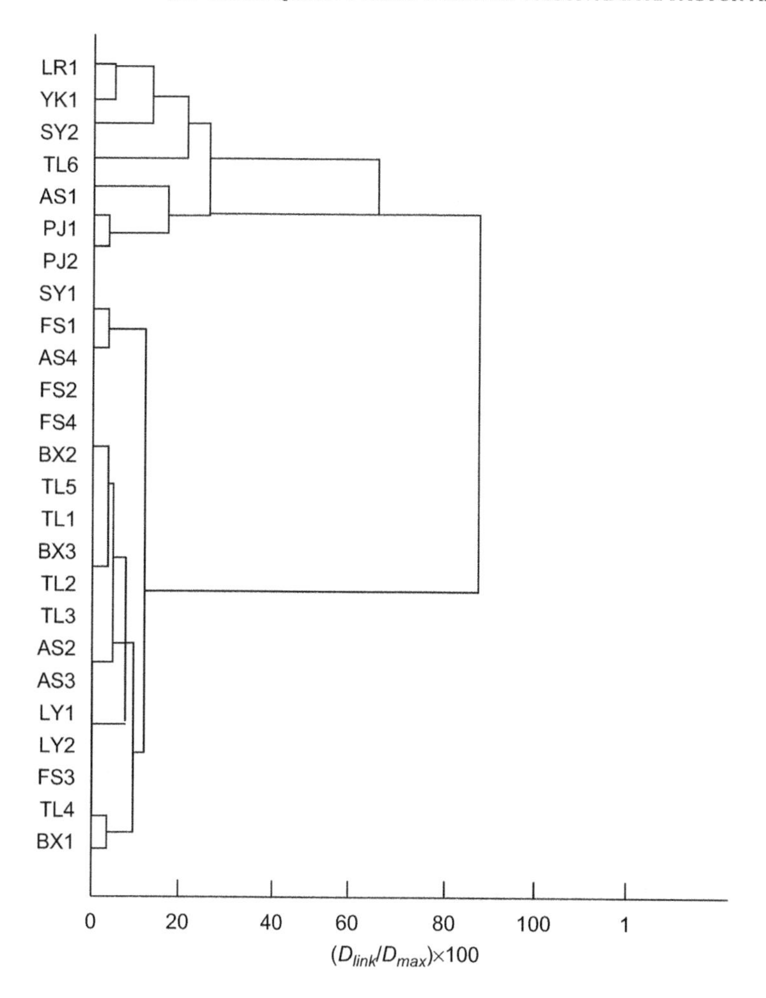

FIGURE 5.1 Dendogram showing the clustering of monitoring sites (Meng et al., 2009); D_{link}/D_{max} represents the quotient of the linkage distance by the maximum distance.

established and the assessed sites were categorised into 9 'healthy' and 'subhealthy' sites and 8 'subsick' and 'sick' sites.

5.9. WATER-QUALITY INDEX BASED ON MULTIVARIATE FACTOR ANALYSIS (COLETTI ET AL., 2010)

The technique of multivariate factorial analysis (MFA) employs a correlation structure of an initial set of "*p*" variables (X_1, X_2 … X_P), replacing it with a smaller set of hypothetical variables, which, being lower in number and with a simpler structure, explain most of the variation in the original variables.

This technique permits learning the behaviour of data from the reduction of the parameters' original space dimension, thus permitting the selection of the most representative variables for the water resource being analysed (De Andrade et al., 2007). Factorial analysis thus permits the identification of more sensitive among the indicators; this can facilitate a monitoring programme as well as the evaluation of changes in the quality of water resources. Toledo and Nicolella (2002) earlier demonstrated that the use of indices based on the factorial analysis technique is useful when one intends to evaluate changes that have taken place in watersheds.

Building upon this, Coletti et al. (2010) used water-quality data of eight parameters – electrical conductivity, pH, ammoniacal nitrogen, ammonia, nitrate, total phosphorus, suspended solids and turbidity – to develop a MFA-based WQI and apply it to Rio das Pedras Watershed, Brazil.

The development of the WQI required three steps: a) preparation of the correlation matrix; b) extraction of the common factors and the possible reduction of space and c) the rotation of axes related to the common factors, aiming for a simple and easily interpreted solution.

A basic matrix was first built without any gaps, a requirement demanded by the MFA technique being employed. Using the matrix from the original data, a correlation matrix was obtained using the Spearman correlation coefficient, which revealed a linear dependency among the variables being studied.

Common factors were then found using factor analysis. Their numbers were determined based on the percent of total variance in the variables.

With the intent of obtaining a simpler factorial load matrix, the axes were rotated using the Varimax procedure. But the results were not satisfactory; hence, the axes without rotation were used in subsequent steps.

The statistical model that formalised factorial analysis is given by

$$z_{ij} = \sum_{p=1}^{m} a_{jp}F_{pi} + u_j Y_{ji}(i = 1, 2, K, N; j = 1, 2, K, n) \quad (5.17)$$

where $a_{jp}F_{pi}$ is the contribution of the common p factor to the linear combination and $u_j Y_{ji}$ the residual error in the representation of the observed z_{ij} measurement.

The model was adjusted using the average of the results from three simple samples collected from six points distributed over the watershed, for the eight parameters over a thirteen-month period.

Application of the index revealed deteriorating conditions of water quality in the Rio das Pedras Watershed due to agricultural activities. The authors thought that the index appeared to be a promising tool in felicitating the monitoring of water-resource availability in the region.

5.10. STUDY OF ANTHROPOGENIC IMPACTS ON KANDLA CREEK, INDIA

Shirodkar et al. (2010) have reported the use of several statistical techniques in conjunction with the 'overall index of pollution' developed earlier by Sargaonkar and Deshpande (2003; described in Chapter 3) to assess the impact of human activities on the water quality of Kandla creek, India.

The authors employed multivariate ordination techniques to segregate data matrices of three different seasons, containing 20 variables. This reduced the number of dimensions in data sets and allowed better detection of major trends or underlying patterns of variation in the data. Redundancy analysis (RDA), a constrained form of principal-component analysis (PCA), was used to explore the relationships between the measured physico-chemical and biological variables where the ordination axes are constrained to be linear combinations of environmental variables. Factor analysis was carried out separately for the data set of each season. The observed screen plot was used to identify the number of principal components (PCs) to be retained in order to comprehend the underlying data structure. The loadings of

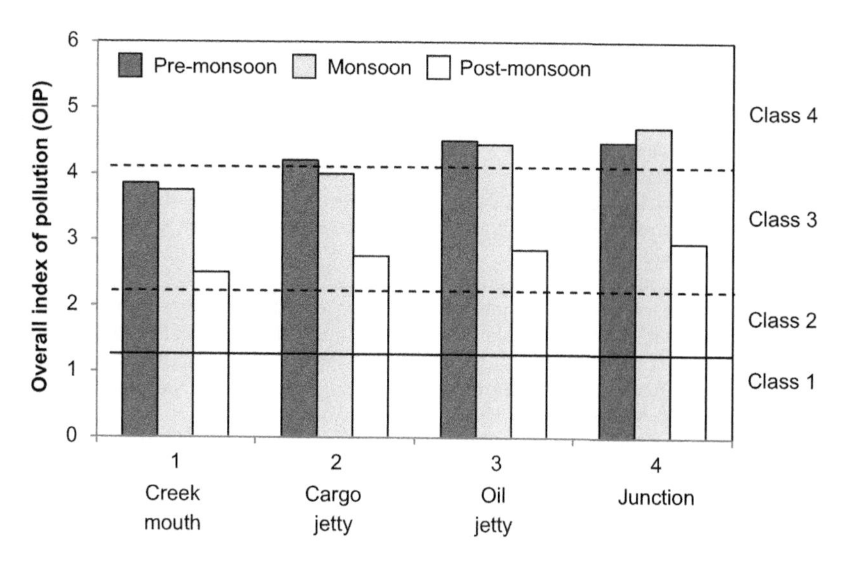

FIGURE 5.2 Overall index of pollution (OIP) at different sampling locations during the three seasons at kandla creek (Shirodkar et al., 2010). The water either belonged to class 3 ('slightly polluted,' index score 2–4) or class 4 ('polluted,' index score > 4).

PCs were considered as strong (>0.75), moderate (0.5–0.75) and weak (0.4–0.5). The authors then employed the Varimax normalised rotation (VNR) which distributes the PC loadings such that their dispersion is maximised by minimising the number of large and small coefficients. The reliability of the data for factor analysis was tested with Kaiser–Meyer–Olkin (KMO), which measures the sampling adequacy. Spearman's correlation matrix was used to identify the significantly correlated variables within the measured environmental data.

The 'overall index of pollution' was then generated following the aggregation procedure earlier reported by Sargaonkar and Deshpande (2003). The findings brought out the impact of seasons on the water quality of different locations within the creek (Figure 5.2).

References

Brown, R.M., McClelland, N.I., Deininger, R.A., Tozer, R.G., 1970. A water quality index – do we dare? Water Sewage Works 117, 339–343.

Coletti, C., Testezlaf, R., Ribeiro, T.A.P., de Souza, R.T.G., Pereira, D.A., 2010. Water quality index using multivariate factorial analysis. Revista Brasileira de Engenharia Agricola e Ambiental 14 (5), 517–522.

Coughlin, Robert E., Hammer, Thomas R., Dickert, Thomas G., Sheldon, Sallie, 1972. Perception and Use of Streams in Suburban Areas: Effects of Water Quality and of Distance from Residence to Stream. Regional Science Research Institute, Philadelphia, PA. Discussion Paper No. 53. March 1972.

De Andrade, E.M., Araujo, L.D.F.P., Rosa, M.F., Disney, W., Alves, A.B., 2007. Surface water quality indicators in low Acaraã basin, Cearãj, Brazil, using multivariable analysis. Engenharia Agricola 27 (3), 683–690.

Dinius, S.H., 1987. Design of an index of water quality. Water Resources Bulletin 23 (5), 833–843.

El-Iskandarani, M., Nasr, S., Okbah, M., Jensen, A., 2004. Principal components analysis for quality assessment of the Mediterranean coastal water of Egypt. Modelling, Measurement and Control C 65 (40575), 69–83.

Harkin, R.D., 1974. An objective water quality index. Journal of the Water Pollution Control Federation 46 (3).

Johnson, R.A., Wichern, D.W., 1998. Applied Multivariate Statistical Analysis. Prentice-Hall Inc.

Joung, H.M., Miller, W.W., Mahannah, C.N., Guitjens, J.C., 1978. A Water Quality Index Based on Multivariate Factor Analysis. Experiment Station J, Series No. 378,

University of Nevada, Nevada Agricultural Experiment Station, Reno, Nevada.

Kendall, M.G., 1955. Rank Correlation Methods. Charles Griffin, London.

Landwehr, J.M., 1979. A statistical view of a class of water quality indices. Water Resources Research 15 (2), 460–468.

Liou, S.-M., Lo, S.-L., Wang, S.-H., 2004. A generalized water quality index for Taiwan. Environmental Monitoring and Assessment 96 (40603), 35–52.

Meng, W., Zhang, N., Zhang, Y., Zheng, B., 2009. Integrated assessment of river health based on water quality, aquatic life and physical habitat. Journal of Environmental Sciences 21 (8), 1017–1027.

Qian, Y., Migliaccio, K.W., Wan, Y., Li, Y., 2007. Surface water quality evaluation using multivariate methods and a new water quality index in the Indian River Lagoon, Florida. Water Resources Research 43 (8).

Sargaonkar, A., Deshpande, V., 2003. Development of an overall index of pollution for surface water based on a general classification scheme in Indian context. Environmental Monitoring and Assessment 89, 43–67.

Sedeno-Diaz, J.E., Lopez-Lopez, E., 2007. Water quality in the RÃo Lerma, Mexico: an overview of the last quarter of the twentieth century. Water Resources Management 21 (10), 1797–1812.

Schaeffer, D.J., Janardan, K.G., 1977. Communicating environmental information to the public a new water quality index. The Journal of Environmental Education 8, 18–26.

Shirodkar, P.V., Pradhan, U.K., Fernandes, D., Haldankar, S.R., Rao, G.S., 2010. Influence of anthropogenic activities on the existing environmental conditions of Kandla Creek (Gulf of Kutch). Current Science 98 (6), 815–828.

Shoji, H., Yomamoto, T., Nakamura, T., 1966. Factor analysis on stream pollution of the Yodo river system. Air and Water Int. Poll. J. 10, 291–299.

Toledo, L.G., Nicolella, G., 2002. Índice de qualidade de água em microbacia sob uso agrícola e urbano Scientia Agrícola 59 (1), 181–186.

CHAPTER

6

Water-Quality Indices Based on Fuzzy Logic and Other Methods of Artificial Intelligence

OUTLINE

Water Quality Indices DOI: 10.1016/B978-0-444-54304-2.00006-3

6.1. INTRODUCTION

Artificial intelligence refers to the capability of certain techniques and tools which enables them to increase the knowledge initially provided to them through a process of inference or 'learning'. The so-called 'knowledge-based' or 'expert' systems are examples of artificial intelligence (AI). The essence of these systems is their ability to acquire heuristic knowledge, usually represented through a set of qualitative conditional expressions with verbal meaning, with the merit of being semantically clear. They are able to increase the initial knowledge base through a process of inference or 'learning'. Due to this, AI systems can be 'trained' to recognise patterns or signals and to respond to them. Fuzzy inference, genetic algorithms, artificial neural networks and self-organising maps are examples of AI techniques that have been employed in developing WQIs. Of these, fuzzy inference has been by far the most extensively utilised, often in conjunction with factor analysis, principal-component analysis and cluster analysis.

6.2. FUZZY INFERENCE

One of the research fields involving AI is fuzzy logic, originally conceived as a way to represent intrinsically vague or linguistic (*more-or-less; if x and y then z, etc*) knowledge. It is based on the mathematics of fuzzy sets (Zadeh, 1965; Ross 2004). The combination of fuzzy logic with expert systems is referred as fuzzy inference (Yager and Filvel, 1994).

The inventor of fuzzy mathematics, Lotfi Zadeh (1965), has noted: "as the complexity of a system increases, our ability to make precise and yet significant statements about its behaviour diminishes until a threshold is reached beyond which precision and significance (or relevance) become almost mutually exclusive characteristics" indicating that real situations are very often uncertain or vague in a number of ways; Zadeh called this vagueness "fuzziness". He combined both the precision and the fuzziness, transforming the traditional dual logic of a yes-or-no type into the more-or-less type of a multi-value logic, a much more flexible concept. Given a certain threshold, multi-value logic can be converged to a yes-or-no-based logic on the basis of membership function or grade of membership – also called 'degree of compatibility' or degree of truth (Zadeh, 1977).

Whereas experimental protocols and subsequent handling of data by conventional techniques fail to accommodate the impact of stochasticity that prevails in nature, as explained in the previous chapter as well as later in this chapter, procedures based on fuzzy rules are able to deal with the uncertainties and inadequacies in knowledge and data. Fuzzy models can represent qualitative aspects of knowledge and human inference processes even in situations where precise quantitative analysis may not be possible. They are less accurate than rigorously applied numerical models, but the gains in simplicity, computational speed and flexibility that result from the use of these models offer heavy compensation for whatever loss in precision that might occur (Bárdossy, 1995).

The use of fuzzy inference offers the following advantages:

1. They can be used to describe a large variety of nonlinear relations.
2. They tend to be simple, since they are based on a set of local simple models.
3. They can be interpreted verbally and this makes them analogous to AI models.
4. They use information that other methods cannot include, such as individual knowledge and experience.
5. Fuzzy logic can deal with, and process, missing data without compromising the final result.

Specifically in indice development, the fuzzy approach has an advantage over the conventional approaches because the former has the ability to expand and combine quantitative and qualitative data that express the ecological status of a water source, making it possible to do without the often misleading 'precision' of crisp approaches. In other words, the ecological complexity of the real world situations (of which a WQI is expected to reflect the overall state) is better captured by fuzzy rule-based models than the ones based on crisp mathematics.

From the time they were introduced in the 1960s (Zadeh, 1965), fuzzy rules have been successfully used to model dynamic systems in several fields of science and engineering. They have been applied to solve water-related complex environmental problems (Sadiq and Rodriguez, 2004; Vemula et al., 2004; Liou and Lo, 2005; McKone and Deshpande, 2005; Ghosh and Mujumdar, 2006) and have come to increasingly dominate WQI development.

In fuzzy inference an input set is mapped into an output set using fuzzy logic. This mapping may be used for decision-making or for pattern recognition. The process involves four main steps (Yen and Langari, 1999; Ross, 2004; Cruz, 2004; Caldeira et al., (2007)):

1. Development of fuzzy sets and membership functions.
2. Fuzzy set operations.
3. Fuzzy logic.
4. Inference rules.

6.3. A PRIMER ON FUZZY ARITHMETIC

The elements of fuzzy arithmetic, in the specific context of its application to WQI development, are presented below (Chang et al., 2001; Ocampo-Duque et al., 2006; and Sadiq and Tesfamariam, 2008).

If an uncertain quantity is one which lies entirely neither in set 'a' nor in set 'b', but belongs to both in varying degrees, fuzzy arithmetic is a generalised form of interval analysis, which is used to address uncertain and/or vague information pertaining to that quantity. A fuzzy number describes the relationship between the uncertain quantity x and a membership function μx, which ranges between 0 and 1.

A fuzzy set is an extension of the classical set theory (in which x is either a member of set A or not) in that an x can be a member set of A with a certain membership function μ_x. Fuzzy sets qualify as fuzzy numbers if they are normal, convex and bounded (Klir and Yuan 1995). Different shapes of fuzzy numbers are possible (e.g., bell, triangular, trapezoidal and Gaussian). Trapezoidal and triangular fuzzy numbers (Figure 6.1) are often used in WQI development. A trapezoidal fuzzy number (TFN) can be represented by four vertices (a, b, c, d) on the universe of discourse (scale X on which a criterion is defined), representing the *minimum, most likely interval and maximum values*, respectively. Triangular fuzzy number (TFN) is a special case of ZFN (trapezoidal fuzzy number), where $b = c$.

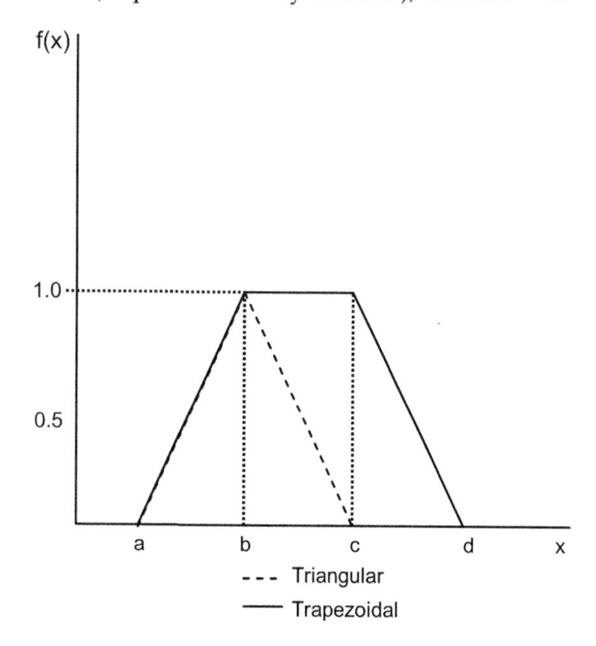

FIGURE 6.1 Trapezoidal and triangular membership functions.

An α-cut of a fuzzy set A is a crisp set (interval) A^{α} that contains all the elements of the universal set X whose membership grades in A are greater than or equal to the specified value of an α, i.e., $A^{\alpha} = \{x | \mu_x \geq \alpha\}$.

Fuzzy arithmetic is based on two premises (Klir and Yuan 1995): *a*) each fuzzy number can fully and uniquely be represented by its α-cut and *b*) α-cuts of each fuzzy number are closed intervals of real numbers for all $\alpha \in (0,1)$. Hence, once the interval is defined, traditional interval analysis can be used (Ferson and Hajagos, 2004). Table 6.1 lists some commonly used interval analysis operations which can be used to carry out fuzzy arithmetic at various predefined α-cut levels, e.g., (0, 0.1, 0.2, ..., 1).

If X is the universe of discourse and its elements are denoted by x, then a fuzzy set A is defined as a set of ordered pairs:

$$A = \{x,\ \mu_A(x) \setminus x \in \mathrm{X}\}$$

where $\mu_A(x)$ is the membership function of x in A. A membership function is an arbitrary curve whose shape is defined by convenience.

The standard fuzzy set operations include union (OR), intersection (AND) and additive complement (NOT). They contain the essence of fuzzy logic. If two fuzzy sets A and B are defined on the universe X, then for a given element x belonging to X, the following operations can be carried out:

$$(\text{Intersection, AND})\ \mu_{A \cap B}(x) = \min(\mu_A(x), \mu_B(x)) \quad (6.1)$$

$$(\text{Union, OR})\ \mu_{A \cup B}(x) = \max(\mu_A(x), \mu_B(x)) \quad (6.2)$$

$$(\text{Additive complement, NOT})\ \mu_{\bar{A}}(x) = 1 - \mu_A(x) \quad (6.3)$$

Another important concept is the inference rule. An if–then rule has the form: "If x is A then z is C", where A and C are linguistic values defined by fuzzy sets in the universes of discourse X and Z, respectively. The if part is called the antecedent, while the then part is called the consequent. The antecedent and the consequent of a rule can have multiple parts.

A fuzzy inference system (FIS) has four parts: fuzzification, weighting, evaluation of inference rules and defuzzification. Fuzzification involves the definition of inputs, outputs as well as their respective membership functions that transform the numerical value of a variable into a membership grade to a fuzzy set, which describes a property of the variable. But not all variables have the same importance; hence, it is necessary to establish a way to guide the influence of each variable in the final score. To do this analytical hierarchy process (AHP) is often applied.

Between the two most frequently used techniques – fuzzy clustering analysis (FCA) and

TABLE 6.1 Some Common Arithmetic Operations Used in Interval Analysis

Operators	Formulae[a]	Results[b]
Summation	$A + B$	$[a_1 + b_1, a_2 + b_2] = [5, 15]$
Subtraction	$A - B$	$[a_1 - b_2, a_2 - b_1] = [1, 5]$
Multiplication	$A \times B$	$[a_1 \times b_1, a_2 \times b_2] = [6, 50]$
Division	A/B	$[a_1/b_2, a_2/b_1] = [0.6, 5]$
Scalar product	$Q \cdot B$	$[Q \cdot b_1, Q \cdot b_2] = [4, 10]$

[a] *The values of A and B are positive; if negative numbers are used, the corresponding min and max values have to be selected.*
[b] $A = [a_1, a_2] = [3, 10]$; $B = [b_1, b_2] = [2, 5]$; $Q = 2$.
$a_1 < a_2$; $b_1 < b_2$; a_i and b_i $(i = 1–2) > 0$; $Q > 0$.

TABLE 6.2 Advantages and Disadvantages of Fuzzy-Based Methods

Advantages	Disadvantages
Easy interpretable by natural language	Not free of eclipsing but can be handled with trial and error process
Can handle complex and vague situation	Cannot incorporate guideline values for water-quality parameters
Can incorporate experts opinion with hard data	Suffer rigidity to some extent (careful selection of parameter can reduce it)
Can describe a large number of nonlinear relationships through simple rules	Easy to manipulate or can be biased due to human subjectivity
Provides a transparent mathematical model	
Able to account interconnection (interdependencies) among parameters	
Capable to handle missing data without influencing the final WQI value	
Free of ambiguity and can represent different water-quality usage if parameters are selected carefully	

Islam et al., 2011.

fuzzy synthetic evaluation (FSE) – the former is applied for clustering the raw data into several categories using the selected operators without respect to any predetermined criteria in relation to each category. Most of the rules designed for FCA are based on the search for centroids or representative objects around which all observations are expected to be clustered on a minimum basis (Selim, 1984; Trauwaert et al., 1991). On the other hand, FSE is designed to group raw data into several different categories according to predetermined quality criteria, which can normally be described using a set of functions that are designed to reflect the absence of sharp boundaries between each pair of adjacent criteria. FSE is considered more relevant than FCA to the assessment of water quality. A well-designed FSE may be capable of covering the uncertainties existing in the sampling and analysis process, comparing the sampling results to the applied quality standards for each parameter and summarising all of the individual parameter values (Otto, 1978; Lu et al., 2000).

Islam et al. (2011) have compiled a list of advantages and disadvantages associated with fuzzy-based methods (Table 6.2) and have compared the virtues and shortcomings of some of the soft computing methods, including fuzzy sets (Table 6.3).

6.4. TOWARDS APPLICATION OF FUZZY RULES IN DEVELOPING WATER-QUALITY INDICES: THE WORK OF KUNG ET AL. (1992)

Among the first to elucidate that several of the shortcomings of conventional WQIs can be overcome by using fuzzy rules, Kung et al. (1992) did not quite develop a fuzzy WQI but came very close to doing it. They, in fact, developed a general methodology based on fuzzy clustering analysis, which could be applied to preexisting WQIs for composite classification of water quality incorporating multiple parameters in a manner, which can to a great degree tide

TABLE 6.3 A Comparison of Virtues and Short Comings of Some of the Soft Computing Methods Used for Water-Quality Assessment

Assessment Criteria	ANN	Fuzz Sets	Evidential Reasoning	Bayesian Network	Rough Sets
Simplicity	L	H	M	M	M
Interpretability	Nil	H	M	H	H
Vagueness	Nil	H	M	M	H
Randomness	Nil	L	H	H	M
Cause and effect	H	H	H	H	M
Redundancy	Nil	H	Nil	Nil	Nil

Islam et al., 2011.

over the lopsidedness and other problems associated with the direct use of the conventional indices based on crisp mathematics.

Besides the problems mentioned in the introduction, there are other difficulties in the technical interpretation of conventional WQIs. For instance, if in a WQI a score of 1.0 or lower is set to indicate acceptable water quality, highly misleading results can be obtained if the index score is calculated from three different possible situations. In the first, all measured constituents may just reach legally permissible standards, causing the unusual and questionable case of 'threshold status.' In the second, the measures of toxic constituents may be far lower than permissible standards but the concentration of eutrophic and scenic parameters may be above permissible standards; yet the index may show only minor deterioration in water quality. Finally, one of the heavy metals may have a higher measure than the legally permissible standard, although the measurements of the other constituents may be significantly lower than legally permissible; yet the index would show a highly polluted status.

Furthermore, in an index like the one mentioned above, a score of 0.99 would indicate a normal status while a score of 1.01 would imply unacceptable water quality. This kind of sharp or 'crisp', 'all-or-nothing' cut-off is unrealistic given the dynamic nature of water quality as explained in the introduction. A lake sampled in the morning might yield an index value of 0.99; the same lake sampled in the afternoon might generate a score of 1.01 but it is not as if the lake water becomes safe by the minute and unsafe by the minute. Moreover, depending on the methods used to develop an index, it is possible to obtain a score of less than 1.00 for the very same levels of water-quality parameters for which an index calculated by a different method may yield a value higher than one!

In this backdrop, Kung et al. (1992) thought that fuzzy mathematics, which had been found to suit many fields in which the complexity and the uncertainty of real-life situations had been captured much more closely by it than by the techniques of crisp mathematics, will also be appropriate for water-quality assessment. They then went on to apply fuzzy clustering to develop water-quality assessment methods that could be combined with preexisting WQIs to obtain more meaningful results from the latter than is possible just with the WQIs.

As elaborated by the authors, classical (crisp) clustering algorithms generate partitions in which each object is assigned to exactly one cluster (Foody, 1992). Very often, however, objects cannot be exclusively assigned to strictly one cluster because they may be lying 'in between' clusters. In those cases, fuzzy clustering methods provide a much more adequate tool for representing real data structures.

A fuzzy relation $\underset{\sim}{R}$ from a set X to a set Y (X, Y are data sets) is a fuzzy subset of $X * Y$ characterised by a membership function $u\underset{\sim}{R}$: $X* Y \rightarrow [0,1]$. For each $x \in X$ and $y \in Y$ (x, y are elements, $x \in X$ and $y \in Y$), $u\underset{\sim}{R}(x, y)$ is referred to as the strength of the relation between x and y. If $X = Y$, then it is said that $\underset{\sim}{R}$ is a fuzzy relation on X. Since the elements of a cluster should be as similar to each other as possible and clusters should be as dissimilar as possible, the clustering process is controlled by use of similarity measures. Hence, a fuzzy cluster is determined by a fuzzy relation. Should a $\underset{\sim}{R}$, the similarity measure matrix, meet the following requirements:

1. Reflexivity:

$\underset{\sim}{R}$ is called reflexive (Zadeh, 1971) if

$$u\underset{\sim}{R}(x,x) = 1 \quad \forall x \in X \tag{6.4}$$

2. Symmetry:

A fuzzy relation $\underset{\sim}{R}$ is called symmetric (Zimmermann, 1985) if

$$u\underset{\sim}{R}(x,y) = u\underset{\sim}{R}(y,x) \quad \forall x,\ Y \in x \tag{6.5}$$

3. Transitivity:

A fuzzy relation $\underset{\sim}{R}$ is called (Max-min) transitive (Zimmermann, 1985) if

$$\underset{\sim}{R}.\underset{\sim}{R} \subseteq \underset{\sim}{R} \tag{6.6}$$

then the fuzzy relation $\underset{\sim}{R}$ is said to be stabilised (Wang, 1983). Furthermore, if a particular cluster represents a set of elements on a specific threshold level, it can be considered a similarity cluster of the level.

In most real-life situations, the fuzzy relation meets the reflexivity and symmetry requirements but not of transitivity. It has to be transformed before it can become part of a clustering chart. This transformation can be done through a max−min self-multiply processing of fuzzy matrix, that is,

$$r_{ij} = \vee(r_{ik} \wedge r_{jk}) = (\underset{\sim}{R}^* \underset{\sim}{R})_{ij} \tag{6.7}$$

where $\vee$ is the maximum operator, $\wedge$ the minimum operator and rij is the element of the ith row and the jth column in the similarity measure matrix $\underset{\sim}{R}$. If, at a certain step, there is

$$\underset{\sim}{R}^* = \underset{\sim}{R}^k = \underset{\sim}{R}^{2k} \tag{6.8}$$

then the fuzzy relation $\underset{\sim}{R}$ is deemed to be stabilised (Wang, 1983). A cluster chart representing the dynamic grouping process can then be worked out, and the sets of elements on a specific threshold level can also be obtained should an appropriate threshold be determined by the experts vis a vis local conditions.

The distinguishing features of the methodology proposed by Kung et al. (1992) are:

1. It shows that fuzzy clustering analysis (FCA) can be used as a complement or an alternative to conventional quantitative methods of water-quality assessment, especially when the WQI score is close to the threshold between normal and abnormal.
2. Use of FCA yields results closer to nature than the conventional common WQI methods, since water-quality assessment is one of those complex and fuzzy fields in which human intellectual activities are most needed.
3. Applying is simpler than using deterministic models in water-quality assessment.
4. Combined with any WQI used in water-quality assessment, the results obtained with FCA cannot only illustrate the water quality in terms of spatial distribution but can also describe characteristics of each class as well as their unique causes.

6.5. ASSESSMENT OF WATER QUALITY USING FUZZY SYNTHETIC EVALUATION AND OTHER APPROACHES TOWARDS DEVELOPMENT OF FUZZY WATER-QUALITY INDICES

Whereas fuzzy clustering analysis classifies samples for unknown standards by relationship, thereby requiring a large amount of data, fuzzy synthetic evaluation (FSE) classifies samples on the basis of known standards. In the former, samples must have a high degree of similarity in addition to categorising factors; therefore, it is better suited to regional classification.

As FSE is a modified version of conventional synthetic evaluation and overcomes some of the defects engrained in binary logic, Lu et al. (1999) carried out water-quality assessment of Fei-Tsui reservoir, Taiwan, using FSE. When comparing the performance of their method with the results of Carlson Index (Carlson, 1977) for identical WQ data, the authors found that Carlson Index was not able to reflect long-term changes in water quality and the overturn phenomena in the reservoir which the FSE method was able to. Later Lu and Lo (2002) used the FSE results to train self-organising maps (SOMs) and obtain deeper insights into the reservoir water-quality data.

A self-organising map (SOM) is a neural-network model and algorithm that implements a characteristic nonlinear projection from the high-dimensional space of sensory or other input signals onto a low-dimensional array of neurons. The term 'self-organising' implies an ability to learn and organise information without being given the associated-dependent output values for the input pattern. The SOM is able to map a structured, high-dimensional signal manifold onto a much lower dimensional network in an orderly fashion. The network organises itself by adjusting the synaptic weights as the input patterns are presented to it; hence, discovery of a new pattern is possible at any instant. Moreover, SOM is noise tolerant, which is a highly desirable attribute when handling data with likely uncertainties.

The successful use of SOM as a complementary tool for FSE by Lu and Lo (2002) further highlights the potential of artificial intelligence in water-quality assessment. A little before this study appeared, Chang et al. (2001) reported the performance of three fuzzy synthetic evaluation techniques in assessing water quality as compared with that of the United States National Sanitation Foundation Water-Quality Index (NSF-WQI). The study employed data collected at seven sampling stations from the Tseng-Wen River, Taiwan. Figure 6.2 summarises the procedure adopted by the authors. WQI was applied in generating the basic outputs, with independent weights selected for each quality parameter. Three fuzzy synthetic evaluation (FSE) techniques – simple fuzzy classification, fuzzy information intensity and defuzzification – were applied sequentially for the water-quality classification. The membership functions were as shown in Figure 6.3.

The authors found that the conventional WQI was 'pessimistic' in the sense that it tended to show the water quality as generally poorer than was indicated by the fuzzy methods. They also found that the fuzzy information intensity method and the defuzzification method gave more self-consistent results than the simple fuzzy classification method. A very similar study has recently been published by Chen et al. (2011).

Haiyan (2002) applied fuzzy comprehensive assessment to assess the quality of air, water and soil in Zhuzhou City, China, based on the monitoring data of 1997 and National Environmental Quality Standards of China. The assessment procedure comprised five steps:

1. Selection of assessment parameters and establishment of assessment criteria.

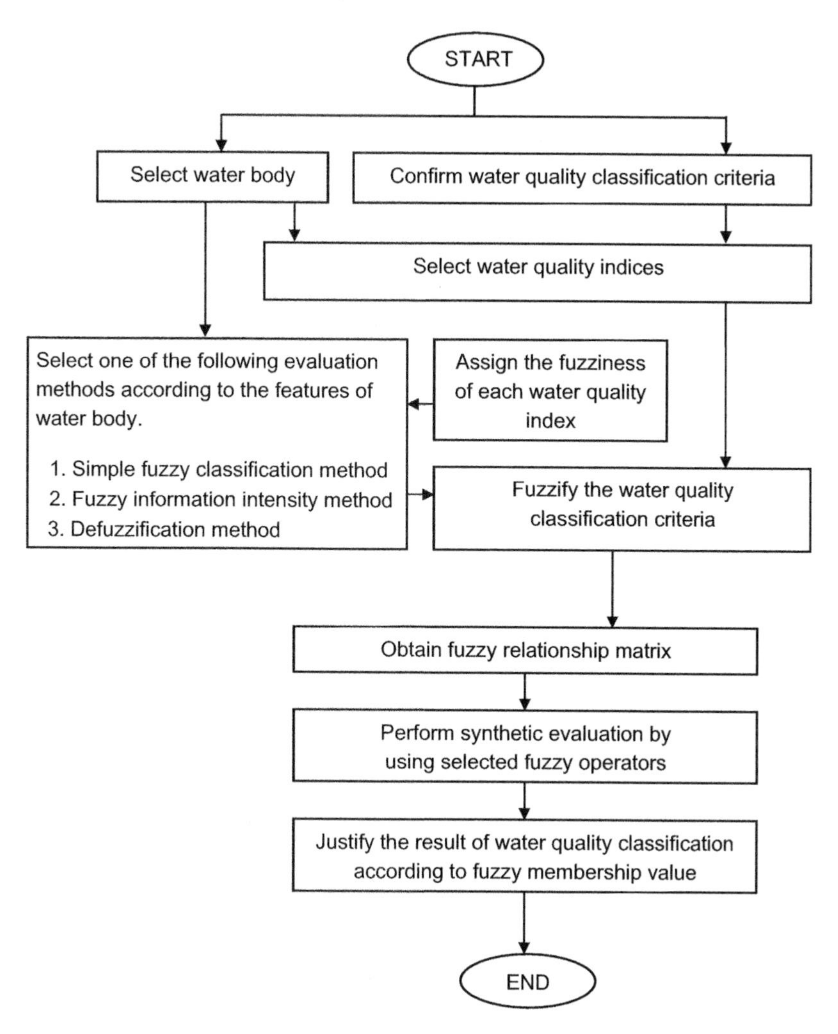

FIGURE 6.2 Flow chart of the procedure used by Chang et al. (2001).

2. Establishment of membership functions of each assessment parameter to assessment criteria at each level.
3. Substitution of the monitoring data of each assessment parameter at each monitoring site and national standards into the membership functions.
4. Allocation of the weights of each assessment parameter at each monitoring site to get a weight matrix.
5. Carrying out the fuzzy algorithm. Reflecting the increasing of application of fuzzy logic in environmental assessment, Adriaenssens et al. (2004) reviewed and assessed applications of fuzzy logic for decision support in ecosystem management, asserting that the application of fuzzy logic seems to be very promising to address issues of sustainability, environmental assessment and predictive models.

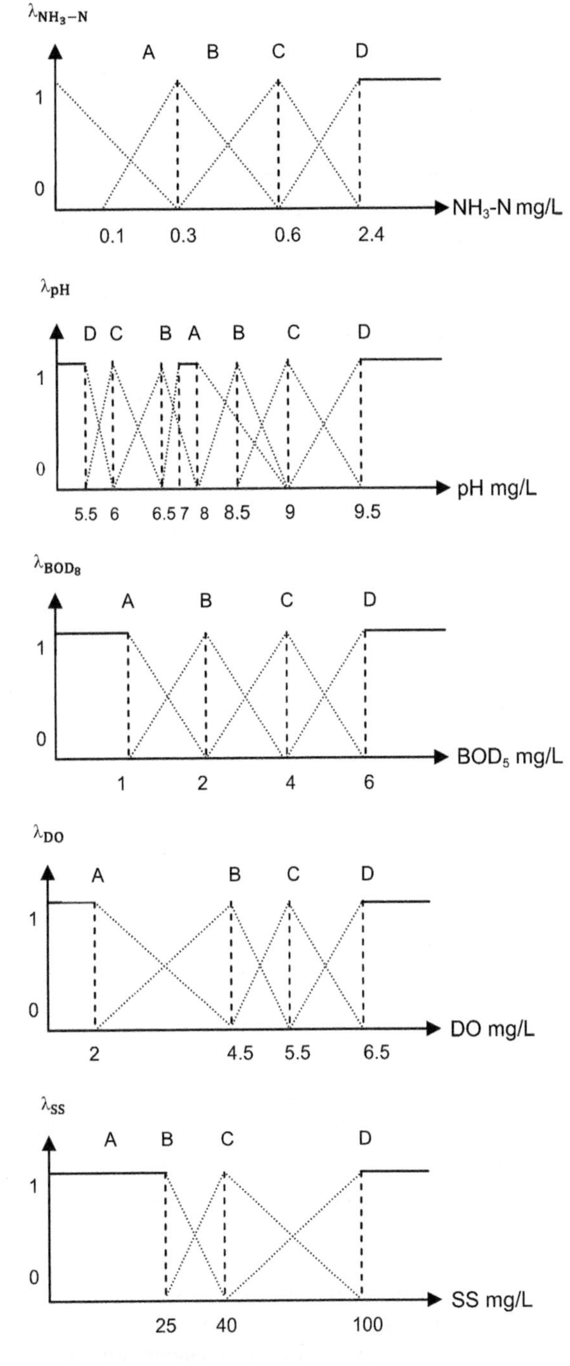

FIGURE 6.3 Membership functions used in the study of Chang et al. (2001).

A fuzzy logic approach for evaluating ecosystem sustainability was proposed by Prato (2005) who stated that the fuzzy logic approach is more appropriate than the conventional crisp sets approach in evaluating the sustainability of an ecosystem.

Continuing the trend, Shen et al. (2005) investigated the status of combined heavy metal and organo-chlorine pesticide pollution and evaluated the soil environmental quality of the Taihu Lake watershed using a fuzzy comprehensive assessment (FCA) method. The evaluation was carried out in six steps:

1. Selection of assessment parameters.
2. Establishment of the membership functions.
3. Calculation of the membership function matrix.
4. Calculation of the weights matrix.
5. Determination of the fuzzy algorithm.
6. Statistical treatment of data.

The authors noted that the FCA method provides a scientific basis for analysing and evaluating the environmental quality of soil.

A fuzzy logic approach was employed by Altunkaynak et al. (2005) to detect and remove the linear trend embedded in the historical records of monthly dissolved oxygen data at Golden Horn. It enabled them to model dissolved oxygen fluctuations in the study area.

6.6. REACH OF FUZZY INDICES IN ENVIRONMENTAL DECISION-MAKING

Silvert (1997, 2000) brought home the advantage of using fuzzy rules in developing environmental indices in a striking fashion when describing the experience he had acquired in the course of assessing impact of fish farms on the seabed, and other studies. He also suggested possible ways in which multi-objective decision-making and consensus building can be facilitated by the use of indices based on fuzzy logic.

It appeared particularly relevant in situations where the perceptions of different stakeholders on acceptability/unacceptability (and their degrees) of environmental conditions not only differ widely but are also, often, in conflict.

Traditionally the symbol μ has been used to represent fuzzy memberships. If x represents the value of an environmental variable, then $\mu(x)$ is the corresponding membership in the set of acceptable conditions and takes a value between zero and one. For example, if a lake becomes hypoxic and all the fish die, then μ would presumably be zero, indicating that this situation is totally unacceptable.

Explaining the philosophy, Silvert (2000) observed that in most situations more than one environmental variable is important, and we can define $\mu_i(x_i)$, which is called the 'partial membership', to represent the acceptability of the ith environmental variable. One then needs to develop ways of combining these partial memberships to obtain a general measure of acceptability. These ways depend on the given application. For example, if the objective is to have a healthy fish population in a lake, and the environmental variables are oxygen level, water temperature, and nutrients, then since adverse levels of any of these can be lethal to fish, the fuzzy intersection, $\mu = \min(\mu_1, \mu_2, \mu_3)$, is appropriate. By definition, the combined acceptability will be as low as the lowest of the three partial acceptabilities. However, there may be some compensatory effects which may cushion the low acceptability of one variable by good values of another. Hence, the choice of combination rule has to be made in the context of the given situation.

In comparison, modelling the relationship between an environmental variable x and its acceptability $\mu(x)$ is fraught with great complications because the notion of acceptability may differ from stakeholder to stakeholder; one man's peach may be other man's poison.

For a large number of water-quality variables, values that are too low or too high are unacceptable; in all such cases it is an intermediate range which is highly acceptable. Moreover, the acceptability of some of the environmental conditions are exceedingly difficult, often impossible, to quantify. For example, there are no commonly accepted techniques for measuring whether an odour is good or bad, and to what degree. Only within specific contexts it may be possible to determine that people find certain smells unacceptable to varying degrees, and one can assign approximate values to μ without having a quantitative measure of x.

It is in situations like this that the great strengths of fuzzy logic come to the fore — it lets one deal with subjective and nonquantitative data which approaches based on crisp mathematics fail to handle. Environmental effects are of concern not just to a few scientists or other stakeholders but to people as a whole, including businessmen, politicians and ordinary citizens. Once one ventures into the social and political realm where the real decisions are made, it simply is not possible to tell people that some of their concerns about loss of scenery, foul smells or noise-induced headaches cannot be considered simply because they cannot be measured by instruments.

Citing the example of a study to assess the impact of fish farms on benthos, Silvert (2000) observes that it is quite fast and easy for a diver to record that there is, for example, 'very little seaweed, a few crabs, and thick patchy bacterial mats', but to actually quantify this in terms of the biomass of seaweed, number of crabs per square metre, and thickness and percentage cover of mats would require special equipment and additional dive time. Other useful data, such as a strong sulphide smell to the cores, are virtually impossible to quantify.

Silvet and co-workers carried out a pilot project to explore the use of fuzzy logic for developing indices of benthic condition under a fish farm in Eilat, Israel, that has been studied by divers for several years (Angel et al., 1998).

Four fuzzy sets were defined, representing nil, moderate, severe and extreme impacts. The use of fuzzy sets was observed to be immediately helpful in resolving one of the major conceptual problems that had made the data difficult to interpret, viz that different types of observations can lead to incompatible results. For example, there may be a healthy benthic community which indicates that the impact is nil or at worst moderate, but extensive patches of thick bacterial mats may point to a severe or even extreme impact. Fuzzy sets allow for nonzero memberships in more than one of these categories, which offers a mechanism for resolving these apparent inconsistencies.

The partial memberships were obtained by a procedure of first assigning membership values to each observation and then combining them. For example, the presence of seagrass was generally considered to be a strong indicator of a healthy seabed (the species dominant in that area, *Halophila stipulacea*, is extremely sensitive to pollution), so if the divers identified the seagrass cover as normal, the partial memberships for nil, moderate, severe and extreme impacts were assigned as $\mu_{NIL} = 0.8$, $\mu_{MOD} = 0.2$ and $\mu_{SEV} = \mu_{xTR} = 0$. These assignments, known as 'association rules', were applied to eight types of observations, namely the extent, thickness and colour of bacterial mats of *Beggiatoa* sp., quantity of seagrass and epi-macro-fauna, degree of bioturbation by macrofauna and by fish, and visibility.

The observations were mainly qualitative, of the sort that could be made by divers without use of measuring instruments; for example, seagrass coverage was reported as 'absent', 'few', or 'normal', and the thickness of the bacterial mats was reported as 'thin', 'thick', or 'massive'. Each variable was also assigned a weight reflecting its importance in the benthic assessment process. The presence of seagrass and the extent of bacterial mats were assigned high weights, as they were considered strong indicators of benthic condition, while the colour of the mats and the presence of visible epi-macro-fauna (which are not always easy to see) were assigned relatively low weights.

The entire procedure was developed to fit the observational data, rather than following the conventional scheme of planning acquisition of data which the planner might have considered potentially useful. In addition to facilitating a simple and practical sampling scheme, of a kind with which a high sampling frequency could be maintained, this also gave the investigators access to an historical database that might have been discarded if they were prepared to accept only rigorously quantitative measurements. Above all else, it made it possible to collect relevant information without the necessity of relying on highly expensive and cumbersome sampling-cum-analysis-based data acquisition programmes.

There were, however, some trade-offs, since simplifying the experimental aspect of the programme was accomplished at the expense of some fairly sophisticated mathematical analysis. Given a large number of benthic observations, the author needed to combine the partial memberships to obtain overall membership values for the four fuzzy sets of nil, moderate, severe and extreme impact. For this he chose to use the method of 'symmetric summation' (Silvert, 1979, 1997) as it makes it easier to avoid value judgements and is hence suitable for the evaluation of environmental impacts.

The fuzzy scores revealed something that had not been detected by other means, namely a sudden worsening of conditions in late summer of 1993. The nil conditions that had been improving over the summer fell to a low level, and the major change was an increase in membership in the moderate impact category, accompanied by a small jump in severe and extreme impacts. The author could not fathom the cause of this deterioration, but it underscored the fact that the use of fuzzy logic to quantify a set of subjective and qualitative observations could dramatically reveal an effect that would have otherwise eluded detection.

The author further observed that the strongest positive feature of fuzzy logic in developing environmental indices is the ability to combine such indices much more flexibly than one can combine discrete measures, which are often simply binary indices corresponding to ordinary ('crisp') sets, such as 'acceptable versus unacceptable'.

This attribute makes fuzzy logic applicable in multi-objective decision-making much more effectively than other techniques. As mentioned earlier, society is not always able to reach consensus on the value of certain components of the environment, so that effects which are acceptable to some segments are far less acceptable (or downright unacceptable) to others. Examples include the abundances of certain birds and marine mammals, which are highly prized for their beauty and entertainment value by recreational users of the environment, but are seen as predators and competitors by fishers and farmers. Many complex issues deal with the marginal (i.e. incremental) value of natural lands, such as the question of how much old-growth forest should remain protected and how much can be exploited. Hence, the concept of acceptability, and of the mathematical concept of a fuzzy set of acceptable conditions, should be based on several acceptability sets, each representing the viewpoint of a different segment of society.

In a coastal region, possible interests of stakeholders may include good recreational use, sustainable wild fisheries and profitable shellfish aquaculture. Relevant variables that impact these interests would then include the abundance of birds and the occurrence of toxic marine algae. From a recreational point of view, birds are usually considered very desirable (except for very common species like seagulls), and the more abundant they are, the higher the degree of satisfaction of the recreational objective. However, fishermen may not share this love for birds; to them most are their 'wage-snatchers', so are attractive birds like oystercatchers and eider ducks for owners of shellfish farms. Toxic algae are merely a nuisance to swimmers and other recreational users, but can cause economic ruin of shellfish farms. Thus, the memberships and weights would differ from stakeholder to stakeholder – for example, the higher the number of eider ducks, the greater the membership (acceptability) for bird watchers but lower the acceptability to mussel farmers. Formalisms of this type can be used to identify key areas of disagreement between stakeholders and may possibly contribute to the resolution of conflict in complex situations by providing a language for quantifying these disagreements.

Silvert (2000) proposed a three-step procedure to deal with conflicts among stakeholders, with the rider that it would be unrealistic to assume that mathematical calculations will be accepted as a decisive means to solve complex social and political issues, but they offer a quantitative expression of the differing values and needs expressed during negotiations, and as such can help clarify the basic underlying issues:

1. Identify environmental variables on which agreement can be reached and reach consensus on the partial memberships and weighting factors. For example, point source air pollution is usually of more concern to nearby residents than to the producers, but everyone agrees that air pollution is undesirable, so it should be possible to arrive at an agreement regarding acceptability levels.
2. This enables the participants to focus on areas where there is real disagreement, such as marine mammals and birds in the situation described above, without being distracted by issues on which consensus is readily achievable.
3. Once the basically different objectives of various groups have been clearly delineated, sets of acceptability scores for the different

objectives can be calculated for different scenarios and used to provide a focus for further discussion.

6.7. A WQI BASED ON GENETIC ALGORITHM

Arguing that conventional aggregation methods often excessively stress the roles of over-the-standard fractional indices and maximal indices, while ignoring the roles of less fractional indices, resulting in evaluations which do not always reflect the actual water quality, Peng (2004) proposed a water-quality index formula in the form of logarithm function of the parameters which relied not on monitoring values of the parameters but, rather, on the values relative to 'base value'. An index was then derived by applying genetic algorithms (GAs) to optimise the parameters. GA was chosen because it has increasingly proved effective as a universal optimum searching algorithm based on evolution theory and possesses the characteristics of ease-of-use, currency, robustness and amenability to parallel disposal (Holland, 1992; Li and Peng, 2000).

A WQI in the form of logarithm function is

$$\mathrm{WQI}_j = a_j + b_j \ \ln(1 + c_j) \tag{6.9}$$

where a_j and b_j are undetermined parameters relative to parameter j and c_j is monitoring value of parameter j.

Considering that the values of acceptable standards of different parameters are greatly different from each other, a_j and b_j are almost always different. But if a 'base value' c_{j0} of j is taken from each individual parameter, and c_j is replaced in Equation (6.9) with x_j, the index value relative to c_{j0} becomes

$$x_j = c_j / c_{j0} \tag{6.10}$$

where 'base value' c_{j0} aims to normalise all parameter values in a way that they correspond to identical level of the corresponding standard.

Consequently, GA can be adopted to optimise a_j and b_j in Equation (6.9), resulting in values a and b applicable to multiple parameters:

$$\mathrm{WQI}_j = a + b \ \ln(1 + x_j) \tag{6.11}$$

The objective function is constructed as:

$$\min f(x) = \frac{1}{KM} \sum_{k=1}^{K} \sum_{j=0}^{N} [\mathrm{WQI}_{jk} - \mathrm{WQI}_{ek}] \tag{6.12}$$

where K is the number of water-quality levels – 5 in the study by Peng (2004), N is the number of the parameters (30) considered in Equation (6.11), WQI_{jk} is the k-level standard index value of parameter j of water quality, calculated from Equation (6.11), and WQI_{ek} is the index objective value of k-level standard, having no relation with parameter j.

Taking K to have values 1, 2, 3, 4, and 5, GA was used to optimise a and b. The Equation (6.12) yielded $a = -0.80$, and $b = 1.56$ which fitted $\min f(x) \leq 0.03$. The subindex then emerged as

$$\mathrm{WQI}_j = -0.80 + 1.56 \ \ln(1 + x_j) \tag{6.13}$$

The averages of N (30) parameters for each level of K (1–5) taken on the basis of WQI_{jk} resulted in the overall WQI:

$$\mathrm{WQI} = \sum_{j=1}^{M} W_j \cdot \mathrm{WQI}_j \tag{6.14}$$

where WQI_j is the single index of parameter j calculated from Equation (6.13) and W_j is the united and weighted value of parameter j.

Considering that water-quality characteristics change in the fashion of S-shaped curves, the roles of WQI_j in lower levels were slightly strengthened and those in higher levels were slightly weakened while calculating comprehensive index. For this purpose, a compromise active function was adopted:

$$W_j' = \begin{cases} (u_j/2)^{1/2} & \text{if } 0 \leq u_j \leq 0.5 \\ 1 - [(1 - u_j)/2]^{1/2} & \text{if } 0.5 \leq u_j \leq 1 \end{cases} \tag{6.15}$$

The author demonstrated the use of his new index by applying it to the water quality of the Qianshan River, Zhuhai.

6.8. THE FUZZY WATER-QUALITY INDEX OF OCAMPO-DUQUE ET AL. (2006)

Describing the step-by-step development of what is arguably one of the first WQIs based on fuzzy logic, Ocampo-Duque et al. (2006) explain the procedure with the hypothetical (and simplified) example of a situation wherein the levels of dissolved oxygen (DO) and organic matter (BOD_5) are deemed sufficient to evaluate water quality (WQ) by means of a fuzzy water-quality (FWQ) index. Choosing 'low', 'medium and 'high' fuzzy sets for inputs, and 'good', 'average' and 'poor' fuzzy sets for the output, trapezoidal membership functions are seen to define these fuzzy sets as in Figure 6.5.

Given that in water-quality assessment, experts often use expressions such as "if the levels of organic matter in a river are low, and the levels of dissolved oxygen are high, then the expected water quality is good", the following rules can be enunciated in fuzzy language for the above mentioned example:

Rule 1. If BOD_5 is low and DO is high then WQ is 'good'.
Rule 2. If BOD_5 is medium and DO is medium then WQ is 'average'.
Rule 3. If BOD_5 is high and DO is low then WQ is 'poor'.

How fuzzy logic-based reasoning is inherently different from the reasoning provided by conventional deterministic approach can be understood by the application of the above rules. For example, if it is required to evaluate water quality at two points in a water-body – *R1* and *R2*, having BOD_5 and DO values of 1.0, 9.0, and 3.3, 6.5, respectively – one can confidently apply Rule 1 to the values at *R1*. However, the same rule cannot be applied with the same degree of sarity to *R2*. Hence when input values are close to boundaries

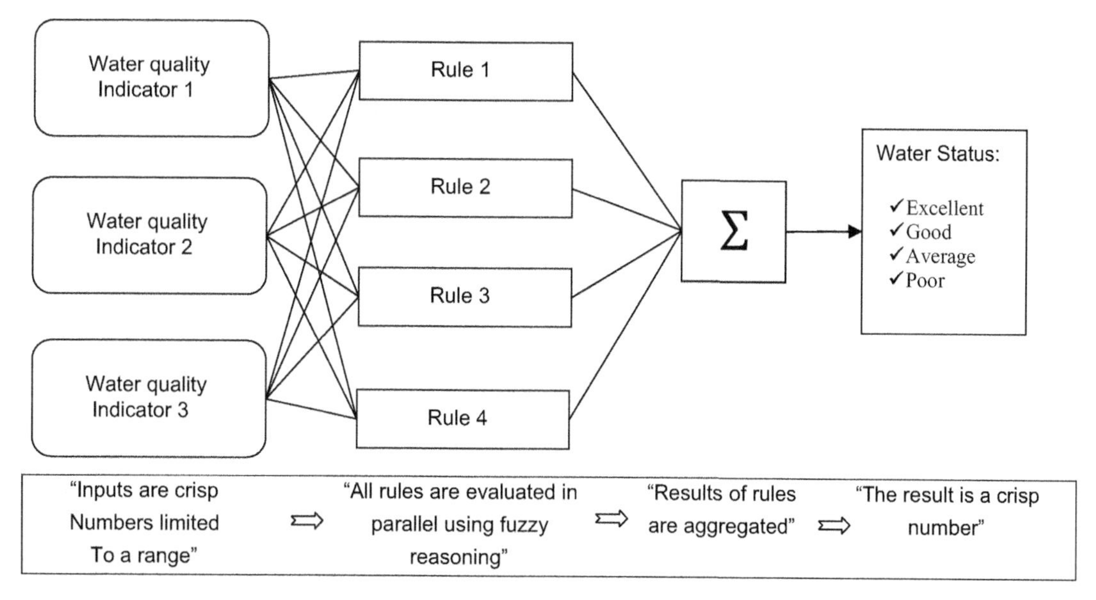

FIGURE 6.4 Input–output mapping in fuzzy logic. *(Adopted from Ocampo-Duque et al., 2006).*

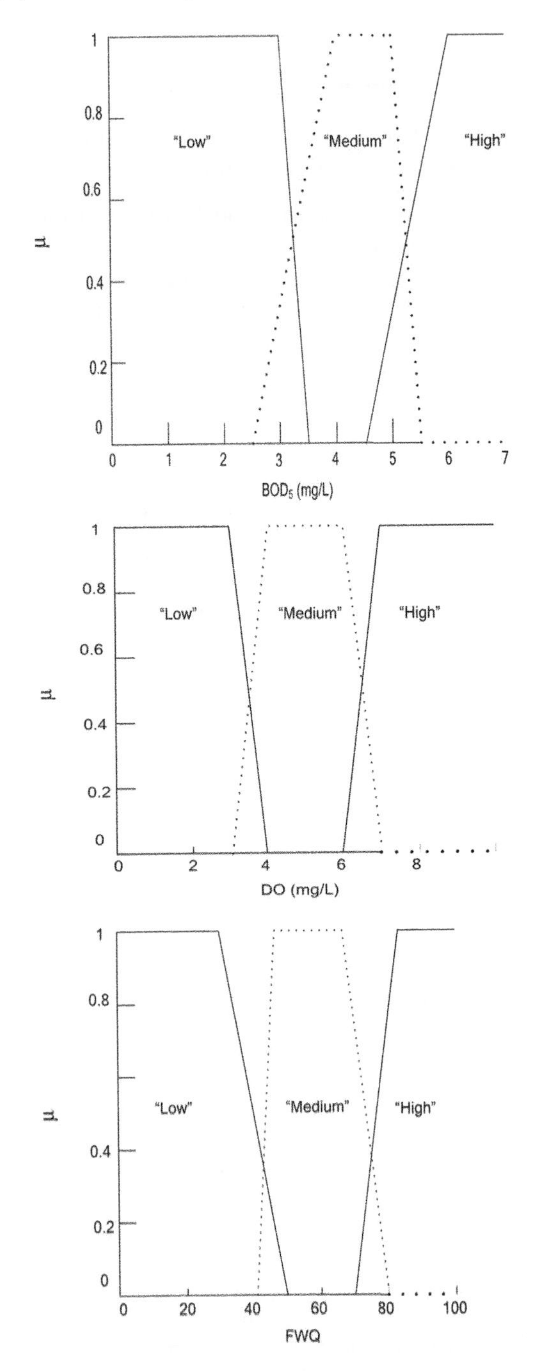

FIGURE 6.5 Membership functions for BOD_5, DO and FWQ parameters in the fuzzy WQI of Ocampo-Duque et al. (2006).

between a fuzzy set and another one, as in *R2*, the output is not so direct, and fuzzy operations should be carried out.

The membership functions in Figure 6.5 indicate that the value of 3.3 for BOD_5 belongs to 'low and 'medium' fuzzy sets, with membership degrees of 0.2, and 0.8, respectively. Similarly, a value of 6.5 for DO belongs to 'high' and 'medium' fuzzy sets, with membership degrees of 0.5 and 0.5, respectively. Hence, the variables belong to more than one set.

As there are multiple parts in the antecedents of the rules, fuzzy logic operations should be applied to give a degree of support for every rule. Applying Equation (2) to the antecedents of the three rules, one gets 0.2, 0.5 and 0.0 degrees of support, respectively.

The degree of support for the entire rule is then used to shape the output fuzzy set. The consequence of a fuzzy rule assigns an entire fuzzy set to the output. This fuzzy set is represented by a membership function that is chosen to indicate the qualities of the consequent. If the antecedent is only partially true, having a value lower than 1, the output fuzzy set is truncated at this value according to the 'minimum implication method' (Ross, 2004). As the degrees of support for the three rules mentioned above were less than 1, the minimum implication method was applicable leading to the deduction that WQ belonged to 'good' fuzzy set, truncated at $\mu = 0.2$, and WQ belonged to 'average' fuzzy set, truncated at $\mu = 0.5$. This is shown in Figure 6.6, where columns refer to the input/output fuzzy sets, and rows are the fuzzy rules.

Since decisions are based on the testing of all the rules in the system, all must be aggregated to make a decision. As depicted in Figure 6.6, output fuzzy sets for each rule are aggregated to a single output fuzzy set. For aggregation the maximum method (Ross, 2004), which unites all the truncated output fuzzy sets, was used.

The final step of defuzzification was then taken. The input for the defuzzification process was the aggregated output fuzzy set. As much

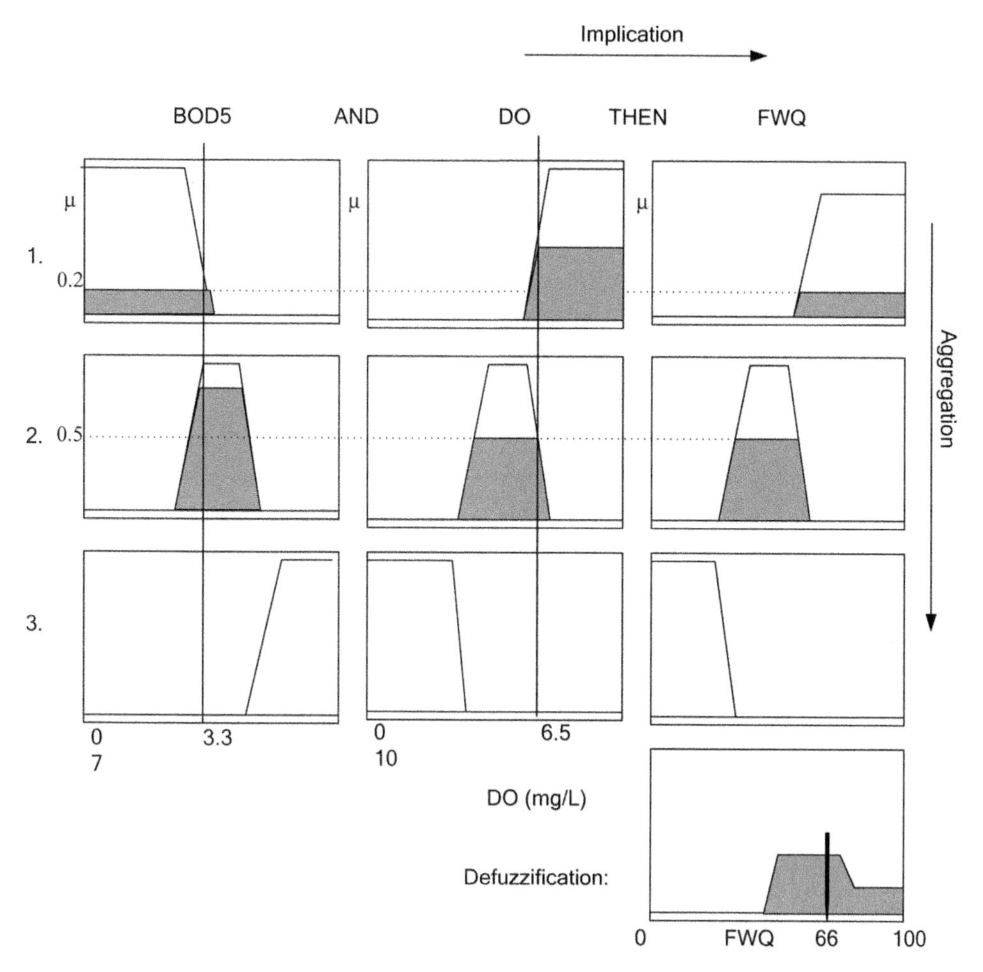

FIGURE 6.6 Fuzzy inference diagram for the water-quality scoring problem with two variables and three rules in the course of the fuzzy WQI development by Ocampo-Duque et al. (2006).

as fuzziness helps the rule evaluation during intermediate steps, the final desired output is a numeric score. The centroid defuzzification method, which is the most preferred and physically appealing of all available methods, was used; it returns the centre of area under the curve formed by the output fuzzy set:

$$Z^* - \frac{\int \mu(z) \cdot z dz}{\int \mu(z) dz} \tag{6.16}$$

By replacing the corresponding membership functions (shown in Figure 6.5) in Equation (5), the water-quality index for the point "R2" is arrived at:

$$\text{FWQ}^* = \frac{\int_{40}^{45}(0.10_z - 4)zdz + \int_{45}^{75}(0.5)zdz + \int_{75}^{78}(-0.10_z + 8)zdz + \int_{78}^{100}(0.2)zdz}{\int_{40}^{45}(0.10_z - 4)dz + \int_{45}^{75}(0.5)dz + \int_{75}^{78}(-0.10_z + 8)dz + \int_{78}^{100}(0.2)dz} = 65.9 \tag{6.17}$$

In actual practice, Ocampo-Duque et al. (2006) used 27 water-quality indicators and 96 rules as explained a little later. The overall procedure is as illustrated in Figure 6.4.

The successful application of a fuzzy inference system (FIS) depends on an appropriated weight assignment to the variables involved in the rules. Weight assignment defines the relative importance and influence of the input parameters in the final score. Even a good FIS can lead to erroneous simulations if the weights are not assigned correctly. Hence, the authors used a comprehensive multi-attribute decision-aiding method based on the analytic hierarchy process (AHP) (Vaidya and Kumar, 2006) to estimate the relative importance of water-quality variables. The AHP is a methodology for solving decision problems by the prioritisation of alternatives. The basis of the AHP is the Saaty's eigenvector method (Saaty, 2003) and the associated consistency index. It is based on the largest eigenvalue and associated eigenvector of $n \times n$ positive reciprocal matrix A. Elements a_{ij} of A are the decision maker's numerical estimates of the preference of n alternatives with respect to a criterion when they are compared pairwise using the 1–9 AHP comparison scale. At the lower limit of the scale, both alternatives are equally preferred; the progressively higher value of the scale to 9 implies that one of the alternatives is preferred.

The singular value decomposition (SVD) method (Gass and Rapcsak, 2004) was utilised in conjunction with AHP to enable the priority of the decision maker to be approximated by the uniquely determined, normalised, positive weight vector w with the values:

$$w_i = \frac{u_i + \dfrac{1}{v_i}}{\sum_{j=1}^{n} u_j + \dfrac{1}{v_j}} \quad i = 1, \ldots, n \tag{6.18}$$

where u and v are the left and right singular vectors belonging to the largest singular value of matrix A, respectively, and n is the number of variables. The consistency measure (CM) of the weight vector was determined on an absolute scale by using the Frobenius norm:

$$\text{CM} = \frac{||A - \tilde{A}||_F}{||A||_F + (41/9)n} \tag{6.19}$$

where matrix $\tilde{A}$ is formed by setting (w_i/w_j) for every pair (i, j). If CM < 0.10, the matrix A is considered to be consistent else the pairwise comparison will have to be made again.

The authors chose 27 WQ parameters to cover a wide range of possible pollutants (Table 6.4). Trapezoidal membership functions were used to represent 'low', 'medium', 'high', 'poor', 'average' and 'good' fuzzy sets:

$$\mu(x; a, b, c, d) = \max\left(\min\left(\frac{x-a}{b-a}, 1, \frac{d-x}{d-c}\right), 0\right) \tag{6.20}$$

where a, b, c and d are membership function parameters (Table 6.4). The ranges shown in the table for fuzzy sets were based on European trend concentrations in river waters (EEA, 2003), guidelines for drinking-water quality (WHO, 2004), toxicity and ecotoxicity parameters, and Spanish available regulations for classifying water in river basins, and setting objectives.

Ninety-six rules were enunciated, three for each indicator and three for each partial score into groups. Each rule had only one antecedent in order to facilitate the weight assignment. The structure of fuzzy rules was: if indicator i is 'Low' then FWQ is 'Good', if indicator i is 'Medium' then FWQ is 'Average', and if indicator i is 'High' then FWQ is 'Poor'. There were exceptions for DO and pH, in whose case rules were: if DO is 'Low' then FWQ is 'Poor', if DO is 'Medium' then FWQ is 'Average', if DO is 'High' then FWQ is 'Good', if pH is 'Low' or pH is 'High' then FWQ is 'Average', and if pH is 'Medium' then FWQ is 'Good'.

A comparison of the performance of the proposed FWQI with the Simplified Water-Quality Index (ISQA) of the Catalan Water Agency (ACA, 2005), and the WQI of Said et al.

TABLE 6.4 Quality Classes Set by Icaga (2007) for his Fuzzy WQI

	Limits of Quality Classes				
	I	*II*	*III*	*IV*	*Observations*
Temperature (T) (°C)	25	25	30	>30	16.75
pH	6.5–8.5	6.5–8.5	6–9	<6–9>	8.40
Dissolved oxygen (DO) (g/m^3)	8	6	3	<3	8.53
Oxygen saturate (OS) (g/m^3)	90	70	40	<40	No data
Chloride (Cl) (g/m^3)	25	200	400	>400	145.74
Sulphat (SO_4) (g/m^3)	200	200	400	>400	499.33
Ammonia (NH_3) (g/m^3)	0.2	1	2	>2	2.108
Nitrite (NO_2) (g/m^3)	0.002	0.01	0.05	>0.05	0.026
Nitrate (NO_3) (g/m^3)	5	10	20	>20	0.930
Total phosphors (g/m^3)	0.02	0.16	0.65	>0.65	No data
Total dissolved solid (TDS) (g/m^3)	500	1500	5000	>5000	1329.08
Color (Pt-co unit)	5	50	300	>300	21.82
Sodium (Na) (g/m^3)	125	125	250	>250	235.88

(2004), which are conventional WQIs, was carried out using the data of Ebro River. It was observed that the WQI and ISQA scores were always over 70, indicating "good water quality". ISQA scores were higher of the two because ISQA does not consider microbiological pollution. In contrast, the FWQI scores indicated the water quality in the lower Ebro as "some portion average and some portion good". FWQI outputs agreed more closely with the real condition reported by the Confederación Hidrográfica del Ebro for the studied zone (CHE, 2004), according to whom Ebro River water quality decreases as it comes closer to the sea.

6.9. ICAGA'S FUZZY WQI

Close on the heels of the fuzzy WQI of Ocampo-Duque et al. (2006) came the one developed by Icaga (2007). His procedure consisted of 6 steps:

1. Setting up of the quality classes of the WQ parameters using observed values. This was done as in Table 6.5.
2. Arrangement of the parameters in their classes to obtain four groups.
3. Use of membership functions to standardise the natural measurement scales of the quality parameter into a measurement of the quality degree (membership grade). In this step, four membership functions were used as follows:

$$mf_1 = \begin{cases} 1 & \text{for } x < a \\ 1 - [(x-a)/(b_1-a)] & \text{for } a \leq x < b_1 \\ 0 & \text{for otherwise} \end{cases} \tag{6.21}$$

$$mf_2 = \begin{cases} (x-a)/(b_1-a) & \text{for } a \leq x < b_1 \\ 1 & \text{for } b_1 \leq x < b_2 \\ 1 - [(x-b_2)/(c_1-b_2)] & \text{for } b_2 \leq x < c_1 \\ 0 & \text{for otherwise} \end{cases} \tag{6.22}$$

TABLE 6.5 Limits of Membership Functions in the Procedure of Icaga (2007)

	The Parameters of the Membership Functions							
			b			*c*		
	a	b_1[a]		b_2[b]	c_1[a]		c_2[b]	*d*
Temperature (T) (°C)	17.5		22.5			27.5		32.5
pH ≥ 7.5	7.5		7.75			8.75		9.25
pH < 7.5	5.75		6.25			6.75		7.5
Dissolved oxygen (DO) (g/m^3)[c]	9		7			4.5		1.5
Chloride (Cl) (g/m^3)	0	50		100		300		500
Sulphat (SO_4) (g/m^3)	50		150		250		350	450
Ammonia (NH_3) (g/m^3)	0		0.4			1.5		2.5
Nitrite (NO_2) (g/m^3)	0		0.004			0.03		0.07
Nitrate (NO_3) (g/m^3)	2.5		7.5			15		25
Total dissolved solid (TDS) (g/m^3)	0		1000			3250		6250
Color (Pt-co unit)	0		27.5			175		425
Sodium (Na) (g/m^3)	31.25		93.75		156.3		218.8	281.3
Output membership function	12.5		37.5			62.5		87.5

[a] *First upper corner of the trapezoidal membership function.*
[b] *Second upper corner of the trapezoidal membership function.*
[c] *The number membership functions in reverse order.*

$$mf_3 = \begin{cases} (x - b_2)/(b - c_1) & \text{for } b \le x < c_1 \\ 1 & \text{for } c_1 \le x < c_2 \\ 1 - [(x - c_2)/(d - c_2)] & \text{for } c_2 \le x < d \\ 0 & \text{for otherwise} \end{cases} \quad (6.23)$$

$$mf_4 = \begin{cases} 1 - [(x - c_2)/(d - c_2)] & \text{for } c_2 < x \le d \\ 1 & \text{for } d < x \\ 0 & \text{for otherwise} \end{cases} \quad (6.24)$$

where mf_i is the ith membership function; x the observed value; and a, b, c, d are the limits of the related membership functions (Tables 6.5 and 6.6); in case the membership function is in trapezoid shape, then the b_1 and c_1 are the first top levels of the trapezoids and b_2 and c_2 are the second top levels of the trapezoids otherwise $b_1 = b_2$ and $c_1 = c_2$. The resulting membership functions had shapes as depicted in Figures 6.7 and 6.8.

4. Use of the four rule bases:

if $QP_1 = I$ or $QP_2 = I$ or ... or $QP_N = I$ then $QP = I$
if $QP_1 = II$ or $QP_2 = II$ or ... or $QP_N = II$ then $QP = II$
if $QP_1 = III$ or $QP_2 = III$ or ... or $QP_N = III$ then $QP = III$
if $QP_1 = IV$ or $QP_2 = IV$ or ... or $QP_N = IV$ then $QP = IV$

where QP_i is the ith quality parameter; I, II, III, IV are the quality classes in traditional classification; and N is the number of quality parameters. In the rule bases the "or" operators were used to obtain maximum values. Following this, the rule bases

TABLE 6.6 Water-Quality Parameters and Their Membership Functions Used for Developing Fuzzy WQI by Ocampo-Duque et al. (2006)

			'Low'			'Medium'				'High'			
	Indicator	**Units**	*a* = *b*	*c*	*d*	*a*	*b*	*c*	*d*	*a*	*b*	*c* = *d*	**Range**
Primary	DO	mg/L	0	3	4	3	4	7	8	7	8	12	0–2
	pH	–	0	6	7.5	6	7	8	9	7.5	9	14	0-4
	CON	μS/cm	0	600	700	600	700	800	900	800	900	1400	0–1400
	SS	mg/L	0	11	13	11	13	15	17	15	17	24	0–24
Organic matter	BOD_5	mg/L	0	2.5	3.5	2.5	3.5	4	5	4	5	10	0–10
	TOC	mg/L	0	3	4	3	4	7	8	7	8	10	0–10
Microbiology	TC	MPN/100 mL	0	250	500	250	500	4000	5000	3750	5000	10,000	0–10,000
	FC	MPN/100 mL	0	100	200	100	200	1600	2000	1500	2000	4000	0–4000
	Sa	Presence - 1 Absence-0 in 1 L	0	0.2	0.4	0.2	0.4	0.6	0.8	0.6	0.8	1	0–1
	FS	MPN/100 mL	0	100	200	100	200	800	1000	800	1000	2000	0–2000
Anions and ammonia	PO_4	mg/L	0	0.2	0.4	0.2	0.4	0.6	0.8	0.6	0.8	1	0–1
	NO_3	mg/L	0	10	20	10	20	30	40	30	40	50	0–50
	NH_4	mg/L	0	0.07	0.14	0.07	0.14	0.18	0.24	0.18	0.24	0.5	0–0.05
	SO_4	mg/L	0	75	100	75	100	125	150	125	150	250	0–250
	Cl	mg/L	0	50	100	50	100	150	200	150	200	250	0–250
	F	mg/L	0	0.3	0.6	0.3	0.6	0.9	1.2	0.9	1.2	1.5	0–1.5
Priority substance	Atr	ng/L	0	80	160	80	160	240	320	240	320	500	0–500
	BTEX	μg/L	0	40	80	40	80	120	160	120	160	200	0–200
	Ni	μg/L	0	10	15	10	15	20	25	20	25	50	0–50
	Sim	ng/L	0	80	160	80	160	240	320	240	320	500	0–500
	TCB	μg/L	0	4	8	4	8	12	16	12	16	20	0–20
	Cr	μg/L	0	10	20	10	20	30	40	30	40	50	0–50
	HCBD	μg/L	0	0.4	0.8	0.4	0.8	1.2	1.6	1.2	1.6	3	0–3
	PAH	ng/L	0	20	40	20	40	60	80	60	80	100	0–100
	As	μg/L	0	15	25	15	25	35	45	35	45	60	0–60
	Pb	μg/L	0	15	25	15	25	35	45	35	45	60	0–60
	Hg	μg/L	0	0.2	0.4	0.2	0.4	0.6	0.8	0.6	0.8	1	0–1
FWQ scores		–	'Poor'			'Average'				'Good'			Range
			0	40	50	40	50	70	80	70	80	100	0–100

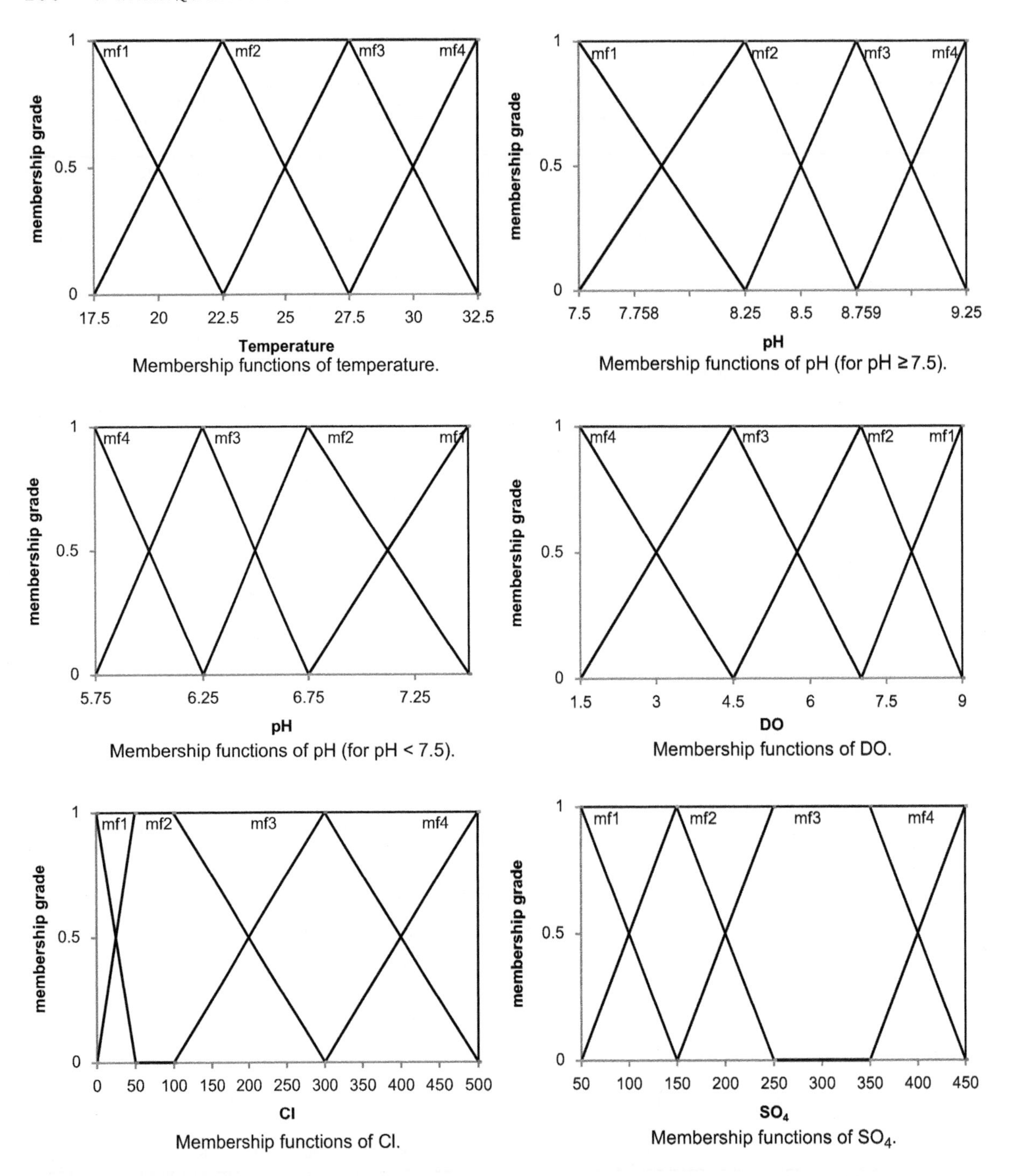

FIGURE 6.7 Membership functions of temperature, pH, DO, Cl and SO_4, in the fuzzy WQI of Icaga (2007).

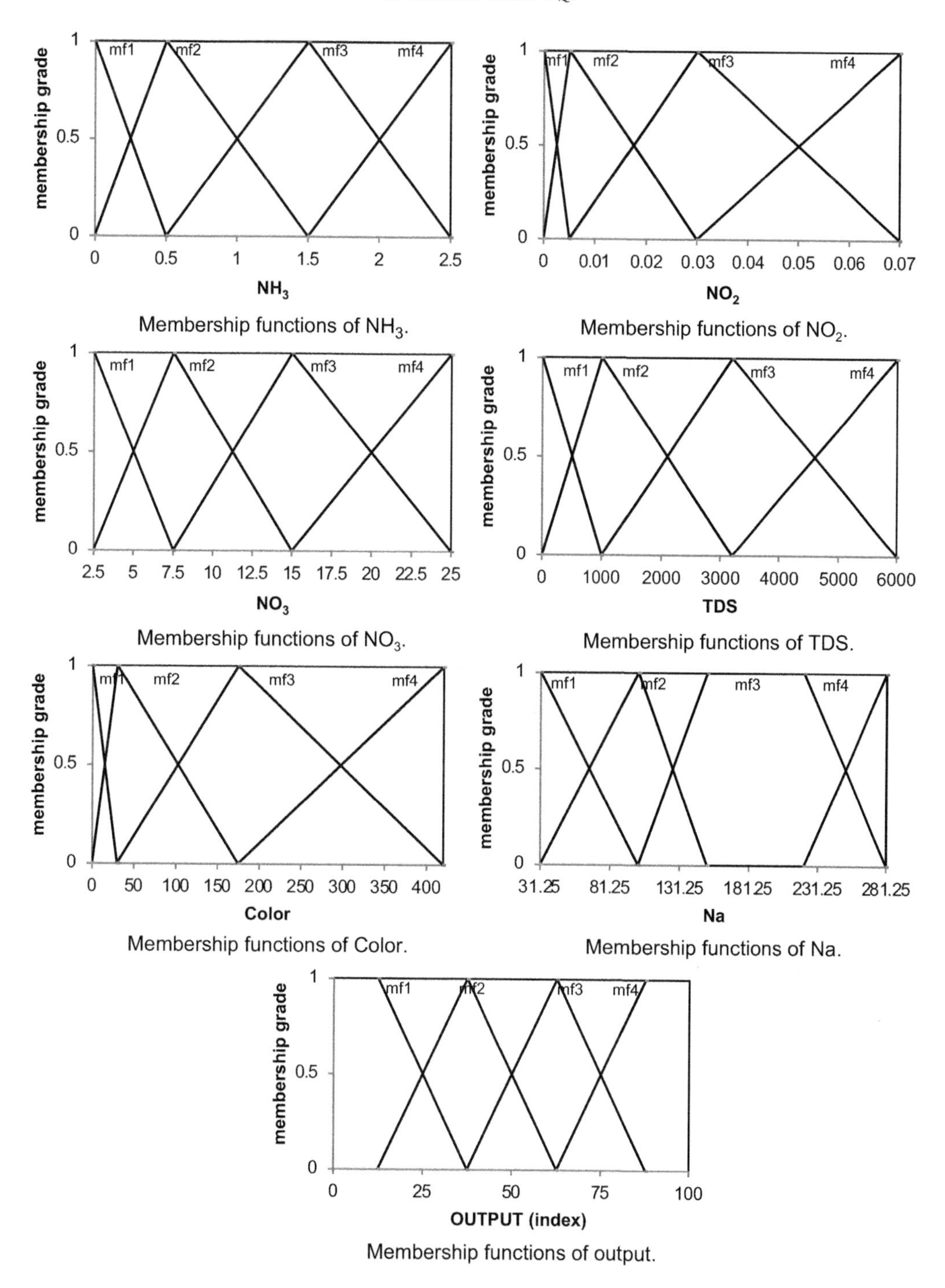

FIGURE 6.8 Membership functions of NH_3, NO_2, NO_3, TDS, colour and general output in the fuzzy WQI of Icaga (2007).

were arranged in the pattern of 'if T = I or DO = I or NO_3 = I, then Output = I′ for all classes.

5. Use of the fuzzy algorithm. In it fuzzy inferences of the groups were determined using grades of membership functions of the parameters.
6. Defuzzification of the four fuzzy inferences of the groups to obtain an index number between 0 and 100 using centroid methods (Bandemer and Gottwald, 1996).

For the Eber lake water, studied by the author, the FWQI was found to be 41.2.

6.10. USE OF ORDERED WEIGHTED AVERAGING (OWA) OPERATORS FOR AGGREGATION

In an attempt to simultaneously reduce the problems of *ambiguity* (*exaggeration*) and *eclipsing*, Sadiq and Tesfamariam (2008) have used *ordered weighted averaging* (OWA) operators in developing a fuzz WQI. The OWA operator (Yager, 1988) provides flexible aggregation that is bounded between the *minimum* and *maximum* operators. The OWA weight generation incorporates decision maker's attitude or tolerance, which can also be related to perceived importance of the environmental system under study.

Even though numerous reports exist on the applications of OWA operators in the civil and environmental Engineering (Sadiq and Tesfamariam, 2007; Makropoulos and Butler, 2006; Smith, 2006), in order to incorporate the fuzzy 'acceptability' measures, a fuzzy number OWA (FN-OWA) was proposed by the authors so that it can handle fuzzy or linguistic values and deal with uncertainty by specifying 'acceptable' environmental quality.

6.10.1. Ordered Weighted Averaging Operators

Most multi-criteria decision analysis problems do not touch the extremes for mutually exclusive probabilities corresponding to multiplication (*AND*-gate) and summation (*OR*-gate) of a fault tree analysis. In other words, these problems require neither strict "*ANDing*" of the *t*-norm (*minimum*) nor strict "*ORing*" of the *s*-norm (*maximum*). To generalise this idea, Yager (1988) introduced a new family of aggregation techniques, called the ordered weighted averaging (OWA) operators, which form a general mean type operator. The OWA operator provides a flexibility to utilise the range of "*ANDing*" (or "*ORing*") to include the attitude of a decision maker in the aggregation process. The OWA operation involves three steps: (1) reorder the input arguments, (2) determine weights associated with the OWA operators and (3) aggregate.

The OWA operator of dimension n is a mapping of $R_n \rightarrow R$, which has an associated n weighting vector $w = (w_1, w_2, \ldots, w_n)^T$, where $w_j, \in [0, 1]$ and $\sum_{j=1}^{n} w_j = 1$. Hence, for a given n-input parameters vector $(\tilde{x}_1, \tilde{x}_2, \ldots, \tilde{x}_n)$, the OWA operator determines the environmental index as follows:

$$\text{Environmental Index (EI)} = OWA(\tilde{x}_1, \tilde{x}_2, \ldots, \tilde{x}_n) = \sum_{j=1}^{n} w_j \tilde{y}_j \quad (6.25)$$

where $\tilde{y}_i$ is the jth largest number in the vector $(\tilde{x}_1, \tilde{x}_2, \ldots, \tilde{x}_n)$, and $(\tilde{y}_1 \geq \tilde{y}_2, \ldots, \geq \tilde{y}_n)$. Therefore, the weights w_j of OWA are not associated with any particular value $\tilde{x}_j$; rather, they are associated to the 'ordinal' position of $\tilde{y}_j$. The linear form of the OWA equation aggregates n-input parameters vector $(\tilde{x}_1, \tilde{x}_2, \ldots, \tilde{x}_n)$ and provides a nonlinear solution (Yager and Filev, 1994).

The range between minimum and maximum values can be determined through the concept of *ORness* (β), which is defined as follows (Yager, 1988):

$$\beta = \frac{1}{n-1} \sum_{i=1}^{n} w_i (n - i) \quad \text{and} \quad \beta \in [0, 1] \quad (6.26)$$

The *ORness* characterises the degree to which the aggregation is like an *OR* operator. The $\beta=0$ refers to a scenario that vector w becomes (0, 0, …, 1), i.e., an input parameter with the minimum value in the n-input parameters vector $(\tilde{x}_1, \tilde{x}_2, \ldots, \tilde{x}_n)$ is assigned the full weight, which implies that the OWA becomes a *minimum* operator. When $\beta=1$, the OWA vector w becomes (1, 0,…,0), i.e., an input parameter with a maximum value in the n-input parameters vector $(\tilde{x}_1, \tilde{x}_2, \ldots, \tilde{x}_n)$ is assigned complete weight, which implies that the OWA collapses to *maximum* operator. Similarly, when $\beta=0.5$, the OWA vector w becomes $(1/n, 1/n, \ldots, 1/n)$, i.e., an arithmetic mean of the input parameter vector $(\tilde{x}_1, \tilde{x}_2, \ldots, \tilde{x}_n)$.

The OWA weights can be generated using a regularly increasing monotone (RIM) quantifier $Q(r)$ as follows:

$$w_i = Q\left(\frac{i}{n}\right) - Q\left(\frac{i-1}{n}\right) \quad i=1,2,\ldots,n \quad (6.27)$$

distribution, i.e., equal weights are assigned to $(\tilde{x}_1, \tilde{x}_2, \ldots, \tilde{x}_n)$, and becomes an arithmetic mean, i.e., $w_i=1/n$. For $\delta>1$, the RIM function leans towards right, i.e., "AND-type" operators manifesting negatively skewed OWA weight distributions. Similarly, for $\delta<1$, the RIM function leans towards left, i.e., "OR-type" operators manifesting positively skewed OWA weight distributions. Discussion on the selection of an appropriate δ value is provided in later sections.

The OWA operators were transformed into FN-OWA for the n-fuzzy input parameters, by describing them as a set of fuzzy numbers $(\tilde{x}_1, \tilde{x}_2, \ldots, \tilde{x}_n)$. If required the reordering and ranking of an n-fuzzy input parameters vector $(\tilde{x}_1, \tilde{x}_2, \ldots, \tilde{x}_n)$.

The 'middle of maximums' (MoM) method was used for *defuzzification* to determine the ordering of fuzzy n-input parameters in a vector $(\tilde{x}_1, \tilde{x}_2, \ldots, \tilde{x}_n)$:

$$SM(EI,k) = 1 - DM(EI,K) = 1 - \frac{[W_1^{SM}|a_{EI}-a_k| + W_2^{SM}|b_{EI}-b_k| + W_3^{SM}|C_{EI}-C_k| + W_4^{SM}|d_{EI}-d_k|]}{[d_{k=5}-a_{k=1}]}$$

Yager (1996) defined a parameterised class of fuzzy subsets, which provide families of RIM quantifiers that change continuously between $Q^*(r)$ and $Q^*(r)$:

$$Q(r) = r^\delta \quad r \geq 0 \quad (6.28)$$

For $\delta=1$; $Q(r)=r$ (a linear function) is called the unitor quantifier; for $\delta\to\infty$; $Q^*(r)$ is the universal quantifier (and-type); and for $\delta\to 0$; $Q^*(r)$ is the existential quantifier (or-type).

Therefore, Equation (3) can be generalised as

$$w_i = \left(\frac{i}{n}\right)^\delta - \left(\frac{i-1}{n}\right)^\delta \quad i=1,2,\ldots,n \quad (6.29)$$

where δ is a degree of a polynomial function. For $\delta=1$, the RIM function is like a uniform

To estimate *EI*, the authors proposed five linguistic constants ($k=1$, 2, …, 5) over the universe of discourse (a quantitative scale on which *EI* are defined), namely, *very poor* (VP), *poor* (P), *fair* (F) and *good* (G) to *very good* (VG) (Table 6.7). The estimated *EI* and linguistic constants were superimposed over the universe of discourse. Based on the maximum similarity between a linguistic constant and the estimated EI, a linguistic constant was assigned which can be related to a specific decision action. The smaller the distance between the estimated *EI* and a particular linguistic constant is, the higher will be the similarity measure *(SM)*, and vice versa.

The FN-OWA methodology was applied to three attitudinal scenarios: $\delta=1/3$ (*OR-type*), $\delta=1$ (*neutral*) and $\delta=3$ (*AND-type*). The

TABLE 6.7 Transformation of Raw Water-Quality Data into Fuzzy Subindices

i (1)	WQ Indicators (2)	Obs. Values (q_i) (3)	Transformation Function[a] (4)	Parameters for the Transformation Function[b] (5)	Fuzzy Sub-Indices (*si*) (6)
1	BOD_5 (mg/L)	20	UDS $(\overline{m}^1, \overline{q}_c^1)$	$\overline{m}^1 = (2.1, 3, 3,.9)$; $\overline{q}_c^1 = (10, 20, 30)$	(0.014, 0.13, 0.34)
2	Fecal coliforms (MPN/100 mL)	66	UDS $(\overline{m}^2, \overline{q}_c^2)$	$\overline{m}^2 = (0.21, 0.3, 0.39)$; $\overline{q}_c^2 = (2, 4, 6)$	(0.25, 0.43, 0.59)
3	DO (proportion)	0.6	US $(q*^3, \overline{n}^3, \overline{p}^3, r^3)$	$\overline{n}^3 = (1.5, 3, 4.5)$; $\overline{p}^3 = (0.9, 1, 1.1)$	(0.41, 0.62, 0.83)
				$q*^3 = 1$; $r^3 = 0$ (used as crisp values)	
4	Nitrates (mg/L)	25	UDS $(\overline{m}^4, \overline{q}_c^4)$	$\overline{m}^4 = (2.1, 3, 3.9)$; $\overline{q}_c^4 = (20, 44, 60)$	(0.04, 0.23, 0.48)
5	PH	7.8	US $(q*^5, \overline{n}^5, \overline{p}^5, r^5)$	$\overline{n}^5 = (1.6, 4, 6.4)$; $\overline{p}^5 = (5.4, 6, 6.6)$	(0.79, 0.87, 0.94)
6	Phosphates (mg/L)	2	UDS $(\overline{m}^6, \overline{q}_c^6)$	$q*^5 = 7$; $r^5 = 0$ (used as crisp values)	(0.08, 0.25, 0.47)
				$\overline{m}^6 = (0.7, 1, 1.3)$; $\overline{q}_c^6 = (0.34, 0.67, 1.01)$	
7	Temperature (°C)	32	US $(q*^7, \overline{n}^7, \overline{p}^7, r^7)$	$\overline{n}^7 = (0.25, 0.5, 0.75)$; $\overline{p}^7 = (6.3, 7, 7.7)$	(0.25, 0.40, 0.63)
				$q*^7 = 20$, $r^7 = 0$ (used as crisp values)	
8	Total solids (mg/L)	1,000	US $(q*^8, \overline{n}^8, \overline{p}^8, r^8)$	$\overline{n}^8 = (0.5, 1, 1.5)$; $\overline{p}^8 = (0.9, 1, 1.1)$	(0.10, 0.17, 0.37)
				$q*^8 = 75$; $r^8 = 0.8$ (used as crisp values)	
9	Turbidity (JTU)	70	UDS $(\overline{m}^9, \overline{q}_c^9)$	$\overline{m}^9 =$, $\overline{q}_c^9 = (25, 50, 75)$	(0.07, 0.27, 0.50)

[a] *Two types of transformation functions are used: UDS* $\overline{m}^i, \overline{q}_c^i = \left(1 + \frac{qi}{\overline{q}_c^i}\right)^{-\overline{m}}$; $US(q*^i, \overline{n}^i, \overline{p}^i, r^i) = \dfrac{\overline{p}^i r^i + (\overline{n}^i + \overline{p}^i)(1 - r^i)\left(\frac{q^i}{q*^i}\right)^{\overline{ni}}}{\overline{p}^i + \overline{n}^i(1 - r^i)\left(\frac{q^i}{q*^i}\right)^{\overline{ni}+\overline{p}^i}}$

[b] *The bar over the transformation parameters triangular fuzzy numbers.*
UDS uniform decreasing subindices, *US* unimodal subindices.
Modified after Swamee and Tyagi, 2000.

corresponding *ORness* values were computed as $\beta = 0.76$, 0.5 and 0.22, respectively. The results of WQI were as plotted in Figure 6.9. With an increase in δ value (or decrease in *ORness* β), the EI values became smaller and vice versa. Therefore, the smaller *ORness* values ($\beta < 0.5$) account for a pessimistic decision maker's attitude and larger *ORness* values ($\beta > 0.5$) represent an optimistic decision maker's attitude. With neutral and and-type decision attitude, the WQI was classified as *poor*, whereas, with OR-type decision-making attitude, the WQI was classified as *fair*. Assuming that these three scenarios are surrogates for the intended use of a river water, for recreational, irrigation, and drinking, respectively, and considering that the adverse consequences of even a marginal fall in quality of source water for drinking water supply can be quite high, the decision maker would be conservative (or pessimistic) selecting smaller *ORness* value ($\beta < 0.5$). Conversely, if the primary use of source water is recreational, the larger *ORness* value ($\beta > 0.5$) would be adopted.

6.11. FUZZY WATER-QUALITY INDICES FOR BRAZILIAN RIVERS (LERMONTOV ET AL., 2008, 2009; ROVEDA ET AL., 2010)

Setting out to develop a water-quality index based on fuzzy rules, and to compare its performance with three pre-existing conventional WQIs on the assessment of Brazilian river watersheds, Lermontov et al., (2008, 2009) developed and tested their fuzzy WQI (FWQI) on the Ribeira do lguape River.

In their index, and following the usual fuzzy rules, fuzzy sets were defined in terms of a membership function that maps a domain of interest to the interval [0, 1]. Curves were used to map the membership function of each set. They showed to which degree a specific value belonged to the corresponding set:

$$\mu A: X \rightarrow [0, 1]$$

Trapezoidal and triangular membership functions (Figure 6.1) were used by the authors for the same nine parameters that form the basis of the official CETESB WQI of the SaoPaulo

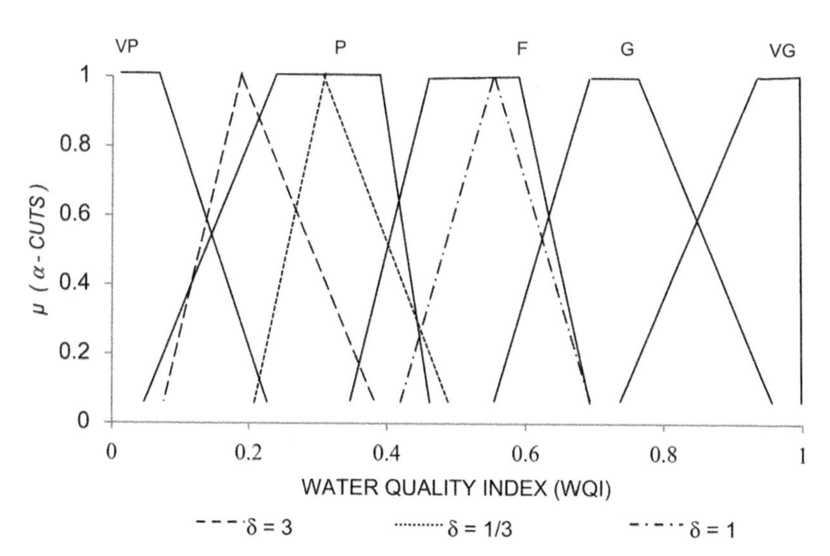

FIGURE 6.9 Estimating water-quality index (WQI) values using results of FNOWA (Sadiq and Tesfamariam, 2008).

State, Brazil, so that their FWQI can be compared with CETESB WQI and validated. In turn, the CETESB WQI has been patterned on the WQI of the US National Sanitation Foundation (NSF, 2007). The water-quality data were used according to the following equations to create the fuzzy sets:

Trapezoidal: $f(x; a, b, c, d)$

$$= \begin{cases} 0 & x < a \text{ OR } d < x \\ \dfrac{(a-x)}{(a-b)} & a \le x \le b \\ 1 & b \le x \le c \\ \dfrac{(d-x)}{(d-c)} & c \le x \le d \end{cases} \quad (6.30)$$

Triangular: $f(x; a, b, c)$

$$= \begin{cases} 0 & x < a \text{ OR } c < x \\ \dfrac{(a-x)}{(a-b)} & a \le x \le b \\ \dfrac{(c-x)}{(c-b)} & b \le x \le c \end{cases} \quad (6.31)$$

The sets were then named according to a perceived degree of quality that ranged from 'very excellent' to 'very bad'. For the parameters' temperature and pH, two sets for each linguistic variable were used because temperature and pH sets have the same linguistic terms above and under the 'very excellent' point while distancing from it. The trapezoidal function was used only for the 'very excellent' linguistic variable and the triangular for all others. The linguistic model of fuzzy inference was used wherein the input data set – the water-quality variables (the antecedents) – was processed using linguistic *if/then* rules to yield an output data set, the consequents.

The flow graph of the process used by the authors is shown in Figure 6.10. The individual quality variables were processed by inference systems, yielding several groups normalised between 0 and 100. The groups were then

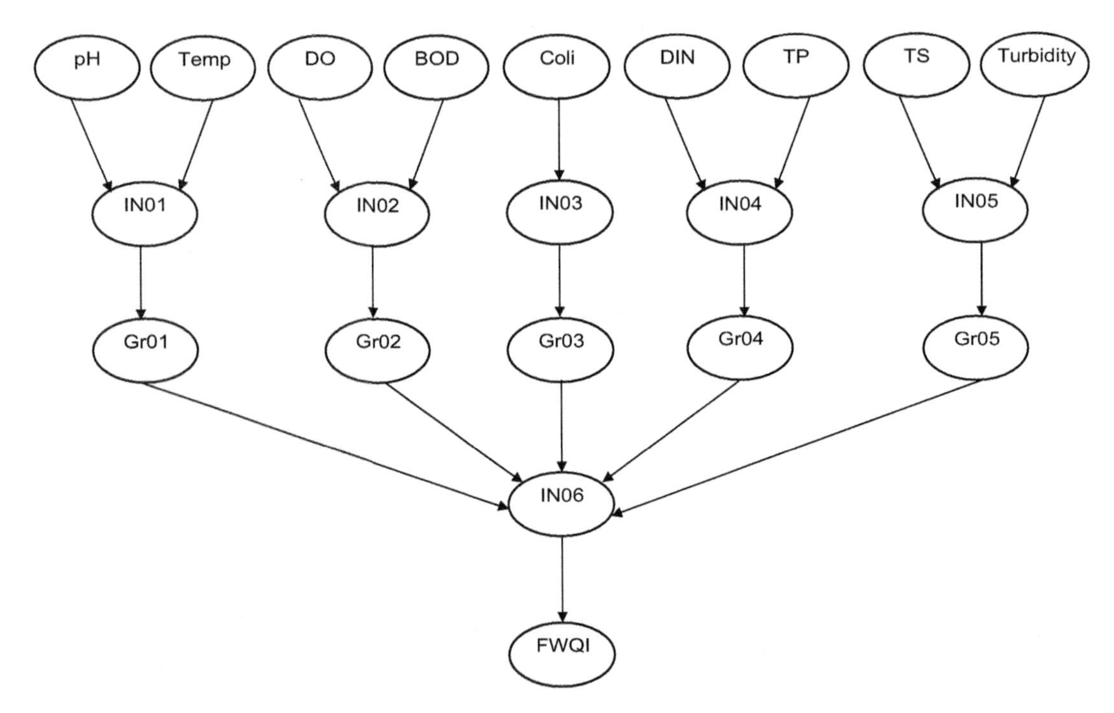

FIGURE 6.10 Flow graph of the process used by Lermontov et al. (2009).

processed for a second time, using a new inference, and the end result was the fuzzy water-quality index – FWQI.

Whereas in the traditional methods for obtaining WQI, parameters are normalised with the help of tables or curves and weight factors (Conesa, 1995; Mitchell and Stapp, 1996; Pesce and Wunderlin, 2000; CETESB, 2004–2006; NSF, 2007) and then calculated by conventional mathematical methods, in the fuzzy rule-based method of these authors, parameters were normalised and grouped through a fuzzy inference system that used the fuzzy sets in Figure 6.11 to obtain the FWQI.

The authors found that their FWQI correlated well with the CETESB WQI (correlation coefficient 0.794).

The FWQI of Lermontov et al. (2009) was used by Roveda et al. (2010) in the study of Sorocaba Rivor in SaoPaulo State, Brazil. They, too, compared the performance of FWQI with that of CETESB WQI and found that the FWQI provided 'a more rigorous classification of water quality at the investigated sampling locations'.

Out of the 107 index values calculated by Roveda et al. (2010), 78 (73 %) gave the same category using both systems, while 29 values (27 %) were in different categories. Of these 29 values, 20 (69 %) were classified by FWQI in lower categories than those obtained using the CETESB WQI. The other 9 gave higher values, albeit often near the lower limit of the better category.

Correlation between the two indices yielded a value of $R = 0.71$.

6.12. A HYBRID FUZZY – PROBABILITY WQI

Nikoo et al. (2010) have presented a hybrid probabilistic water-quality index (PWQI) by utilising fuzzy inference systems (FIS), Bayesian networks (BNs) and probabilistic neural networks (PNNs). The proposed methodology can provide the probability mass function (PMF) of the quality of a water body based on the existing water-quality-monitoring data. The methodology was applied by the authors to Jajrood River, Iran which showed its effectiveness in river water-quality assessment and zoning.

The flow chart of the methodology is presented in Figure 6.12. At first, the water-quality indicators were selected considering the temporal and spatial variations of the water-quality variables and the main characteristics of pollution loads in the study area. With the

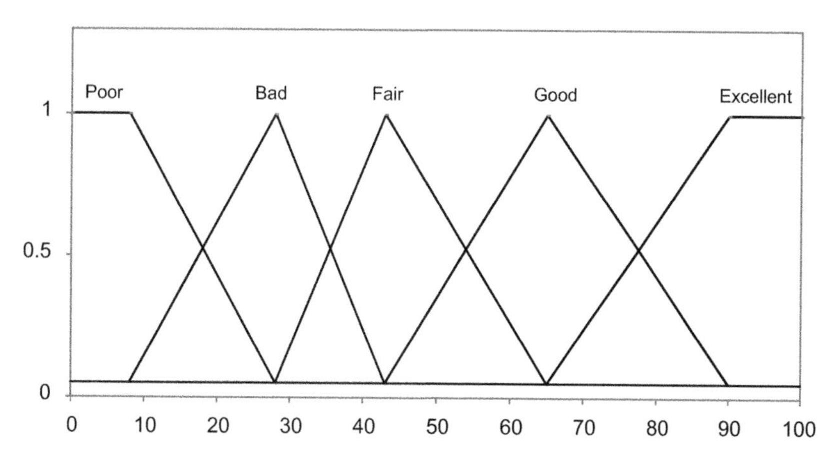

FIGURE 6.11 Graph of fuzzy set functions used by Lermontov et al. (2009).

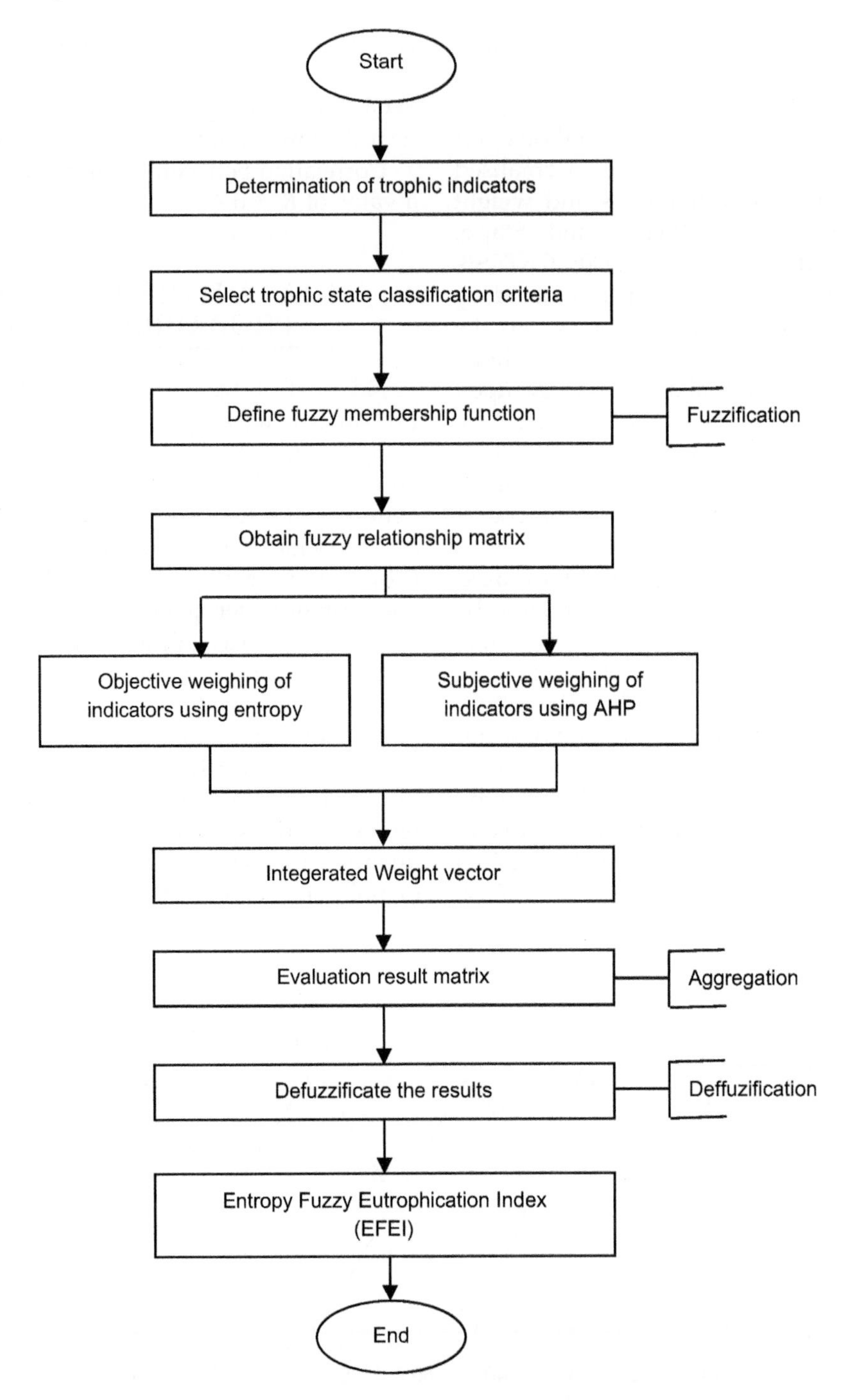

FIGURE 6.12 Flow chart of the entropy – based fuzzy eutrophication index proposed by Taheriyoun et al. (2010).

water-quality values of those parameters (as measured in the Jajrood River), scores of two well-known indices were calculated: the US National Sanitation Foundation WQI (NSF-WQI) and the Condition Council of Ministers of Environment WQI (CCME-WQI). Based on the index values, five classes of water quality were set: excellent, good, fair, marginal and poor. These classes, and corresponding fuzzy membership functions, were considered for inputs (NSF-WQI and CCME-WQI) and output (classes of the quality of water) of the FIS. Since the two indices yield different scores for the same set of water-quality parameters, it was necessary to establish a way to guide the influence of each index in the final score. Using the method of Karamouz et al. (2003), the pairwise comparison matrix in the AHP was used for weight assignment to NSF-WQI and CCME-WQI. The relative weights were found to be 0.37 and 0.63. The analytic hierarchy process (AHP) was used for weight assignment to the inputs of the FIS.

In the next step, the fuzzy inference rules were developed using experts' opinion about the relationship between the inputs and output of the FIS. To construct the PWQI, the trained FIS was utilised in a Monte Carlo analysis. For this purpose, the values of NSF-WQI and CCME-WQI were considered as random variables with uniform distribution and the outputs of the FIS in different simulations were saved. The BN and PNN were then trained and validated using the results of the Monte Carlo analysis. The trained PNN and BN were used as a PWQI for water-quality assessment and zoning in river systems. The accuracy of the trained BN was assessed by comparing its output with the output of FIS. The average relative error in the use of the trained BN was 7.8%, which showed that the trained BN can be accurately used for probabilistic water-quality assessment.

A major feature of PWQI is that it is flexible as the number of FIS inputs can be increased. For example, variables or indices that show the sediment quality or the concentration of pesticides in water or sediment can also be considered as the inputs of the FIS.

A combination of fuzzy inference and artificial neural network has been used by Safavi (2010) for water-quality predictions when studying the Zayandehroud River, Iran.

6.13. AN ENTROPY-BASED FUZZY WQI

In the fuzzy synthetic evaluation (FSE) method, the weights of water-quality parameters are determined empirically or by using the pairwise comparison method. These methods are dependent on expert's judgements and biases, and carry the possibility of misjudgements (Chowdhury and Husain, 2006).

The entropy theory, which has its roots in thermodynamics, was introduced to the information theory by Shannon (1948). In information theory, the degree of disorder of a system is measured. The larger the value of entropy in an information system, the more randomness it possesses, hence less information expressed by the data (Zeleny, 1982). In other words, entropy can be a measure of uncertainties in data and the extent of useful information provided by the data. Therefore, entropy can serve as an objective means of defining the weights of water-quality parameters or indicators based on the extent of useful information contained in the available data.

Chowdhury and Husain (2006) developed a methodology for health risk management of different water treatment technologies using the entropy and fuzzy set theories applied in a multi-attribute decision-making space. In their study, the weights of the attributes were determined based on a combined approach using the analytic hierarchy procedure (AHP) as a mean of measuring the subjective weight and using entropy for determining the objective weight of each attribute.

Building upon the work of Chowdhury and Husain (2006), Taheriyoun et al. (2010) have developed an entropy-based fuzzy synthetic evaluation method to capture the randomness and uncertainties in the input water-quality data. With it they have developed an index to analyse the trophic status of reservoirs. Fuzzy membership functions were defined for selected water-quality indicators and the weights of the indicators were determined using the entropy and AHP. The procedure is as sketched in Figure 6.13.

As the weighing procedure distinguishes the WQI of Taheriyoun et al. (2010) from other fuzzy WQIs, its salient points are described below.

The m set of data for n indicators was used to form the evaluation matrix:

$$EM = \begin{bmatrix} X_1 & X_2 & \dots & X_{1n} \\ X_2 & X_2 & \dots & X_{2n} \\ \vdots & \vdots & \ddots & \vdots \\ X_{m1} & X_{m2} & \dots & X_n \end{bmatrix} \tag{6.32}$$

Element X_{ik} of the evaluation matrix represents the ith set of data for the kth indicator ($i = 1, 2, \ldots, m$; $k = 1, 2, \ldots, n$) (Zeleny, 1982; Hwang and Yoon, 1981).

The following procedure was used for determining weights of indicators:

1. Normalisation of the elements of the original evaluation matrix:

For the sake of having the same scale of measurement for the n indicators, it was assumed that all the initial entry values in the matrix are in the range from 0 to 1. This was achieved by normalising the elements of the initial matrix:

$$r_{ik} = \begin{cases} \dfrac{X_{ik}}{\max\{X_{ik}\}_k} & : \text{for maximum criterion} \\[2ex] \dfrac{\max\{X_{ik}\}_k}{X_{ik}} & : \text{for minimum criterion} \end{cases} \tag{6.33}$$

The maximum criterion refers to the indicator with preferred higher value and thus each element was divided by the maximum value of kth indicator (maximum value in column k in Equation (6.32)). On the other hand, the minimum criterion means the indicator with preferred lower value and the minimum value is divided by each element. It follows that

$$r_{ik} \in [0, 1]$$

2. Calculating the probability of the criterion to occur, p_{ik}:

$$p_{ik} = \frac{r_{ik}}{\sum_{k=1}^{m} r_{ik}}$$

3. Measurement of the entropy of the kth criterion (indicator):

$$E_k = -C \sum_{i=1}^{m} [p_{ik} \cdot Ln, p_{ik}]$$

where c represents a constant, defined as

$$C = \frac{1}{Ln\, \text{m}}$$

4. Calculating the objective importance of the indicators as the weight of entropy:

Considering that a weight assigned to an attribute (indicator or criteria) is directly related to the average intrinsic information generated by a given set of data in addition to its subjective assessment, the degree of diversification (d_k) of the information provided by k^{th} indicator was defined as the complementary of entropy value:

$$d_k = 1 - E_k \tag{6.34}$$

Therefore, the objective importance of kth criteria was evaluated as

$$W_k = \frac{d_k}{\sum_{k=1}^{n} d_k} \tag{6.35}$$

Equations (9) and (10) reflect that lesser the entropy of the indicators higher the information

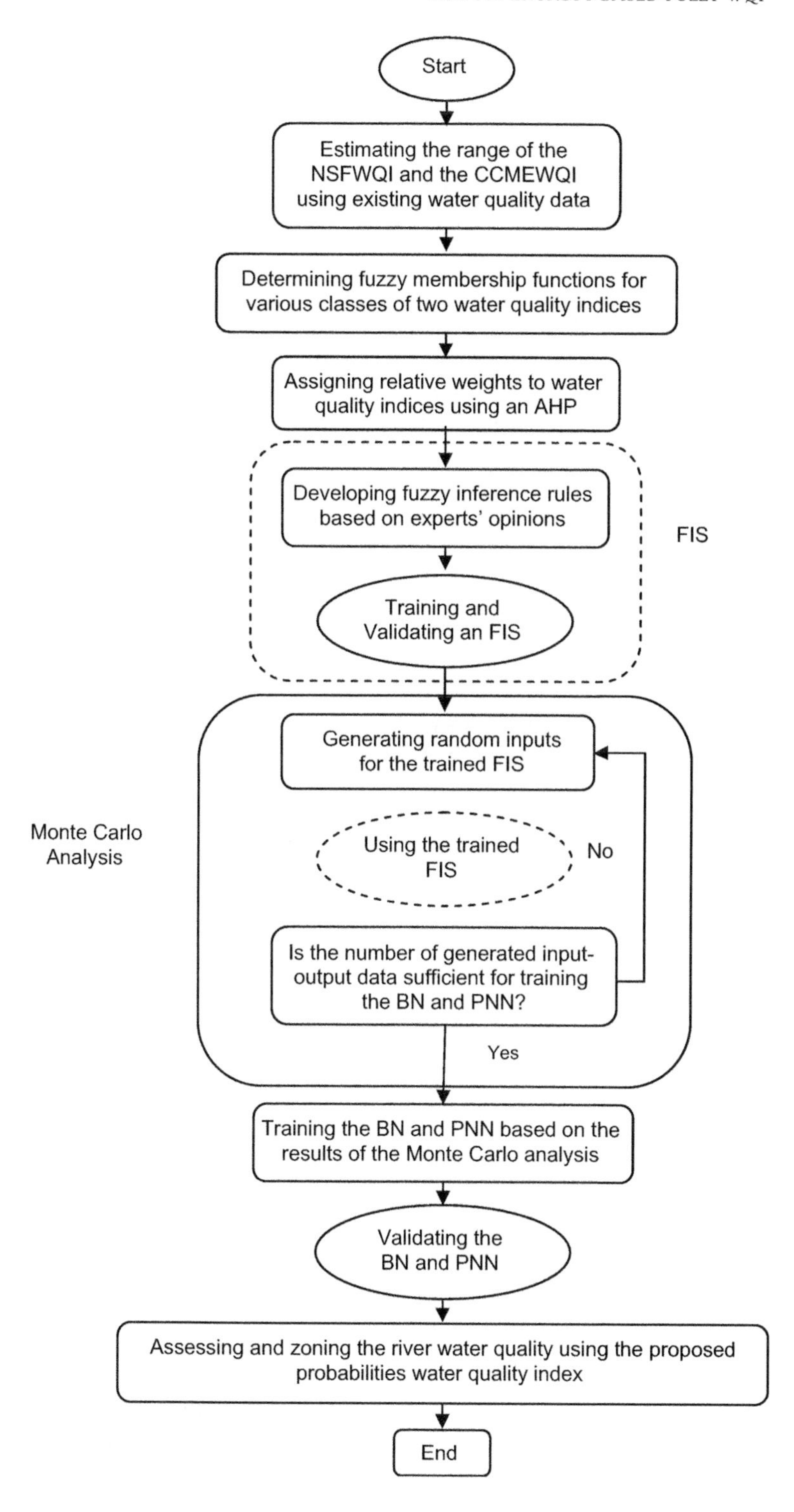

FIGURE 6.13 Water-quality assessment methodology based on a hybrid fuzzy-probabilities WQI of Nikoo et al. (2010).

content. Thus, appropriate weights can be assigned.

5. Combining the objective importance w_k with the subjective importance λ_k to evaluate the integrated importance of the kth indicator parameter W_k^*:

$$W_k^* = \frac{\lambda_k \cdot W_k}{\sum_{j=1}^{n} \lambda_j W_j}$$

The aggregation was done in a manner similar to the one described by Chang et al. (2001). In other words, the aggregation of the results was achieved by multiplying the matrix of fuzzy membership and weighted vector of the n indicators (Lu et al., 1999):

$$B = W^* \times R = [b_e b_m b_o]$$

in which

$$W^* = [W^*_C W^*_P W^*_{SD} W^*_{HO}]$$

$$R = \begin{bmatrix} \mu_e & \mu_n & \mu_o \\ \mu_p & \mu_{pn} & \mu_p \\ \mu_{SDe} & \mu_{SDm} & \mu_{SDo} \\ \mu_{HOe} & \mu_{HOm} & \mu_{HOo} \end{bmatrix}$$

where b_e, b_m and b_o are the components that show the membership degree of the reservoir trophic status corresponding to eutrophic, mesotrophic and oligotrophic, respectively.

Application of the index revealed that chlorophyll a contained the highest amount of useful information with the least entropy content and the secchi disk depth had the least amount of available information and the highest entropy in the data set. Trophic levels were determined based on the indicators with the least degree of uncertainty and the highest level of information.

The authors believe that their method will be particularly useful when dealing with missing data or unreliable information as it enables measuring the entropy content of indicator values. In this way, the best indicators for differentiating the trophic levels can be chosen and their impact on the process can be proportionally weighted. As a consequence, the evaluation process is likely to be more accurate.

6.14. A FUZZY RIVER POLLUTION DECISION SUPPORT SYSTEM

Nasiri et al. (2007) have proposed what they call a fuzzy 'multiple-attribute decision support expert system' to compute a water-quality index and to provide an outline for the prioritisation of alternative plans based on the extent of improvements in WQI. The steps to fuzzy water-quality index development are similar to the ones described in Chapter 6. Nasiri et al. (2007) have then added the following steps:

1. *Development of reach quality index.* For each river reach, a water-quality index was computed by aggregating the water-quality attributes with respect to their weights via the Yager max–min multiple-attribute decision-making model (Yager, 1980). In doing so, for reach r, a quality index set $\tilde{QI}_r$ was defined as follows:

$$\tilde{QI}_{r_{r=1,\ldots,R}} = \bigcap_{i=1,2,\ldots,I} \tilde{S}_{i,r}^{w^*i,r} = \left\{ \left(S, \mu_{\tilde{QI}_r}(S) \right) \middle| \mu_{\tilde{QI}_r}(S) = \mathrm{Min}_{i=1,\ldots,I}\left(\left(\mu_{\tilde{S}_{i,r}}(S) \right)^{w^*i,r} \right) \right\} \quad (6.36)$$

Then, based on the Yager algorithm, the element with the maximum membership value showed the decision point in terms of reach quality index QI_r^*:

$$QI_r^* = \left\{ s \middle| \mu_{\tilde{QI}_r}(s) = \mathrm{Max}(\mu_{\tilde{QI}_r}) \right\} \quad (6.37)$$

2. *Prioritising per reach.* A major objective of the reach water-quality index is to allow the evaluation of the benefits associated with different alternative water-quality management plans in river reaches. The index is correlated directly with the water-quality classifications and provides a direct measure of the relative water quality of a reach within the boundaries of water uses. Thus, a water-quality management plan that can potentially decrease a reach quality index represents an improvement in water quality of that reach, which can be a basis for comparing the alternatives to prioritise them for every reach. Therefore, for each plan the effectiveness indicator was defined in reach r as

$$+QI_r^* = \begin{cases} 1 - \left(\dfrac{QI_r^*}{QI_r^*(0)}\right) & QI_r^* < QI_r^*(0) \\ 0 & \text{Otherwise} \end{cases} \tag{6.38}$$

where QI_r^* and $QI_r^*(0)$ are estimations for the water-quality index of reach r, after and before a water-quality management plan, respectively.

3. *Reach importance degree.* Given that in terms of water quality, a reach is defined via its associated quality factors, Nasiri et al. (2007) assume that the criticality of a reach – in terms of water quality – is derived from the criticalities of water-quality factors in that reach (which, in turn, are defined in relation to the reach water uses). Hence, the reach importance is governed by importance degrees of its quality factors:

$$\tilde{W}_r = \{\tilde{W}_{1,r}, \tilde{W}_{2,r}, \ldots, \tilde{W}_{I,r}\} \tag{6.39}$$

Therefore in a fuzzy sense,

$$\tilde{W}_r = \cup_{i=1,2,\ldots,I} \tilde{W}_{i,r} = \left\{ \left(W, \mu_{\tilde{W}_r}(w)\right) \middle| \mu_{\tilde{W}_r}(w) = \mathrm{Max}_i \left(\mu_{\tilde{W}_{i,r}}(w)\right) \right\} \tag{6.40}$$

In this way, the fuzzy distribution for a reach importance contained all fuzzy distributions for criticalities of water-quality factors that reach $\tilde{W}_{i,r} \subseteq \tilde{W}_r$.

Each reach can have a similar weight set that covers and depicts all possible importance degrees and criticalities involved in that reach. The reach with a larger number of quality factors with higher criticalities shall have higher importance.

4. *River-quality index.* The river-quality index was obtained by aggregating the water-quality indices for all reaches of the river computed in Equation (6.39) with respect to the reach weights computed via Equation (6.40). As the reach indices are crisp values, an arithmetic aggregation has to be used instead of set theoretic operators (such as intersection, union, etc.). Using a simple additive weighting method, and considering the representative values of reach importance sets w_r^*, led to

$$WQI^* = \sum_{r=1}^{R} W_r^* QI_r^* \Big/ \sum_{r=1}^{R} W_r^* \tag{6.41}$$

5. *Prioritising per river quality.* The plan effectiveness indicator for the river, similar to the reach indicator, which would show the amount of improvement that can be provided in the quality of river water via a water-quality management plan, is defined as

$$+QI^* = \begin{cases} 1 - \left(\dfrac{QI^*}{QI^*(0)}\right) & QI^* < QI^*(0) \\ 0 & \text{Otherwise} \end{cases} \tag{6.42}$$

where QI^* and $QI^*(0)$ are the estimations for the water-quality index of the river (Equation (6.41)) after and before a water-quality management plan, respectively.

By comparing these plan effectiveness indicators, a rank for each water-quality plan can be found, as a higher value of indicator means a higher rank and priority in terms of effectiveness.

The authors have demonstrated the applicability of this methodology by a case study of Tha Chin River basin, Thailand.

6.15. A FUZZY INDUSTRIAL WQI

A fuzzy industrial water-quality index of Soroush et al. (2011), which operates on 0–100 scale, was applied to the Zayandehrud River, located in Isfahan province, Iran. It used six parameters (pH, TH, TA, SO_4^{2-}, Cl^-, and TDS) and appeared a convenient tool for continuous monitoring of river water for industrial purposes.

6.16. IMPACT OF STOCHASTIC OBSERVATION ERROR AND UNCERTAINTY IN WATER-QUALITY EVALUATION

When evaluating water quality, the influence of physical weight of the observed index is normally taken into account, but the influence of stochastic observation error (SOE) is not adequately considered. Using Monte Carlo simulation, combined with Shannon entropy, the principle of maximum entropy (POME) and Tsallis entropy, Wang et al. (2009) investigated the influence of stochastic observation error (SOE) for two cases of the observed index: small observation error and large observation error. As randomness and fuzziness represent two types of uncertainties that are deemed significant and should be considered simultaneously when developing or evaluating water-quality models, the authors used three models: two of the models, named as model I and model II, considered both the fuzziness and randomness, and another model, considered only fuzziness. The results from three representative lakes in China showed that for all three models, the influence of SOE on water-quality evaluation was significant irrespective of whether the water-quality index has a small observation error or a large observation error. Furthermore, when there was a significant difference in the accuracy of observations, the influence of SOE on water-quality evaluation increased. The water-quality index whose SOE was minimum determined the results of evaluation.

References

ACA, Agencia Catalana del Agua (Catalonia, Spain). (2005). Available at: <http://www.mediambient.gencat.net/aca/ca/inici.jsp> [Accessed October 2005].

Adriaenssens, V., De Baets, B., Goethals, P.L.M., De Pauw, N., 2004. Fuzzy rule-based models for decision support in ecosystem management. Science of the Total Environment 319, 1–12.

Altunkaynak, A., Ozger, M., Cakmakci, M., 2005. Fuzzy logic modeling of the dissolved oxygen fluctuations in Golden Horn. Ecological Modelling 189, 436–446.

Angel, D., Krost, P., Silvert, W., 1998. Describing benthic impacts of fish farming with fuzzy sets: theoretical background and analytical methods. Journal of Applied Ichthyology 14, 1–8.

Bandemer, H., Gottwald, S., 1996. Fuzzy Sets, Fuzzy Logic Fuzzy Methods with Appications. John Wiley & Sons, Chichester. 239.

Bárdossy, A.D., 1995. Fuzzy Rule-based Modeling with Applications to Geophysical Biological and Engineering Systems. CRC Press, Boca Raton, New York, London, Tokyo.

Caldeira, A.M., Machado, M.A.S., Souza, R.C., Tanscheit, R., 2007. Inteligência Computacional aplicada a administração economia e engenharia em Matlab. Thomson Learning, São Paulo.

Carlson, R.E., 1977. A trophic state index for lakes. Limnol Oceanogr 22 (2), 361–369.

Chang, N.-B., Chen, H.W., Ning, S.K., 2001. Identification of river water quality using the fuzzy synthetic evaluation approach. Journal of Environmental Management 63 (3), 293–305.

CHE, Confederación Hidrográfica del Ebro. (2004). Memoria de la Confederación Hidrográfica del Ebro del año 2003. Available at: <http://www.oph.chebro.es/> [Accessed October 2005].

Chen, D., Ji, Q., Zhang, Fuzzy, H., 2011. Synthetic evaluation of water quality of Hei River system 280–283.

Chowdhury, S., Husain, T., 2006. Evaluation of drinking water treatment technology: an entropy-based fuzzy application. Journal of Environmental Engineering, ASCE 132 (10), 1264–1271.

Companhia de Tecnologia de Saneamento Ambiental (CETESB), 2004, 2005 and 2006 Relatório 57 de Qualidade das Águas Interiores do Estado de São Paulo, São Paulo.

Conesa, F.V.V., 1995. Guía Metadológica para la Evaluacón del Impacto Ambiental. Ed. Mundi-Prensa.

Cruz, A.J.de.O, 2004. Lógica Nebulosa. Notas de aula. Universidade Federal do Rio de Janeiro, Rio de Janeiro.

EEA, 2003. Air pollution by ozone in Europe in summer 2003: overview of exceedances of EC ozone threshold values during the summer season April–August 2003 and comparisons with previous years. Topic Report No 3/2003. European Economic Association, Copenhagen. http://reports.eea.europa.eu/topic_report_2003_3/en, p. 33.

Ferson, S., Hajagos, J.G., 2004. Arithmetic with uncertain numbers: rigorous and (often) best possible answers. Reliability Engineering and System Safety 85, 135–152.

Foody, G.M., 1992. A fuzzy sets approach to the representation of vegetation continua from remotely sensed data: an example from lowland health. Photogrammetric Engineering and Remote Sensing 58 (2), 221–225.

Gass, S.I., Rapcsak, T., 2004. Singular value decomposition in AHP. European Journal of Operational Research 154, 573–584.

Ghosh, S., Mujumdar, P.P., 2006. Risk minimization in water quality control problems of a river system. Advances in Water Resources 29 (3), 458–470.

Haiyan, W., 2002. Assessment and prediction of overall environmental quality of Zhuzhou City, Hunan Province, China. Journal of Environmental Management. 66, 329–340.

Holland, J.H., 1992. Genetic algorithms[J]. Scientific American 4, 44–50.

Hwang, C., Yoon, K., 1981. Multiple Attribute Decision Making, Methods and Applications, a State-of The-art Survey. Springer-Verlag.

Icaga, Y., 2007. Fuzzy evaluation of water quality classification. Ecological Indicators 7 (3), 710–718.

Islam, N., Sadiq, R., Rodriguez, M.J., Francisque, A., 2011. Reviewing source water protection strategies: a conceptual model for water quality assessment. Environmental Reviews 19 (1), 68–105.

Karamouz, M., Zahraie, B., Kerachian, R., 2003. Development of a master plan for water pollution control using MCDM techniques: A case study. Water International, IWRA 28 (4), 478–490.

Klir, G.J., Yuan, B., 1995. Fuzzy Sets and Fuzzy Logic: Theory and Applications. Prentice-Hall International, Upper Saddle River, NJ.

Kung, H., Ying, L., Liu, Y.C., 1992. A complementary tool to water quality index: fuzzy clustering analysis. Water Resources Bulletin 28 (3), 525–533.

Lermontov, A., Yokoyama, L., Lermontov, M., Machado, M.A.S., 2009. River quality analysis using fuzzy water quality index: Ribeira do Iguape river watershed. Brazil Ecological Indicators 9 (6), 1188–1197.

Lermontov, A., Yokoyama, L., Lermontov, M., MacHado, M.A.S., 2008. Aplicação da lógica nebulosa na parametrizaç ão de um novo índice de qualidade das águas. Engevista 10 (2), 106–125.

Li, Z.Y., Peng, L.H., 2000. Damage index formula of air quality evaluation based on optimum of genetic algorithms. China Environmental Science 20 (4), 313–317.

Liou, Y.T., Lo, S.L., 2005. A fuzzy index model for trophic status evaluation of reservoir waters. Water Research 39 (7), 1415–1423.

Lu, R.K., Shi, Z.Y., 2000. Features and recover of degraded red soil Soil 4, 198–209. In Chinese.

Lu, R.S., Lo, S.L., 2002. Diagnosing reservoir water quality using self-organizing maps and fuzzy theory. Water research 36 (9), 2265–2274.

Lu, R.S., Lo, S.L., Hu, J.Y., 1999. Analysis of reservoir water quality using fuzzy synthetic evaluation. Stochastic Environmental Research and Risk Assessment 13 (5), 327–336.

Lu, R.S., Lo, S.L., Hu, J.Y., 2000. Analysis of reservoir water quality using fuzzy synthetic evaluation. Stochastic Environmental Research and Risk Assessment 13 (5), 327.

Makropoulos, C.K., Butler, D., 2006. Spatial ordered weighted averaging: incorporating spatially variable attitude towards risk in spatial multicriteria decision-making. Environmental Models and Software 21 (1), 69–84.

McKone, T.E., Deshpande, A.W., 2005. Can fuzzy logic bring complex environmental problems into focus? Environmental Science & Technology 39, 42A–47A.

Mitchell, M.K., Stapp, W.B., 1996. Field Manual for Water Quality Monitoring: An Environmental Education Program for Schools, vol. 277. Thomson-Shore, Inc, Dexter, Michigan.

Nasiri, F., Maqsood, I., Huang, G., Fuller, N., 2007. Water quality index: a fuzzy river-pollution decision support expert system. Journal of Water Resources Planning and Management 133, 95–105.

Nikoo, M.R., Kerachian, R., Malakpour-Estalaki, S., Bashi-Azghadi, S.N., Azimi-Ghadikolaee, M.M., 2010. A probabilistic water quality index for river water quality assessment: a case study. Environmental Monitoring and Assessment, 1–14.

NSF, National Sanitation Foundation International. (2007). Available in: <http://www.nsf.org> (Accessed on October of 2007).

Ocampo-Duque, W., Ferre-Huguet, N., Domingo, J.L., Schuhmacher, M., 2006. Assessing water quality in rivers with fuzzy inference systems: a case study. Environment International 32 (6), 733–742.

Otto, W.R., 1978. Environmental Indices: Theory and Practice. Ann Arbor Science Publishers Inc, Ann Arbor, MI.

Peng, L., 2004. A Universal Index Formula Suitable to Multiparameter Water Quality Evaluation. Numerical Methods for Partial Differential Equations 20 (3), 368–373.

Pesce, S.F., Wunderlin, D.A., 2000. Use of water quality indices to verify the impact of Cordoba City (Argentina) on Suquia River. Water Research 34 (11), 2915–2926.

Prato, T., 2005. A fuzzy logic approach for evaluating ecosystem sustainability. Ecological Modelling 187, 361–368.

Ross, T.J., 2004. Fuzzy Logic with Engineering Applications. John Wiley & Sons, New York.

Roveda, S.R.M.M., Bondanca, A.P.M., Silva, J.G.S., Roveda, J.A.F., Rosa, A.H., 2010. Development of a water quality index using a fuzzy logic: A case study for the sorocaba river IEEE World Congress on Computational Intelligence. WCCI 2010.

Saaty, T.L., 2003. Decision-making with the AHP: why is the principal eigenvector necessary? European Journal of Operational Research 145, 85–91.

Sadiq, R., Rodriguez, M.J., 2004. Fuzzy synthetic evaluation of disinfection by-products – A risk-based indexing system. Journal of Environmental Management 73 (1), 1–13.

Sadiq, R., Tesfamariam, S., 2007. Probability density functions based weights for ordered weighted averaging (OWA) operators: an example of water quality indices. European Journal of Operational Research 182 (3), 1350–1368.

Sadiq, R., Tesfamariam, S., 2008. Developing environmental indices using fuzzy numbers ordered weighted averaging (FN-OWA) operators. Stochastic Environmental Research and Risk Assessment 22 (4), 495–505.

Safavi, H.R., 2010. Prediction of river water quality by adaptive neuro fuzzy inference system (ANFIS). Journal of environmental studies 36 (53), 1–10.

Said, A., Stevens, D.K., Sehlke, G., 2004. An innovative index for evaluating water quality in streams. Environmental Management 34 (3), 406–414.

Selim, S.Z., 1984. Soft clustering of multi-dimensional data: a semi-fuzzy approach. Pattern Recognition 17 (5), 559–568.

Shannon, C., 1948. A mathematical theory of communication. The Bell System Technical Journal 27, 379–423.

Shen, G., Lu, Y., Wang, M., Sun, Y., 2005. Status and fuzzy comprehensive assessment of combined heavy metal and organo-chlorine pesticide pollution in the Taihu Lake region of China. Journal of Environmental Engineering 76, 355–362.

Silvert, W., 1979. Symmetric summation: A class of operations on fuzzy sets IEEE Trans. Syst. Man, Cyber. SMC-9. 657–659.

Silvert, W., 1997. Ecological impact classification with fuzzy sets. Ecological Modelling 96, 1–10.

Silvert, W., 2000. Fuzzy indices of environmental conditions. Ecological Modelling 130 (1–3), 111–119.

Smith, P.N., 2006. Flexible aggregation in multiple attribute decision making: application to the Kuranda Range road upgrade. Cybernetics and Systems 37 (1), 1–22.

Soroush, F., Mousavi, S.F., Gharechahi, A., 2011. A fuzzy industrial water quality index: Case study of Zayandehrud River system. Iranian Journal of Science and Technology, Transaction B: Engineering 35 (1), 131–136.

Swamee, P.K., Tyagi, A., 2000. Describing water quality with aggregate index. Journal of Environmental Engineering Division ASCE 126 (5), 450–455.

Taheriyoun, M., Karamouz, M., Baghvand, A., 2010. Development of an entropy-based Fuzzy eutrophication index for reservoir water quality evaluation. Iranian Journal of Environmental Health Science and Engineering 7 (1), 1–14.

Trauwaert, E., Kaufman, L., Rousseeuw, P., 1991. Fuzzy clustering algorithms based on the maximum likelihood principle. Fuzzy Sets and Systems 42, 213–227.

Vaidya, O.S., Kumar, S., 2006. Analytic hierarchy process: an overview of applications. European Journal of Operational Research 169, 1–29.

Vemula, V.R., Mujumdar, P.P., Gosh, S., 2004. Risk evaluation in water quality management of a river system. Journal of Water Resources Planning and Management-ASCE 130, 411–423.

Wang, Pei-zhuang, 1983. Theory of Fuzzy Sets and Its Application. Shanghai Science and Technology Publishers, Shanghai, China.

Wang, D., Singh, V.P., Zhu, Y.-s., Wu, J.-c., 2009. Stochastic observation error and uncertainty in water quality evaluation. Advances in Water Resources 32 (10), 1526–1534.

World Health Organisation (2004). Guidelines for drinking water quality. Health Criteria and Other Supporting Information (2nd ed.). Geneva, 2, 231–233.

Yager, R.R., 1980. On a general class of fuzzy connectives. Fuzzy Sets and Systems 4 (3), 235–322. FSSYD8 0165-011 10.1016/0165-011(80)90013-5.

Yager, R.R., 1988. On ordered weighted averaging aggregation in multicriteria decision making. IEEE Transactions on Systems, Man, and Cybernetics 18, 183–190.

Yager, R.R., 1996. Quantifier guided aggregation using OWA operators. International Journal of General Systems 11, 49–73.

Yager, R.R., Filev, D.P., 1994. Parameterized "andlike" and "orlike" OWA operators. International Journal of General Systems 22, 297–316.

Yager, R.R., Filvel, D.P., 1994. Essentials of Fuzzy Modeling and Control. John Wiley & Sons, New York.

Yang, S.M., Shao, D.G., Shen, X.P., 2005. Quantitative approach for calculating ecological water requirement of seasonal water-deficient rivers. Shuili Xuebao/Journal of Hydraulic Engineering 36 (11), 1341–1346.

Yen, J., Langari, R., 1999. Fuzzy Logic: Intelligence. Control and Information. Prentice-Hall, Inc.

Zadeh, L.A., 1965. Fuzzy sets. Information and Control 8, 338–353.

Zadeh, L.A., 1971. Similarity Relations and Fuzzy Orderings. Information Science 3, 177–200.

Zadeh, L.A., 1977. Fuzzy Sets and Their Application to Pattern Recognition and Clustering Analysis. Classification and Clustering, San Francisco, California. 251–299.

Zeleny, M., 1982. Multiple Criteria Decision Making. McGraw-Hill, New York.

Zimmermann, H.J., 1985. Fuzzy Set Theory — and Its Applications. Kluwer Nijhoff Publishing, Norwell, Massachusetts.

CHAPTER

7

Probabilistic or Stochastic Water-Quality Indices

OUTLINE

7.1. INTRODUCTION

Whereas fuzzy theory deals with uncertainty in knowledge vis a vis *truth* – and tries to make good sense of partial or relative truths (fuzziness) – probability theory deals with uncertainty in knowledge vis a vis occurrence of *events*. In other words, probability theory strives to make predictions about *events* from a state of partial knowledge. At a particular sampling station not all samples at all times will pass or fail a given criteria of compliance. Likewise, one or more parameters may be within or outside permissible limits at different times or at different locations. But any water-quality monitoring programme can only generate partial knowledge, limited by the number of sampling stations that can be put and the frequency of sampling that can be managed within resource constraints. Based on these inputs, it is required to assess the probability whether the water quality of a water body at any given instant is *likely* to be favourable or unfavourable, and in what aspects. Hence, the challenge is to obtain information on the pattern of compliance–noncompliance of a large number of water-quality variables as a function of time and space from a limited (and necessarily incomplete) volume of data. The information then has to be utilised in getting an idea about the likely overall quality of the water. 'Probabilistic' or 'stochastic' water-quality indices aim to do that.

Landwehr (1979) had noted that water-quality constituents upon which all indices depend are random variables. It follows that regardless of their construct, all indices are also random variables.

Water Quality Indices DOI: 10.1016/B978-0-444-54304-2.00007-5

If x denotes a water-quality constituent (which is a random variable) and $f(x)$ denotes the probability density function of x, then, in a sense, $f(x)$ is itself an indicator of water-quality conditions, and therefore the parameters characterising $f(x)$ are measures of water quality. In terms of the parameters which characterise $f(x)$, changes in water quality may be regarded as absolute or relative. For example, if negligible values of x imply good quality, as is the case with respect to toxic and carcinogenic chemicals, then a decrease in μ, the mean of x, implies an improvement in quality in an absolute sense. However, if there is some desirable range for the values of x, as might be the case if x represents pH, then whether an increase or decrease in μ implies improvement or deterioration of quality depends upon whether μ is greater or less than the limits of the desired range.

The variability in the values of x, measured by σ^2, the variance of x, provides another measure, although a relative one, of water quality. For example, since it is easier to detect a trend or to design a treatment process when σ^2 is small, it might be said that the water is of better or preferable quality when the variability is low.

Most countries in the world try to manage their water resources on the basis of water-quality standards. These standards are used to determine whether or not the quality of a water body violated a standard relative to some water-use category. The interpretation is based on crisp cut-offs, for example, if a water-quality standard says that for a water to be suitable for drinking it should have 0.05 mg/L or lower of copper, all samples containing copper above 0.05 mg/L will be deemed 'unfit' and the ones containing 0.05 mg/L or lower concentration of copper will be deemed 'fit'.

This compliance/noncompliance concern can be viewed conceptually as a very simple index (Landwehr, 1979). The rating curves $\phi_i(x_i)$, $i = 1,\ldots n$, where n is the number of constituents specified in the guideline, are defined as step functions which take on a value of 1, good quality, when the value of x is in compliance and 0, bad or unacceptable quality, otherwise.

Then, the 'index' is defined as

$$I = \prod_{i=1}^{n} (\phi_i(x))^{1/n} \tag{7.1}$$

Since $\phi_i(x) = 0$ or 1, I will be 0 or 1. In fact, I will be equal to 1, indicating a good-quality situation, if and only if each constituent is in compliance with the guideline. The exponent $1/n$ is computationally superfluous but is included to strictly maintain the functional form of the index.

The intent of the law is to have all regulated bodies be in compliance, that is, ideally, I should be 1 for all locations. But since x_i, $i = l,\ldots, n$, are random variables, it is difficult to evaluate the success of the policy without some knowledge of the distribution of x_i. Furthermore, since the standards or guidelines represent a goal to be met, one must manipulate the distribution function of the x_i to bring the water quality into compliance with this goal. If the x_i of concern may be considered to be independent, then the x_i may be treated individually; if not, the problem becomes very complex. It must be recalled here that correlative behaviour among constituents, except occasionally in relation to temperature, is rarely discussed in the statement of water-quality standards; effectively, they are treated as independent variables.

Regulatory bodies have tried to go around the 'all-or-nothing' nature of water-quality standards in relation to the stochasticity inherent in most natural phenomena (including water quality) by stipulations such as "the water quality will be acceptable if all parameters comply in 95% of samples taken from a point at least n_1 number of times over a period covering n_2 number of days." But there are problems associated with such stipulations which probabilistic indices have tried to address, as described in the following sections.

7.2. A 'GLOBAL' STOCHASTIC INDEX OF WATER QUALITY

Beamonte et al. (2005) have proposed a stochastic water-quality index that takes into account the uncertainty surrounding the acceptability level when a set of water-quality data is assessed for the purpose. Their stochastic index is built with the probability classification vector of each parameter. In order to obtain those vectors, a mixed-lognormal model has been introduced and its statistical analysis developed. The methodology was applied to the data observed in the La Presa station, one of the sampling points of the Spanish surface water-quality network.

The index attempts to address the problems in using the administrative classification procedure of water quality given by an EU (then EEC) directive. According to the directive, water shall be assumed to conform to the parameters in Table 7.1 if samples of this water taken at regular intervals at the same sampling point show that it complies with the parametric values for the water quality in question in 95% of the samples. Also, in the samples which do not comply, the water quality should not deviate from the parametric values by more than 50%.

Once all samples taken during a specific period of time at the same sampling point have been observed, the administrative classification procedure is then applied wherein each parameter is classified into the quality level its 95 sampling percentile belongs to. Then, the sampling point is classified in the same level as its worst quality parameter.

But the simplistic procedure can give misleading interpretations. For example, if a mere 6% of the analytical determinations of a parameter in a sampling point do not comply with category A3 (Table 7.1), the sampling point is classified as +A3, although all other parameters may belong to category A1. Moreover, this system is very sensitive to the existence of outliers, especially so when the number of data points is not sufficiently high, as is usually the case. For example, if there are no more than 20 analytical determinations of each parameter in a sampling point, and if even just one of them belongs to +A3, its related parameter as well as the sampling point will get classified as +A3.

Beamonte et al. (2005) first developed a water-quality index involving the EEC administrative classification of sampling points. For it, they defined the administrative quality vector of the sampling point as a vector (a, b, c, d) showing the number of parameters in each quality level, where a is the number of parameters with quality A1, b is the number of parameters with quality A2 and so on. They then proposed comparing two sampling points through their administrative quality vectors (a_1, b_1, c_1, d_1) and (a_2, b_2, c_2, d_2) in the following manner: if d_1 is less than d_2, the first point has better quality than the second one, but in the case of a tie, one proceeds to compare c_1 and c_2 and so on.

The order introduced in that way in the administrative quality vectors allowed the authors to arrange all of them from the worst one, given by the vector (0, 0, 0, k), where k is the number of parameters used, to the best one, given by the vector (k, 0, 0, 0). The rank assigned to a sampling point by this arrangement was used as its water-quality index as Beamonte et al. (2004) had earlier proved that the rank of an administrative quality vector (a, b, c, d) is given by

$$I(a,b,c,d) = \frac{1}{6}\left(s_1^3 + 3s_1^2 + 2s_1\right) + \frac{1}{2}\left(s_2^2 + s_2\right) + a + 1, \tag{7.2}$$

where $s_1 = a + b + c$ and $s_2 = a + b$. The authors called $I(a, b, c, d)$ the 'administrative quality index' because it was consistent with the EEC's administrative classification procedure.

The worst value of this index will always be 1, but its best value would depend on

TABLE 7.1 Classification of Different Water-quality Parameters, and the Overall Quality, on the Basis of the Probabilistic and the Deterministic ('Unitary Vector') Indices (Beamonte et al., 2005).

	Probability Vector				Unitary Vector			
Parameter	**A1**	**A2**	**A3**	**+A3**	**A1**	**A2**	**A3**	**+A3**
Color	.962	.038	.000	.000	1	0	0	0
Temperature	.961	.000	.000	.039	1	0	0	0
Nitrates	1.000	.000	.000	.000	1	0	0	0
Fluoride	1.000	.000	.000	.000	1	0	0	0
Dissolved iron	.977	.023	.000	.000	1	0	0	0
Copper	.885	.115	.000	.000	1	0	0	0
Zinc	1.000	.000	.000	.000	1	0	0	0
Arsenic	.987	.000	.011	.002	1	0	0	0
Cadmium	.996	.000	.000	.004	1	0	0	0
Total chromium	.990	.000	.000	.010	1	0	0	0
Lead	.944	.000	.000	.056	1	0	0	0
Selenium	.877	.000	.000	.123	1	0	0	0
Mercury	.991	.000	.000	.009	1	0	0	0
Barium	.447	.553	.000	.000	1	0	0	0
Cyanide	.997	.000	.000	.003	1	0	0	0
Sulphates	.000	.000	.000	1.000	0	0	0	1
Phenols	.514	.377	.107	.002	1	0	0	0
Dissolved hydrocarbons	.142	.837	.020	.001	0	1	0	0
Aromatic hydrocarbons	.903	.000	.096	.001	1	0	0	0
Total pesticides	.876	.107	.015	.002	1	0	0	0
Ammonia	1.000	.000	.000	.000	1	0	0	0
Overall	17.450	2.050	.250	1.250	19	1	0	1

$k = a + b + c + d$, the number of parameters under study, and takes the value $(k+3)(k+2)(k+1)/6$. With a simple linear transformation, this index can be translated to a scale from 0 to 100.

7.2.1. The Stochastic WQI

The shortcoming of the procedure described above is that it classifies each parameter in a quality level, the level its 95 percentile belongs to, and then proceeds as if no uncertainty about the parameter classification were present. Hence, we could represent each parameter through a unitary vector: a vector with a component equal to one, the component associated to the classification level obtained and their other three components equal to zero. The administrative classification vector will then be the sum of those k unitary vectors. However, those unitary vectors give a false sense of certainty; actually, due to sample variability, the estimate of the percentile could be in a different quality level than the true percentile, giving an erroneous classification.

In order to deal with the uncertainty inherent in the problem the authors proposed the following procedure leading to the stochastic WQI. Taking η to be the 95 percentile of the distribution of a new analytical result of a parameter, the posterior distribution about η was obtained following the Bayesian paradigm, and the classification probability vector $p = (p_1, p_2, p_3, p_4)$, where p_i is the posterior probability that η belongs to the interval representing the i quality level.

The authors proposed to use this classification probability vector instead of the unitary vector used in Equation (7.2). The probability vector p is expected to have one component close to 1 and the others negligible if there is no uncertainty about the quality level the parameter belongs to; otherwise, some of the components of p will have significant values.

The sum of the probability vectors associated with each one of the k parameters included in the study forms the *stochastic quality vector*, an alternative to the administrative quality vector. The components of the stochastic quality vector are real, though not integers, but not negative either and add to k.

Given a stochastic quality vector (v_1, v_2, v_3, v_4), obtained at a specific sampling point, the *stochastic water-quality index* is given by

$$SI(v_1, v_2, v_3, v_4) = \frac{1}{6}\left(s_1^3 + 3s_1^2 + 2s_1\right) + \frac{1}{2}\left(s_2^2 + s_2\right) + v_1 + 1, \quad (7.3)$$

where $s_1 = v_1 + v_2 + v_3$ and $s_2 = v_1 + v_2$.

The statistical procedure used to obtain the probability classification vector $p = (p_1, p_2, p_3, p_4)$ would depend on the probabilistic model used to represent the behaviour of the data. The authors considered normal distribution and a 'mixed' lognormal distribution for the purpose.

The normal model is based on the supposition that the available data on a given parameter follow a normal distribution with unknown mean μ and variance σ^2. The 95 percentile, which is the quantity of interest for the SWQI, is then $\eta = \mu + 1.64\sigma$. In order to obtain the probability classification vector $p = (p_1, p_2, p_3, p_4)$, for the posterior distribution of η, a Monte Carlo approach was adopted involving these steps: a) obtain a vector (μ_1, σ_1^2) generated from the posterior distribution of (μ, σ^2); b), compute $\eta_1 = \mu_1 + 1.64\ \sigma_1$ and repeat those steps N times to obtain a random sample of size N from the posterior distribution of η, $\{\eta_1, \ldots, \eta_N\}$.

The probability that η belongs to the interval associated to quality level A1 can now be estimated as $p_1 = R_1/N$, where R_1 is the number of points in the sample $\{\eta_1, \ldots, \eta_N\}$ belonging to that interval. The choice of N can be made through an upper bound of the standard error of $p_1 = R_1/N$, which is $p_1(1 - p_1)/\sqrt{N}$. The other components of the probability classification vector $p = (p_1, p_2, p_3, p_4)$ can be estimated in a similar way.

But the normal statistical model is unsuitable for most parameters of water quality and can be used only in limited number of cases after an adequate transformation of the data. A logarithmic transformation is more effective but since water-quality parameters are non-negative and some of the parameters can take the value of zero frequently, either because the parameter is really not present in the water or because it is present in such a small amount that it is unobservable, any reasonable statistical model has to allow for a positive mass of probability, $\pi > 0$, at the point zero.

Hence, the authors have proposed a mixed probabilistic model, neither discrete nor continuous, which they have called mixed-lognormal and denoted by MLN$(Y \mid \pi, \mu, \sigma^2)$. This model is such that $P(Y = 0) = \pi > 0$ and log Y, given that Y is positive, is a normal variable with mean μ and variance σ^2.

Given a random sample of size m from the MLN(Y |π, μ, σ^2) model, the sufficient statistics are the number of nonzero observations, n, and the sample mean and variance of the logarithms of the nonzero data, $\bar{x}$ and s_2.

The classification probability p_1 is given by the posterior probability $P(\eta \leq c_1)$, where c_1 is the mandatory limiting value for A1. In a similar way, through the posterior cumulative distribution function of η, the whole probability classification vector $p = (p_1, p_2, p_3, p_4)$ can be computed.

The posterior distribution of η is a mixed distribution because η is zero for any π greater than 0.95; hence, $P(\eta \leq c) = P(\eta = 0) + P(\eta \leq c|\eta \neq 0)\ (1 - P(\eta = 0))$. But $P(\eta = 0) = P(\pi > .95) = 1 - F_\pi(0.95)$, where $F_\pi(\cdot)$ is the cumulative distribution function of the posterior distribution of π.

$F_\pi(0.95)$ is easy to compute because the posterior distribution of π is Beta: besides, $P(\eta \leq c|\eta = 0)$ can be estimated by simulation: if $\{\eta_1, \ldots, \eta_N\}$ is a Monte Carlo sample generated from the continuous part of the posterior distribution of η, and R is the number of points ηi smaller than c, R/N is a Monte Carlo estimate of $P(\eta \leq c|\eta \neq 0)$. Hence, $1 - F_\pi(.95) + F_\pi(.95)$ R/N is an estimate of $P(\eta \leq c)$.

How the unitary and probability vectors for individual parameters, as well as the overall quality vector (the corresponding indices), differ can be seen from Table 7.1 which is based on the treatment of the water-quality data of La Presa monitoring station, near Valencia, Spain, by the authors.

As may be seen, for 13 of the parameters the probability that they fall under one of the classifications is greater than 0.95. Another 6 parameters are less, but reasonably, certain to fall under one of classifications 80% to <95% times. But for barium and phenols the probabilities that they may belong to Class A1 or A2 are fairly close to each other. As per deterministic (unitary vector) index, barium falls under Class A1 but the SWQI puts it more decisively in Class A2.

7.3. A MODIFICATION IN THE GLOBAL STOCHASTIC INDEX BY CORDOBA ET AL. (2010)

A slight modification has been done in the above-mentioned SWQI of Beamonte et al. (2005) by Cordoba et al. (2010) with the dual purpose of using it to study the water quality of 12 rivers coming under the care of Spain's Confederation Hydrographic del Jucar (CHJ), and for comparing its performance with a conventional WQI. The performance of SWQI was also compared with a dissolved oxygen (DO)-based index and a few biological indices.

The modification involved reconciling the SWQI with a 'general' water-quality index (GPI) of Provencher and Lamontagne (1977); the latter being a conventional WQI hitherto used by CHJ. GHQ has the following form, similar to one of numerous other conventional indices (Chapter 3):

$$\mathrm{GQI} = \sum_{i=1}^{n} Q_i P_i \tag{7.4}$$

where n is the number of physical and chemical variables that are analysed; Q_i represents an equivalence function that transforms the concentration of variable i into a quality level that ranges between 0 and 100 (with 0 corresponding to the worst level and 100 to the ideal level, according to the water's intended use) and P_i is the weighting of variable i. The sum of all weightings must add up to 1.0 so that the calculated index will be between 0 and 100.

According to this scale, the quality level associated with variable i is 'excellent' if $Q_i = 100$; 'very good' if $100 > Q_i \geq 85$; 'good' if $85 > Q_i \geq 75$; 'usable' if $75 > Q_i \geq 60$; 'bad' (i.e., correction is required) if $60 > Q_i > 0$ and 'unacceptable' if $Q_i = 0$. To calculate GQI, the values of P_i and Q_i were obtained following the procedure used by the Spanish Hydrographic Federations to determine the 'general quality index'.

Cordoba et al. (2010) transformed the values of Q_i used to calculate GQI, obtained from data, so that the measurement scale for both indices became similar. The transformed values of Q_i used to define the quality levels associated with the variables and the quality intervals used to calculate PWQI were chosen to reflect the measurement sense of GQI. The PWQI was then expressed as

$$\mathrm{PWQI} = \frac{1}{6}s_1^3 + (3s_1^2 + 2s_1) + \frac{1}{2}(s_2^2 + s_2) + v_1 + 1 \quad (7.5)$$

where s_1, s_2 and v_1 are values obtained according to the probability that the different physical and chemical characteristics belong to prearranged quality intervals that have been deduced from the corresponding 95th percentile for that variable, and by attending to the concretely probabilistic model that represents the data's behaviour. In other words, the PWQI was adapted to the levels of quality defined in the GQI.

To compare PWQI with GQI and other indices, data on nine water-quality variables recorded from 22 stations on 12 rivers by CHJ over a 15-year period (1990–2005) were utilised.

As should be expected, the classification obtained with the two indices is quite similar for the stations with extreme behaviours. That is, stations with a very good or very bad quality were classified as such in equally clear fashion by both the indices. But for intermediate water quality, the output of PWQI differed from that of GWQI.

For the entire study period, GQI correlated positively with PWQI, in terms of the Kendall correlation coefficient as well as the Spearman correlation coefficients, yielding 0.636 ($p < 0.001$) and 0.768 ($p < 0.001$), respectively. Similar results were obtained with the other indices used in the comparison.

The results also revealed that the mean value of the PWQI did not change significantly when any one of the following variables was excluded from the calculation: total phosphates, nitrates, dissolved oxygen and suspended matter. In contrast, the PWQI scores changed significantly if any of the following variables were excluded: total coliforms, conductivity, biochemical oxygen demand, chemical oxygen demand and pH. Moreover, omitting the latter variables increased the mean value of the index, indicating better water quality.

References

Landwehr, J.M., 1979. A statistical view of a class of water quality indices. Water Resources Research 15 (2), 460–468.

Beamonte, E., Bermúdez, J.D., Casino, A., Veres, E., 2004. Un Indicator Global para la Calidad del Agua. Aplicación a las Aguas Superficiales de la Comunidad Valenciana, Estadística Española 156, 357–384.

Beamonte, E., Bermúdez, J.D., Casino, A., Veres, E., 2005. A global stochastic index for water quality: the case of the river Turia in Spain. Journal of Agricultural, Biological, and Environmental Statistics 10 (4), 424–439.

Cordoba, E.B., Martinez, A.C., Ferrer, E.V., 2010. Water quality indicators: comparison of a probabilistic index and a general quality index. The case of the Confederación Hidrográfica del Jú car (Spain). Ecological Indicators 10 (5), 1049–1054.

Provencher, M., Lamontagne, M.P., 1977. Méthode De Determination D'un Indice D'appréciation De La Qualité Des Eaux Selon Différentes Utilisations. Québec, Ministère des Richesses Naturelles, Service de la qualité des eaux.

CHAPTER

8

'Planning' or 'Decision-Making' Indices

OUTLINE

Water Quality Indices DOI: 10.1016/B978-0-444-54304-2.00008-7

8.1. INTRODUCTION

8.1.1. The 'Planning' or 'Decision-Making' Indices

As we have detailed in the preceding chapters, each and every WQI can be used as a tool in (a) assessing water quality of different sources, (b) comparing the water quality of different sources and linking the findings with the impacting factors, (c) evolving measures to improve water quality based on the previous step and (d) undertaking resource allocation and ecomanagement exercises accordingly, etc. In other words, each and every WQI can be a tool for planning and decision making.

Nevertheless, some indices have been specifically reported as 'planning tools'. This chapter describes the significant ones among them.

8.2. WATER-QUALITY MANAGEMENT INDICES

Close on the heels of the US National Sanitation Foundation's water-quality index (NSF—WQI; Brown et al., 1970), the US Environmental Protection Agency (USEPA) took up the initiative to put in place water-quality management indices. The USEPA's Office of Water Programs, in collaboration with The MITRE Corporation developed a set of three indices: Index of Water Quality or the 'prevalence, duration and intensity (PDI) of pollution, the National Planning Priorities Index (NPPI) and the Priority Action Index (PAI).

The indices were then evaluated for some 1000 planning areas (metropolitan areas, subbasins, etc.) across the USA. The work was done in 1971 but was reported four years later (Truett et al., 1975). Hence, this set of three indices may well be called the first-ever indices for water-quality planning and management.

8.2.1. Index of Water Quality or the PDI Index

The index is given by

$$V = \frac{P \times D \times I}{M} \tag{8.1}$$

Here, prevalence (P) is meant to denote the number of stream miles within the planning area in which the water is not in compliance with established Federal/State water-quality standards or other legal criteria of water quality. Such noncompliance (as established by stream sampling and analysis) is deemed to exist when the in-stream water quality deviates from any one of the several legal criteria.

The duration (D) of a pollution situation was supposed to be measured in terms of the number of quarter-year periods or seasons in which it occurred; with weightage as follows:

1. 0.4 for violations occurring in a single quarter,
2. 0.6 for violations occurring in two quarters,

3. 0.8 for violations occurring in three quarters,
4. 1.0 for violations occurring in all four quarters.

Intensity of pollution (*I*) is an indicator of the severity of pollution in a zone, expressed in terms of its effects rather than of water-quality parameters. It permits the evaluator to avoid the process of explicitly forming judgements concerning the seriousness of each criterion deviation and to move directly to the question of impacts on human welfare and ecology.

Values of *I* range from 0 to 1 and represent the simple addition of the values assigned to three component measures which classify impacts according to ecological, utilitarian and aesthetic considerations.

The value (*V*) of the PDI index for a given planning area or 'block development unit' (BDU) is the product of the measure of prevalence, duration and intensity of pollution, divided by the total stream miles (*M*) within the BDU.

8.2.2. National Planning Priorities Index (NPPI)

The index was intended to form the basis for EPA to initiate planning efforts in specific areas and to rank those areas in terms of priority. In turn, these priorities were expected to be established in such a way that, for the whole nation, the resulting order of allocation of funds would ensure that:

1. funds are granted and used in a cost-effective manner for the planned municipal water-treatment projects (plus other types of water quality control measures).
2. the maximum percentage of the nation's population benefits from improved water quality.
3. the maximum percentage of water bodies meets or exceeds required water-quality standards.

In order that the NPPI reflects the actual situation relative to the need and priority for water-quality management planning in each planning area (BDU), the list of parameters included in the index was modified and expanded to incorporate the following:

1. Current Area Population
2. Downstream Affected Population
3. Planned Investment
4. Controllability
5. Required Planning Level – an integer scale from 1 to 4 was used with a value of 1 indicating essentially no planning and 4 indicating a fully developed and implementable plan for highly complex areas. Values of 2 and 3 indicate intermediate planning levels.
6. Delta Planning Level – defined as the difference between current planning level and required planning level, to show the need for more planning.
7. PDI Index – The PDI index used as an input for NPPI differed slightly from the formulation given above, in that it did not express the number of polluted stream miles as a ratio to the total stream miles within the BDU, and it permitted the pollution zone to be associated with pollution sources.
8. Per Capita Planning Cost

After the parameters to determine the NPPI were identified, a value function was formed to represent how each parameter was going to affect the value of the index (Figure 8.1). A weight was assigned to each parameter to indicate the relative importance of the parameter with respect to the Planning Priority Index (Table 8.1). The forms of the value functions were determined by joint participation of personnel from Office of Water Programs and MITRE; the values of the weights were assigned by Office of Water Programs personnel and represented an averaging of the independent judgements of ten experienced water-quality personnel.

The method of computing the index involved a summation of the weighted value of each parameter after the value had been transformed

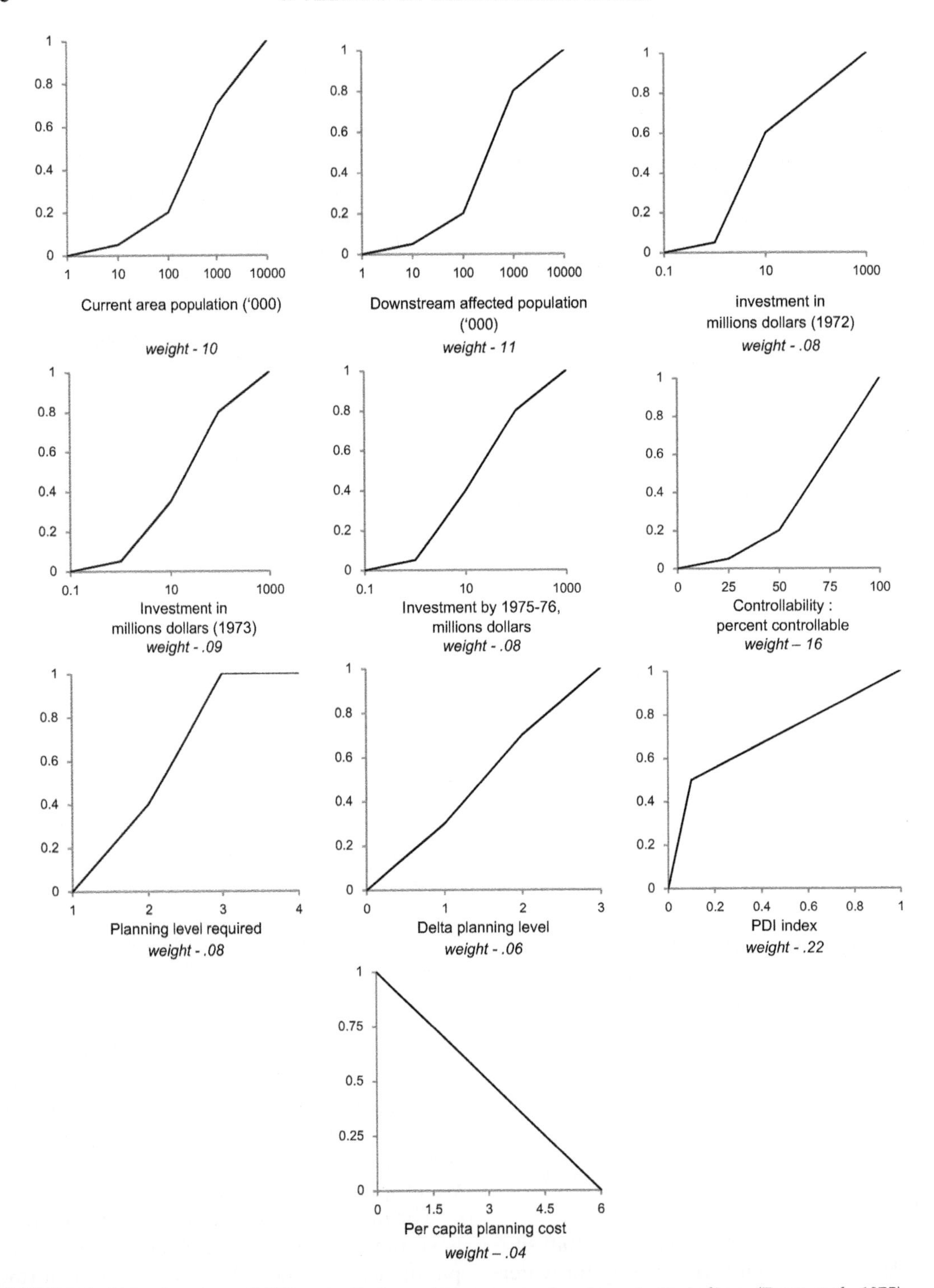

FIGURE 8.1 Value functions of different criteria used to compute planning priority indices. *(Truett et al., 1975)*

TABLE 8.1 Weights Assigned to Different Parameters in the NPPI.

Criterion	Weight
Current Area Population	.10
Downstream Affected Population	.11
Investment FY 72	.08
Investment FY 73–74	.09
Investment FY 75–76	.08
Controllability	.16
Required Planning Level	.06
Delta Planning Level	.06
PDI Index	.22
Per Capita Planning Cost	.04
Total	1.00

(Truett et al., 1975)

by the respective functions shown in Figure 8.1. To wit,

$$NPPI_i = \sum_j a_j f_j(x_{ij})$$

where

i designates particular planning area (or BDU),

$NPPI_i$ is the index value for that BDU,

j designates a particular parameter,

a_j is the "importance weighting" assigned to that parameter, $\sum a_j = 1$,

x_{ij} is the value of the jth parameter for the ith BDU and

f_j is the transform (or value function) for the jth parameter, as given in Figure 8.1.

Transformation of a given parameter value produces the "mapped value" needed for the computation.

8.3. DEE'S WQI-BASED ENVIRONMENTAL EVALUATION SYSTEM

Dee et al. (1972, 1973) proposed a system for evaluating the environmental impact of large-scale water resources projects. The system included a water-quality index, which was represented by 12 common water-quality variables (such as DO pH, turbidity and faecal coliforms), besides pesticides and toxic substances. The subindices of various water-quality variables were similar to those in the Brown's NSF–WQI (Brown et al. 1970).

The index was calculated with and without considering the proposed water resources project. The difference between the two scores provided a measure of the environmental impact (EI) of the project:

$$EI = \sum^{78} W_i I_i(\text{with}) - \sum^{78} W_i I_i(\text{without}) \tag{8.2}$$

8.4. ZOETEMAN'S POLLUTION POTENTIAL INDEX (PPI)

This index was developed by Zoeteman (1973) as a planning tool based not on observed water-quality variables but on indirect factors assumed to be responsible for pollution. It was based on the size of the population within a given drainage area, the degree of economic activity and the average flow rate of the river:

$$PPI = NG/Qx10^{-6} \tag{8.3}$$

where

N is the number of people living in a drainage area,

G is the average per capita gross National Product (GNP) and

Q is the yearly average flow rate (m^3/sec)

Zoeteman applied the PPI to 160 river sites throughout the world, comparing PPI values with the pollutant variables for which more than 40 observations were available. The PPI ranged from 0.01 to 1000 for these rivers. The PPI was also applied to the Rhine River (1973).

8.5. ENVIRONMENTAL QUALITY INDEX PRESENTED BY INHABER (1974)

An environmental quality index, specifically developed for Canada, was presented by Inhaber (1974). It was a composite of four indices, representing air, water, land and miscellaneous aspects of environmental quality.

The air-quality index comprised of three major sections (Figure 8.2), dealing with air quality in urban areas, air quality around these areas and air quality outside major urban areas.

The water-quality index comprised of two major sections. The first pertained to industrial and municipal discharges of wastes into waters and the second with the actual measured water quality and some secondary aspects of it (Figure 8.2).

In the land-quality index, six subindices were combined. These included characteristics of forests, overcrowding in cities, erosion, access to parkland, strip mining and sedimentation (Figure 8.2).

In the index for miscellaneous aspects of environmental quality, subindices on pesticides and radioactivity were combined (Figure 8.2).

Some of the subindices in each of the four main sections were themselves made up of sub-subindices. For example, the subindex on forestry had three sub-subindices.

To combine the indices or subindices, the root-mean-square method was used in the belief that it combines the advantages of simplicity with a greater sensitivity to extreme values of environmental conditions than ordinary linear averaging.

A system of weights, to underscore the fact that some parts of the environment are more important than others, was also used. These weights were assigned on the advice of experts, but, the author was quick to point out, the weights were by no means certain, adding that, in fact, determining a more equitable method of preferential weighting, one which also takes into account the judgement of the public, could be an important task if future EQIs are constructed.

A second type of weighting was also used with respect to population. Since many parts of the EQI describe environmental conditions for people, indices were weighted, whenever possible, according to the population that was likely to be affected. For example, if a given city had twice the population of another, it had twice the mathematical emphasis placed on it.

Since air, water and land are all important components of environmental quality, and since no data were available to assess the relative importance of each, it was decided to weight them equally. Since the index for miscellaneous aspects of environmental quality was much less comprehensive on average than the other three, it was given a lower weight. Hence, for a weight 1 given to total environmental quality, the weights assigned to the air, water, land and miscellaneous indices were 0.3, 0.3, 0.3 and 0.1, respectively. The Environmental Quality Index (EQI) was given by:

$$\mathrm{EQI} = \left\{\left[0.3(I_{\mathrm{air}})^2 + 0.3(I_{\mathrm{water}})^2 + 0.3(I_{\mathrm{land}})^2 + 0.1(I_{\mathrm{misc}})^2\right]/1\right\}^{1/2} \quad (8.4)$$

The values of I_{air}, I_{water}, I_{land} and I_{misc} were 0.99, 0.73, 0.54 and 0.088, respectively, leading to EQI of 0.74.

The author has emphasised that the value 0.74 for the EQI does not represent *the* measure of the state of the Canadian environment, but rather *a* measure, based on the many assumptions that were made. A lower or higher value for any subsequent year in which an EQI is calculated would mean that, on the average, environmental conditions are getting better or worse, respectively.

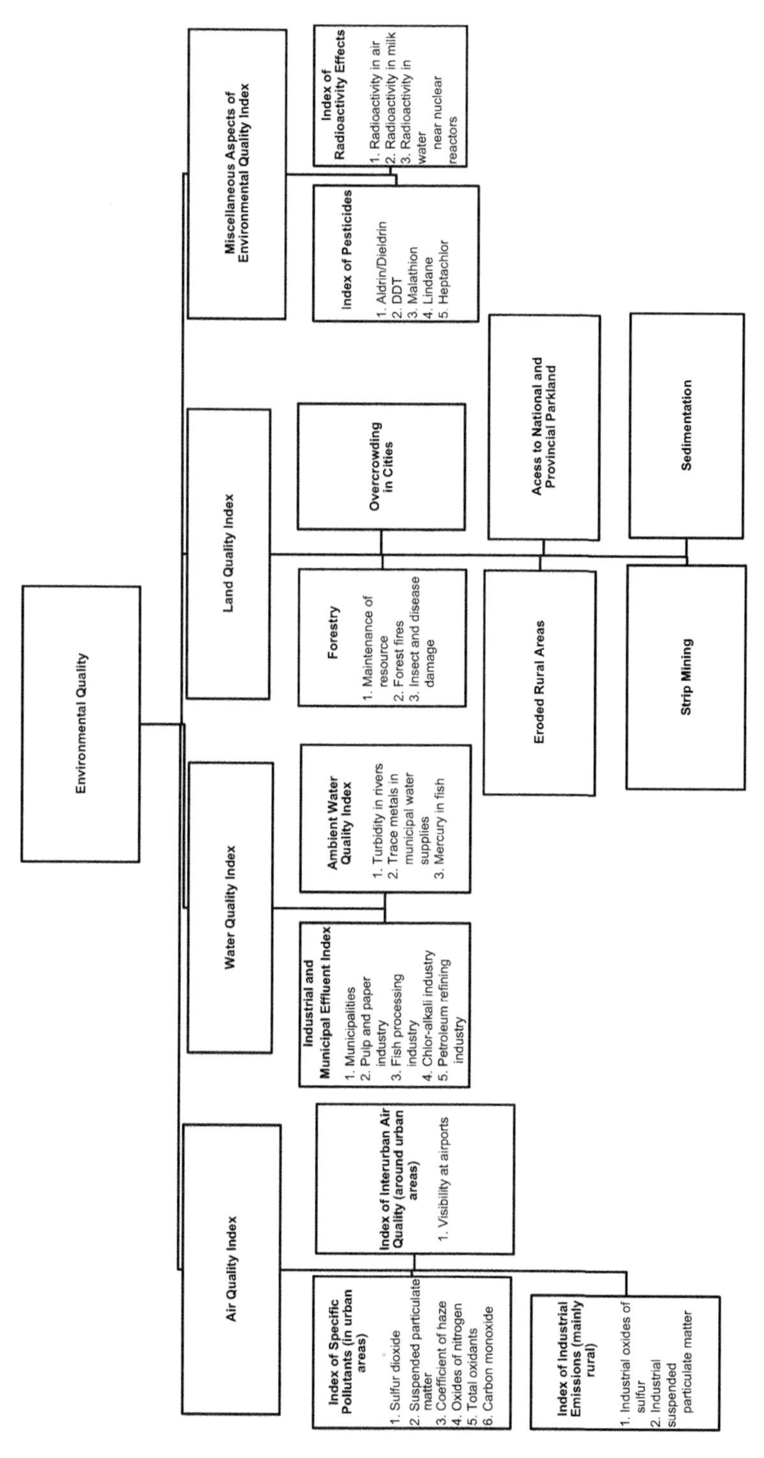

FIGURE 8.2 Schematic diagrams of the indices for air quality, water quality, land quality and miscellaneous aspects of environmental quality described by Inhaber (1974).

8.6. JOHANSON AND JOHNSON'S POLLUTION INDEX

Johanson and Johnson (1976) developed a planning index as a tool to assist in the process of identifying candidate polluted locations. He used the index to screen 652 data sets from water ways across the nation. For each location, Pollution Index (PI) was computed as follows:

$$PI = \sum_{i=1}^{a} W_i C_i \tag{8.5}$$

where

W_i is the weight for pollutant variable i and

C_i is the highest concentration of pollution variable i reported in a location of interest.

For each pollutant i, the weight was based on the reciprocal of the median of observed national concentrations. Using the index, it was possible to scan the data by computer and identify the locations receiving the highest priority for removal of pollutants.

8.7. OTT'S NPPI

Ott (1978) formulated the National Planning Priorities Index (NPPI) as a tool for assigning priorities to different demand sectors in order to ensure that funds are granted and used in a cost-effective manner for the planned water-treatment projects. It was computed as the weighted sum of 10 subindices:

$$NPPI = \sum_{i=1}^{10} W_i I_i \tag{8.6}$$

where each subindex I_i was computed using a segmented linear function.

8.8. WATER-QUALITY INDICES FOR OPERATIONAL MANAGEMENT

House and Ellis (1987) proposed a WQI and three other indices – the potable water supply index (PWSI), the aquatic toxicity index (ATI) and the potable sapidity index (PSI) – for operational management of water resources. They demonstrated the utility of their system of indices by a comparative study with the National Water Council's (NWC) classification of a number of rivers in the Greater London region, UK.

The indices were developed by the procedure given in Figure 8.3. Nine parameters were considered: DO, NH_3–N, BOD, NO_3, Cl, suspended solids, pH, temperature and total coliforms. A scale of 10–100 was subdivided into four classes of water quality:

1. Class I (71–100) to reflect water of high quality suitable for all high value uses at low treatment cost.
2. Class II (51–70) to reflect waters of reasonable quality suitable for high value uses at moderate treatment costs.
3. Class III (31–50) to reflect polluted waters with generally moderate value uses and high treatment costs.
4. Class IV (10–30) to reflect badly polluted waters of low economic value requiring a large investment in treatment facilities.

The rating curves were constructed by equating the concentrations of individual parameters contained within published water-quality directives and criteria to the potential use subdivisions outlined above. The greatest emphasis was placed on criteria contained within the European Economic Commission and the European Inland Fisheries Advisory Commission's water-quality proposals. Where quality criteria were given in the form of maximum desirable or 'guideline' concentrations, the median water-quality rating (WQR) for that use was ascribed. However, where these criteria were maximum permissible or 'mandatory' concentrations, use limiting WQRs where applied. In this way, the derived index score was expected to not only classify water quality and indicate possible water use but also locate a water body accurately within that class, thus

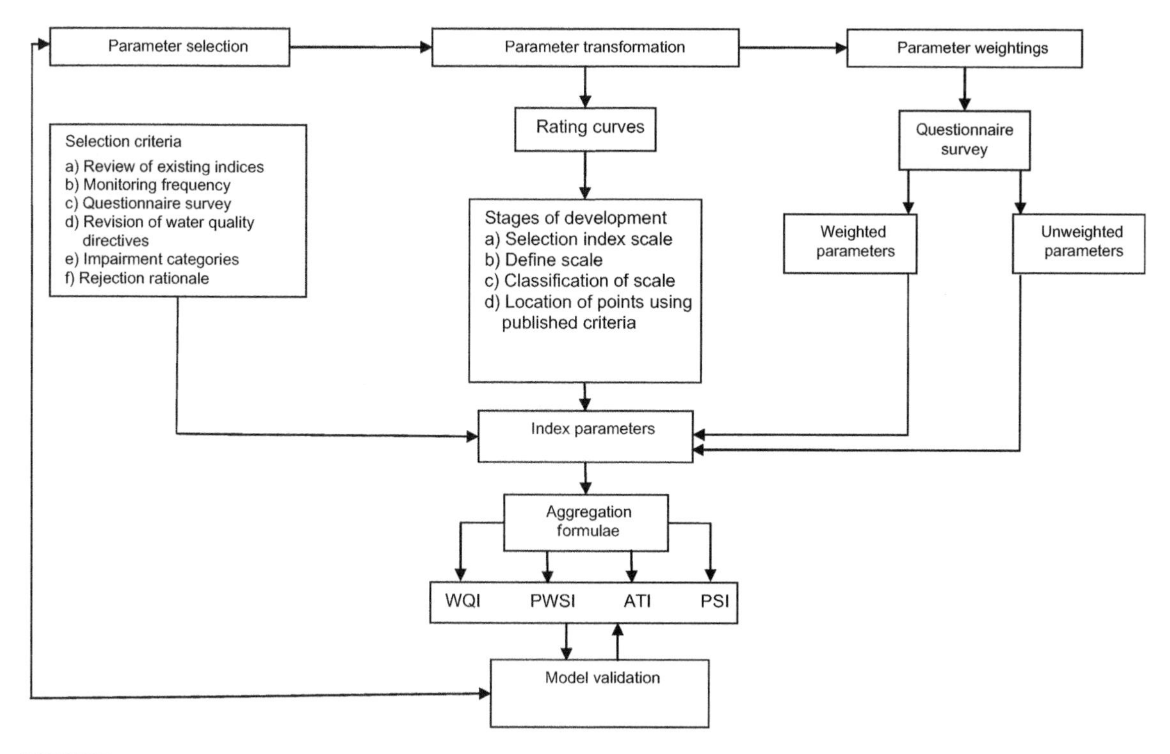

FIGURE 8.3 Procedure used in the development of four water-quality indices by House and Ellis (1987).

providing the maximum amount of information to the user and affording best management practices. It also made it possible to assess the economic gains or losses in value that might accrue from a change in use due to either an improvement or deterioration in water quality.

Weightings were applied to each parameter by means of a questionnaire study completed by officers from the water authorities and river purification boards of England, Wales and Scotland.

The index was computed with the aggregation function:

$$WQI = \frac{1}{100}\left[\sum_{i=1}^{n} q_i w_i\right] \tag{8.7}$$

where

q_i represents the WQR for the ith parameter,

w_i represents the weighting for the ith parameter and

n represents the number of parameter

Three other indices were also developed following the procedures outlined in Figure 8.3. The first of these – the Potable Water Supply Index (PWSI) – was based on thirteen physio-chemical and biological constituents (Table 8.2), and was intended for application to waters primarily used for the abstraction of drinking water, or in instances where this potential water use is under consideration. The remaining two indices were in essence the indices of toxicity. The first – the Aquatic Toxicity Index (ATI) – was based on nine parameters including toxic heavy metals and phenols (Table 8.2), all of which are potentially harmful to fish and wildlife. The second – the Potable Sapidity Index (PSI) – reflects the suitability of water for use in potable water supply (PWS) on the basis of twelve parameters which include toxic heavy metals, pesticides and hydrocarbons (Table 8.2).

TABLE 8.2 Water-quality Parameters Covered in the Three Companion Indices of House and Ellis (1987)

PWSI	PSI	ATI
Dissolved oxygen	Total copper	Dissolved copper
Ammoniacal nitrogen	Total zinc	Total zinc
Biochemical oxygen demand	Total cadmium	Dissolved cadmium
Suspended solids	Total lead	Dissolved lead
Nitrates	Total chromium	Dissolved chromium
pH	Total arsenic	Total arsenic
Temperature	Total mercury	Total mercury
Chlorides	Total cyanide	Total cyanide
Total coliforms	Phenols	Phenols
Sulphates	Total hydrocarbons	
Fluorides	Polyaromatic hydrocarbons	
Colour	Total pesticides	
Dissolved iron		

8.9. INDEX TO REGULATE WATER-MANAGEMENT SYSTEMS

Pryazhinskaya and Yaroshevskii (1996) proposed a method to simulate functioning water-management systems of river basins. They considered functions of the principal components of the model, and a general description of its relations with other models. The conventional methods of developing the regulations to control the water-management systems were analysed by the authors, and the forms of describing these regulations were revealed in order to standardise their representation when inputted into the simulation. Particular attention was paid to the methods for integrating the results of the simulation experiment into a form suitable for analysis.

8.10. INDEX TO ASSESS THE IMPACT OF ECOREGIONAL, HYDROLOGICAL AND LIMNOLOGICAL FACTORS

Ravichandran et al. (1996) developed a methodology consisting of principal-component analysis (PCA) of 23 features of the geological, geomorphological, basin morphometry and land-use aspects of the Tamiraparani basin, South India, defined in terms of 63 micro-basins.

The authors used the PCA scores calculated on five components to cluster the micro-basins into groups based on a similarity measure. The groups identified in the analysis were traced on the drainage map to delineate nine ecoregions. A water-quality survey of the identified ecoregions was then carried out, and the pH, EC, DO, TDS, major ions and nutrients were estimated in 278 water samples. A PCA of the water-quality data revealed that three processes appear to be particularly important for water quality in this basin: the geological origin of ionic richness variables, nutrient leaching from agricultural operations and the carbonate system.

The authors then compared the spatial ability of the ecoregions to account for regional variations in water quality using two existing classification methods (hydrological and limnological). Water samples were grouped in terms of ecoregional, hydrological and limnological classifications, based on their location in the basin. Visual examination of the box plots of water-quality variables showed that ecoregions had less within-region variation with statistically significant differences between the group means than either hydrological or limnological classification.

Discriminant analysis was also performed which displayed the relatively better ability of ecoregions to account for spatial water-quality variations than the other groups in the space defined by the first two discriminant functions representing the ionic richness and nutrient variability of the water samples.

8.11. A WATERSHED-QUALITY INDEX

In order to characterise the spatial and temporal variability of surface water quality in the Chilblain River watershed, Central Chile, a WQI was calculated from nine physico-chemical parameters, periodically measured at 18 sampling sites, by Debels et al. (2005). On the basis of the results from a principal-component analysis (PCA), modifications were introduced into the original WQI to reduce the costs associated with its implementation.

Two of the versions of the WQI, WQIDIR1 and WQIDIR2, which are both based on one laboratory analysis (chemical oxygen demand) and four field measurements (pH, temperature, conductivity and dissolved oxygen),were seen to adequately reproduce the most important spatial and temporal variations observed with the original index.

8.12. INDEX FOR WATERSHED POLLUTION ASSESSMENT

Sanchez et al. (2007) have proposed the water-quality index (WQI) and its correlation with the dissolved oxygen deficit (D) as indicators of the environmental quality of watersheds. The water-quality index was derived from the following empirical equation (Pesce and Wunderlin, 2000):

$$\mathrm{WQI} = k\frac{\sum_i C_i P_i}{\sum_i P_i} \qquad (8.8)$$

where k is a subjective constant with a maximum value of 1 for apparently good quality water and 0.25 for apparently highly polluted water, C_i is the normalised value of the parameter and P_i is the relative weight assigned to each parameter. The authors did not consider the constant k in order to preclude a subjective evaluation, as was done by others using this equation (Nives, 1999; Hernández-Romero et al., 2004). In relation to the parameter P_i, the maximum value of 4 was assigned to parameters of relevant importance for aquatic life as, for example, DO and TSS, while the minimum value of 1 was assigned to parameters with minor relevance to watershed pollution such as temperature and pH (Table 8.3) The water was classified as "very bad", "bad", "medium", "good" and "excellent", on the basis of scores 0–25, 26–50, 51–70, 71–90 and 91–100, respectively (Jonnalagadda and Mhere, 2001).

The values of the water-quality index (WQI) seemed to correlate with the values of oxygen deficit (D). It was found that when the value of D increased, the value of WQI decreased. The two variables were correlated in a linear equation:

$$\mathrm{WQI} = -6.39D + 93.61 \qquad (8.9)$$

The regression coefficient was 0.91 with $p \leq 0.1$. Hence, D can be used to provide a quick measure of water quality in the study area of the authors, the Las Rozas watershed in Spain. The water quality seemed to decrease in autumn, spring and summer, when compared with winter.

8.13. A GIS-ASSISTED WATER-QUALITY INDEX FOR IRRIGATION WATER

Irrigation water quality is an attribute which is very difficult to define as its effects are interlinked with soil characteristics, plant species, duration of irrigation, etc., in a complex fashion. From a management point of view, it is more meaningful to analyse all related parameters as a combination rather than focussing on one

TABLE 8.3 Values of C_i[a] and P_i for Different Parameters of Water Quality, Based on European Standards (EU 1975) in the Index of Sanchez et al. (2007)

		C_i										
Parameter	P_i	**100**	**90**	**80**	**70**	**60**	**50**	**40**	**30**	**20**	**10**	**0**
Range of Analytical Values												
pH	1	7	7–8	7–8.5	7–9	6.5–7	6–9.5	5–10	4–11	3–12	2–13	1–14
K[b]	2	<0.75	<1.00	<1.25	<1.50	<2.00	<2.50	<3.00	<5.00	<8.00	<12.00	>12.00
TSS	4	<20	<40	<60	<80	<100	<120	<160	<240	<320	<400	>400
Amm.	3	<0.01	<0.05	<0.10	<0.20	<0.30	<0.40	<0.50	<0.75	<1.00	<1.25	>1.25
NO_2^-	2	<0.005	<0.01	<0.03	<0.05	<0.10	<0.15	<0.20	<0.25	<0.50	<1.00	>1.00
NO_3^-	2	<0.5	<2.0	<4.0	<6.0	<8.0	<10.0	<15.0	<20.0	<50.0	<100.0	>100.0
P_T	1	<0.2	<1.6	<3.2	<6.4	<9.6	<16.0	<32.0	<64.0	<96.0	<160.0	>160.0
COD	3	<5	<10	<20	<30	<40	<50	<60	<80	<100	<150	>150
BOD_5	3	<0.5	<2.0	<3	<4	<5	<6	<8	<10	<12	<15	>15
DO	4	≥7.5	>7.0	>6.5	>6.0	>5.0	>4.0	>3.5	>3.0	>2.0	>1.0	<1.0
T	1	21/16	22/15	24/14	26/12	28/10	30/5	32/0	36/−2	40/−4	45/−6	>45<−6

[a] *All values, except pH, in mg/l*
[b] *Conductivity in mS/cm.*

or the other isolated parameter. With this aspect in view, Simsek and Gunduz (2007) have proposed a GIS-integrated tool to evaluate the quality of irrigation waters with regards to potential soil and crop problems. It is mainly an index method that addresses five types of hazards: (a) salinity hazard; (b) infiltration and permeability hazard; (c) specific ion toxicity; (d) trace-element toxicity and (e) miscellaneous impacts on sensitive crops.

A linear combination of these groups was formulated by the authors to form their irrigation water-quality (IWQ) index. The method was applied to assess the irrigation water quality of the Simav Plain located in western Anatolia, Turkey.

The water-quality parameters for the IWQ index were selected on the basis of the guidelines presented by Ayers and Westcot (1985) as given in Tables 8.4–8.6. These parameters are expected to not only best characterise the associated hazard but also combine with others to form a general pattern of water quality for the particular resource.

The IWQ index was calculated as

$$IWQ = \sum_{i=1}^{5} G_i \tag{8.10}$$

where i is an incremental index and G represents the contribution of each one of the five hazard categories included in Table 8.4.

The first category is the salinity hazard that is represented by the EC value of the water and is formulated as

$$G_1 = W_1 r_1 \tag{8.11}$$

where W_1 is the value of the weight of this hazard group and r is the rating value of the parameter as given in Table 8.4.

TABLE 8.4 Classification of the IWQ Index Parameters by Simsek and Gunduz (2007)

Hazard	Weight	Parameter	Range	Rating	Suitability for Irrigation
Salinity hazard	5	Electrical conductivity (μS/cm)	EC < 700	3	High
			700 ≤ EC ≤3000	2	Medium
			EC > 3000	1	Low
Infiltration and permeability hazard	4	See Table IV for details			
Specific ion toxicity	3	Sodium adsorption ratio (−)	SAR < 3.0	3	High
			3.0 ≤ SAR ≤ 9.0	2	Medium
			SAR > 9.0	1	Low
		Boron (mg/l)	B < 0.7	3	High
			0.7 ≤ B ≤ 3.0	2	Medium
			B > 3.0	1	Low
		Chloride (mg/l)	CI < 140	3	High
			140 ≤ CI ≤ 350	2	Medium
			CI > 350	1	Low
Trace-element toxicity	2	See Table V for details			
Miscellaneous effects to sensitive cops	1	Nitrate–Nitrogen (mg/l)	NO_3 −N < 5.0	3	High
			5.0 ≤ NO_3–N ≤ 30.0	2	Medium
			NO_3–N > 30.0	1	Low
		Bicarbonate (mg/l)	HCO_3 < 90	3	High
			90 ≤ HCO_3 ≤ 500	2	Medium
			HCO_3>500	1	Low
		pH	7.0 ≤ pH ≤ 8.0	3	High
			6.5 ≤ pH < 7.0 and 8.0 < pH ≤ 8.5	2	Medium
			pH < 6.5 or pH > 8.5	1	Low

TABLE 8.5 Classification for Infiltration and Permeability Hazard (Simsek and Gunduz, 2007)

	SAR						
	<3	3–6	6–12	12–20	>20	Rating	Suitability
EC	>700	>1200	>1900	>2900	>5000	3	High
	700–200	1200–300	1900–500	2900–1300	5000–2900	2	Medium
	<200	<300	<500	<1300	<2900	1	Low

TABLE 8.6 Rating for Trace-element Toxicity in the IWQ Index of Simsek and Gunduz (2007)

Element	Range of concentration	Rating	Suitability for Irrigation
Aluminium (mg/l)	Al < 5.0	3	High
	5.0 ≤ Al ≤ 20.0	2	Medium
	Al > 20.0	1	Low
Arsenic (mg/l)	As < 0.1	3	High
	0.1 ≤ As ≤ 2.0	2	Medium
	As > 2.0	1	Low
Beryllium (mg/l)	Be < 0.1	3	High
	0.1 ≤ Be ≤ 0.5	2	Medium
	Be > 0.5	1	Low
Cadmium (mg/l)	Cd < 0.01	3	High
	0.01 ≤ Cd ≤ 0.05	2	Medium
	Cd > 0.05	1	Low
Chromium (mg/l)	Cr < 0.1	3	High
	0.1 ≤ Cr ≤ 1.0	2	Medium
	Cr > 1.0	1	Low
Cobalt (mg/l)	Co < 0.05	3	High
	0.05 ≤ Co ≤ 5.0	2	Medium
	Co > 5.0	1	Low
Copper (mg/l)	Cu < 0.2	3	High
	0.2 ≤ Cu ≤ 5.0	2	Medium
	Cu > 5.0	1	Low
Fluoride (mg/l)	F < 1.0	3	High
	1.0 ≤ F ≤ 15.0	2	Medium
	F > 15.0	1	Low
Iron (mg/l)	Fe < 5.0	3	High
	5.0 ≤ Fe ≤ 20.0	2	Medium
	Fe > 20.0	1	Low
Lead (mg/l)	Pb < 5.0	3	High
	5.0 ≤ Pb ≤ 10.0	2	Medium
	Pb > 10.0	1	Low

(*Continued*)

TABLE 8.6 Rating for Trace-element Toxicity in the IWQ Index of Simsek and Gunduz (2007) (*cont'd*)

Element	Range of concentration	Rating	Suitability for Irrigation
Lithium (mg/l)	Li < 2.5	3	High
	2.5 ≤ Li ≤ 5.0	2	Medium
	Li > 5.0	1	Low
Manganese (mg/l)	Mn < 0.2	3	High
	0.2 ≤ Mn ≤ 10.0	2	Medium
	Mn > 10.0	1	Low
Molybdenum (mg/l)	Mo < 0.01	3	High
	0.01 ≤ Mo ≤ 0.05	2	Medium
	Mo > 0.05	1	Low
Nickel (mg/l)	Ni < 0.2	3	High
	0.2 ≤ Ni ≤ 2.0	2	Medium
	Ni > 2.0	1	Low
Selenium (mg/l)	Se < 0.01	3	High
	0.01 ≤ Se ≤ 0.02	2	Medium
	Se > 0.02	1	Low
Vanadium (mg/l)	V < 0.1	3	High
	0.1 ≤ V ≤ 1.0	2	Medium
	V > 1.0	1	Low
Zinc (mg/l)	Zn < 2	3	High
	2 ≤ Zn ≤ 10	2	Medium
	Zn > 10.0	1	Low

The second category is the infiltration and permeability hazard that is represented by EC−SAR combination and is formulated as

$$G_2 = W_2 r_2 \qquad (8.12)$$

where W_2 is the weight value of this hazard group and r_2 is the rating value of the parameter as given in Table 8.5.

The third category is the specific ion toxicity that is represented by SAR, chloride and boron ions in the water and is formulated as a weighted average of the three ions:

$$G_3 = \frac{W_3}{3}\sum_{j=1}^{3} r_j \quad (8.13)$$

where j is an incremental index, W_3 is the weight value of this group as given in Table 8.4 and r is the rating value of each parameter as given in Table 8.5.

The fourth category is the trace-element toxicity and is formulated as a weighted average of all the ions available for analysis:

$$G_4 = \frac{W_4}{N}\sum_{k=1}^{N} r_k \quad (8.14)$$

where k is an incremental index, N is the total number of trace element available for the analysis, W_4 is the weight value of this group and r is the rating value of each parameter as given in Table 8.6.

The fifth and the final category is the miscellaneous effects to sensitive crops that is represented by nitrate–nitrogen and bicarbonate ions and the pH of the water, and is formulated as a weighted average:

$$G_5 = \frac{W_5}{3}\sum_{m=1}^{3} r_m \quad (8.15)$$

where m is an incremental index, W is the weight value of this group and r is the rating value of each parameter as given in Table 8.4.

8.14. A SYSTEM OF INDICES FOR WATERSHED MANAGEMENT

To investigate relationships between water quality and land use in watersheds serving public water supply in the Northeast of Goiás, a set of four indices were developed and tested by Bonnet et al. (2008).

Principal-component analysis (PCA) was performed on 174 data points covering for the annual average, drought and flood. The Index of water quality (IQA) was obtained as follows:

$$\mathrm{IQA_n} = \sum (p_1 \cdot c_n : p_6 \cdot c_n) \quad (8.16)$$

where p is value of each of the six parameters and c is a coefficient (weight) of the eigenvector.

A weighted IQA was then obtained as the respective percentage of total variance:

$$\mathrm{IQA_p} = \sum (\mathrm{IQA_1}.\%_1 : \mathrm{IAQA_6}.\%_6) \quad (8.17)$$

An index of seasonal contrast (ISC) was computed as follows:

$$\mathrm{ISC} = (\mathrm{IQA_1\ full} - \mathrm{IQA_1\ dry})/\mathrm{IQA_1\ full} \quad (8.18)$$

A normalised vegetation index (NRVI) was computed by mapping land cover into two categories: "use," including the annual crop classes, culture centre pivot and pasture, and "remaining native vegetation", comprising submontane deciduous forest and montane; semideciduous alluvial forest, lowland, submontane and montane; pioneer formations, river and/or lake, wooded savanna, woodland, woody park and grassy savanna park. Based on the area and perimeter of each catchment, as well as each polygon of use and remaining within the basins, the NRVI was obtained as

$$\mathrm{NRVI} = \frac{\mathrm{area_{remaining}} - \mathrm{area_{use}}}{\mathrm{area_{remaining}} + \mathrm{area_{use}}} \quad (8.19)$$

The authors found that the water quality was below the standards in 62.43% of the cases. Acceptable levels of NRVI were achieved in only 31.52% of the assessed basins.

8.15. A FUZZY WQI FOR WATER-QUALITY ASSESSMENT OF SHRIMP FORMS

Following a methodology similar to the one used by Ocampo-Duque et al. (2006), Carbajal and Sanchez (2008) developed a fuzzy WQI (FWQI) which handles temperature, salinity, DO and pH as inputs to monitor habitat quality in shrimp farms. The authors compared the performance of their FWQI with that of the Canadian Council of Ministers of Environment (CCME) WQI to obtain a pattern summarised in Figure 8.4. As may be seen, even though the two indices follow a consistent order vis a vis their scores – CCME–WQI < FWQI – the FQI appears to be more sensitive to changes in the values of temperature, DO and salinity.

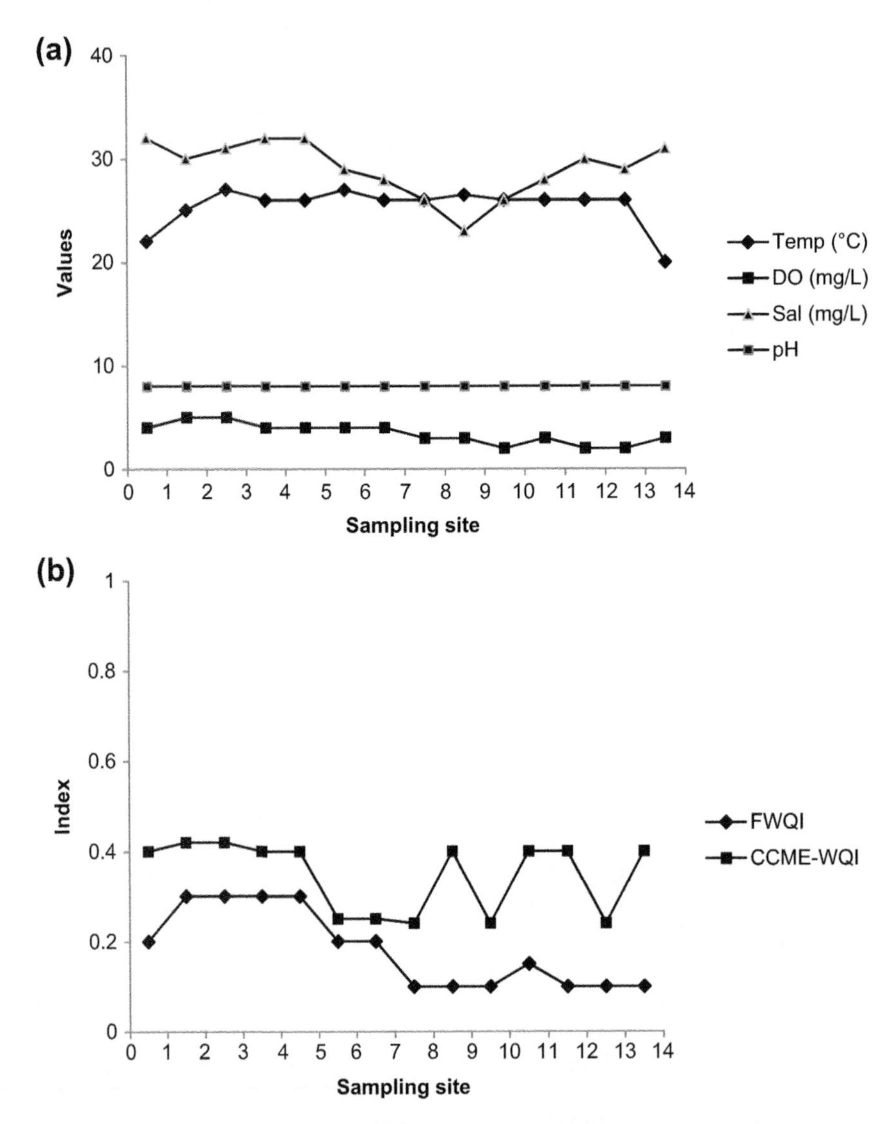

FIGURE 8.4 Variation in water-quality parameters (above) and in the corresponding scores of the two indices. *(Carbajal and Sanchez, 2008).*

8.16. AN INDEX TO ASSESS ACCEPTABILITY OF RECLAIMED WATER FOR IRRIGATION

To assess the possible toxicity to plants when using treated wastewater, Deng and Yang (2009) have proposed an 'irrigation security' index.

The toxicity of a pollutant present in an irrigation water, i, to a plant is related to the exposure doze, x_i, by the expression

$$y_i = a x_i^b \tag{8.20}$$

where y_i is the toxicity of the ith pollutant to a plant species, x_i is the exposure dose of ith pollutant and a, b are constants; $b > 1$. Higher the probability of a pollutant occurring at harmful levels in an irrigation water, greater the risk to the receiving plants. Hence,

$$ps_i = (d_i/Std_i) \times (W_i + CV_i)^k \tag{8.21}$$

where ps_i is irrigation risk from i^{th} pollutant, a dimensionless number; d_i is the mean concentration of the harmful substance in reclaimed water, mg/L; Std_i is the acceptable limit of ith pollutant; W_i is the coefficient of damage by the ith pollutant and CV_i is coefficient of variation of the ith pollutant concentration in the water and k is a control ratio, more than 0.

It can be deduced that the irrigation risk would be considerable if ps_i is more than 0.3, the concentration of the ith pollutant is large or its coefficient of variation is very large. When ps_i is more than 1, the irrigation risk would go beyond acceptable limit. Using this reasoning the authors have grouped water-quality constituents on the basis of their potential toxicity (when occurring beyond certain levels) to plants as follows:

BOD, COD, Cr, SS, TKN, TP, temperature, pH < LAS, total salt, petroleum, Cl^-, S^{2-}, Cu, Zn, Sn, B < Hg, Cd, As, Cr^{6+}, Pb, fluoride, CN^-, volatile phenol, benzene, chloral, ascaris eggs acrolein, fecal coliform

8.17. AN INDEX FOR IRRIGATION WATER-QUALITY MANAGEMENT

An irrigation water-quality index that expresses possible alterations in water quality for irrigation, thereby facilitating evaluation/mitigation of possible problems caused to irrigated soil and plants, has been proposed by Meireles et al. (2010).

The index was developed in two steps. In the first step, parameters that contribute to most variability in irrigation water quality were identified using principal-component analysis (PCA) and factor analysis (FA). In the second step, a definition of quality measurement values (q_i) and aggregation weights (w_i) was established. Values of q_i were estimated based on each parameter value, according to irrigation water-quality parameters proposed by the University of California Committee of Consultants and by the criteria established by Ayers and Westcot (1999), shown in Table 8.7. Water-quality parameters were represented by a nondimensional number; the higher the value, the better the water quality (Table 8.7).

Values of q_i were calculated on the basis of the tolerance limits shown in Table 8.7 and water-quality results determined in laboratory:

$$\mathbf{q}_i = \mathbf{q}_{i\max} - \left[\left(\mathbf{X}_{ij} - \mathbf{X}_{\inf}\right)^* \left(\mathbf{q}_{i\mathrm{amp}}\right) / \mathbf{X}_{\mathrm{amp}}\right] \tag{8.22}$$

where $\mathbf{q}_{i\max}$ is the maximum value of $\mathbf{q}_i$ for the class; $\mathbf{X}_{ij}$ is the observed value for the parameter; $\mathbf{X}_{\inf}$ is the corresponding value to the lower limit of the class to which the parameter belongs; $\mathbf{q}_{i\mathrm{amp}}$ is class amplitude; $\mathbf{X}_{i\mathrm{amp}}$ is class amplitude to which the parameter belongs.

In order to evaluate $\mathbf{X}_{\mathrm{amp}}$ of the last class of each parameter, the upper limit was considered to be the highest value determined in the

TABLE 8.7 Parameter Limiting Values Used in the WQI of Meireles et al. (2010)

q_i	EC ($dS\ cm^{-1}$)	SAR° ($(mmol_c\ L^{-1})^{1/2}$)	Na^+ ($mmol_c\ L^{-1}$)	Cl^- ($mmol_c\ L^{-1}$)	HCO_3 ($mmol_c\ L^{-1}$)
85–100	$0.20 \leq CE < 0.75$	$2 \leq SAR° < 3$	$2 \leq Na < 3$	$1 \leq Cl < 4$	$1 \leq HCO_3 < 1.5$
60–85	$0.75 \leq CE < 1.50$	$3 \leq SAR° < 6$	$3 \leq Na < 6$	$4 \leq Cl < 7$	$1.5 \leq HCO_3 < 4.5$
35–60	$1.50 \leq CE < 3.00$	$6 \leq SAR° < 12$	$6 \leq Na < 9$	$7 \leq Cl < 10$	$4.5 \leq HCO_3 < 8.5$
0–35	$EC < 0.20$ or $EC \geq 3.00$	$SAR° < 2$ or $SAR° \geq 12$	$Na < 2$ or $Na \geq 9$	$Cl < 1$ or $Cl \geq 10$	$HCO_3 < 1$ or $HCO_3 \geq 8.5$

physico-chemical and chemical analysis of the water samples.

Each parameter weight used in the WQI was obtained from the PCA/FA, by the sum of all factors multiplied by the explainability of each parameter. Then w_i values were normalised such that their sum equals one:

$$w_i - \frac{\sum_{j=1}^{k} F_j A_{ij}}{\sum_{j=1}^{k} \sum_{i=1}^{n} F_j A_{ij}} \quad (8.23)$$

where w is the weight of the parameter for the WQI; F is the autovalue of component; A_{ij} is the explainability of parameter i by factor j; i is the number of physico-chemical and chemical parameters selected by the model, ranging from 1 to n and j is the number of factors selected in the model, varying from 1 to k.

The water-quality index was calculated as

$$WQI = \sum_{i=1}^{n} q_i w_i \quad (8.24)$$

where WQI is in the range 0–100; q_i is the quality of the ith parameter, a number from 0 to 100 which is a function of its concentration and w_i is the normalised weight of the ith parameter which is a function of its importance (in explaining the global variability in water quality).

In order to identify the most significant interrelation of water-quality parameters in the Acarau River, for each resulting factor of PC, a matrix rotation procedure was adopted by Meireles et al. (2010) using the Varimax method. This method minimises the contribution of parameters with a lower significance in the factor such that the parameters present loads close to one or zero, eliminating the intermediate values which make interpretation of the results difficult (Vega et al., 1998; Helena et al., 2000; Wunderlin et al., 2001).

The water-management guidelines based on the range of WQI were drawn by the authors (Table 8.8) and applied to the study of Acarau River basin, Brazil.

8.18. INDEX FOR THE ANALYSIS OF DATA GENERATED BY AUTOMATED SAMPLING (CONTINUOUS MONITORING) NETWORKS

Automated sampling networks are able to continuously monitor certain water-quality variables (such as electrical conductivity, pH, temperature, DO, turbidity, ammonia and nitrate) which can be 'sensed' by instruments without time delay but are not able to monitor some other oft-used variables in water quality assessment (such as BOD_5 and coliform) for which no instantaneous sensors are available. Hence, the WQIs which may be adequate for assessing discretely monitored data may not necessarily be appropriate for continuous monitoring networks.

Moreover, due to their high acquisition frequency (sometimes with minute resolution), very large databases (samples and variables repeated in time) can be obtained in one day

TABLE 8.8 Water-Use Guidelines Provided by Meireles et al. (2010) Based on their WQI

		Recommendation	
WQI	**Water-Use Restrictions**	**For Soil**	**For Plants**
85 ≤ 100	No restriction (NR)	May be used for the majority of soils with low probability of causing salinity and sodicity problems, being recommended leaching within irrigation practices, except for in soils with extremely low permeability.	No toxicity risk for most plants
70 ≤ 85	Low restriction (LR)	Recommended for use in irrigated soils with light texture or moderate permeability, being recommended salt leaching. Soil sodicity in heavy texture soils may occur, being recommended to avoid its use in soils with high clay levels 2:1.	Avoid salt sensitive plants
55 ≤ 70	Moderate restriction (MR)	May be used in soils with moderate to high permeability values, being suggested moderate leaching of salts.	Plants with moderate tolerance to salts may be grown
40 ≤ 55	High restriction (HR)	May be used in soils with high permeability without compact layers. High-frequency irrigation schedule should be adopted for water with EC above 2.000 dS m^{-1} and SAR above 7.0.	Should be used for irrigation of plants with moderate to high tolerance to salts with special salinity control practices, except water with low Na, Cl and HCO_3 values
0 ≤ 40	Severe restriction (SR)	Should be avoided its use for irrigation under normal conditions. In special cases, may be used occasionally. Water with low salt levels and high SAR require gypsum application. In high saline content water soils must have high permeability, and excess water should be applied to avoid salt accumulation.	Only plants with high salt tolerance, except for waters with extremely low values of Na, Cl and HCO_3.

for subsequent analysis. With this attribute an appropriate index can make it possible to detect point episodes, which, although they might have had a significant impact on the quality of the water body, would normally remain hidden when monitoring is dependent on only discontinuous data.

Based on this premise, Terrado et al. (2010) evaluated the suitability of five preexisting WQIs for use with continuous monitoring networks:

1. The Catalan Water Agency's WQI: ISQA
2. WQI of Pesce and Wunderlin (2000)
3. River-pollution index (Liou et al., 2004)
4. NSF–WQI (Brown et al. 1970)
5. WQI of the Canadian Council of Ministers of the Environment (CCME–WQI; CCME, 2001)

The criteria Terrado et al. (2010) set to base their evaluation is summarised in Table 8.9. They assigned more significance to those indices which were predominantly based on the variables which are obtained continuously (e.g., pH, conductivity, turbidity, dissolved oxygen, water temperature, ammonia and, in some cases, nitrates, chlorides and phosphates). They attached lower significance to those indices which required several other parameters. A large

TABLE 8.9 Criteria on which Terrado et al. (2010) have Shortlisted Five WQIs

Criteria	Indices				
	ISQA	Pesce and Wunderlin	Liou	NSF-WQI	CCME-WQI
Parameters measured using continuous sampling	Fair	Good	Fair	Bad	Good
Adaptability to different uses of water body	Bad	Bad	Bad	Bad	Good
Existing guidelines to define objectives	Good	Good	Good	Good	Fair
Experience of real application	Fair	Bad	Bad	Fair	Good
Consideration of the amplitude (amount by which the objectives are not met)	Fair	Fair	Fair	Fair	Good
Programming difficulty	Good	Good	Good	Fair	Good
Tolerance to missing data	Bad	Bad	Bad	Bad	Fair
Need of synchorinized data	Bad	Bad	Bad	Bad	Good
Tolerance to wrong data	Bad	Bad	Bad	Bad	Bad

Good | Fair | Bad

number of WQIs that have not been developed to work with data of high resolution and use average values to calculate the index score, with the subsequent loss of relevant information, were omitted from contention. In addition, they preferred flexible indices that allowed use of different parameters, selection of a specific time period and establishment of different objectives depending on various water uses. Whenever these objectives were attained, they gave the index a higher score in evaluating it against the set criteria.

Finally, they assigned higher weightage on simplicity in programming, tolerance to missing or erroneous data and the possibility of the index working with non-synchronised data.

The assessment led to the selection of CCME–WQI as the most suitable tool for categorising water bodies using data generated by automated sampling stations.

The authors conducted a sensitivity analysis and identified several advantages and disadvantages of CCME–WQI (Table 8.10). They have also made the following recommendations:

1. For monthly quality evaluation of a water body, daily calculation of the index should be done before taking a monthly average, instead of directly calculating the monthly index from the raw data.
2. Variables of dubious relevance, those with most test values fulfilling the established objectives, and those presenting a strong correlation, should be excluded from the analysis in order to avoid index bias.
3. Depending on the context, different weights should be assigned to different variables, so as to highlight the significance of those having a greater impact on water quality and decrease the contribution of those having a lesser impact. This step may introduce a certain degree of subjectivity, hence should be taken with great caution.

TABLE 8.10 Advantages and Disadvantages of CCME–WQI as Identified by Terrado et al. (2010)

Advantages	Disadvantages
1. Flexibility in the selection of input parameters and objectives	1. Missing guidelines about the variables to be used for the index calculation
2. Adaptability to different legal requirements and different water uses	2. Missing guidelines about the objectives specific to each location and particular water use
3. Statistical simplification of complex multivariate data	3. Easy to manipulate (biased)
4. Clear and intelligible diagnostic for managers and the general public	4. The same importance is given to all variables
5. Suitable tool for water-quality evaluation in a specific location	5. No combination with other indicators or biological data
6. Easy to calculate	6. Only partial diagnostic of the water quality
7. Tolerance to missing data	7. F_1 not working appropriately when too few variables are considered or when too much convariance exists among them. The factor has too much weight in calculating the index.
8. Suitable for analysis of data coming from automated sampling	
9. Experience in implementation	
10. Considers amplitude (of difference from the objective)	

8.19. AN INDEX OF DRINKING-WATER ADEQUACY FOR THE ASIAN COUNTRIES

Kallidaikurichi and Rao (2010) argue that an index for water use ought to reflect the key variables determining the availability of an adequate amount of drinking water of the right quality. Such an index, according to the authors, should address the following issues:

1. Ensuring adequate drinking water is facilitated if a nation has its own resources.
2. The quality of access mechanism is such that it ensures supply.
3. The capacity to purchase water,
4. Adequacy of the actual use of water
5. Water of good quality

Based on this rationale, the authors surmised that an index of drinking-water adequacy (IDWA) should reflect resources, access, capacity, use and quality. Accordingly, they proceeded to develop five constituent indices of the IDWA.

8.19.1. Resource Indicator

Estimates of renewable internal freshwater resources per capita were taken from the World Development Indicators (WDI, 2006) and converted to a log scale. Taking the resource per capita, R_j, for country j, the first constituent index was written as

$$\text{Resource Indicator } (RI) = \left[\log R_j - \log R_{\min}/\log R_{\max} - \log R_{\min}\right] \times 100 \tag{8.25}$$

The $R_{\max}$ value belongs to that of Papua New Guinea (PNG) at 138,775 m^3 in 2004. The $R_{\min}$ was assumed to be 1 m^3, which in log form is zero.

8.19.2. Access Indicator

The maximum possible access value is availability of safe water for 100% of the population. The minimum cannot be zero, since it is a demographic–geographic impossibility, given that people need at least some minimum drinking water in any settlement. As Ethiopia had the lowest access rate of 22% in 2004, the index took the form

$$\text{Access Indicator } (AI) \text{ for country } 'j' = [(A_j - 22)/(100 - 22)] \times 100 \tag{8.26}$$

8.19.3. Capacity Indicator

The index is given by

Capacity Indicator (CI) for country 'j'

$$= \left[\text{Log } C_j - \text{Log } C_{\min}/\text{Log } C_{\max} - \text{Log } C_{\min}\right] \times 100 \quad (8.27)$$

8.19.4. Use Indicator

To obtain the index of use based on the estimate of per capita consumption, the authors took the minimum as 70 LPCD, as prescribed by the Indian Government. For the maximum, the authors chose the data of Singapore, where water conservation is combined with guaranteed continuous supply of water that can be safely consumed straight from the tap. The 1995–2002 average of per capita domestic consumption in Singapore was found to be 167 LPCD. The index took the form

Use Indicator (UI) for country 'j'

$$= [(U_j - 70)/(167 - 70)] \times 100 \quad (8.28)$$

8.19.5. Quality Indicator

In the absence of reliable national data on water quality, the authors used diarrhoeal death rate (DR) which is expressed as diarrhoeal deaths per 100,000 people. World Health Organization (WHO) data indicate a maximum DR close to 100 in Lao PDR and a minimum of 0.5 in the Republic of Korea. The index was computed by taking the difference between 100 and the country value, as an indirect measure of water quality:

Quality Indicator (QI) for country 'j'

$$= [(100 - \text{DR}_j)/(100 - 0)] \times 100$$

$$= 100 - \text{DR}_j \quad (8.29)$$

IDWA was then calculated as a simple average of the five constituent indices:

$$\frac{RI + AI + CI + UI + QI}{5} \quad (8.30)$$

The authors have also proposed an IDWA-II, in which the single national access indicator has been replaced with urban and rural house connection rates, thus raising the number of component indices from 5 to 6, and focussing on the need to supply drinking water to urban and rural areas on the same footing, if possible.

Application of IDWA and IDWA-II on 23 Asian countries indicated that the two indices were strongly correlated yet were able to bring out the subtle differences between the drinking-water adequacies of different countries due to difference in the extents of success of their rural water supply schemes.

8.20. INDICES FOR THE PREDICTION OF STREAM OF QUALITY IN AN AGRICULTURAL SETTING

Shiels (2010) has attempted to link landscape indices to stream water quality in a predominantly agricultural landscape, using remotely sensed and geospatial data assisted by geographical information system (GIS).

8.20.1. Natural Cover Index (I^{NC})

To obtain this index, the area of land in natural cover (NC) is divided by the total area (TA) in the subwatershed:

$$I^{NC} = \frac{\text{NC}}{\text{TA}} \quad (8.31)$$

Here, NC is defined as areas where significant human activity is limited to nature observation, hunting, fishing, timber harvest, etc., and where vegetation is allowed to grow for many years. It was calculated by selecting for all forested, shrub, grassland and wetland land use/land cover classes from the 2001 National Land Cover Database of the USA.

8.20.2. River–Stream Corridor Integrity Index (I^{RSCI})

For this index, the area of land in natural cover within 100 m on each side of the stream, the corridor natural cover (CNC), was divided by the total (100 m on each side) corridor area (TCA):

$$I^{RSCI} = \frac{CNC}{TCA} \tag{8.32}$$

8.20.3. Wetland Extent Index (I^{WE})

The present extent of wetlands (PEWs) was divided by the historic extent of wetlands (HEWs) to get an approximate wetland acreage remaining in the watershed:

$$I^{WE} = \frac{PEW}{HEW} \tag{8.33}$$

Inverse distance weights have been used because the distance from a land use (point source) in a watershed to a stream affects the impact that land use has on a stream (Hale et al., 2004).

8.20.4. Extent of Drained Land Index (I^{EDL})

Drained area of land (DA) was divided by the total land area (TA) in each watershed:

$$I^{EDL} = \frac{DA}{TA} \tag{8.34}$$

8.20.5. Percent of Agriculture on Slopes (I^{PAGS})

The area of agriculture land (row crops) in three slope ranges – 3–10% (PAS1), 10–30% (PAS2) and 30% + (PAS3) – was divided by the total agricultural area (TAA):

$$I^{PAG3} = PAG1 \times 0.25 + PAG2 \times 0.5 + PAG3 \times \frac{1}{TAA} \tag{8.35}$$

8.20.6. Proximity of CAFOs to Streams Index (I^{PCS})

The number of confined animal feeding operations (CAFOs) points was found within a 0–100 m (PC1), 101–500 m (PC2), 501–1000 m (PC3) or a 1000+ m (PC4) range from a river–stream course, and each range was divided by the highest CAFOs in watershed (HCW) value:

$$I^{PCI} = \left(\frac{PC1}{HCW} \times 1 + \frac{PC2}{HCW} \times 0.5 + \frac{PC3}{HCW} \times 0.25 + \frac{PC4}{HCW} \times 0.125\right) \times 1.75 \tag{8.36}$$

Here a multiplier of 1.75 has been used to increase the index value to a maximum of 1.0. Typical individual indicator scores and the numerical values used to calculate the scores can be seen in Table 8.11.

8.21. AN INDEX TO ASSESS EXTENT OF WASTEWATER TREATMENT

In order to provide an overall picture of the water-quality level achieved by different wastewater-treatment sequences and to compare their performances, a new index, called the '*wastewater polishing index*', has been devised by Verlicchi et al. (2011). It is based on six parameters most often monitored for wastewaters discharged into surface water bodies and for reuse: BOD_5, COD, SS, P_{tot}, NH_4 and *Escherichia coli*.

To develop rating curves for the six parameters, two key points have been assumed for each indicator corresponding to the expected limits of each specific range (Table 8.12): a minimum value equal to 0 and a maximum value equal to the Italian legal limits for effluent discharge into surface water bodies $C_{i,\,law}$, with i = BOD_5, COD, SS, P_{tot}, NH_4 and *E. coli*.

TABLE 8.11 An Illustrative set of Scores for the System of Indices Used by Shiels (2010)

Index	Code	Calculation	Score
Natural cover	I^{NC}	1231 ha of natural vegetation/10,103 ha of land in watershed	0.12
River–stream corridor	I^{RSCI}	505 ha of natural vegetation/2562 ha of buffer	0.20
Wetland extent	I^{WE}	196 ha of present wetlands/3542 ha of hydric soils	0.06
Extent of drained land	I^{EDL}	2766 ha of drained land/10,103 ha of land in watershed	0.27
Percent of agriculture on slopes: 3–10%, 10–30% and >30%	I^{PAGS}	(4370 ha of agriculture on 3–10% slopes + 1019 ha of agriculture on 10–30% slopes + 26 ha of agriculture on 30% + slopes)/7155 ha of agriculture land in watershed	0.23
Proximity of CAFO's to streams	I^{PCS}	(0 CAFOs within 100 m/15 max CAFOs × 1 + 5 CAFOs within 500 m/15 max CAFOs × 0.5 + 3 CAFOs within 1000 m/15 max CAFOs × 0.25 + 0 CAFOs outside 1000 m/15 max CAFOs × 0.125) × 1.75	0.38

For all six variables, the assumed rating curves are straight lines connecting minimum with the $C_{i\ law}$. In this way, six subindices have been defined.

If a measured concentration C_i happens to be greater than $C_{i,\ law}$, the linear correlation can be extrapolated resulting in a subindex value greater than 100.

The new index is given by

$$\text{WWPI} = \frac{\sum_i I_i^{n_i}}{\sum_i 100^{n_i}} \times 100 = \frac{I^1_{BOD_5} + I^1_{COD} + I^1_{SS} + I^1_{NH_4} I^1_P + I^{1.4}_{E.coli}}{5 \times 100^1 + 100^{1.4}} \times 100 \quad (8.37)$$

where I_i is the subindex corresponding to the basic parameter i = BOD_5, COD, SS, P, NH_4 and *E. coli* and n_i are equal to 1 for all parameters, with the exception of *E. coli* for which it is 1.4.

The higher value for *E. coli* subindex is to give greater weight to the disinfection ability of the polishing sequence under study.

The WWPI can indicate at one glance how much the achieved effluent quality is below the level allowed by the Italian law for discharge into a surface water body because for an adequately treated wastewater the index value should be $\leq$100.

A sensitivity analysis by the authors revealed that *E. coli* is the most influential parameter, as even a small increase/decrease in its value causes a significant percentage variation in the corresponding WWPI. For instance, an increase equal to 20% in its value results in a WWPI increase of about 20%, while the same increase in the value of one of the other five parameters causes an increase in the range of 0.5–3%.

TABLE 8.12 Range of Values of Six Water Quality Variables, Used in Developing Subindices by Verlicchi et al. (2011)

Limit	BOD_5 (mg L^{-1})	COD (mg L^{-1})	SS (mg L^{-1})	NH_4 (mg L^{-1})	P_{tot} (mg L^{-1})	*E. coli* (CFU/100 mL)
Minimum	0	0	0	0	0	0
$C_{i\ law}$	25	125	35	15	1	5000

The influence of different variables on the final WWPI values increases in the following order: $NH_4 < COD < BOD_5 < SS < P_{tot} << E. coli$.

8.22. USE OF INDICES FOR PRIORITISING PACEMENT OF WATER-QUALITY BUFFERS TO CONTROL NONPOINT POLLUTION

A comparison of five physically based, spatially distributed empirical indices was carried out by Dosskey and Qiu (2011) to assess the degree to which they identified the same or different locations in watersheds where vegetative buffers would function better for reducing agricultural nonpoint source pollution.

All the five indices were calculated on a 10 m × 10 m digital elevation grid on agricultural land in the 144 km^2 Neshanic River watershed in New Jersey, USA. The indices include the topography-based Wetness Index (WI) and Topographic Index (TI) and three soil survey-based indices: sediment trapping efficiency (STE), water trapping efficiency (WTE) and groundwater interaction (GI).

The WI was calculated for each grid cell by

$$\mathrm{WI} = \ln(\alpha/\tan\beta) \tag{8.38}$$

where α is the catchment area draining to the grid cell and β is the slope angle of the cell in degrees.

The TI refines the saturation excess interpretation of WI by accounting for the water-storage capability of soil in the grid cell (Walter et al., 2002). It was calculated by

$$\mathrm{TI} = \ln(\alpha/\tan\beta) - \ln(K_{sat}D) \tag{8.39}$$

where K_{sat} is the saturated hydraulic conductivity of the soil profile having depth D above a layer that restricts percolation such as a bedrock or compact soil layer.

An empirical sediment factor was calculated based on soil survey attributes by

$$\text{Sediment factor} = D_{50}/\mathrm{RKLS} \tag{8.40}$$

where D_{50} is the median particle diameter (mm) of the surface soil by texture class, and R, K, L and S are rainfall erosivity, soil erodibility, slope length and slope steepness factors, respectively, from the revised universal soil loss equation in English units (Renard et al., 1997).

Sediment trapping efficiency (STE), as a percent of input load, was estimated in a buffer under otherwise standard conditions by

$$\mathrm{STE} = 100 - 85\,e^{-1320(\text{Sediment Factor})} \tag{8.41}$$

An empirical infiltration factor was calculated by

$$\text{Infiltration factor} = K_{sat}^2/\mathrm{RLS} \tag{8.42}$$

where K_{sat} is the saturated hydraulic conductivity (in $hour^{-1}$) of the surface soil layer.

Water trapping efficiency (WTE), in percent of input volume, in a buffer under the standard conditions described above was calculated by

$$\mathrm{WTE} = 97\,(\text{Infiltration Factor})^{0.26} \tag{8.43}$$

Results showed that each index associated higher pollution risk and mitigation potential to a different part of the landscape. For watersheds where pollutant loading was generated by both saturation-excess (emphasised by TI and WI) and infiltration-excess processes (emphasised by STE and WTE), the indices were complementary.

References

Ayers, R.S., Westcot, D.W., 1985. Water Quality for Agriculture, 174 pp. Rome, FAO, Paper 29. rev. 1.

Ayers, R.S., Westcot, D.W.A., 1999. Qualidade da Águana Agricultura, second ed., Campina Grande: UFPB. 218 p. (Estudos FAO: Irrigacao e Drenagem, 29).

Bonnet, B.R.P., Ferreira, L.G., Lobo, F.C., 2008. Water quality and land use relations in Goias: A watershed scale analysis. Revista Arvore 32 (2), 311–322.

Brown, R.M., McClelland, N.I., Deininger, R.A., Tozer, R.G., 1970. A water quality index – do we dare? Water Sewage Works 117, 339–343.

Canadian Council of Ministers of the Environment (CCME), 2001. Canadian water quality index 1.0 technical report and user's manual. Canadian Environmental Quality Guidelines Water Quality Index Technical Subcommittee, Gatineau, QC, Canada.

Carbajal, J.J., Sanchez, L.P., 2008. Classification based on fuzzy inference systems for artificial habitat quality in shrimp farming. In: 7th Mexican International Conference on Artificial Intelligence – Proceedings of the Special Session. MICAI 2008, pp. 388–392.

Debels, P., Figueroa, R., Urrutia, R., Barra, R., Niell, X., 2005. Evaluation of water quality in the ChillÃ¡n River (Central Chile) using physicochemical parameters and a modified Water Quality Index. Environmental Monitoring and Assessment 110 (40603), 301–322.

Dee, N., Baker, J., Drobny, N., Duke, K., Fahringer, D., 1972. Environmental evaluation system for water resource planning (to Bureau of Reclamation, U.S. Department of Interior). Battelle Columbus Laboratory, Columbus, Ohio. January, 188 pages.

Dee, N., Baker, J., Drobny, N., Duke, K., Whitman, I., Fahringer, D., 1973. An environmental evaluation system for water resource planning. Water Resource Research 9 (3), 523–535.

Deng, J., Yang, L., 2009. Irrigation Security of Reclaimed Water Based on Water Quality in Beijing. Environ. Sci. & Eng. Coll., Huangshi Inst. of Technol., Huangshi, China. CORD Conference Proceedings, 1–4.

Dosskey, M.G., Qiu, Z., 2011. Comparison of indexes for prioritizing placement of water quality buffers in agricultural watersheds1. JAWRA Journal of the American Water Resources Association 47 (4), 662–671.

European Union (EU), 1975. Council Directive 75/440/EEC of 16 June 1975 concerning the quality required of surface water intended for the abstraction of drinking water in the Member States. Official Journal L 194 25/07/1975, 0026–0031.

Hale, S.S., Paul, J.F., Heltshe, J.F., 2004. Watershed landscape indicators of estuarine benthic condition. Estuaries and Coasts 27, 283–295.

Helena, B., Pardo, R., Vega, M., Barrado, E., Fernandez, J.M., Fernandez, L., 2000. Temporal evaluation of groundwater composition in an alluvial aquifer (Pisuerga River, Spain) by principal component analysis. Water Research 34 (3), 807–816.

Hernández-Romero, A.H., Tovilla-Hernández, C., Malo, E.A., Bello-Mendoza, R., 2004. Water quality and presence of pesticides in a tropical coastal wetland in southern Mexico. Marine Pollution Bulletin 48 (11–12), 1130–1141.

House, M.A., Ellis, J.B., 1987. The development of water quality indices for operational management. Water Science and Technology 19 (9), 145–154.

Inhaber, H., 1974. Environmental quality: outline for a national index for Canada. Science (New York, N.Y.) 186 (4166), 798–805.

Johanson, E.E., Johnson, J.C., 1976. Contract, (68-01-2920). USEPA, Washington DC, USA.

Jonnalagadda, S.B., Mhere, G., 2001. Water quality of the odzi river in the Eastern Highlands of Zimbabwe. Water Research 35 (10), 2371–2376.

Kallidaikurichi, S., Rao, B., 2010. Index of drinking water adequacy for the Asian economies. Water Policy 12 (S1) 135–154.

Liou, S.-M., Lo, S.-L., Wang, S.-H., 2004. A generalized water quality index for Taiwan. Environmental Monitoring and Assessment 96 (40603), 35–52.

Meireles, A.C.M., de Andrade, E.M., Chaves, L.C.G., Frischkorn, H., Crisostomo, L.A., 2010. A new proposal of the classification of irrigation water. Revista Ciencia Agronomica 41 (3), 349–357.

Nives, S.G., 1999. Water quality evaluation by index in Dalmatia. Water Research 33, 3423–3440.

Ocampo-Duque, W., Ferré-Huguet, N., Domingo, J.L., Schuhmacher, M., 2006. Assessing water quality in rivers with fuzzy inference systems: a case study. Environment International 32 (6), 733–742.

Ott, W.R., 1978. Environmental Indices: Theory and Practice. Ann Arbor Science Publishers Inc, Ann Arbor, MI.

Pesce, S.F., Wunderlin, D.A., 2000. Use of water quality indices to verify the impact of Cordoba City (Argentina) on Suquia River. Water Research 34 (11), 2915–2926.

Pryazhinskaya, V.G., Yaroshevskii, D.M., 1996. A conception of developing a system to simulate functioning water management systems of river basins with allowance made for water quality indices. Water Resources 23 (4), 449–456.

Ravichandran, S., Ramanibai, R., Pundarikanthan, N.V., 1996. Ecoregions for describing water quality patterns in Tamiraparani basin, South India. Journal of Hydrology 178 (40634), 257–276.

Renard, K.G., Foster, G.R., Weesies, G.A., McCool, D.K., Yoder, D.C., 1997. Predicting Soil Erosion by Water: a Guide to Conservation Planning with the Revised Universal Soil Loss Equation (RUSLE). Agric. Handb. no. 703. USDA, Washington, DC.

Sanchez, E., Colmenarejo, M., Vicente, J., Rubio, A., Garcia, M., Travieso, L., Borja, R., 2007. Use of the water quality index and dissolved oxygen deficit as simple indicators of watersheds pollution. Ecological Indicators 7 (2), 315–328.

Shiels, D.R., 2010. Implementing landscape indices to predict stream water quality in an agricultural setting: an assessment of the Lake and River Enhancement (LARE) protocol in the Mississinewa River watershed,

East-Central Indiana. Ecological Indicators 10 (6), 1102–1110.

Simsek, C., Gunduz, O., 2007. IWQ Index: A GIS-integrated technique to assess irrigation water quality. Environmental Monitoring and Assessment 128 (40603), 277–300.

Terrado, M., Barceló, D., Tauler, R., Borrell, E., Campos, S.D., 2010. Surface-water-quality indices for the analysis of data generated by automated sampling networks. TrAC – Trends in Analytical Chemistry 29 (1), 40–52.

Truett, J.B., Johnson, A.C., Rowe, W.D., Feigner, K.D., Manning, L.J., 1975. Development of water quality management indices. Journal of the American Water Resources Association 11 (3), 436–448.

Vega, M., Pardo, R., Barrado, E., Deban, L., 1998. Assessment of seasonal and polluting effects on the quality of river water by exploratory data analysis. Water Research 32 (12), 3581–3592.

Verlicchi, P., Masotti, L., Galletti, A., 2011. Wastewater polishing index: A tool for a rapid quality assessment of reclaimed wastewater. Environmental Monitoring and Assessment 173 (1-4), 267–277.

Walter, M.T., Steenhuis, T.S., Mehta, V.K., Thongs, D., Zion, M., Schneiderman, E., 2002. Refined conceptualization of TOPMODEL for shallow subsurface flows. Hydrological Processes 16, 2041–2046.

World Development Indicators (WDI), Swanson, E., 2006. http://www-wds.worldbank.org/external/default/WDSContentServer/IW3P/IB/2010/04/21/000333037_20100421010319/Rendered/PDF/541650WDI0200610Box345641B01PUBLIC1.pdf.

Wunderlin, D.A., Díaz, M.P., Amé, M.V., Pesce, S.F., Hued, A.C., Bistoni, M.A., 2001. Pattern recognition techniques for the evaluation of spatial and temporal variations in water quality. A case study: Suquía River Basin (Córdoba – Argentina). Water Research 35, 2881–2894.

Zoeteman, B.C.J., 1973. Potential pollution index as a tool for river water. Quality Management, 336–350.

CHAPTER

9

Indices for Assessing Groundwater Quality

OUTLINE

Water Quality Indices DOI: 10.1016/B978-0-444-54304-2.00009-9

9.1. INTRODUCTION

The water-quality standards which are set for different purposes – drinking, swimming, irrigation, industrial use, etc., – apply to all water sources, terranian as well as subterranian. As water-quality indices (WQIs) are either developed with reference to standards, or are meant to assist in the setting of standards, the approaches to developing indices of groundwater quality are very similar to the ones used for surface water quality. The difference lies mainly in parameter selection.

Whereas parameters such as coliforms and BOD feature in most of the surface WQIs, they are rarely covered in ground WQIs. On the other hand, most ground WQIs incorporate minerals such as boron and arsenic which are rarely included in surface WQIs.

9.2. THE WQI OF TIWARI AND MISHRA (1985)

The WQI of Tiwari and Mishra (1985) was not specific to groundwater but over the years it has been used extensively for groundwater-quality assessment, mainly in different regions of India, but outside as well (Ketata et al., 2011). The steps associated with the development of this index are as follows:

1. *Parameter selection*: This has been mostly done subjectively by different authors, choosing parameters which past experience had indicated to be of importance in their regions. For example, in areas where groundwater was known to be high in one or more elements such as boron and iodine, those elements were included.
2. *Assignment of weightage*: This is done using the equations

$$W_i = \frac{K}{O_i} \tag{9.1}$$

$$K = \frac{1}{(\sum 1/O_i)} \tag{9.2}$$

where W_i is the weight, K is a constant and O_i corresponds to WHO (World Health Organization) or ICMR (Indian Council of Medical Research) standards of the parameters.

The quality rating Q_i is given by

$$q_i = \frac{(V_{\text{actual}} - V_{\text{ideal}})}{(V_{\text{standard}} - V_{\text{ideal}})} \times 100 \tag{9.3}$$

where q_i is the quality rating of the ith parameter for a total of n water samples, V_{actual} is the value of the water-quality parameter obtained from the laboratory analysis of the sample and V_{standard} is the value of the water-quality parameter obtained from the water-quality standard. The value of V_{ideal} is 7 for pH and zero for all other parameters. O_i corresponds to the WHO/ICMR standard value of the parameters.

3. *Aggregation:* The index is obtained with the function:

$$\text{WQI} = \text{antilog} \sum_{i=1}^{n} w_i \log q_i \tag{9.4}$$

The authors who have used this procedure in recent years to develop ground WQIs include Ramachandramoorthy et al. (2010) and Srivastava et al. (2011).

9.3. ANOTHER OFT-USED GROUNDWATER-QUALITY INDEX DEVELOPMENT PROCEDURE

Another procedure of developing a groundwater-quality index has been used by several authors, for example, Soltan (1999), Ramakrishnaiah et al. (2009), Banoeng-Yakubo et al. (2009), Vasanthavigar et al. (2010), Banerjee and Srivastava (2011) and others. Giri et al. (2010) have used it to develop a 'metal pollution

TABLE 9.1 Assignment of Weights in the Ground Water WQI of Vasanthavigar et al. (2010)

Water Quality Parameters	The Water Quality Standard (BIS 10500 1991)	Weight (w_i)	Relative Weight
Total dissolved solids	500	5	0.116
Bicarbonate (mg/L)	–	1	0.023
Chloride (mg/L)	250	5	0.116
Sulphate (mg/L)	200	5	0.116
Phosphate (mg/L)	–	1	0.023
Nitrate (mg/L)	45	5	0.116
Fluoride (mg/L)	1	5	0.116
Calcium (mg/L)	75	3	0.070
Magnesium (mg/L)	30	3	0.070
Sodium (mg/L)	–	4	0.093
Potassium (mg/L)	–	2	0.047
Silicate (mg/L)	–	2	0.047
		$\Sigma w_i = 41$	$\Sigma W_i = 0.953$

index' (MPI) and applied it to the study of Bagjata mining area in India. It is described below with the example of the version developed by Vasanthavigar et al. (2010).

The authors chose 12 parameters (TDS, HCO_3, Cl, SO_4, PO_4, NO_3, F, Ca, Mg, Na, K and Si) and assigned weights (w_i) according to their perceived importance in the overall quality of water for drinking purposes (Table 9.1). The Bureau of Indian Standards (BIS) stipulations formed the reference point.

A maximum weight of 5 was assigned to nitrate, total dissolved solids, chloride, fluoride and sulphate considering that these often influence groundwater quality the most. Bicarbonate and phosphate were given the minimum weight of 1 as the two rarely play a significant role in groundwater quality. Calcium, magnesium, sodium and potassium were assigned weight ranging between 1 and 5. The relative weights (W_i) of each parameter (Table 9.1) were then computed from the following equation:

$$W_i = w_i / \sum_{i=1}^{n} w_i \tag{9.5}$$

where W_i is the relative weight;

w_i is the weight of the ith parameter ($i = 1 \ldots n$) parameter and
n is the number of parameters.

In the third step, a quality rating scale (q_i) for each parameter was assigned using the equation:

$$q_i = (C_i/S_i) \times 100 \tag{9.6}$$

where q_i is the quality rating, C_i is the concentration of each chemical parameter in each water sample in milligrams per litre, S_i is the drinking-water standard for each chemical parameter (mg/L) according to the guidelines of the BIS 10500 (1991).

TABLE 9.2 Index Scores and the Water Quality They Represent (Vasanthavigar et al., 2010)

Range	Type of Water
<50	Excellent water
50–100.1	Good water
100–200.1	Poor water
200–300.1	Very poor water
>300	Water unsuitable for drinking

For computing the WQI, the SI_i was first determined for each chemical parameter, which was then used to determine the WQI:

$$SI_i = W_i \times q_i \qquad (9.7)$$

$$\text{WQI} = \sum SI_i \qquad (9.8)$$

Where SI_i is the subindex of the ith parameter;

q_i is the rating based on concentration of ith parameter and
n is the number of parameters.

The WQI range and the corresponding type of water were set as in Table 9.2.

Studies on 148 samples by the authors revealed that their WQI followed the trend of chloride and EC. This indicted that chloride and EC can serve as indicators of groundwater pollution.

9.4. INDEX OF AQUIFER WATER QUALITY (MELLOUL AND COLLIN, 1998)

An index of aquifer water quality (IAWQ) has been presented by Melloul and Collin (1998) for assessing empirical regional groundwater quality, simultaneously utilising data values of a number of chemical parameters characterising salinity and pollution.

In order to transform raw chemical data into rating values (Y) as regards standards, each value of a parameter, P_{ij} (field data value parameter i in cell j) was related to its desired standard value P_{id} vis a vis drinking, irrigation and other water purposes (WHO, 1993). Each relative value, X_{ij}, was estimated as

$$X_{ij} = P_{ij}/P_{id} \qquad (9.9)$$

To express X_{ij} as a corresponding index rating value, related to groundwater quality, Y_i was assigned to each X_{ij} value as follows:

- for high water quality, with X_{ij} equal to $0{\cdot}1$, the corresponding index rating value would be around 1;
- for acceptable water quality, with X_{ij} equal to 1 (the raw value of the parameter P_i equal to its standard desired value), the corresponding index rating value of such water would be 5 and
- for unacceptable groundwater quality, with X_{ij} equal to or higher than 3.5 (the initial value of the parameter P_i equal to or higher than 3.5 times its standard desired value), the corresponding index rating value would be 10.

Based on the operational hydrological experience, the authors thought that $Y_1 = 1$ for $X_1 = 0.1$; $Y_2 = 5$ for $X_2 = 1$ and $Y_3 = 10$ for $X_3 = 3.5$.

For any parameter i in any cell j an adjusted parabolic function of rates $Y_{ij} = f(X_{ij})$ was determined for each cell as:

$$Y_i = -0.712X_i^2 + 5{\cdot}228X_i + 0.484 \qquad (9.10)$$

From this equation the corresponding rating Y_i can be estimated for any value of X_i. Thus, after this transformation of the field data the index formula will involve only Y values, representing input data for the next step in the development of the indexation formula.

The proposed IAWQ formula to numerically assess any groundwater-quality situation was stated as a summation of weights multiplied by respective ratings of various parameters i for each cell j as follows:

$$IAWQ_j = C/n\left[\sum_{i=1}^{n}(Wri.Yri)\right] \quad (9.11)$$

where C is a constant, used to ensure desired range of numbers (in this case, $C = 10$); i, n *is the* number of chemical parameters involved ($i = 1, \ldots, n$). This value has been incorporated in the denominator to average the data; Wri is the relative value of $W_i/Wmax$, where W_i is a weight for any given parameter and $Wmax$ is the maximum possible weight. Lower numerical values define lower pollution potential, whilst higher values define heightened pollution potential, hence a W_i value would be larger if a given parameter were toxic or hazardous to groundwater quality.

The values of $Wmax$ and $Ymax$ were incorporated into Equation (9.3) to represent W and Y values as related to a reference level, as well as to ensure that the ultimate IAWQ value remains within a scale of 1–10, in order to assess the relative level of salinisation and pollution. With this, the IAWQ values can be more readily compared from one site to the other, while providing a means of determining the relative influence of additional parameters upon groundwater quality.

In this manner, to pinpoint a source of contamination in areas potentially vulnerable to pollution, applicable 'fingerprint' or 'indicator' chemical parameters can be utilised to identify such specific sources of pollution as industrial and solid-waste sites. For example, in an industrial region, such chemical parameters as heavy metals, organics, etc., could serve as indicators of industrial activity. However, due to the ubiquitous availability of data on indicator parameters Cl and NO_3, these parameters can be applied in an initial step towards IAWQ formula application in order to gain a broad appreciation of the water quality in a region.

The equation, for estimating IAWQ when considering only Cl and NO_3 for a cell (j), reduces to

$$IAWQ_j = C/n[(W_{Clr}.Y_{Clr}) + (W_{NO_3r}.Y_{NO_3r})]j \quad (9.12)$$

The application of the index was tested by the authors in the assessment of impacts of extensive irrigation, water drawl and salinity intrusion in the Sharon area.

9.5. GROUNDWATER-QUALITY INDEX OF SOLTAN (1999)

Soltan (1999) proposed a WQI based on nine water-quality parameters including heavy metals (NO^{3-}, PO^{3-}, Cl^-, TDS, BOD, Cd, Cr, Ni and Pb) to assess the water quality from artesian wells located near Dakhla oasis, Egypt. The indices for individual parameters were calculated as follows:

$$WQI = \sum_{i=1}^{n} q_i \quad (9.13)$$

where

$$q_i = 100 \times \frac{V_i}{S_i} \quad (9.14)$$

The average water-quality index for n parameters was calculated using the expression

$$AWQI = \frac{\sum_{i=1}^{n} q_i}{n} \quad (9.15)$$

where n is the number of parameters and q_i is quality rating for the ith parameter. V_i is the observed value of the ith parameter and S_i is the water-quality standard for the ith parameter.

The permissible or critical pollution index value was set at 100. The AWQI has a value of 0 when all pollutants are absent, and a value of 100 when all pollutants reach their permissible limits. AWQI values exceeding 100 indicate that the water sample may suffer from serious pollution problems.

9.6. A GROUNDWATER CONTAMINATION INDEX

In order to aid the depiction of level of groundwater contamination at various regions on a map, a contamination index, (C_d), has been developed to provide a general view of the extent of groundwater contamination (Backman et al. 1998). The parameters chosen for the calculation of the index are the ones that are considered potentially harmful. Two indices are calculated, one based on those parameters that pose a health risk and the other based on technical—aesthetic aspects.

For the calculation of the contamination index for health-risk aspects, F^-, NO_3^-, UO_2^{2-}, As, B, Ba, Cd, Cr, Ni, Pb, Rn and Se have been considered. For the other index based on technical—aesthetic aspects, the parameters considered are T. D. S, SO_4^{2-}, Cl^-, F^-, NO^{3-}, NH_4^+, Al, As, Ba, Cd, Cr, Cu, Fe, Hg, Mn, Pb, Sb, Se and Zn.

Either index is given by

$$C_d = \sum_{i=1}^{n} C_{fi} \tag{9.16}$$

where

$$C_{fi} = \frac{C_{Ai}}{C_{Ni}} - 1 \tag{9.17}$$

where C_{fi} is the contamination factor for the ith component,

C_{Ai} is the analytical value of the ith component and

C_{Ni} is the upper permissible concentration for the ith component, N representing the normative value.

9.7. AN INDEX FOR SURFACE WATER AS WELL AS GROUNDWATER QUALITY

An index for Dalmatia region, Croatia, developed by Stambuk-Glijanovik (1999) addresses the issues of surface water quality as well groundwater quality.

The index is given by the equation:

$$WQI = \frac{WQE}{WQE_{MAC}} \tag{9.18}$$

In it the tested water-quality evaluation (WQE) is divided by the water-quality evaluation WQE_{MAC} which satisfies the maximum admissible concentration (MAC) of first-class water according to the standard for drinking water.

The Croatian standard consists of four classes. The first one includes underground and surface waters which can be used in the natural state or after being disinfected for drinking water. The second one includes underground and surface waters which cannot be used for drinking purposes without first being treated. The third group includes water which cannot be used either in its natural state or after being treated. The fourth class is not used.

WQE is calculated by summing up individual quality ratings (q_i) and weighting these parameters in total quality evaluation (w_i):

$$WQE = \sum_{i=1}^{n} q_i w_i \tag{9.19}$$

where

$\sum_{i=1}^{n} q_i w_i$ is the weighed sum,

q_i is the water-quality score of parameter i; w_i is the weighting factor of parameter i and n is the number of parameters. Nine water-quality parameters were used by the author to determine the water-quality index (WQI): temperature, mineralisation, corrosion coefficient, $K = (SO_4 + Cl)/HCO_3$, dissolved oxygen, biochemical oxygen demand, total nitrogen, protein, nitrogen, total phosphorus and total coliform (MPN coli/100 ml).

After determining the nine parameters, the results were recorded and transferred to a WQE table which contained the range of possible results of the parameters and their score values (Table 9.3). By summing up all parameters the water-quality evaluation was obtained.

The score values of each particular parameter as well as its weights were arbitrarily estimated, on the basis of a survey.

The grades present the weights, i.e., the percentage of *approximately* 100% water quality and do not depend on the MAC value.

If mineralisation, coefficient K and protein N are not determined, the index is evaluated in relation to the C_{80} concentration which includes 80% of the results and is calculated from the equation:

$$C_{80} = \overline{C} + t\sigma \tag{9.20}$$

in which $\overline{C}$ is the mean value, σ is the standard deviation and t is the value of a Student t-test for 80% of probability level.

The oxygen coefficient is calculated from

$$\alpha \text{ saturation } O_2 = MAC/(C - t\sigma)$$

The reciprocal coefficient value is employed because the decrease in oxygen content deteriorates the water quality.

9.8. USE OF GROUNDWATER-QUALITY INDEX, CONTAMINATION INDEX AND CONTAMINATION RISK MAPS FOR DESIGNING WATER-QUALITY MONITORING NETWORKS

Three assessment parameters – areas of high vulnerability to groundwater pollution (such as urban clusters and industrial belts), presence of potential sources of contamination and contamination indices – have been used by Ramos Leal et al. (2004) to develop a methodology for the design of groundwater-quality-monitoring networks.

The vulnerability was assessed using DRASTIC method (which covers Depth to water table, net Recharge, Aquifar media, Soil media, Topographic slope, Impact of vadose zone media and hydranlic Conductivity) and another geohydrological procedure SINTACS (Aller et al., 1985; Civita and De Maio, 1997).

The water-quality index had the form

$$ICA = k\frac{\sum C_i P_i}{\sum P_i} \tag{9.21}$$

where C_i is function percentage value assigned to the parameters according to the concentrations, P_i is the weight assigned to each parameter and k is a constant that takes the values shown in Table 9.4. The constant, k, is related to aesthetic characteristics of the sample, such as the water's appearance and odour.

The minimum value ICA can take is 0 and the maximum is 100. A smaller ICA represents lower water quality and a higher ICA represents higher water quality.

The contamination index (Backman et al., 1998) used by the authors has the form

$$C_d = \sum_{i=1}^{n} C_{fi} \tag{9.22}$$

where

$$C_{fi} = \frac{C_{Ai}}{C_{Ni}} - 1 \tag{9.23}$$

C_{fi} is contamination factor for the ith component, C_{Ai} is analytic value of the ith component and C_{Ni} is the permissible highest concentration of the ith component.

The higher the value of C_d, the higher the contamination of the resource.

TABLE 9.3 Selection of Parameters, the Possible Range of Investigations of Parameters and Their Scores for Water Quality Evaluation as Used by Stambuk-Glijanovik (1999)

q_iw_i	Temperature (°C)	Mineralisation (mg/L)	(Cl + SO_4)/alk.	Dissolved Oxygen % Saturation	BOD5 (mg/L)	Total N (mg/L)	Protein N (mg/L)	Total P (mg/L)	MPN coli/100 ml	q_iw_i
	7	7	6	16	10	16	10	12	16	
16				90–105		0.0–0.06			0–50	16
15				90–87; 105–115		0.06–0.1			50–200	15
14				87–84; 115–125		0.10–0.15			200–400	14
13				84–80; 125–130		0.15–0.2			400–600	13
12				80–77; 130–135		0.2–0.3		0–0.03	600–900	12
11				77–74; 135–140		0.3–0.4		0.03–0.06	900–1200	11
10				74–71; 140–145	0–1.2	0.4–0.5	0.0–0.03	0.06–0.1	1200–1800	10
9				71–66; 154	1.2–2.0	0.5–0.6	0.03–0.05	0.1–0.15	1800–2500	9
8				66–63	2.0–2.5	0.6–0.7	0.05–0.08	0.15–0.18	2500–3500	8
7	8–12	350		63–60	2.5–3.0	0.7–0.8	0.08–0.10	0.18–0.22	3500–5000	7
6	6–8; 12–15	350–500	0–0.25	60–55	3.0–3.5	0.8–0.9	0.10–0.13	0.22–0.26	5000–7000	6
5	4–6; 15–17	500–800	0.25–0.4	55–50	3.5–4.0	0.9–1.0	0.13–0.16	0.26–0.3	7000–10000	5
4	−4; 17–19	800–1200	0.4–0.65	50–45	4.0–4.5	1.0–1.2	0.16–0.20	0.3–0.35	10000–15000	4
3	19–21	1200–1500	0.65–1.5	45–35	4.5–5.5	1.2–1.5	0.20–0.25	0.35–0.4	15000–20000	3
2	21–23	1500–2000	1.5–3	35–25	5.5–6.5	1.5–2	0.25–0.50	0.4–0.5	20000–28000	2
1	23–25	2000–3000	3–6	25–10	6.5–8.0	2–3	0.50–1	0.5–1	28000–50000	1
0	+25	+3000	+6	−10	+8	+3	+1	+1	+50000	0

TABLE 9.4 Illustrative Example of the Standardisation Procedure Used by Stigter et al. (2006)

Sample no.	Concentration of NO_3^-, mg/L	NO_3^-		
		≤GL*	GL–MAC@	>MAC
1	31	0	1	0
2	135	0	0	1
3	6	1	0	0

*guide level, @ = maximum admissible concentration.

9.9. ATTRIBUTE REDUCTION IN GROUNDWATER-QUALITY INDICES BASED ON ROUGH SET THEORY

Xiong et al (2005) have applied rough set (RS) theory, which is a mathematical tool for analysing uncertainty and imprecision, for attribute reduction of groundwater-quality indices. For this, the significance of the attributes to groundwater-quality index and the groundwater-quality class has been explored. A discernibility matrix is used to arrive at the attribute reduction. The authors feel that the RS theory is simple, precise and practical for attribute reduction in the groundwater-quality indices.

9.10. INDEX DEVELOPMENT USING CORRESPONDENCE FACTOR ANALYSIS

Stigter et al. (2006) distributed groundwater-quality data obtained from two study areas in Portugal among three classes for each parameter involved, on the basis of drinking-water guidelines. The first class had concentrations below the guide level (GL), whereas the third class has concentrations above the maximum admissible concentration (MAC) for each parameter, as defined by the former drinking water directive of the European Union (80/778/EEC). The second class had concentrations in between the two guideline classes. The standardisation procedure was then performed by applying a simple binary codification: 1 if the sample belonged to a class, 0 if not. Table 9.5 illustrates

TABLE 9.5 Characterisation of the Data Set Used for the Selection of the Variables by Stigter et al. (2006) for Incorporation in the GWQI

	pH	EC (μS/cm)	Na^+ (mg/L)	K^+ (mg/L)	Mg^{2+} (mg/L)	Ca^{2+} (mg/L)	NH_4^+ (μg/L)	Cl^- (mg/L)	HCO_3^- (mg/L)	SO_4^{2-} (mg/L)	NO_3^- (mg/L)	PO_4^{3-} (μg/L)
Minimum	6.4	773	27	0	6	43	0	50	77	16	5	0
1st Quartile	6.9	1170	72	2	28	137	31	119	307	65	50	19
2nd Quartile	7.1	1560	115	3	34	172	51	184	370	119	79	52
3rd Quartile	7.3	2150	166	5	47	248	120	339	425	202	151	114
Maximum	8.5	6400	702	60	141	566	250	2077	570	437	581	5100
PV 98	6.5–9.5	2500	200	–	–	–	500	250	–	250	50	–
MAC 80	9.5[a]	–	150	12	50	200[b]	500	–	–	250	50	6691
GL 80	6.5–8.5	400	20	10	30	100	50	25	–	25	25	535
% Exceed PV	0%	15%	29%	8%	21%	33%	0%	35%	–	15%	75%	0%

PV 98 = parametric value in 1998 EU Directive; MAC 80, GL 80 = maximum admissible concentration and guide level in 1980 EU Directive.
[a] *Defined in Portuguese legislation.*
[b] *Based on Portuguese MAC for hardness.*

the procedure for nitrate, whose GL and MAC are 25 and 50 mg/L, respectively.

In the next step, two standard water samples were defined of extremely high and low quality; the first class (≤GL) was assigned to all parameters for the high-quality sample, whereas the low-quality sample was located in the third class (>MAC). The values were then aggregated by subjecting the standard and real samples to correspondence factor analysis (CFA).

CFA was used as it is able to identify the underlying pattern of relationships within a data set. This is basically done by rearranging the data into a small number of uncorrelated 'components' or 'factors' that are extracted from the data by statistical transformations. Such transformations involve the diagonalisation of some kind of similarity matrix of the parameters, such as a correlation or variance–covariance matrix (Brown, 1998; Pereira and Sousa, 2000). Each factor describes a certain amount of the statistical variance of the analysed data and is interpreted according to the intercorrelated parameters. The main advantage of CFA is that symmetry is conferred to the data matrix (Benzecri, 1977; Pereira and Sousa, 2000), thus permitting the simultaneous study of correlations within and between variables and samples.

Stigter et al. (2006) performed the diagonalisation on the similarity matrix of the two standard samples, as this resulted in the extraction of a single eigenvector explaining 100% of the data variance and diametrically opposing the high- and low-quality samples.

Subsequently, the real water samples were orthogonally projected on the extracted factor, in order to define the degree of association between those and the two quality standards. The resulting scores corresponded to the final index values, which ranged between −1 (high quality) and 1 (low quality) and were discrete rational numbers. The authors believe that although the process may appear complicated at first sight, its application is actually rather straightforward and involves a relatively low extent of data processing. The entire CFA routine can be run in a single software, and the orthogonal projection (and index calculation) can be mathematically expressed by the equation

$$F_i = \frac{1}{p\sqrt{\lambda}} \sum_{j=1}^{m} \delta_j L_j \tag{9.24}$$

where F_i is sample i's factor score, p is the number of parameters involved in the index construction, λ is the factor eigenvalue, δ_j is the Boolean code, ($\delta_j = 1$ if sample belongs to parameter class j, $\delta_j = 0$ if not), L_j is the factor loading of class j, m is number of classes ($=3p$).

TABLE 9.6 Water Quality Variables Which Were Subjected to Principal-Component Analysis as a Prelude to WQI Formulation by Mohamad Roslan et al. (2007)

Heavy Metal	Nonmetals	Physical Characteristic	Aggregate Indicator
Copper	Dissolved oxygen	Temperature	BOD
Zinc	pH	Salinity	COD
Iron	Ammoniacal nitrogen	Electric conductivity	Phenol
Lead	Nitrate	Turbidity	
Chromium trivalent	Nitrite	Total suspended solid	
Chromium hexavalent	Phosphate	Total dissolved solid	
Nickel	Sulphate		
Cobalt	Sulphide		
Manganese	Free chlorine		
Silver	Cyanide		
Tin	Arsenic		
Aluminium	Boron		
Mercury			

The characterisation of the data done by the authors for use in the GWQI is illustrated in Table 9.6.

9.11. INDICES FOR GROUNDWATER VULNERABILITY ASSESSMENT

A set of three indices – the intrinsic vulnerability index (IVI), the contaminant source index (CSI) and the well capture zone and receptor index (WI) – together with the IAWQ of Melloul and Collin (1998) have been used by Nobre et al. (2007) to develop a procedure for groundwater vulnerability and risk mapping (Figure 9.1).

The IAWQ has been described earlier in this chapter; Nobre et al. (2007) used chloride and nitrate as 'fingerprint' parameters to compute the IAWQ in their procedure. As may be seen from Figure 9.1, the system of indices has been used in conjunction with source–pathway–receptor risk chain analysis using GIS and process-based modelling. This methodology aims to integrate the controlling features that interfere along the contaminant pathway from source to receptor such as recharge, natural attenuation, soil, aquifer media and wells distribution. It uses a fuzzy hierarchy model to evaluate source hazard potential, enabling subjective reasoning to be incorporated and to rank the importance of each class of contaminant to groundwater impact.

9.12. GROUNDWATER-QUALITY INDEX TO STUDY IMPACT OF LANDFILLS

Mohamad Roslan et al. (2007) have created a ground WQI to study the impact of landfills

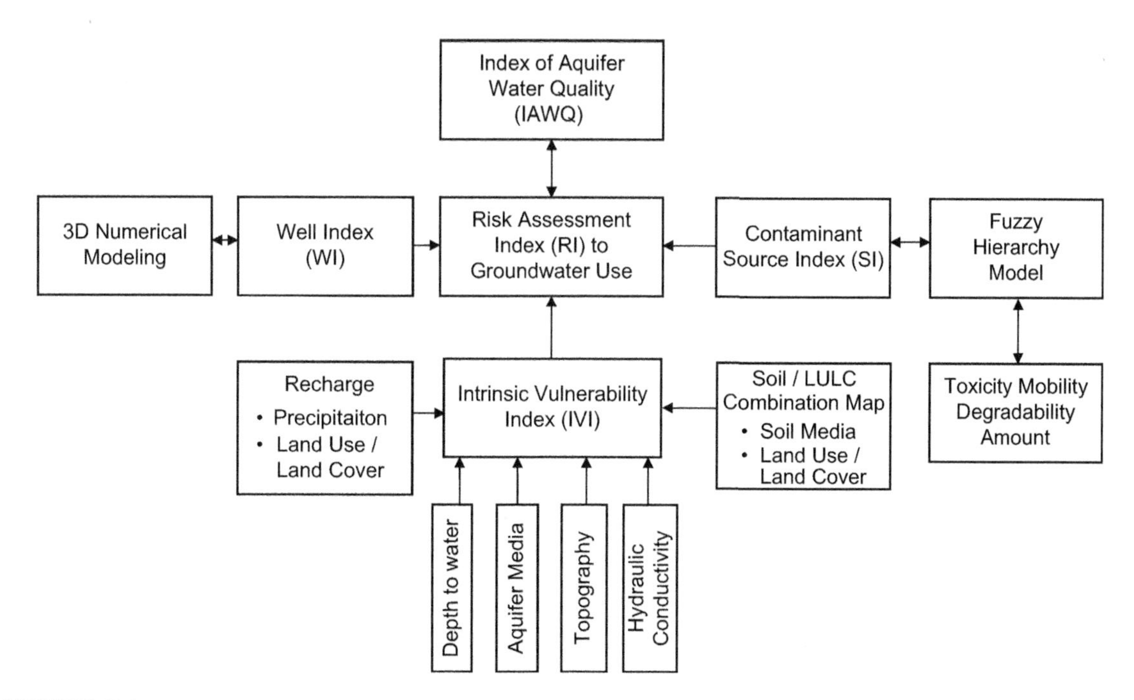

FIGURE 9.1 A procedure for groundwater vulnerability and risk mapping presented by Nobre et al. (2007) using a system of indices.

TABLE 9.7 'Benchmark' Scale Values Assigned in the Ground WQI of Mohamad Roslan et al. (2007)

Variable	Concentration Range	Benchmarking Scale Value
Electric conductivity	$x \leq 40$	10
	$40 < x < 20000$	$\frac{(\log 20000 - \log x)}{(\log 20000 - \log 40)} \times 10$
	$x \geq 2000$	0
Total dissolved solid	$x \leq 50$	10
	$50 < x < 1500$	$\frac{(\log 1500 - \log x)}{(\log 1500 - \log 50)} \times 10$
	$x \geq 1500$	0
Salinity 1	$x \leq 1$	10
	$1 < x < 20$	$\frac{(\log 20 - \log x)}{\log 20} \times 10$
	$x \geq 20$	10
Nitrate	$x \leq 1$	10
	$1 < x < 10$	$(1 - \log x) \times 10.0$
	$x \geq 10$	0
Nitrite	$x = 0$	10
	$0 < x < 1$	$(-\log x) / 3.001$
	$x \geq 1$	0
COD	$x \leq 1$	10
	$1 < x < 10$	$(1 - \log x) \times 10$
	$x \geq 10$	0
Ferum	$x = 0$	10
	$0 < x < 1$	$\frac{-\log x}{2.01} \times -10$
	$x \geq 1$	0

and have applied it to Sabak area in Malaysia.

The authors used water-quality data covering 32 variables (Table 9.7), collected from six sites for three years. Using principal-component analysis (PCA), they shortlisted 7 parameters (Table 9.8) which seemed to be influencing the water quality the most. A 'benchmarking' was then done by the authors with reference to the raw water-quality standard of the Government of Malaysia (Table 9.8).

The benchmarking encompassed a 0–10 scale; when the concentrations surpass the maximum limit set by the standards the value given is zero.

From the benchmarking range, 'radar plots' were drawn to determine the shape of the polygon within the range of each variable (Figure 9.2).

The groundwater-quality index corresponds to the percentage of the polygon area, which was calculated using the equation

TABLE 9.8 Values of k Corresponding to Physical Characteristics of Contamination (Conesa, 1993), Used in the Methodology of Ramos Leal et al. (2004)

k	Corresponding Aesthetic Impact
1.0	For clear water without apparent contamination.
0.75	For water with slight colour, scum, and cloudy, not natural appearance
0.50	For water with polluted appearance and strong odour
0.25	For black waters with evident fermentation and odours

$$A = \sum [0.5 \times \sin (360/7) \times \text{left value} \times \text{right value}] \tag{9.25}$$

where left and right refer to the sides of triangles within the polygon. For the study area, the index value was 26.67, representing an overall poor water quality.

Sensitivity analysis of the index revealed that for every drop of 0.1 in benchmarking value in any variable, the index would be decreased by 0.3.

9.13. INDICES FOR OPTIMISING GROUNDWATER-QUALITY MONITORING NETWORK

Yeh et al. (2008) have reported an attempt at developing a cost-effective programme for monitoring the quality of groundwater, which involves sampling from only a fraction of the existing monitoring wells. The aim is to obtain, at lesser costs than usual, sufficient information to determine the ambient water quality in the main aquifers.

The authors have used a multivariate geostatistical method to select a well network for

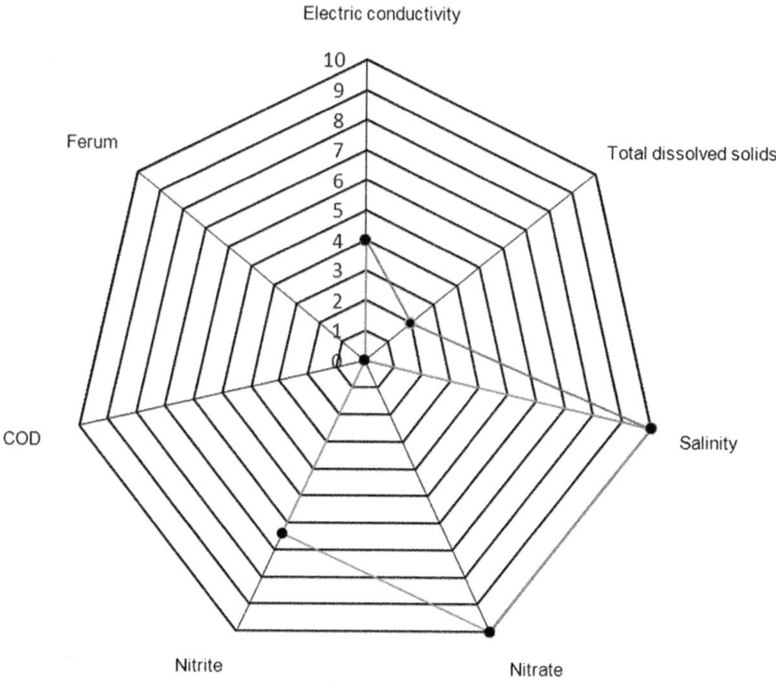

FIGURE 9.2 Radar plot for computation of the ground WQI of Mohamad Roslan et al. (2007).

monitoring groundwater quality on the basis of indices while considering multiple variables of multi-scale geostatistical structures. The study yields a rank of original groundwater-level monitoring wells for selection of least number of wells needed to obtain information of requisite representativeness.

9.14. ECONOMIC INDEX OF GROUNDWATER QUALITY BASED ON THE TREATMENT COST

Queralt et al. (2008) have presented an economic quality index for groundwater integrating various weighted analytical parameters to produce a value between 0 and 100. The index is useful in evaluating the state of and the variations in the quality of groundwater due to anthropogenic and natural causes, and incorporates estimation of costs likely in making the water potable.

9.15. THE INFORMATION-ENTROPY-BASED GROUNDWATER WQI OF PEI-YUE ET AL. (2010)

Following a logic similar to the one used by Taheriyoun et al. (2010) in developing a surface-water-quality index (chapter 6), Pei-Yue et al. (2010) have used the concept of information entropy in assigning weights for a WQI. The index was used for groundwater-quality assessment in Pengyang County, China, towards rational development of local groundwater resources and groundwater protection.

As a first step in developing their WQI, the authors assigned weights to each of the chosen 14 parameters using the concept of information entropy.

For m water samples taken to evaluate the water quality ($i = 1, 2, \ldots, m$), and for each sample having 'n' evaluated parameters ($j = 1, 2, \ldots, n$), eigen value matrix X was constructed:

$$X = \begin{bmatrix} x_{11} & x_{12} & \cdots & x_{1n} \\ x_{21} & x_{22} & \cdots & x_{2n} \\ \vdots & \vdots & \ddots & \vdots \\ x_{m1} & x_{m2} & \cdots & x_{mn} \end{bmatrix} \tag{9.26}$$

In order to eliminate the influence caused by the difference in different units of characteristic indices and different quantity grades, data pretreatment was done as follows.

According to the attributes of each index, the feature indices may be divided into four types: efficiency type, cost type, fixed type and interval type. For the efficiency type, the construction function of normalisation is

$$y_{ij} = \frac{x_{ij} - (x_{ij})_{\min}}{(x_{ij})_{\max} - (x_{ij})_{\min}} \tag{9.27}$$

For the cost type, the construction function of normalisation is

$$y_{ij} = \frac{(x_{ij})_{\max} - x_{ij}}{(x_{ij})_{\max} - (x_{ij})_{\min}} \tag{9.28}$$

After transformation, the standard-grade matrix Y is obtained:

$$Y = \begin{bmatrix} y_{11} & y_{12} & \cdots & y_{1n} \\ y_{21} & y_{22} & \cdots & y_{2n} \\ \vdots & \vdots & \ddots & \vdots \\ y_{m1} & y_{m2} & \cdots & y_{mn} \end{bmatrix} \tag{9.29}$$

The ratio of the index value of the jth index in the ith sample is

$$P_{ij} = y_{ij} / \sum_{i=1}^{m} y_{ij} \tag{9.30}$$

The information entropy is given by

$$e_j = -\frac{1}{\ln m} \sum_{i=1}^{m} P_{ij} \ln P_{ij} \tag{9.31}$$

The smaller the value of e_j, the bigger the effect of jth index. Then the entropy weight can be calculated as

TABLE 9.9 Classification of Groundwater in the Information-Entropy-Based WQI of Pei-Yue et al. (2010)

WQI	Rank	Water Quality
<50	1	Excellent water quality
50~100	2	Good water quality
100~150	3	Medium or average water quality
150~200	4	Poor water quality
>200	5	Extremely poor water quality

$$w_j = \frac{1 - e_j}{\sum_{j=1}^{n} (1 - e_j)} \quad (9.32)$$

Here w_j is defined as the entropy weight of jth parameter. The weights obtained by the authors are given in Table 9.9. In the next step for calculating WQI, assignment of quality-rating scale (q_j) for each parameter was done:

$$q_j = \frac{C_j}{S_j} \times 100 \quad (9.33)$$

where C_j is the concentration of each chemical parameter in each water sample (mg/L), and S_j is the Chinese Water Quality standard for drinking water of each parameter (mg/L). The WQI was calculated as:

$$\text{WQI} = \sum_{j=1}^{n} \omega_j q_j \quad (9.34)$$

Based on WQI, scores, groundwater was classified into five ranks, ranging from 'excellent' to 'extremely poor' (Table 9.10).

9.16. A WQI FOR GROUNDWATER BASED ON FUZZY LOGIC

Jinturkar et al. (2010) have studied groundwater quality at Chikhli, India, on the basis of a WQI developed by them using fuzzy logic. The index was developed along the lines similar to one used by Ocampo-Duque et al. (2006) and others for surface water fuzzy WQIs (Chapter 6). The basic steps are as in Figure 9.3.

The authors covered 7 water-quality parameters and developed triangular membership functions for all of them (Figure 9.4).

Using fuzzy sets associated with each parameter, and with the water-quality

TABLE 9.10 Index Range and the Findings of Water Quality Survey (Jinturkar et al., 2010)

The FWQI Range	Clarification of Water	Description	Percent of Samples Complying Hand Nos.
0.8–1	Excellent	All measurements are within objectives virtually all of the time	28
0.6–0.8	Good	Conditions rarely depart from natural or desirable levels	44
0.4–0.6	Fairly Good	Conditions sometimes depart from natural or desirable levels	28
0.2–0.4	Fair	Conditions often depart from natural or desirable levels	0
0–0.2	Poor	Conditions usually depart from natural or desirable levels	0

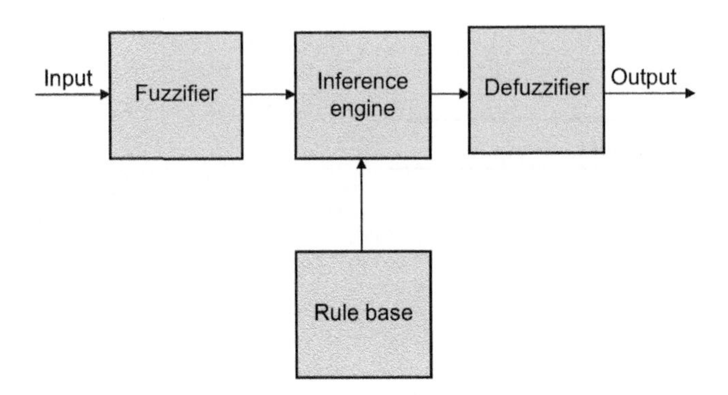

FIGURE 9.3 Main steps in the development of fuzzy WQI used by Jinturkar et al. (2010).

standards of the World Health Organization (WHO) and Indian Council of Medical Research (ICMR) as the guidelines, general IF–THEN rules were developed as exemplified below:

Rule #1: IF (pH is low) and (TDS is excellent) and (Ca is low) and (Mg is low) and (Hardness is low) and (Cl is low) THEN (water-quality index is poor).

Rule #76: IF (pH is medium) and (TDS is fair) and (Ca is low) and (Mg is low) and (Hardness is low) and (Cl is high) THEN (water-quality index is good).

In Rule #1, water-quality index is dictated by TDS as excellent, while pH, Ca, Mg, hardness and Cl are classified as low. For Rule #76, on the other hand, the pH, TDS and Cl are medium, fair and high, respectively, while Ca, Mg and hardness are low to yield 'good' level water-quality index. In this way every parameter was analysed for the development of rule base. A total of 82 rules were framed for testing the rule base.

9.17. USE OF WQI AND GIS IN AQUIFER-QUALITY MAPPING

Ketata et al. (2011) assigned a weight (w_i) to each of the ten parameters (pH, TDS, Cl, SO_4, HCO_3, NO_3, Ca, Mg, Na and K) based on their perceived effects on primary health. A maximum weight of 5 was assigned to dissolved solids, chloride, sulphate and nitrate due to their in controlling groundwater quality in most situations. Bicarbonate was given the minimum weight of 1 as it plays an insignificant role in groundwater quality assessment. Other parameters calcium, magnesium, sodium and potassium were assigned weights between 1 and 5 depending on their importance shaping water quality for drinking purposes.

The rest of the procedure for index development was identical to the one used by several authors earlier as described earlier in this chapter with reference to the WQI of Vasanthavigar et al. (2010).

The authors have integrated the WQI with geographic information systems (GIS) as depicted in Figure 9.5 to generate maps of groundwater quality of the El Khairat aquifer, Tunisia. Earlier reports on groundwater-quality assessment using WQI and GIS include the assessment of pesticides impacts in pine plantations, Australia (Pollock et al. (2005), effect of irrigation water quality in Turkey (Simsek and Gunduz, 2007) and assessment of groundwater quality in a highland village in Kerala, India (Hatha et al., 2009).

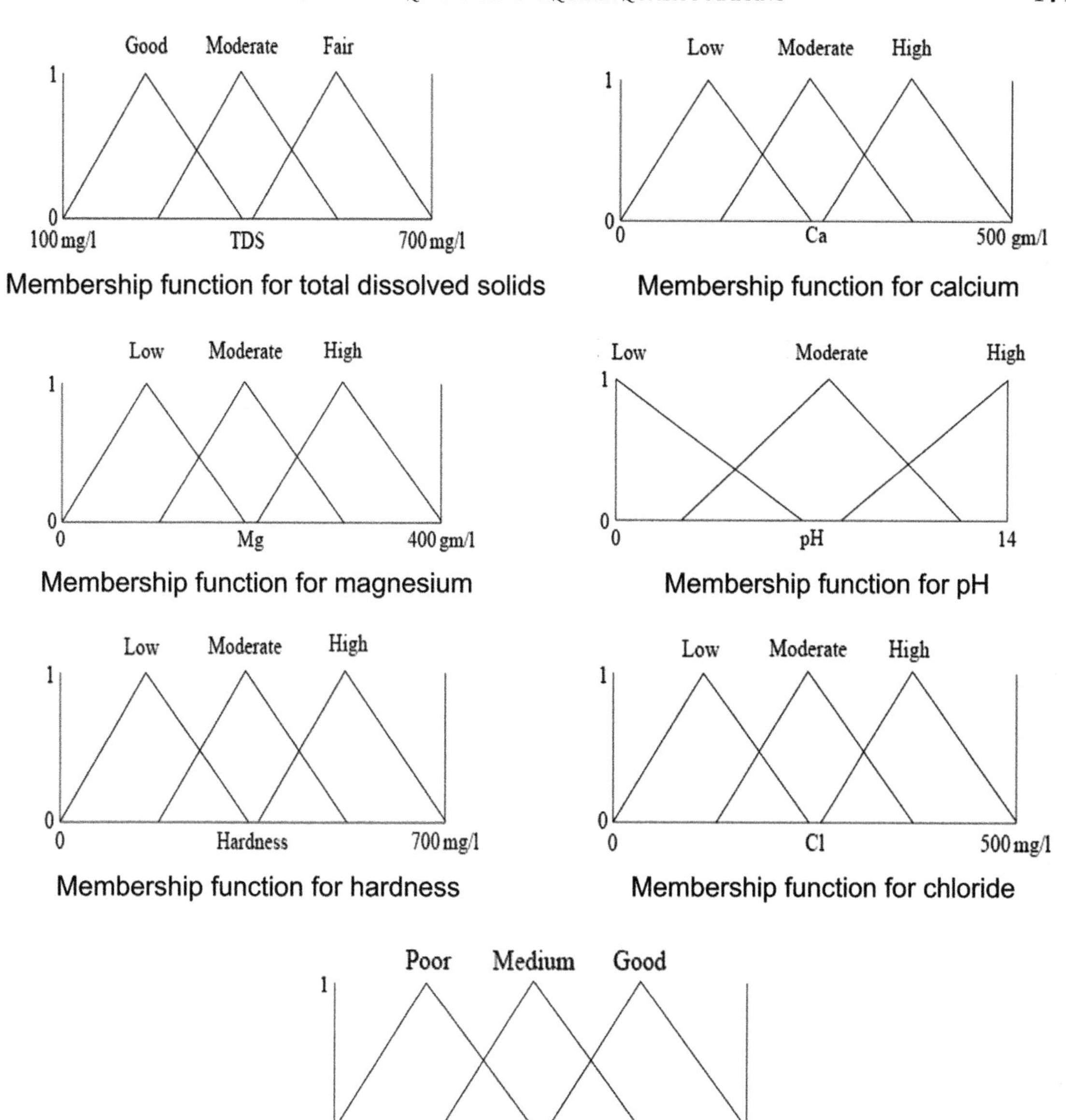

FIGURE 9.4 Membership functions used by Jinturkar et al. (2010) for their fuzzy ground water WQI.

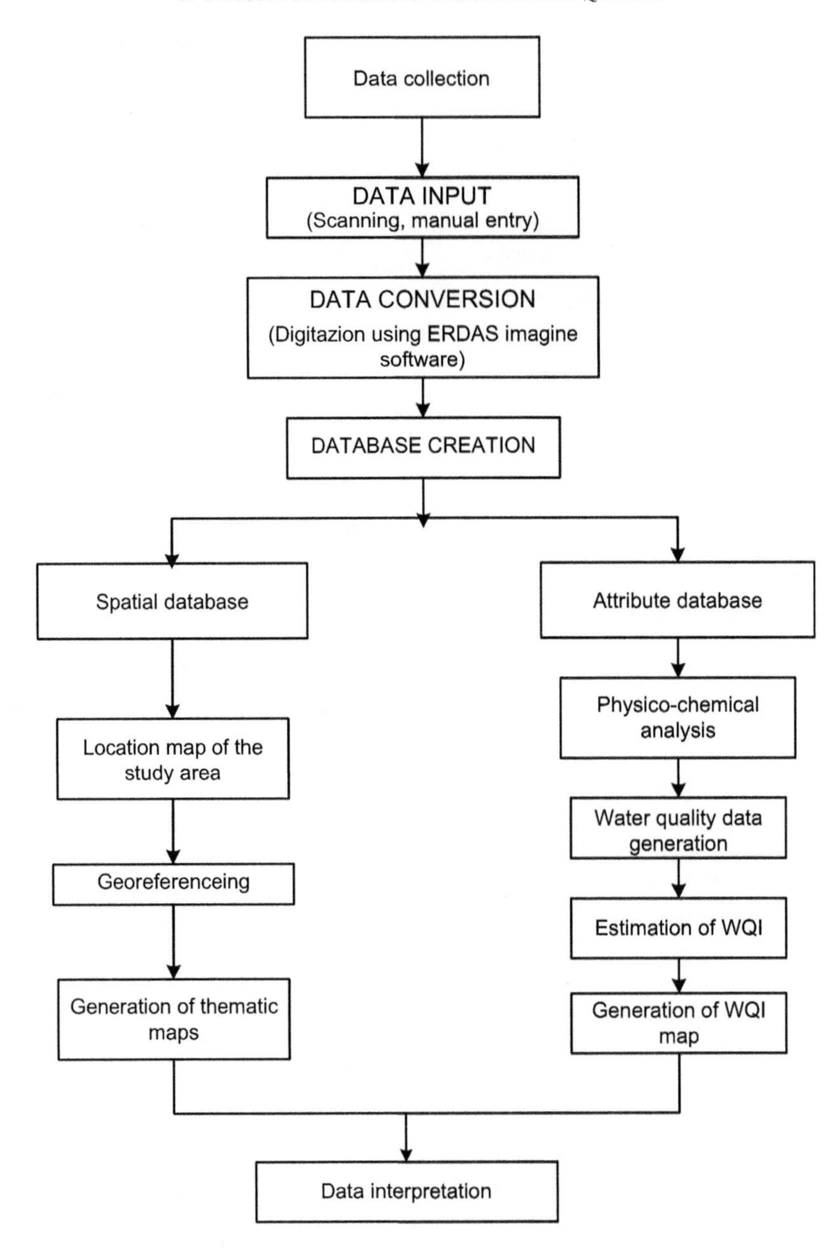

FIGURE 9.5 Use of WQI along with GIS in aquifer-quality mapping by Ketata et al. (2011).

References

Aller, L., Bennett, T., Lehr, J.H., Petty, R.J., 1985. DRASTIC: A Standardized System for Evaluating Ground Water Potential Using Hydrogeological Settings, Environmental Research Laboratory. US Environmental Protection Agency, Ada Oklahoma.

Backman, B., Bodiš, D., Lahermo, P., Rapant, S., Tarvainen, T., 1998. Application of a groundwater contamination index in Finland and Slovakia. Environmental Geology 36 (1–2), 55–64.

Banerjee, T., Srivastava, R.K., 2011. Evaluation of environmental impacts of Integrated Industrial Estate-Pantnagar through application of air and water quality indices. Environmental Monitoring and Assessment 172 (1–4), 547–560.

Banoeng-Yakubo, B., Yidana, S.M., Emmanuel, N., Akabzaa, T., Asiedu, D., 2009. Analysis of groundwater quality using water quality index and conventional graphical methods: the Volta region, Ghana. Environmental Earth Sciences 59 (4), 867–879.

Benzécri, J.P., 1977. L'Analyse des correspondances. Les Cahiers de l'Analyse des Données. II (2), 125–142.

Brown, J.R., 1998. Recommended chemical soil test procedures for the north central region. Missouri Agric. Exp. Stn. North Central Regional Res. Publ. no. 221 (Revised). SB 1001. Columbia.

Civita, M., De Maio, A., 1997. SINTACS. Un Sistema Paramétrico per la Valutazione e la Cartografia della Vulnerabilitá Degli Acquiferi All'inquinamento. Metodologia & Automatizzazione. 191. Pitagora Editrice Bologna.

Conesa Fdez.-Vitora, V., 1993. Methodological Guide for Enviromental Impact Evaluation (Guia Metodológica para la Evaluación de Impacto Ambiental), first ed. Mundi Prensa, Madrid, p. 276.

Giri, S., Singh, G., Gupta, S.K., Jha, V.N., Tripathi, R.M., 2010. An Evaluation of Metal Contamination in Surface and Groundwater around a Proposed Uranium Mining Site, Jharkhand, India. Mine Water and the Environment 29 (3), 225–234.

Hatha, A.A.M., Rejith, P.G., Jeeva, S.P., Vijith, H., Sowmya, M., 2009. Determination of groundwater quality index of a highland village of Kerala (India) using geographical information system. Journal of Environmental Health 71 (10), 51–58.

Jinturkar, A.M., Deshmukh, S.S., Agarkar, S.V., Chavhan, G.R., 2010. Determination of water quality index by fuzzy logic approach: A case of ground water in an Indian town. Water Science and Technology 61 (8), 1987–1994.

Ketata, M., Gueddari, M., Bouhlila, R., 2011. Use of geographical information system and water quality index to assess groundwater quality in El Khairat deep aquifer (Enfidha, Central East Tunisia). Arabian Journal of Geosciences, 1–12.

Melloul, A.J., Collin, M., 1998. A proposed index for aquifer water-quality assessment: the case of Israel's Sharon region. Journal of Environmental Management 54 (2), 131–142.

Mohamad Roslan, M.K., Mohd Kamil, Y., Wan nor Azmin, S., Mat Yusoff, A., 2007. Creation of a ground water quality index for an open municipal landfill area. Malaysian Journal of Mathematical Sciences 1 (2), 181–192.

Nobre, R.C.M., Rotunno Filho, O.C., Mansur, W.J., Nobre, M.M.M., Cosenza, C.A.N., 2007. Groundwater vulnerability and risk mapping using GIS, modeling and a fuzzy logic tool. Journal of Contaminant Hydrology 94, 277–292.

Ocampo-Duque, W., Ferré-Huguet, N., Domingo, J.L., Schuhmacher, M., 2006. Assessing water quality in rivers with fuzzy inference systems: A case study. Environment International 32 (6), 733–742.

Pei-Yue, L., Hui, Q., Jian-Hua, W., 2010. Groundwater quality assessment based on improved water quality index in Pengyang County, Ningxia, Northwest China. E-Journal of Chemistry 7 (Suppl. 1), S209–S216.

Pereira, H.J., Sousa, A.J., 2000. Análise de Dados para o Tratamento de Quadros Multidimensionais: Textos de Apoio ao Curso Intensivo de Análise de Dados, 1988-2000. CVRM, Instituto Superior Técnico, Lisbon, Portugal. p. 105.

Pollock, D.W., Kookana, R.S., Correll, R.L., 2005. Integration of the pesticide impact rating index with a geographic information system for the assessment of pesticide impact on water quality. Water, Air, and Soil Pollution: Focus 5 (1–2), 67–88.

Queralt, E., Pastor, J.J., Corp, R.M., Galofré, A., 2008. Economic index of quality for ground water (IEQAS) based on the potabilisation treatment cost. Practical application to aquifers in Catalonia. Índice económico de calidad para las aguas subterráneas (IEQAS) basado en el coste del tratamiento de potabilización. Aplicación práctica en los acuíferos de Cataluña 28 (293), 89–94.

Ramachandramoorthy, T., Sivasankar, V., Subramanian, V., 2010. The seasonal status of chemical parameters in shallow coastal aquifers of Rameswaram Island, India. Environmental Monitoring and Assessment 160 (1-4), 127–139.

Ramakrishnaiah, C.R., Sadashivaiah, C., Ranganna, G., 2009. Assessment of water quality index for the groundwater in Tumkur taluk, Karnataka state, India. E-Journal of Chemistry 6 (2), 523–530.

Ramos Leal, J.A., 2002. Validación de mapas de vulnerabilidad acuífera e Impacto Ambiental, Caso Río Turbio, Guanajuato. Tesis de Doctorado. Instituto de Geofísica, UNAMUR.

Ramos Leal, J.A., Barrón Romero, L.E., Sandoval Montes, I., 2004. Combined use of aquifer contamination risk maps and contamination indexes in the design of water quality monitoring networks in Mexico. Geofísica Internacional 43 (4), 641–650.

Simsek, C., Gunduz, O., 2007. IWQ Index: A GIS-integrated technique to assess irrigation water quality. Environmental Monitoring and Assessment 128 (1–3), 277–300.

Soltan, M.E., 1999. Evaluation of groundwater quality in Dakhla Oasis (Egyptian Western Desert). Environmental Monitoring and Assessment 57, 157–168.

Srivastava, P.K., Mukherjee, S., Gupta, M., Singh, S.K., 2011. Characterizing monsoonal variation on water quality index of river mahi in India using geographical information system. Water Quality, Exposure and Health 2 (3–4), 193–203.

Stambuk-Giljanovic, N., 1999. Water quality evaluation by index in Dalmatia. Water Research 33 (16), 3423–3440.

Stambuk-Giljanovik, N., 2003. Comparison of Dalmation water evaluation indices. Water Environment Research 75, 388–405.

Stigter, T.Y., Ribeiro, L., Carvalho Dill, A.M.M., 2006. Evaluation of an intrinsic and a specific vulnerability assessment method in comparison with groundwater salinisation and nitrate contamination levels in two agricultural regions in the south of Portugal. Hydrogeology Journal 14 (1–2), 79–99.

Taheriyoun, M., Karamouz, M., Baghvand, A., 2010. Development of an entropy-based Fuzzy eutrophication index for reservoir water quality evaluation. Iranian Journal of Environmental Health Science and Engineering 7 (1), 1–14.

Tiwari, T.N., Mishra, M., 1985. A preliminary assignment of water quality index to major Indian rivers. Indian Journal of Environmental Protection 5 (4), 276–279.

Vasanthavigar, M., Srinivasamoorthy, K., Vijayaragavan, K., Rajiv Ganthi, R., Chidambaram, S., Anandhan, P., Manivannan, R., Vasudevan, S., 2010. Application of water quality index for groundwater quality assessment: Thirumanimuttar sub-basin, Tamilnadu, India. Environmental Monitoring and Assessment 171 (1–4), 595–609.

WHO, 1993. Guidelines for Drinking-Water Quality: Recommendations, second ed. World Health Organisation, Geneva, p. 188.

Xiong, J.Q., Li, Z.Y., Zou, C.W., 2005. Attribute reduction of groundwater quality index based on the rough set theory. Shuikexue Jinzhan/Advances in Water Science 16 (4), 494–499.

Yeh, M.S., Shan, H.Y., Chang, L.C., Lin, Y.P., 2008. Establishing index wells for monitoring groundwater quality using multivariate geostatistics. Journal of the Chinese Institute of Civil and Hydraulic Engineering 20 (3), 315–330.

CHAPTER

10

Water-Quality Indices of USA and Canada

OUTLINE

10.1. INTRODUCTION

In the preceding chapters several water-quality indices (WQIs) have been described, among others, which have been developed by investigators working in the USA or Canada. Then why a separate chapter on the WQIs of these two countries? The main reason is that USA and Canada have had several official WQIs of their individual States/provinces and also have national WQIs.

In contrast, most other countries have only national indices, or none at all (Khan and Abbasi, 1997a,b; 1998a,b; 1999a,b, 2000a,b; 2001; Abbasi, 2002; Sargaonkar and Deshpande, 2003).

Water Quality Indices DOI: 10.1016/B978-0-444-54304-2.00010-5

The other reason is that the US National Sanitation Foundation's WQI (NSF−WQI) and the Canadian Council of Ministers of the Environment WQI (CCME−WQI) have served as models for several official WQIs of other countries. These indices have also been used often as frames of references by investigators when developing or validating new WQIs (Stojda and Dojlido, 1983; Soltan 1999; Bordalo et al., 2006; Abrahao et al., 2007; Sedeno-Diaz and Lopez-Lopez, 2007. Bordalo and Savva-Bordalo, 2007. Avvannavar and Srihari 2008; Carbajal and Sanchez, 2008; Chaturvedi and Bhasin; 2010; Nikoo et al., 2010; Thi Minh Hanh et al., 2011).

This chapter covers the various official WQIs – of state/provincial level as well national level – of USA and Canada. It also describes the work of Said et al. (2004) which was aimed to develop an index which could do with lesser number of parameters than the NSF−WQI and CCME−WQI.

10.2. WQIs OF CANADA

10.2.1. The Alberta Index

The Alberta index uses a 'performance indicator' for water quality, which essentially computes the fraction of samples which do not comply with water-quality standards from among the total sampled. It is represented as

$$A = (n_{\text{exceed}}/n_{\text{measure}}) \times 100 \qquad (10.1)$$

A is, therefore, the percentage of the samples, taken at a site where one or more of the variables exceed the objectives for a designated use. The designated uses in Alberta index include recreation, agriculture and aquatic life.

As may be seen, the index can give meaningful results only if a statistically significant number of samples (>30) are taken randomly from a site. With progressive lesser number of samples it may give increasingly biased results.

10.2.2. The Centre St Laurent Index

The WQI developed at Centre St Laurent (CSL) comprises of three variables (CCME, 1999):

$$\text{WQI} = \left[\sum (A_i \times F_i)\right] \Big/ n \qquad (10.2)$$

where A_i is the mean level of exceedance for parameter i for guideline i. When a parameter value exceeds a guideline for that variable, the ratio 'exceeding value/guideline value' is calculated. These ratios are summed and then divided by the number of times they occur. F_i is the frequency of values that exceed a guideline for a parameter relative to the total number of values obtained for that parameter ($F_i = F_{\text{exceed}}/F_{\text{total}}$ for parameter i). n is the number of parameters. Depending on the water use under consideration, CSL calculates different water-quality indices.

This index is more sophisticated than the Alberta Index but is not able to account for intermediate situations – for example, a water-quality parameter being within acceptable level but very close to the limit compared to another situation wherein the parameter may be well below the limit. In the former case, the probability (or the risk) that a parameter may 'cross over' to an unacceptable level is much higher than the latter case, but the index is not sensitive to this intermediate situation.

10.2.3. The British Columbia Water Quality Index (BCWQI)

The British Columbia water quality index (BCWQI) includes one extra factor not considered in other indices (CCME, 1999):

$$\text{WQI} = (F_1^2 + F_2^2 + F_3^2)^{1/2} \qquad (10.3)$$

where F_1 is the percentage of water-quality guidelines exceeded; F_2 is the percentage of measurements in which one or more of the guidelines were exceeded; and F_3 is the maximum (normalised to 100) extent by which any of the guidelines were exceeded. Two of its factors are analogous to components of other indices: F_2 is similar to the Alberta index, while F_3 is similar to the Centre St Laurent index. F_1 was not used in any of the other indices (CCME, 1999).

The British Columbia Water Quality Index (BCWQI) is based on a variety of objectives depending upon the designated water use: drinking, recreation, irrigation, livestock watering, wildlife and aquatic life. Separate rankings were published based for each use. BCWQI actually uses an inverted scale, where an index value of 0–5 denotes excellent water quality while an index value 60–100 denotes poor water quality (Husain et al., 1999). BCWQI is the state index for Manitoba and has also strongly influenced the national index — the Canadian Water Quality Index (WQI) — as may be seen in Section 2.7.

10.2.4. The Manitoba Adaptation of the BCWQI

The state of Manitoba has adopted the British Columbia approach. Based on its evaluation of four years of data on eight sites in Manitoba, it concluded that the BCWQI appeared to give reasonable results for Manitoba.

10.2.5. The Ontario Index

Ontario index uses BCWQI with modified F_3 so that it represents the average normalised exceedance rather than the maximum. It is done to avoid saturation problem in F_3 as noticed when large exceedances of objectives occurred (CCME, 1999).

10.2.6. The Quebec Index

The Quebec index is based on an approach originally developed in New Zealand (Smith, 1990). It is based on the minimum of a number of subindices, which are calculated for each of the water-quality parameters measured (CCME, 1999):

$$\mathrm{WQI} = \mathrm{minimum}(\mathrm{I}_1, \mathrm{I}_2, \ldots, \mathrm{I}_n) \qquad (10.4)$$

This approach considers use of 'Delphi curves' for the calculation of the subindices. Delphi curves are based on expert opinions as to the significance of a particular level of a water-quality component. These curves are nonlinear and represent the aggregated opinion of a particular level of specific parameters in terms of the designated water use. The Quebec index represents the 'worst case impairment' of any of the variables measured (CCME, 1999).

10.2.7. The Canadian Council of Ministers of Environment Water Quality Index (CCME–WQI)

The CCME–WQI is an adaptation of the BCWQI. There are three factors in the index, each of which has been scaled between 0 and 100.

In the CCME–WQI, the values of the three measures of variance from selected objectives for water quality are combined to create a vector in an imaginary 'objective exceedance' space. In the index 'objectives' refer to Canada-wide water-quality guidelines or site-specific water-quality objectives. The length of the vector is then scaled to range between 0 and 100, and subtracted from 100 to produce an index which is 0 (or close to 0) for very poor water quality, and close to 100 for excellent water quality.

The CCME–WQI consists of three factors (Figure 10.1):

Factor 1 (F_1) – Scope: This factor is called scope because it assesses the extent of the noncompliance of water-quality guideline over

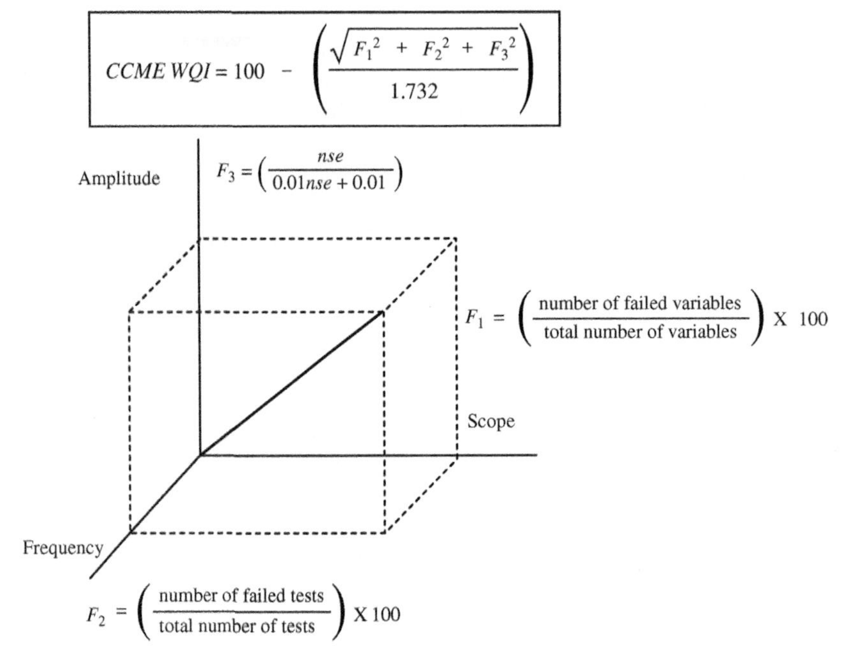

FIGURE 10.1 Graphical representation of the water-quality index (WQI) calculated in a three-dimensional space by summing three factors (F_1, F_2 and F_3) as vectors. The index changes in direct proportion to changes in F_1, F_2 and F_3 (Terrado et al., 2010).

the period of interest. It has been adopted directly from the BCWQI:

$$F_1 = \left(\frac{\text{Number of failed variables}}{\text{Total number of variables}}\right) \times 100 \quad (10.5)$$

where variables indicate those water-quality parameters with objectives which were tested during the time period for the index calculation.

Factor 2 (F_2) – Frequency: It represents the percentage of individual tests that do not meet the objectives ('failed tests'):

$$F_2 = \left(\frac{\text{Number of failed tests}}{\text{Total number of tests}}\right) \times 100 \quad (10.6)$$

The formulation of this factor is also drawn directly from the BCWQI.

Factor 3 (F_3) – Amplitude: It represents the amount by which the failed test values do not meet their objectives, and is calculated in three steps:

1. The number of times by which an individual concentration is greater than (or less than, when the objective is a minimum) the objective is termed an 'excursion' and is expressed as follows. When the test value must not exceed the objective

$$\text{excursion}_i = \left(\frac{\text{Failed Test Value}_i}{\text{Objective}_j}\right) - 1 \quad (10.7)$$

For the cases in which the test value must not fall below the objective:

$$\text{excursion}_i = \left(\frac{\text{Objective}_j}{\text{Failed Test Value}_i}\right) - 1 \quad (10.8)$$

2. The total extent by which individual tests fail to comply is calculated by summing the excursions of individual tests from their objectives and dividing by the total number of tests (those which do and do not meet their objectives). This variable, referred to as the normalised sum of excursions, or *nse*, is calculated as

$$nse = \frac{\sum_{i=1}^{n} \text{excursion}_i}{\text{\# of tests}} \quad (10.9)$$

3. F_3 is then calculated by an asymptotic function that scales the normalised sum of the excursions from objectives (*nse*) to yield a range between 0 and 100:

$$F_3 = \left(\frac{nse}{0.01\,nse + 0.01}\right) \quad (10.10)$$

The CCME–WQI is finally calculated as

$$\text{CCME–WQI} = 100 - \left(\frac{\sqrt{F_1^2 + F_2^2 + F_3^2}}{1.732}\right) \quad (10.11)$$

The factor of 1.732 arises because each of the three individual index factors can range as high as 100. This means that the vector length can reach $\sqrt{100^2 + 100^2 + 100^2} = \sqrt{30,000} = 173.2$ as a maximum. Division by 1.732 brings the vector length down to 100 as a maximum.

It may be seen that the CCME–WQI is closely related to the BCWQI which, in turn, has been found to be extremely sensitive to sampling design and on the chosen water-quality objective (Said et al., 2004). The behaviour of CCME–WQI was studied by way of the application of the CCME–WQI to Canadian rivers by Khan et al. (2003). They found that at none of the three sites the index showed the water to be of quality fit enough for aquatic and drinking use; but for two of the sites it was found to be of excellent quality for agricultural use. A summarised report of the water quality for all sites and for three different water uses found by the authors is presented in Table 10.1.

A time series analysis was also done by Khan et al. (2003) using annual data sets, which indicated that there was a deteriorating trend in drinking water quality (prior to treatment) of two sites (Kejimkujik National Park, NS-EMAN Site, and Bedeque Bay, PEI-EMAN Site). The limited number of records of the parameters significantly affected the time series analysis. A lower number of records generally tended to give a higher WQI (suggesting a better water quality) because less data reduce the probability and proportion of exceedance with respect to the CCME Canadian Water Quality Guidelines and the frequency of exceedance. It was found that total iron, total lead, pH and total cadmium were the parameters that frequently exceed the guidelines.

In a more recent study (Terrado et al., 2010), the suitability of five water-quality indices which included CCME–WQI and NSF–WQI, among others, was evaluated for use in automated sampling (continuous monitoring) networks. The authors found CCME–WQI to be best suited for the purpose. More details of the study have been presented in Chapter 8.

TABLE 10.1 Water Quality of Three Rivers as Indicated by CCME–WQI (Khan et al., 2003)

Sites	Uses	Water Quality
Dunk River (Bedeque Bay EMAN Site)	Drinking	Fair
	Aquatic	Poor
	Agriculture	Good
Mersey River (Kejimkujik National Park EMAN Site)	Drinking	Fair
	Aquatic	Poor
	Agriculture	Excellent
Point Wolfe River (Fundy National Park EMAN Site)	Drinking	Good
	Aquatic	Poor
	Agriculture	Excellent

CCME–WQI continues to be used outside Canada (Carbajal and Sanchez, 2008; Zouabi Aloui and Gueddari, 2009).

10.3. WQIs OF THE USA

10.3.1. The Oregon Water Quality Index (OWQI)

The Oregon Department of Environmental Quality (ODEQ) developed the original Oregon Water Quality Index (OWQI) in 1980. The OWQI is calculated in two steps. The raw analytical results for each variable, having different units of measurement, are transformed into unitless subindex values. These values range from 10 (worst case) to 100 (ideal). These subindices are then combined to give a single WQI value ranging from 10 to 100. The OWQI is integrating measurements of eight water-quality variables (temperature, DO, BOD, pH, ammonium and nitrate nitrogen, total phosphates, total solids and faecal coliform) (Cude, 2001).

10.3.2. The Florida Stream Water Quality Index (FWQI)

The Florida Stream Water Quality Index (FWQI) was developed in 1995 under the Strategic Assessment of Florida's Environment (SAFE). It is an arithmetic average of water clarity (turbidity and total suspended solids), dissolved oxygen; oxygen-demanding substances (biological oxygen demand [BOD], chemical oxygen demand [COD], total organic carbon [TOC]), nutrients (phosphorus and nitrogen), bacteria (total and faecal coliform) and biological diversity (natural or artificial substrate macroinvertebrate diversity and Beck's Biotic Index). The values for this index were determined as follows: 0 to less than 45 represents good quality, 45 to less than 60 represents fair quality and 60 to 90 represents poor quality (SAFE 1995).

10.3.3. The Lower Great Miami's Watershed Enhancement Programme Water Quality Index (WEPWQI)

In 1996, the Lower Great Miami Watershed Enhancement Program (WEP) in Dayton, Ohio, developed a water-quality index and a river index. This river index is calculated in two steps: the WEPWQI, which consists of chemical, physical and biologic variables, and then the River Index, which consists of the water-quality variables plus measurements of water flow and clarity (turbidity). Both indices are expressed as a rating scaled from excellent, good, fair, to poor (WEP 1996).

10.3.4. The National Sanitation Foundation Water Quality Index (NSF–WQI)

This has been described in Chapter 3. The NSF–WQI or its variants have been used, with or without adoptive changes, in several countries including India (Bhargava, 1985); Brazil (Abrhao et al., 2007); Mexico (Sedeno-Diaz and Lopez-Lopez, 2007); Guinea-Bissan (Bordalo and Savva-Bordalo, 2007); Poland (Stojda and Dojlido, 1983); Egypt (Soltan, 1999); Portugal (Bordalo et al., 2006); Italy (Giuseppe and Guidice, 2010), etc. Studies from Croatia suggest the superiority of the geometric form of NSF–WQI (Stambuk-Giljanovic 1999, 2003).

The procedure on which NSF–WQI is based has also been used in developing indices for ground water quality (Chaturvedi and Bhasin, 2010).

10.4. THE WQI OF SAID ET AL. (2004)

Said et al. (2004) have argued that even though there are several water-quality indices that have been developed in USA and Canada to evaluate water quality, all of these indices have eight or more water-quality variables.

However, as per Said et al. (2004), most watersheds do not have long-term and continuous data for these variables. Therefore, there is a need to develop a new WQI that uses fewer variables and that can be used to compare the status of different sites.

The authors have introduced an index which is based on only 5 parameters and yet aims to be as representative as other indices which are based on larger number of parameters. The development of their WQI equation consists of two steps. The first is ranking water-quality variables according to their significance. The variables included in their WQI are DO, total phosphates, faecal coliform, turbidity and specific conductivity. In the second step, the authors have 'short-listed' parameters which have the most pronounced impact on the water quality and which influence other parameters as well. Based on tests, they have given DO the highest weight followed by faecal coliform and total phosphorus. The temperature, turbidity and specific conductance have been given the least influence. A final form was then selected that keeps the index in a simple equation and a 'reasonable' numerical range. The logarithm is used to give small numbers that can be easily appreciated by the decision makers, the stakeholders and the general public. A sensitivity analysis was performed to test the performance of the index and to verify that the index is responsive to influential water-quality variables, as shown in Figure 10.2.

In the final form, the powers of the variables have been chosen for the WQI based on the effect of each variable on water conditions. For example, higher concentrations of faecal coliform and total phosphorus are likely to be very harmful for health and aquatic life. Hence, the forms of the faecal coliform and total phosphorus in the index formula have been chosen to give strong responses to these effects.

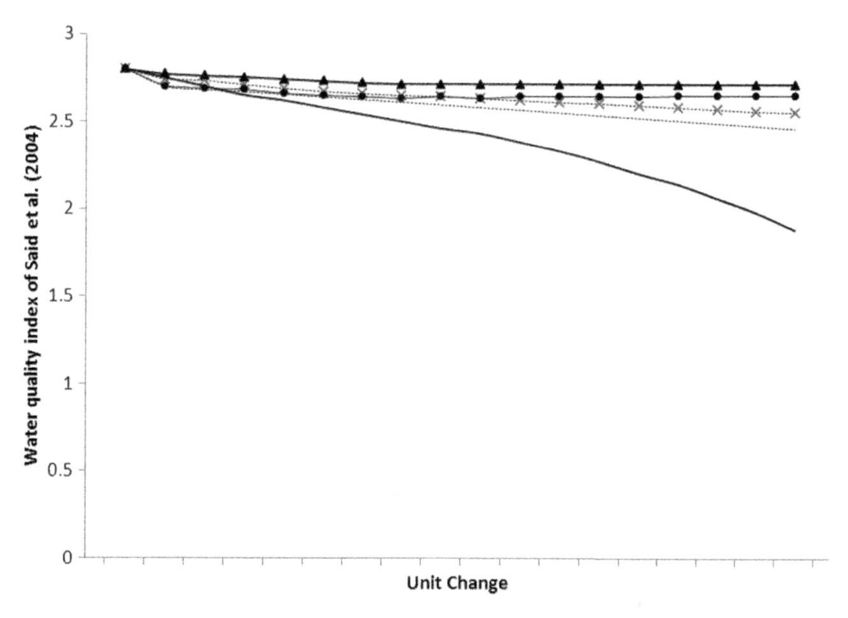

FIGURE 10.2 The sensitivity of the water-quality index of Said et al. (2004) to unit change in water-quality variables. The dissolved oxygen (DO) has the most rapid effect on WQI followed by faecal coliform (F-coli), total phosphorus (TP), specific conductance (SC) and turbidity. DO ——— (10–100% Salt); F-Coli ········ (0–3000 no/100 mL); TP ·····×····· (0–2 mg/L); SC —•— (0–5000 uS/cm); Turbidity—▲— (0–1000 NTU).

Said et al. (2004) felt that turbidity and specific conductance have linear effects, which have lesser influence on the values of the variables, in the index formula. This is because, for example, turbidity would not be very dangerous unless it is associated with a higher level of disease-causing micro-organisms that will make faecal coliform higher as well in the formula.

According to the authors, there is no need to standardise the variables for calculating their index. The calculations are further simplified through the elimination of subindices which are part of most other sophisticated indices.

The index of Said et al. (2004) takes the form

$$WQI = \log \frac{(DO)^{1.5}}{(3.8)^{TP}(Turb)^{0.15}(15)^{FCol/10000} + 0.14(SC)^{0.5}} \quad (10.12)$$

where DO is the dissolved oxygen (% oxygen saturation) Turb is the turbidity (Nephelometric turbidity units [NTU]), TP is the total phosphates (mg/L), FCol is the faecal coliform bacteria (counts/100 mL) and SC is the specific conductivity in (MS/cm at 25 °C). The index has been designed to range from 0 to 3. The maximum or ideal value of this index is 3. In very good waters that have 100% dissolved oxygen, no TP, no faecal coliform, turbidity less than 1 NTU and specific conductance less than 5 MS/cm, the value of this index will be 3. At lower scores up to 2 the water is deemed acceptable; less than 2 represents marginal case, needing some remediation. If one or two variables have deteriorated, the value of this index will be less than 2. If most of the variables have deteriorated, the index would fall below 1, which would mean poor water quality.

10.4.1. Comparison with Other Indices

The main attribute of the WQI of Said et al. (2004), viz, its reliance on lesser number of parameters, which makes it simpler to compute and use, can also be its major limitation. Although it has the advantage of reducing water-quality variables to a minimum subset, it cannot always show the impact of random short-term changes, such as a spill, except if it occurs repeatedly or for a long time. The best results with this index can be obtained in natural conditions and natural measurement sites (not downstream of river outfall). The index can be used to assess water quality for general uses, but cannot be used in making regulatory decisions or to indicate water quality for specific uses. Localised changes in water quality may also not be immediately reflected. Another change not necessarily reflected in the index is the stream habitat. In addition, the index cannot be used to indicate contamination from trace metals, organic contaminants or other toxic substances. Not all of these limitations are, however; unique to this WQI − other indices are also only as widely applicable as the number of parameters they cover.

Table 10.2 shows a sample calculation using the NSF−WQI (Mitchell and Stapp 1996) and the WQI of Said et al. (2004). The NSF−WQI includes nine water-quality variables (DO, faecal coliform, pH, BOD, temperature, TP, nitrate, turbidity and TS) and its score is 77.9, which lies on the good water classification region, so the water is considered good. To get the NSF−WQI score, it is necessary that the Q-value should be determined for each variable. Also, a weighting factor has to be assigned to each variable. In contrast, the WQI of Said et al. (2004) gives a value of 2.22, which indicates that the water is good, in just one simple step. How the two WQIs change with changes in three water-quality variables − DO, turbidity and total P − is illustrated in Figure 10.3.

The relative complexity of other indices is seen in a sample calculation (Table 10.3) with reference to the WEPWQI (WEP 1996), described earlier in Section 3.3. The index value is 54, which lies in the good region according to the ranking criteria of this index. This index

TABLE 10.2 Comparison of the results of the index of Said *et al.* (2004) with the National Sanitation Foundation Water Quality index (NSF–WQI)

Variable	Result	Unit	Q-value	Weight Factor	Subtotal
DO	82	% Saturation	90	0.17	15.3
Fecal coliform	12	#/100 mL	72	0.16	11.52
pH	7.67		92	0.11	10.12
BOD	2	mg/L	80	0.11	8.8
Change in temp.	5	°C	72	0.10	7.2
Total phosphate (PO4)	0.5	mg/L	60	0.10	6
Nitrates	5	mg/L NO_3	67	0.10	6.7
Turbidity	5	NTU	85	0.08	6.8
TS	150	mg/L	78	0.07	5.46
NSF–WQI					77.9
WQI of Said *et al.* (2004)	2.22				

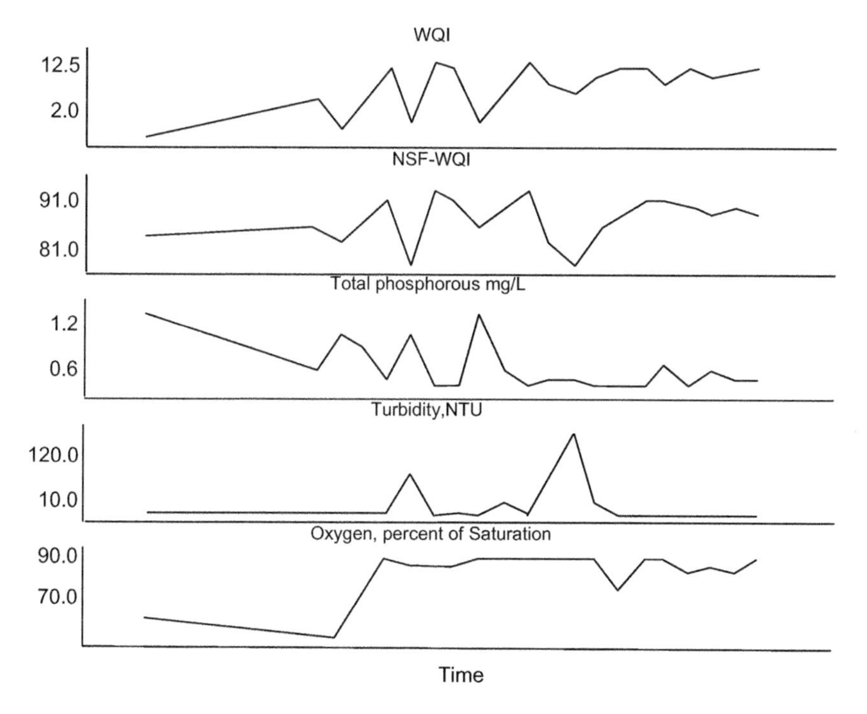

FIGURE 10.3 Pattern of variation of the WQI of Said et al. (2004) and the NSF–WQI as functions of variation in three water-quality variables.

TABLE 10.3 Complexity of Pre-Existing Indices Such the Watershed Enhancement Program Water Quality Index (WEPWQI) in Comparison to the Simpler WQI of Said et al. (2004)

Variable	Averaged Value	Rating	WQ Weighting Factor	WQ Weighted Subtotals	WQ Weighting Maxima
Total ammonia	3.24	1	2	2	8
Atrazine	0.02	4	–	–	–
Chlorpyrifos	0.01	4	–	–	–
Pesticides	(4 + 4)/2 = 4	2	8	8	
Dissolved oxygen	10.02	4	3	12	12
Escherichia coli	80.00	4	–	–	–
Faecal coliform	105.00	4	–	–	–
Pathogens	(4 + 4)/2 = 4	2	8	8	
Fish toxicity	69.00	2	3	6	12
Nitrate	6.97	1	2	2	8
PAH	0.25	4	2	8	8
pH	8.33	3	1	3	4
Specific conductivity	0.66	2	1	2	4
(Water) temperature	16.82	3	1	3	4
WEPWQI subtotals				54	72
WQI	2.11				

needs 15 water-quality variables to get the river index. In contrast, the WQI of the authors gets a score of 2.11, which also indicates that the water is good, with much lesser effort.

Table 10.4 shows an example of the new WQI for a number of water-quality conditions and for the same data as presented in Tables 10.2 and 10.3. It highlights the disparity in rating of water by the two indices. Although a water with 77.9 of NSF–WQI (Table 10.2) is a 'good' water, another water with a lesser score, 72 of WEPWQI is an 'excellent' water (Table 10.3). To lay persons, this situation may be highly confusing. This, feel Said et al. (2004), raises the need for a new general index with a unique simple range from 1 to 3. Table 10.3 shows that most stream waters have values of this index between 2 and 3 except when total phosphorus is more than 0.5 mg/L, turbidity more than 50 NTU, faecal coliform more than 200 organisms/100 mL or specific conductance more than 750-S/cm. In this table, the lower limit for good waters can be a water that has 50% saturation of DO, 200 organisms/100 mL for a single water sample, 0.05 mg/L TP, turbidity of 50 NTU and SC of 100 ms/cm as shown in row 5.

Figure 10.3, which shows a comparison between the new WQI and NSF–WQI, gives some interesting observations. On a day when the NSF–WQI gives a value of 84%, indicating 'very good' water, the proposed index gives a value of only 1.83, which is reflective of a water that needs remedial action, such as a TMDL. The reason for this difference in interpretation of

TABLE 10.4 Range of the WQI of Said et al. (2004) as Influenced by the Levels of Different Variables

DO (% saturation)	Turbidity (NTU)	Fcol (col/100 mL)	Fcol (col/100 mL/10,000)	TP (mg/L)	SC (MS/cm)	WQI
90	1	100	0.01	0.02	1	2.85
70	10	200	0.02	0.7	20	2.12
60	50	500	0.05	1	90	1.71
90	80	1000	0.1	1.4	270	1.66
50	50	200	0.02	0.05	100	2.01
90	100	3000	0.3	0.5	270	1.89
100	200	5000	0.5	0.5	100	1.74
100	270	4000	0.4	0.2	300	1.94
82	5	12	0.0012	0.5	3	2.43
100	0.5	0	0	0	0.5	3.00
60	200	200	0.02	0.1	300	1.96
82	5	12	0.0012	0.5	75	2.30
100	20.35	105	0.0105	0.7	660	2.11

water quality by the two indices is the elevated phosphorus level of 1.5 mg/L. Whereas the index of Said et al. (2004) reckons water with phosphorus concentration of 1.5 mg/L as not 'very good', NSF–WQI does so. In general, the index of Said et al. (2004) is more sensitive to the elevated values of phosphorus and turbidity or decreased values of DO, than NSF–WQI.

Said et al. (2004) feel that their WQI has advantages over the others: it is very simple, fast, does not need to standardise the water-quality variables or to calculate subindices, and it needs much lesser number of water-quality variables to evaluate the water-quality situation. But for the index to be representative, the measurements should not be performed downstream of a wastewater treatment plant or in areas where large amounts of animal or untreated human waste is deposited into the stream. In general, this index gives results similar to those calculated using NSF–WQI or WEPWQI while using fewer variables.

References

Abbasi, S.A., 2002. Water Quality Indices. State of the art report, Scientific Contribution No.-INCOH/SAR-25/2002. INCOH, National Institute of Hydrology, Roorkee.

Abrahão, R., Carvalho, M., Da Silva Jr., W.R., Machado, T.T.V., Gadelha, C.L.M., Hernandez, M.I.M., 2007. Use of index analysis to evaluate the water quality of a stream receiving industrial effluents. Water SA 33 (4), 459–465.

Avvannavar, S.M., Shrihari, S., 2008. Evaluation of water quality index for drinking purposes for river Netravathi, Mangalore, South India. Environmental Monitoring and Assessment 143 (1–3), 279–290.

Bhargava, D.S., 1985. Water quality variations and control technology of Yamuna River. Environmental Pollution Series A: Ecological and Biological 37 (4), 355–376.

Bordalo, A.A., Savva-Bordalo, J., 2007. The quest for safe drinking water: an example from Guinea-Bissau (West Africa). Water Research 41 (13), 2978–2986.

Bordalo, A.A., Teixeira, R., Wiebe, W.J., 2006. A water quality index applied to an international shared river basin: the case of the Douro River. Environmental Management 38 (6), 910–920.

Carbajal, J.J., Śanchez, L.P., 2008. Classification based on fuzzy inference systems for artificial habitat quality in shrimp farming, 388–392.

CCME (1999). Canadian Environmental Qualities Guidelines, Canadian Council of Ministers of the environment, Manitoba Statuary Publications, Winnipeg, Canada.

Chaturvedi, M.K., Bhasin, J.K., 2010. Assessing the water quality index of water treatment plant and bore wells, in Delhi, India. Environmental Monitoring and Assessment 163 (1–4), 449–453.

Cude, C.G., 2001. Oregon water quality index: a tool for evaluating water quality management effectiveness. Journal of the American Water Resources Association 37 (1), 125–137.

Giuseppe, B., Guidice, R.L., 2010. Application of two water quality indices as monitoring and management tools of rivers. Case study: the Imera Meridiopnale River Italy. Environmental Management 45, 856–867.

Husain, T., Khan, A.A., Mukhtasor, A., 1999. Final Report on Water Quality Index for Northwest Territories. Water Management Planning Section, Yellowknife. 12–13.

Khan, F., Husain, T., Lumb, A., 2003. Water quality evaluation and trend analysis in selected watersheds of the Atlantic region of Canada. Environmental Monitoring and Assessment 88 (40603), 221–242.

Khan, F.I., Abbasi, S.A., 1997a. Accident hazard index: a multiattribute method for process industry hazard rating. Process Safety and Environment 75, 217–224.

Khan, F.I., Abbasi, S.A., 1997b. Risk analysis of a chloralkali industry situated in a populated area using the software package MAXCRED-II. Process Safety Progress 16, 172–184.

Khan, F.I., Abbasi, S.A., 1998a. Multivariate hazard identification and ranking system. Process Safety Progress 17, 157–170.

Khan, F.I., Abbasi, S.A., 1998b. DOMIFFECT (DOMIno eFFECT): user-friendly software for domino effect analysis. Environmental Modelling and Software 13, 163–177.

Khan, F.I., Abbasi, S.A., 1999a. Assessment of risks posed by chemical industries: application of a new computer automated tool MAXCRED-III. Journal of Loss Prevention Process 12, 455–469.

Khan, F.I., Abbasi, S.A., 1999b. The world's worst industrial accident of the 1990s: what happened and what might have been – a quantitative study. Process Safety Progress 18, 135–145.

Khan, F.I., Abbasi, S.A., 2000a. Analytical simulation and PROFAT II: a new methodology and a computer automated tool for fault tree analysis in chemical process industries. Journal of Hazardous Materials 75, 1–27.

Khan, F.I., Abbasi, S.A., 2000b. Towards automation of HAZOP with a new tool EXPERTOP. Environmental Modelling and Software 15, 67–77.

Khan, F.I., Abbasi, S.A., 2001. An assessment of the likelihood of occurrence, and the damage potential of domino effect (chain of accidents) in a typical cluster of industries. Journal of Loss Prevention in the Process Industries 14, 283–306.

Lower Great Miami watershed enhancement program. (1996). Miami valley river index. Available at: http://www.mvrpc.org/wq/wep.htm.

Mitchell, M.K., Stapp, W.B., 1996. Field Manual for Water Quality Monitoring: An Environmental Education Program for Schools, p. 277.

Nikoo, M.R., Kerachian, R., Malakpour-Estalaki, S., Bashi-Azghadi, S.N., Azimi-Ghadikolaee, M.M., 2010. A probabilistic water quality index for river water quality assessment: a case study. Environmental Monitoring and Assessment, 1–14.

Said, A., Stevens, D.K., Sehlke, G., 2004. An innovative index for evaluating water quality in streams. Environmental Management 34 (3), 406–414.

Sargaonkar, A., Deshpande, V., 2003. Development of an overall index of pollution for surface water based on a general classification scheme in Indian context. Environmental Monitoring and Assessment 89, 43–67.

Sedeno-Diaz, J.E., Lopez-Lopez, E., 2007. Water quality in the Rão Lerma, Mexico: An overview of the last quarter of the twentieth century. Water Resources Management 21 (10), 1797–1812.

Smith, D.G., 1990. A better water quality indexing system for rivers and streams. Water Research 24 (10), 1237–1244.

Soltan, M.E., 1999. Evaluation of ground water quality in Dakhla Oasis (Egyptian Western Desert). Environmental Monitoring and Assessment 57 (2), 157–168.

Stambuk-Giljanovic, N., 1999. Water quality evaluation by index in Dalmatia. Water Research 33 (16), 3423–3440.

Stambuk-Giljanovic, N., 2003. The water quality of the Vrgorska Matica River. Environmental Monitoring and Assessment 83 (3), 229–253.

Stojda, A., Dojlido, J., 1983. A study of water quality index in Poland. WHO. Water Quality Bulletin 1, 30–32.

Strategic assessment of Florida's environment. (1995). Florida stream water quality index, statewide summary. Available at: http://www.pepps.fsu.edu/safe/environ/swq1.html

Terrado, M., Barceló, D., Tauler, R., Borrell, E., Campos, S.D., 2010. Surface-water-quality indices for the analysis of data generated by automated sampling networks. TrAC – Trends in Analytical Chemistry 29 (1), 40–52.

Thi Minh Hanh, P., Sthiannopkao, S., The Ba, D., Kim, K.W., 2011. Development of water quality indexes to identify pollutants in Vietnam's surface water. Journal of Environmental Engineering 137 (4), 273–283.

Zouabi Aloui, B., Gueddari, M., 2009. Long-term water quality monitoring of the Sejnane reservoir in northeast Tunisia. Bulletin of Engineering Geology and the Environment 68 (3), 307–316.

CHAPTER

11

WQI-Generating Software and a WQI-based Virtual Instrument

OUTLINE

11.1. INTRODUCTION

The first, and apparently the only one so far, water-quality index (WQI)-generating software has been reported by Sarkar and Abbasi (2006). The software is also capable of reporting the index values in the form of a virtual 'water-quality meter'. The latter has the appearance of a speedometer with a dial pointing at the WQI score and the areas covering different ranges of index value given colours like red, yellow and green to indicate polluted, passable and good quality.

The software/virtual instrument has been named QUALIDEX (water QUALIty inDEX).

11.2. THE BASIC ARCHITECTURE OF QUALIDEX

QUALIDEX comprises of four modules, namely – the database module, the index generation module, the water-quality comparison module and the report generation module. The basic architecture of QUALIDEX is depicted in Figure 11.1.

11.2.1. WQI Generation Module

The software enables the user to generate the following well-known water-quality indices for any water source of the water-quality use:

- The Oregon Water Quality Index (OWQI)
- Aquatic Toxicity Index (ATI)
- Diniu's Water Quality Index (DWQI)
- Overall Index of Pollution (OIP) of National Environmental Engineering Research Institute (NEERI), India
- Water-quality index of Central Pollution Control Board (CPCB), India

Water Quality Indices DOI: 10.1016/B978-0-444-54304-2.00011-7

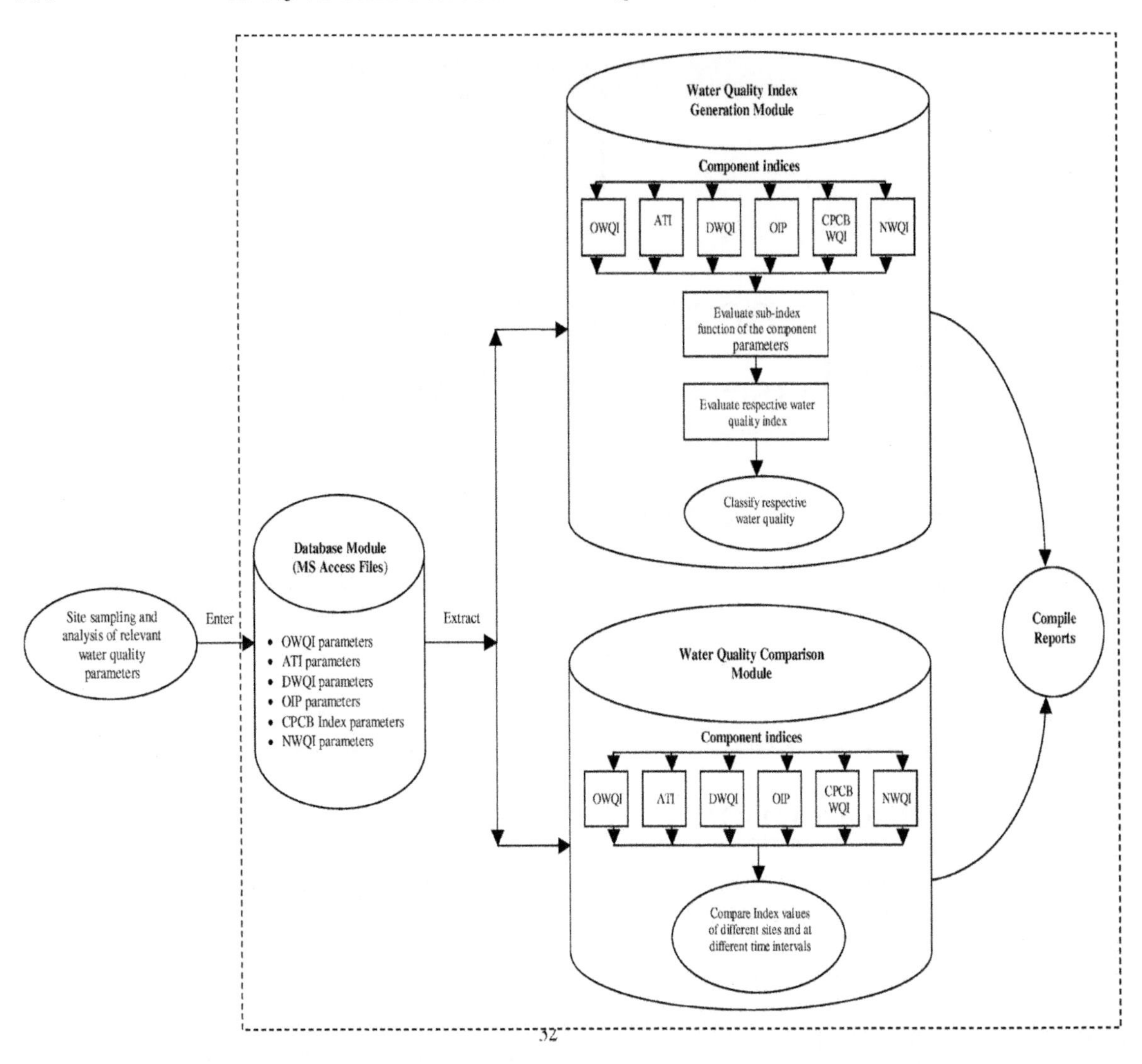

FIGURE 11.1 Conceptual framework of QUALIDEX.

In addition, the software contains the *New Water Quality Index Submodule* with which new index may be generated on the basis of the parameters chosen by the user, and the concerned weightage, applicability range, subindex and aggregation function defined by the user.

The indices described above have been coded within this module. Table 11.1 lists the features of each of these indices. To evaluate any particular water-quality index, the user needs to first specify the site as well as the date and time at which he/she wants to analyse the water quality and subsequently extract the respective values of the component water-quality parameters for the site. Individual dialogue boxes have been created for the detailed assessment of the status of each of the water-quality parameters

TABLE 11.1 Subindex Functions and the Weightages of the Component Index of QUALIDEX

Oregon Water Quality Index (OWQI)					
Parameter	**Weightage**	**Range Applicable (y)**	**Subindex Function (SI_i)**	**Aggregation Function**	**Water-Quality Classification**
Temperature (°C)	–	$T \leq 11$	$SI_T = 100$		
		$11 < T < 29$	$SI_T = 76.54 + 4.172\,T - 0.1623T^2 - 2.0557E - 3T^3$		
		$T > 29$	$SI_T = 10$		
DO_C (mg/l) DO_S (% saturation)	–	$DO_C < 3.3$	$SI_{DO} = 10$		
		$3.3 < DO_C < 10.5$	$SI_{DO} = -80.29 + 31.88DO_C - 1.401DO_c^2$		
		$DO_C \geq 10.5$	$SI_{DO} = 100$		
		$100\% < DO_S \leq 275\%$	$SI_{DO} = 100\exp((DO_S - 100)^* - 1.197E - 2)$		
		$DO_S > 275\%$	$SI_{DO} = 10$		
BOD, 5 day (mg/l)	–	$BOD \leq 8$	$SI_{BOD} = 100\exp(BOD^* - 0.1993)$		
		$BOD > 8$	$SI_{BOD} = 10$		
pH	–	$pH < 4$	$SI_{pH} = 10$		10–59 – Very poor
		$4 \leq pH < 7$	$SI_{pH} = 2.628\exp(pH^*0.5200)$		60–79 – Poor
		$7 \leq pH \leq 8$	$SI_{pH} = 100$	$WQI = \sqrt{\frac{n}{\sum_{i=1}^{n}\frac{1}{s_i^2}}}$	80–84 – Fair
		$8 < pH \leq 11$	$SI_{pH} = 100\exp(pH - 8)^* -0.5188)$		85–89 – Good
		$pH > 11$	$SI_{pH} = 10$		90–100 – Excellent
Total solids (mg/l)	–	$TS < 40$	$SI_{TS} = 100$		
		$40 < TS \leq 220$	$SI_{TS} = 142.6\exp(TS - 8.862E - 3)$		
		$TS > 220$	$SI_{TS} = 10$		
Ammonia+NO_3–N	–	$N \leq 3$	$SI_N = 100\exp(N^* - 0.4605)$		
		$N > 3$	$SI_N = 10$		
Total Phosphorus	–	$P \leq 0.25$	$SI_P = 100–299.5P - 0.1384P^2$		
		$P > 0.25$	$SI_P = 10$		
		$FC \leq 50$	$SI_{FC} = 98$		
Faecal Coliform (nos./100 ml)	–	$50 < FC \leq 1600$	$SI_{FC} = 98\exp((FC - 50)^* - 9.9178E - 4)$		
		$FC > 1600$	$SI_{FC} = 10$		

(Continued)

TABLE 11.1 Subindex Functions and the Weightages of the Component Index of QUALIDEX (*cont'd*)

Aquatic Toxicity Index (ATI)					
Parameter	**Weightage**	**Range Applicable (x)**	**Subindex Function (y)**	**Aggregation Function**	**Water-Quality Classification**
DO (mg/l)	–	$0 \leq DO \leq 5$	$y = 10x$		
		$5 < DO \leq 6$	$y = 20x - 50$		
		$6 < DO \leq 9$	$y = 10x + 10$		
		$DO > 9$	$y = 100$		
pH	–	–	$y = 98 \exp[-(x - 8.16)^2 \cdot (0.4)]$		
			$+17 \exp[-(x - 5.2)^2 \cdot (0.5)]$		
			$+15 \exp[-(x - 11)^2 \cdot (0.72)] + 2$		
Manganese	–	–	$y = 0.115 \exp^{-0.05} \cdot \exp^{0.0013x} + 5$		
Nickel	–	–	$y = -28 \ln(x - 10) + 211$		
Fluoride	–	–	$y = -71 \ln(0.001(x + 2.5)) - 235$		
Chromium	–	–	$y = -40 \ln(0.1(x + 150)) + 210$		60–100 – Suitable for all fish life
Lead	–	–	$y = -27 \ln(0.1(x - 30))$	$I = \frac{1}{100}\left(\frac{1}{n}\sum_{i=1}^{n} q_i\right)^2$	51–59 – Suitable only for hardy fish species
Ammonium		$NH_4^+ \geq 0.02$	$y = 100$		0–50 – Totally unsuitable for normal fish life
		$0.02 < NH_4^+ \leq 0.062$	$y = -500x + 110$		
		$0.062 < NH_4^+ \leq 0.05$	$y = 40/(x + 0.65)^2$		
		$NH_4^+ > 0.5$	$y = -5.8x + 32.5$		
Copper	–	–	$y = -26 \ln(x - 18) + 180$		
Zinc	–	–	$y = -22 \ln(0.001(x - 20)) + 16$		

Orthophosphates	–	–	$y = 100\ \exp^{-2.4x}$		
Potassium	–	–	$y = 150\ \exp^{-0.02x} + 8$		
Turbidity	–	–	$y = -220\ \ln(0.001\ \ln(x) + 30) - 689$		
Total dissolved salts	–	–	$y = 117\ \exp^{-0.00068x} - 7$		
Dinius Water Quality Index					
Parameter	**Weightage**	**Range Applicable (x)**	**Subindex Function (y)**	**Aggregation Function**	**Water-Quality Classification**
DO (% saturation)	0.019	–	$0.82DO + 10.56$		–
BOD, 5 day (mg/l)	0.097	–	$108\ (BOD)^{-0.3494}$		
Coli (MPN Coli/100 ml)	0.090	–	$136\ (Coli)^{-0.1311}$		
E. Coli (Coli/100 ml)	0.116	–	$106\ (E.\ Coli)^{-0.1286}$		
Alkalinity (ppm $CaCO_3$)	0.063	–	$110\ (Alk)^{-0.1342}$		
Hardness (ppm $CaCO_3$)	0.065	–	$552\ (Ha)^{-0.4488}$		
Chloride (mg/l)	0.074	–	$391\ (Cl)^{-0.3480}$	$IWQ = \sum_{i=1}^{n} I_i w_i$	
Sp. Conductance (micro-mhos/cm)	0.079	–	$306\ (Sp.C)^{-0.3315}$		
pH	0.077	<6.9	$10^{0.6803+0.1856(pH)}$		
		6.9–7.1	1		
		>7.1	$10^{3.57-0.2216(pH)}$		
Nitrate (mg/l)	0.090	–	$125(N)^{-0.2718}$		
Temperature (°C)	0.077	–	$10^{2.004-0.0-382(Ta\text{-}Ts)}$		
Colour (Colour units – Pt std)	0.063	–	$127(C)^{-0.2394}$		

(Continued)

TABLE 11.1 Subindex Functions and the Weightages of the Component Index of QUALIDEX (*cont'd*)

Overall Index of Pollution (OIP)					
Parameter	**Weightage**	**Range Applicable (y)**	**Subindex Function (P_i)**	**Aggregation Function**	**Water-Quality Classification**
Turbidity	–	≤5	$P_i = 1$		
		5–10	$P_i = (y/5)$		
		10–500	$P_i = (y + 43.9)/34.5$		
pH	–	7	$P_i = 1$		
		7	$P_i = 1$		
		>7	$P_i = \exp((y - 7)/1.082)$		
		<7	$P_i = \exp((7 - y)/1.082)$		
Colour	–	10–150	$P_i = (y + 130)/140$		
		150–1200	$P_i = y/75$		
% DO	–	<50	$P_i = \exp(-(y - 98.33)/36.067)$		
		50–100	$P_i = (707.58 - y)/14.667$		
		≥100	$P_i = (y - 79.543)/19.054$		
		<2	$P_i = 1$		
BOD	–	2–30	$P_i = y/1.5$		
TDS	–	≤500	$P_i = 1$		
		500–1500	$P_i = \exp((y - 500)/721.5)$		
		1500–3000	$P_i = (y - 1000)/125$		
		3000–6000	$P_i = y/375$		0–1 – Excellent

Hardness	–	≤75	$P_i = 1$	$OIP = \frac{\sum_i P_i}{n}$	1–2 – Acceptable
					2–4 – Slightly polluted
		75–500	$P_i = \exp(y + 42.5)/205.58$		4–8 – Polluted
		>500	$P_i = (y + 500)/125$		8–16 – Heavily polluted
Cl	–	≤150	$P_i = 1$		
		150–250	$P_i = \exp((y/50) - 3)/1.4427)$		
		>250	$P_i = \exp((y/50) + 10.167)/10.82$		
NO_3	–	≤20	$P_i = 1$		
		20–50	$P_i = \exp((145 - y)/76.28)$		
		50–200	$P_i = y/65$		
SO_4	–	≤150	$P_i = 1$		
		150–2000	$P_i = ((y/50 + 0.375)/2.5121)$		
		50–5000	$P_i = (y/50)$**.0301		
		5000–15000	$P_i = ((y/50) - 50)/16.071$		
		>15000	$P_i = (y/15000) + 16$		
As	–	≤0.005	$P_i = 1$		
		0.005–0.01	$P_i = y/0.005$		
		0.01–0.1	$P_i = (y + 0.015)/0.0146$		
		0.1–1.3	$P_i = (y + 1.1)/0.15$		
F	–	0–1.2	$P_i = 1$		
		1.2–10	$P_i = ((y/1.2) - 0.3819)/0.5083$		

(Continued)

TABLE 11.1 Subindex Functions and the Weightages of the Component Index of QUALIDEX (*cont'd*)

Ved Prakash Index (CPCB WQI)					
Parameter	**Weightage**	**Range Applicable (x)**	**Subindex Function (I_i)**	**Aggregation Function**	**Water-Quality Classification**
DO (% saturation)	0.31	0–40%	$I_{DO} = 0.18 + 0.66x$		
		40–100%	$I_{DO} = -13.5 + 1.17x$		
		100–140%	$I_{DO} = 163.34 - .62x$		
BOD (mg/l)	0.19	0–10	$I_{BOD} = 96.67 - 7.0x$	$WQI = \sum_{i=1}^{p} w_i I_i$	63–100—Good to excellent
		10–30	$I_{BOD} = 38.9 - 1.23x$		50–63—Medium to good
pH	0.22	2–5	$I_{PH} = 16.1 + 7.35x$		38–50—Bad
		5–7.3	$I_{PH} = -47.61 + 20.09x$		<38—Bad to very bad
		7.3–10	$I_{PH} = 316.96 - 29.85x$		
		10–12	$I_{PH} = 96.17 - 8.0x$		
Faecal Coliform	0.28	10^3–10^5	$I_{coli} = 42.33 - 7.75 \log_{10}x$		
		$>10^5$	$I_{coli} = 2$		

MS Access Table																		
SiteName	Latitude	Longitude	Date	Time	Turbidity	pH	Color	% DO	BOD	TDS	Hardness	Chloride	Nitrate	Sulfate	Coli	Arsenic	Fluoride
Site 1	11.930 N	79.820 E	1/5/04	9:00	2.5	6.8	12	88	1	350	67.5	120.5	15.2	120.5	45	0.0045	1
Site 2	11.910 N	79.825 E	2/5/04	9:00	7.5	6.2	120	80	2.8	1000	100	210	35.5	200.5	450	0.0075	1.35
Site 3	11.925 N	79.818 E	3/5/04	9:00	50	5.5	200	68	5.5	2000	240	505.5	48.5	310.5	4000	0.025	1.75
Site 4	11.875 N	79.800 E	4/5/04	9:00	150.5	4.8	500	35	10	2800	450	700	75	600	7500	0.075	5.5
Site 5	11.900 N	79.810 E	5/5/04	9:00	200	5.2	1000	18	18	4500	650	900	150	1200	12000	1.12	8.5
Site 6	11.915 N	79.815 E	6/5/04	9:00	8	7.8	68.5	120	1.85	1200	85	230	41.5	205.5	300	0.006	1.35
Site 7	11.850 N	79.822 E	7/5/04	9:00	175	9.3	450	140	7.5	2500	410	750	75	678	7800	0.08	4.5

FIGURE 11.2 Typical data set keyed into the Ms access database module of the Overall Index of Pollution (OIP).

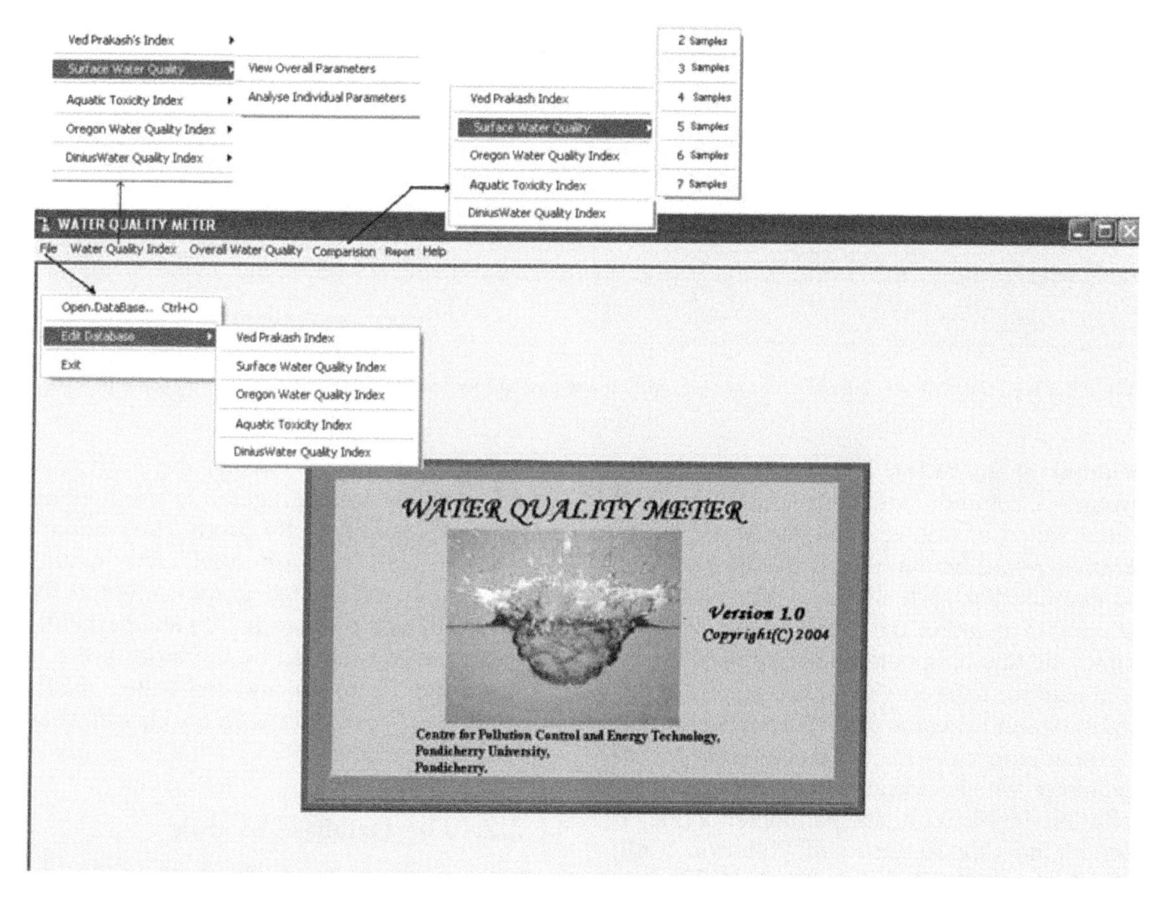

FIGURE 11.3 Main menu of graphic user interface of QUALIDEX.

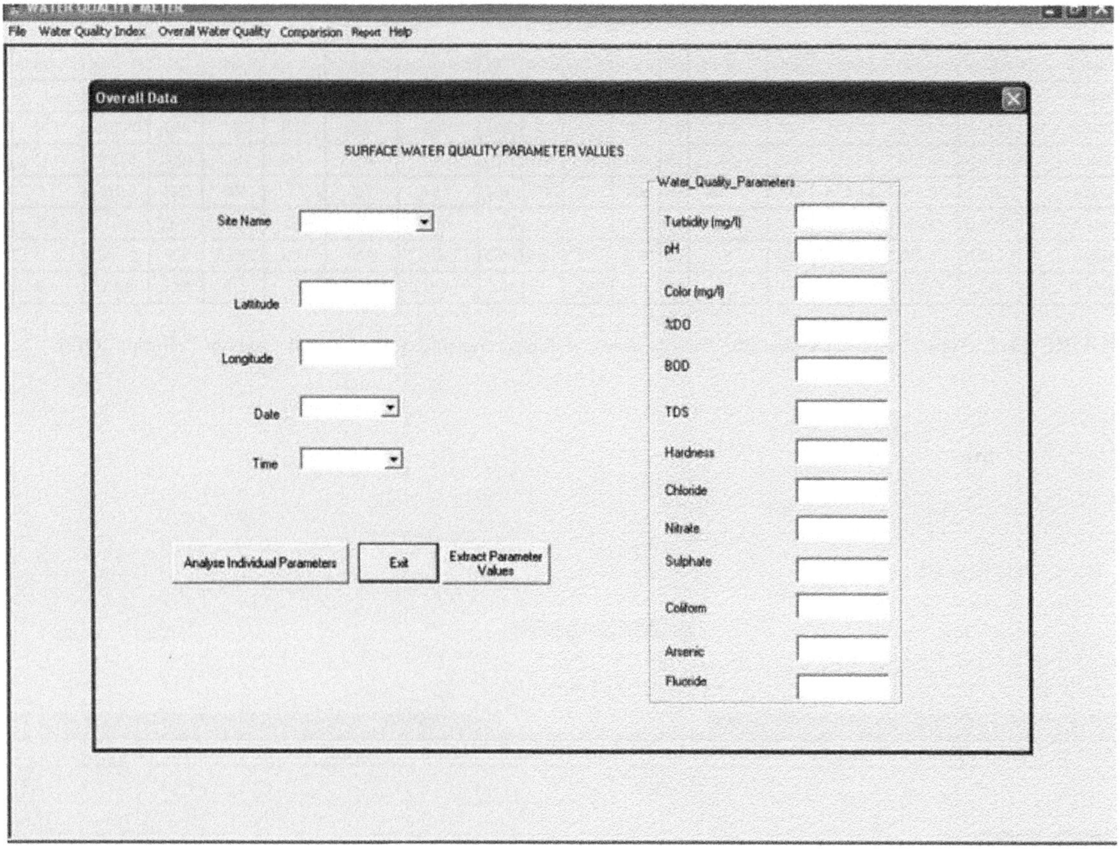

FIGURE 11.4 User-friendly interface for entering, editing and saving raw water-quality data into the database module.

included in an index. This can be accessed through a common interactive interface from where the user has to navigate through the parameter-specific dialogue boxes in a sequential manner. For each water-quality parameter included in an index, the user is expected to first extract the raw parameter value of the site from the database. The software subsequently evaluates the subindex value of the parameter. Option is available to view the subindex curve for the parameter which indicates the variation of the pollution levels with the parameter value. In case of the Overall Index of Pollution (OIP), there is the option for the classification of the water quality with respect to each of the individual parameters. Finally, the individual subindex values are aggregated to produce the overall water-quality index score. The comparisons of the state of individual water-quality parameters are automated graphically and the overall status of the water quality at a particular site and time is indicated by the index score in the water-quality meter and the water quality is classified in accordance with the classification scheme of the index.

11.2.2. The Database Module

The water analysis data covering the parameters needed for each of the five indices

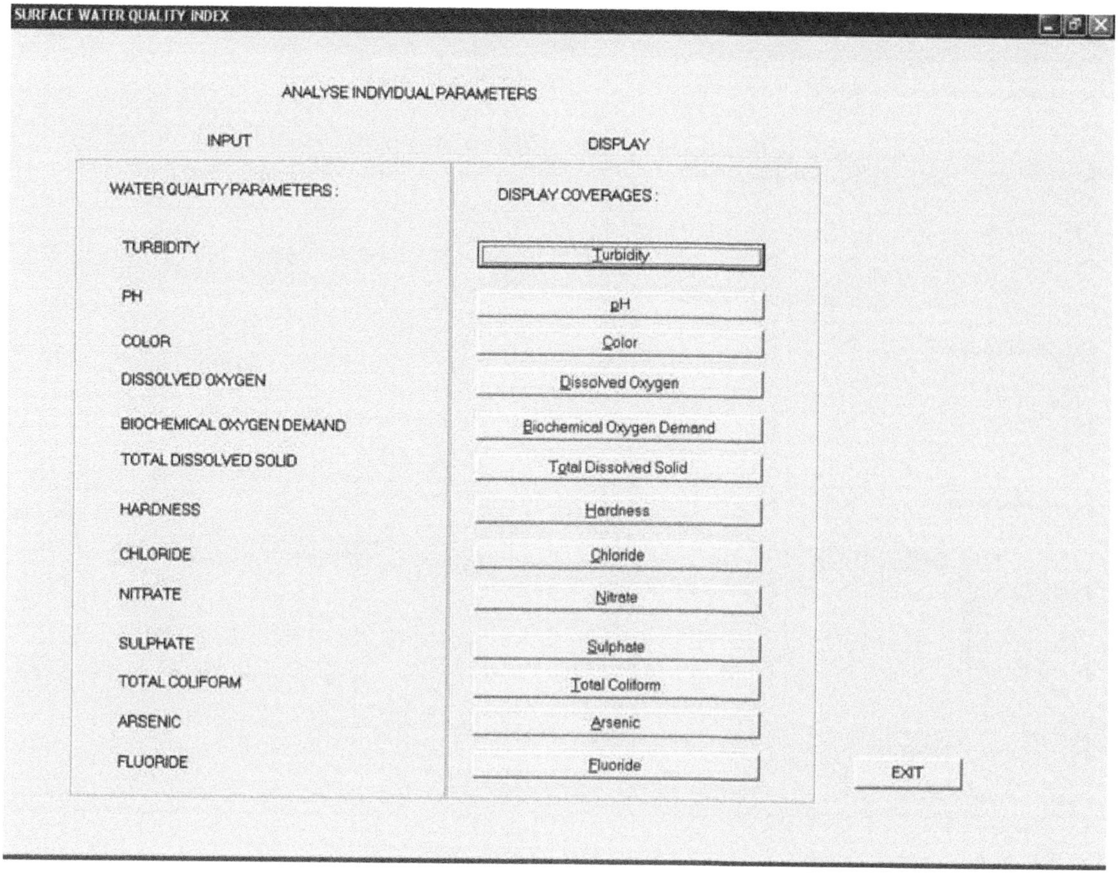

FIGURE 11.5 Common interface to analyse the individual parameters of the Overall Index of Pollution (OIP).

available in the QUALIDEX are stored in the database module. The module comprises of five MS Access files, one for each of the component index. A typical data set as keyed into this module is depicted in Figure 11.2. For each of the index, the corresponding parameter values of a specific site at a particular date and time are stored in the respective MS Access spreadsheet. The MS Access files have been connected to the software through the Open Database Connectivity (ODBC) data source administrator of windows. A user-friendly interface has been developed within the software itself to enable the users to enter, save and edit raw water-quality data as shown in Figure 11.4.

11.2.3. The New Water Quality Index (NWQI) Submodule

This submodule enables the users to generate their own water-quality index and compare the results with the other well-known indices presently included in QUALIDEX. It provides flexibility to the users to choose the component water-quality parameters which they feel have a significant influence upon a specific water quality, associate appropriate subindex

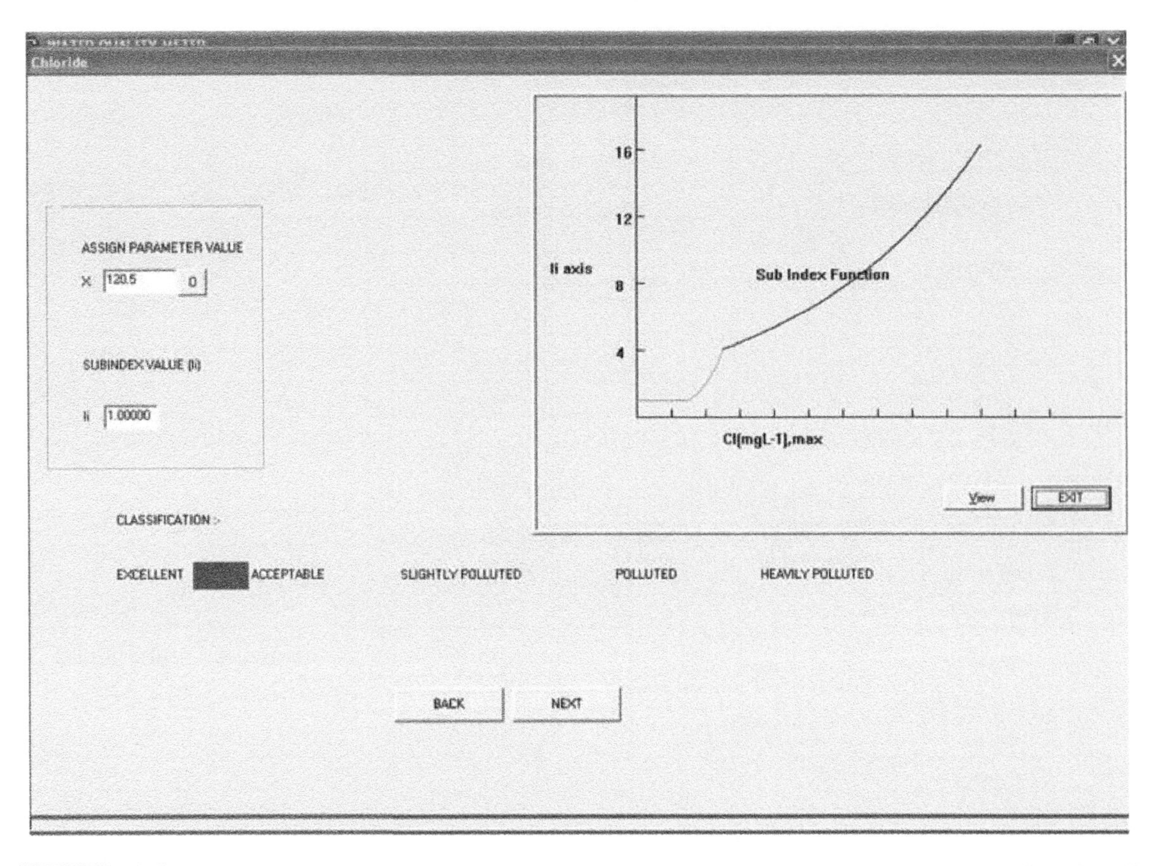

FIGURE 11.6 Sample dialogue box for the evaluation of subindex function of a component parameter of the Overall Index of Pollution (OIP).

functions with them and thereby assess the overall status of the water quality on the basis of their perception of the water pollution problem in their local areas. The minimum operator aggregation function has been employed so as to identify the parameter with the lowest subindex score that plays a most significant role in depleting the water quality. Consequently, after identifying the parameter that contributes maximum to the pollution, appropriate counter measures may be taken to manage the pollution. A sample screenshot of this module has been shown in Figure 11.9, while a typical work flow for operating the New Water Quality Index has been depicted in Figure 11.10.

11.2.4. Water-Quality Comparison Module

Monitoring the variations in the quality of water of a region helps in gaining invaluable insights into the underlying causative factors. A comprehensive assessment of the spatio-temporal variations of water quality is of considerable importance to the policy makers in developing appropriate policies and mitigation measures. QUALIDEX has a water-quality

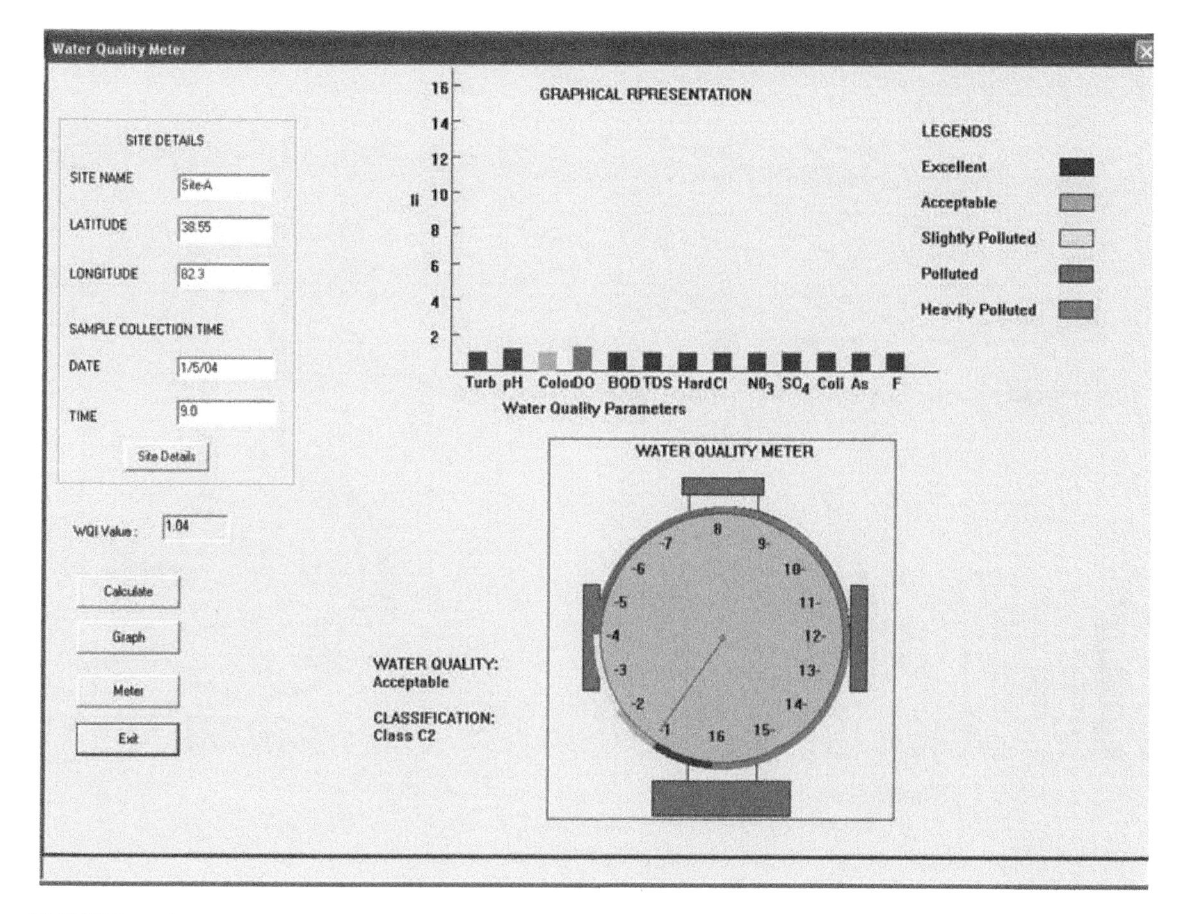

FIGURE 11.7 Sample dialogue box for the overall water-quality assessment for the Overall Index of Pollution (OIP).

comparison module that has been developed primarily with an aim to:

1. Assess the spatial variations in water quality (i.e., the variations at different sites)
2. Assess the temporal variations in the water quality of a site (i.e., the variations at different time intervals)

There is provision for the comparison of water quality with respect to each of the five water-quality indices included in the software. Through this module, the user has the option to compare the water quality of up to seven samples. In order to assess the spatial variations in the water quality of the samples collected from different sites, the user has to choose the respective sites, date and time of collection of the samples from the drop-down menu. The software subsequently extracts the parameter values of the sites from the database module and evaluates the overall index value for the selected sites. In case the user wants to assess the temporal variations in the water quality of a given site, (s)he needs to select the particular site as well as the dates and times for which (s)he wants the comparisons. The software will extract the parameter values and evaluate the overall index values for the site at different

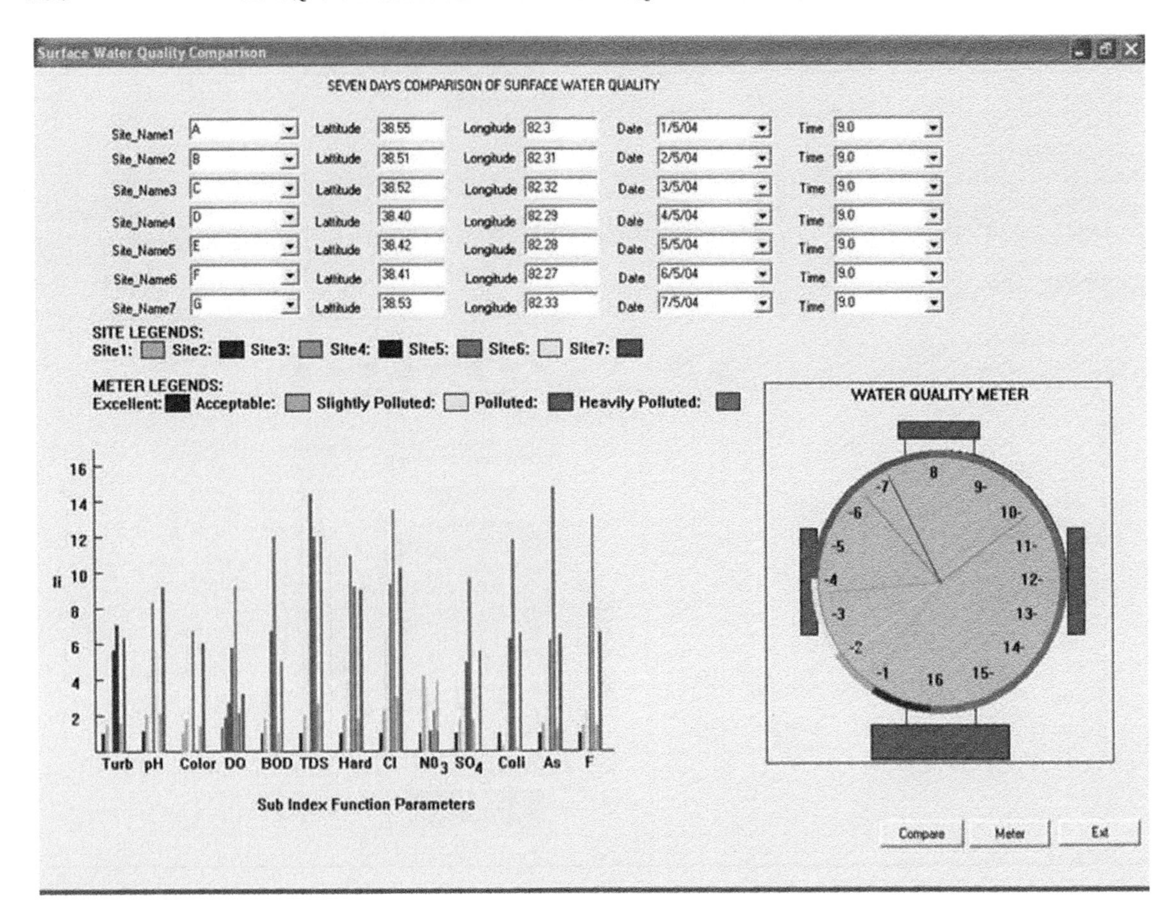

FIGURE 11.8 Sample dialogue box for the water-quality comparison using Overall Index of Pollution (OIP).

time intervals. The final results are automated in the form of user-friendly bar graphs and the water-quality index scores are indicated in the water-quality meter.

11.2.5. The Report Generation Module

The results of the calculations performed in the water-quality index generation module and the water-quality comparison module may be viewed in the form of compact summary tables through this module. For each run, the corresponding summary tables are stored in html files within the software which can be accessed from the report generation menu on the GUI of QUALIDEX. General format of the report generated for the Water Quality Comparison Module has been depicted in Table 11.2. The comparison matrix consists of a number of cells. Each of the cells of the matrix has been divided into two sectors which contain the water-quality parameter value of the site and the corresponding evaluated subindex function value. At the bottom of the matrix, the overall index score is evaluated by aggregating all the subindex function values of component parameters for a particular site. The water-quality classification is

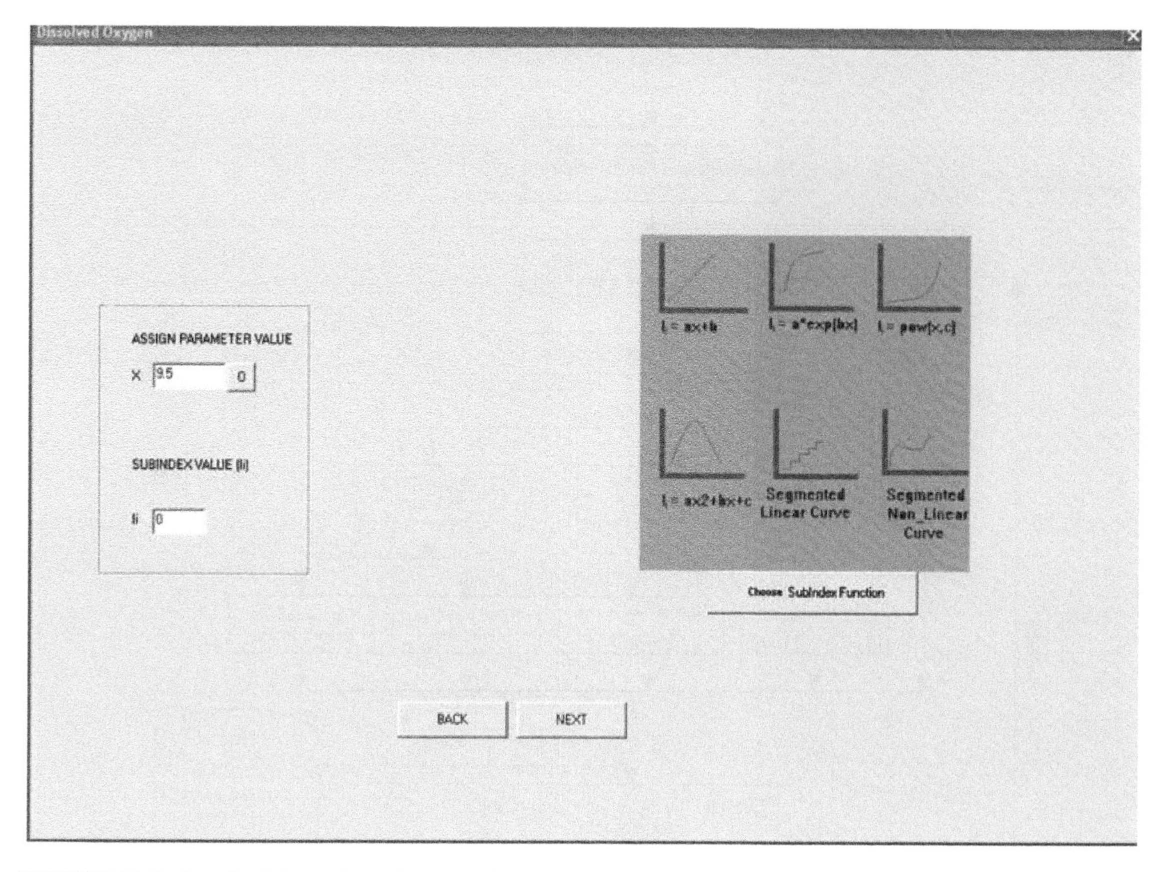

FIGURE 11.9 Sample dialogue box which enables the user to select the subindex function for a parameter in the new water quality index.

also depicted depending on the aggregate index value.

Sample screenshots of QUALIDEX have been shown in Figures 11.3–11.8 for the Overall Index of Pollution (OIP). Figure 11.3 shows the main menu forming the graphic user interface of QUALIDEX from where the corresponding modules of QUALIDEX may be accessed. Figure 11.4 shows the interface developed for entering raw water-quality data, editing data, as well as saving the data into the database module. Figure 11.5 represents the standard interface developed for analysing the component water-quality parameters of an index (Overall Index of Pollution in this case). Figure 11.6 represents a sample dialogue box of the Overall Index of Pollution developed for analysing an individual parameter by generating the subindex function curve and evaluating the subindex function of the parameter. Figure 11.7 depicts the standard format of overall water-quality assessment performed in the Water Quality Index Generation Module. This includes evaluation of the WQI value, a graphical comparison of the water-quality status of the components parameters of the module and classification of the water quality. The figure also shows the virtual instrument 'water quality

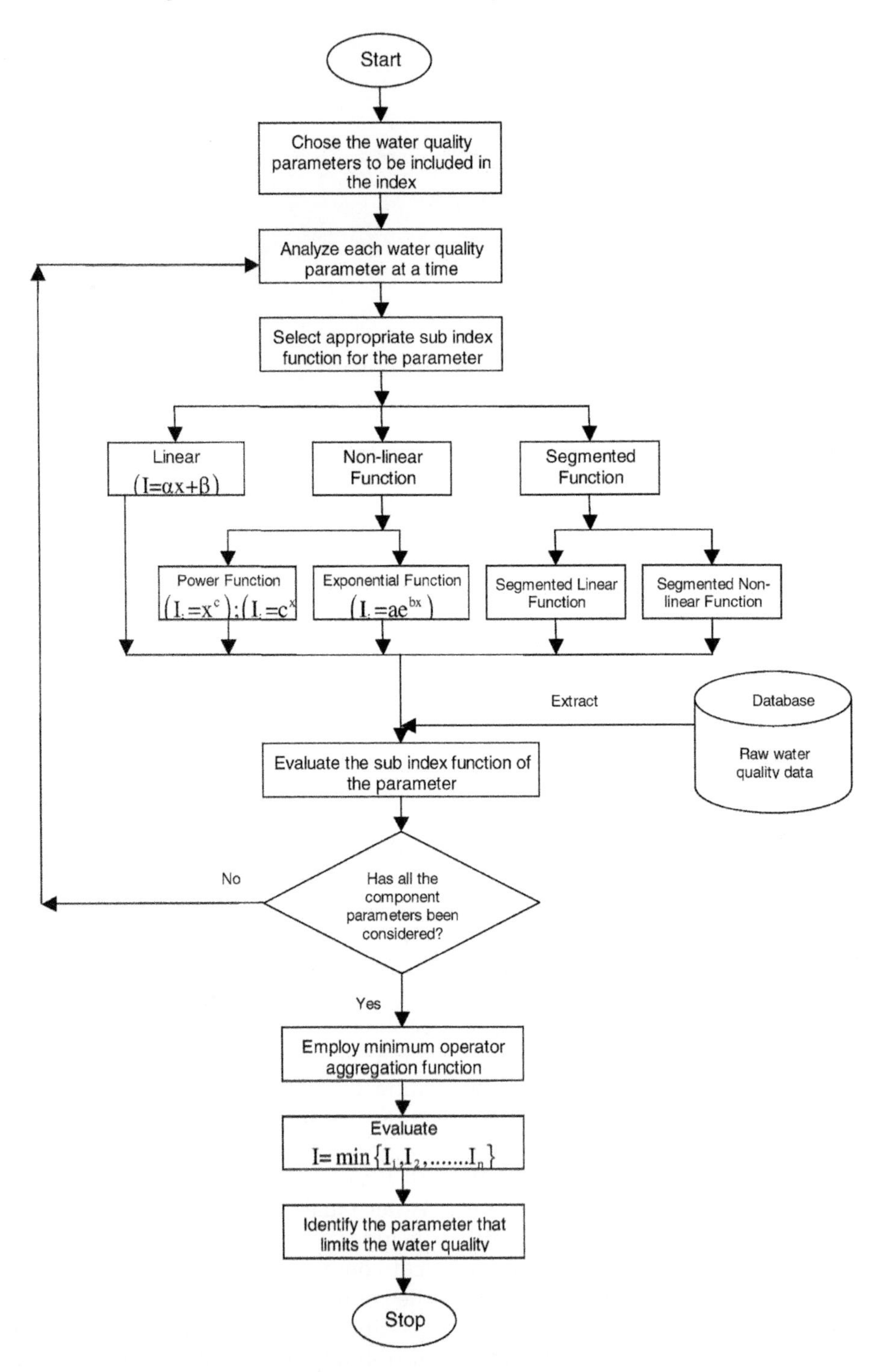

FIGURE 11.10 Sequence of steps for the generation of the new water-quality index submodule.

TABLE 11.2 General Format of the Report Generated for the Water-quality Comparison Module for the Overall Index of Pollution (OIP)

Overall Index of Pollution (OIP)														
Parameters	Parameter values for the seven different sites and corresponding Sub-index values													
	Site 1	S.I	Site 2	S.I	Site 3	S.I	Site 4	S.I	Site 5	S.I	Site 6	S.I	Site 7	S.I
Turbidity	2.5	1.00	7.5	1.50	50.0	2.72	150.5	5.63	200	7.07	8.0	1.60	175	6.34
pH	6.8	1.20	6.2	2.09	5.5	4.0	4.8	7.64	5.2	5.28	7.8	2.09	9.3	8.38
Color	12	1.01	120	1.76	200	2.66	500	6.67	1000	13.33	68.5	1.42	450	6.0
% DO	88	1.33	80	1.88	68	2.7	35	5.79	18	9.27	120	2.12	140	3.17
BOD (mg/l)	1.0	1.00	2.8	1.87	5.5	3.67	10.0	6.67	18	12.0	1.85	1.0	7.5	5.0
TDS (mg/l)	350	1.00	1000	2.0	2000	8.0	2800	14.40	4500	12.0	1200	2.64	2500	12.0
Hardness (mg/l)	67.5	1.00	100	2.0	240.0	3.95	450	10.97	650	9.20	85	1.86	410	9.03
Cl (mg/l)	120.5	1.00	210	2.29	505.5	6.31	700	9.33	900	13.51	230	3.03	750	10.23
NO_3 (mg/l)	15.2	1.00	35.5	4.21	48.5	3.55	75	1.16	150	2.31	41.5	3.89	75	1.15
SO_4 (mg/l)	120.5	1.00	200.5	1.75	310.5	2.62	600	4.93	1200	9.70	205.5	1.79	678	5.55
Coli	45	1.00	450	0.27	4000	2.41	7500	6.22	12000	11.82	300	0.19	7800	6.60
As (mg/l)	.0045	1.00	.0075	1.50	.025	2.74	.075	6.16	1.12	14.80	.006	1.20	.08	6.51
F (mg/l)	1.0	1.00	1.35	1.46	1.75	2.12	5.5	8.26	8.5	13.18	1.35	1.46	4.5	6.62
Overall Index Score	**1.04**		**1.89**		**3.67**		**7.22**		**10.27**		**1.87**		**6.66**	
Water quality Classification	Acceptable		Acceptable		Slightly polluted		Polluted		Heavily polluted		Acceptable		Polluted	

meter' with which changes in the water quality occurring with, say, inflow of a pollutant or seawater can be visually seen as well as recorded.

The subindex values and the corresponding index scores have been compared for the Overall Index of Pollution (OIP) with the help of water quality of seven different sites. The corresponding output generated by the Water Quality Comparison Module has been depicted in Figure 11.8. The figure also displays the readings on the 'water quality meter'.

Reference

Sarkar, C., Abbasi, S.A., 2006. Qualidex – a new software for generating water quality indice. Environmental Monitoring and Assessment 119 (1–3), 201–231.

CHAPTER

12

Water-Quality Indices Based on Bioassessment: An Introduction

OUTLINE

12.1. INTRODUCTION

As can be seen from Part 1 of this book, the water-quality indices (WQI) of the modern and post-modern times have been predominantly based on the assessment of physical and chemical characteristics, most often pH, dissolved oxygen, temperature, turbidity, hardness, total solids, nitrogen, phosphorous, some metals and some pesticides. Among 'biological' characteristics only faecal coliforms and biological oxygen demand (BOD) have featured in most of these indices. Beginning from Horton's Index (Horton, 1965), which is regarded as the first modern WQI, to the post-modern indices (Khan et al., 2003; Parinet et al., 2004; Sarkar and Abbasi, 2006; Kannel et al., 2007), this has been a common feature of the frequently used indices. National water-quality indices, such as the Canadian WQI (CCME, 2001) and the one developed by India's Central pollution Control Board (Sarkar and Abbasi 2006) as also well-known provincial or state indices such as the Oregon WQI (Cude 2001, 2002) and the British Columbia WQI (Khan et al., 2003) also follow this trend. One of the rare exceptions is the Florida Stream WQI (SAFE 1995) which includes micro-invertebrates in its repertoire.

In recent years, increasing concern has been expressed for this near-total reliance on indices based on physcio-chemical parameters; to the neglect of biological parameters. There are two reasons for this concern:

1. Any interpretation of water quality by physico-chemical parameters is restricted to the parameters actually measured. For example, we may find a water sample with its

Water Quality Indices DOI: 10.1016/B978-0-444-54304-2.00012-9 pH, salinity, hardness, BOD, chemical oxygen demand (COD), *etc.*, all within limits for drinking. But that water may contain harmful level of some heavy metal, or some pesticide or even radioactivity! We have dwelt upon these aspects earlier in chapter 4. There are so many natural and anthropogenic chemicals that can be present in water that it is practically impossible to analyse each and every one of the chemicals. The great ability of water to dissolve other chemicals – due to which it is called 'the universal solvent' – adds to the difficulty in analysing any water sample to its full extent. On the other hand, aquatic organisms, specially the *community structure* of organisms such as plankton, macroinvertebrates, fishes and benthos, fairly reflect not only the current water quality but also the overall ecosystem health of a water body (Figure 12.1). It can even give an indication of the direction from which the ecosystem health has come to the point of analysis and where it is likely to go. Moreover, the state of a water body also reflects the state of the environment *around* it (Figure 12.2). This enhances the value of bioassessment even more. In other words, the community structure of the organisms (biota) of an aquatic habitat or an ecosystem integrates and reflects the cumulative effects of the factors impacting that habitat or the ecosystem over time. Indices based predominantly on physico-chemical parameters are unable to do so.

2. It has been recognised that the under-utilisation of bioassessment in the past with over-reliance on physico-chemical assessment has been a major factor responsible for the deterioration of the ecological integrity of river ecosystems (Karr and Chu 1995, 2000; Roux et al., 1999, Dallas 2002). Whereas North America, Australia, New Zealand, Central Europe and Western Europe have had biotic indices, other regions of the world, notably Asia, have been almost exclusively relying on 'abiotic' indices. This is a worrisome situation, made alarming in recent years due to the hitherto unaccounted for, but likely to be very strong, *additional* impact of global warming on aquatic ecosystems (Abbasi and Abbasi, 2011a). There is an urgent need for much more widespread use of biotic indices in conjunction with the indices based on physical and chemical parameters.

12.2. BIOTIC INDICES IN THE CONTEXT OF THE EVOLUTION OF WATER-QUALITY INDICES

The concept of a water-quality Index (WQI) was introduced in its rudimentary form more than 160 years ago, when presence or absence of certain organisms in a water source was used in Germany as an indicator of the fitness or otherwise of that water source. The development of that index came in the wake of the realisation that human activities produced pollution harmful to the biota (Davis and Simon, 1995; Perry & Vanderklein, 1996). Efforts were made to track the extent of biological degradation; the latter was even considered an indicator of the presence of human activities. The first-ever WQI was, thus, a 'biotic' index.

However, by and by, the focus of water-quality evaluation shifted, for much of the first half of the twentieth century, to the effects of chemical contaminants; rarely were connections between chemical criteria and ambient biotic condition documented. A few deviations to this general trend began to occur from 1964 onwards, but were largely restricted to the United States of America (USA) and some parts of Europe. Even now, as mentioned earlier, the use of biotic indices is very sparse, if at all, in most of the developing countries. In 1964, the Trent Biotic Index (TBI) was developed for streams of Florida (USA). In 1981, the first multimetric index – the Index of Biotic Integrity (IBI) – was introduced,

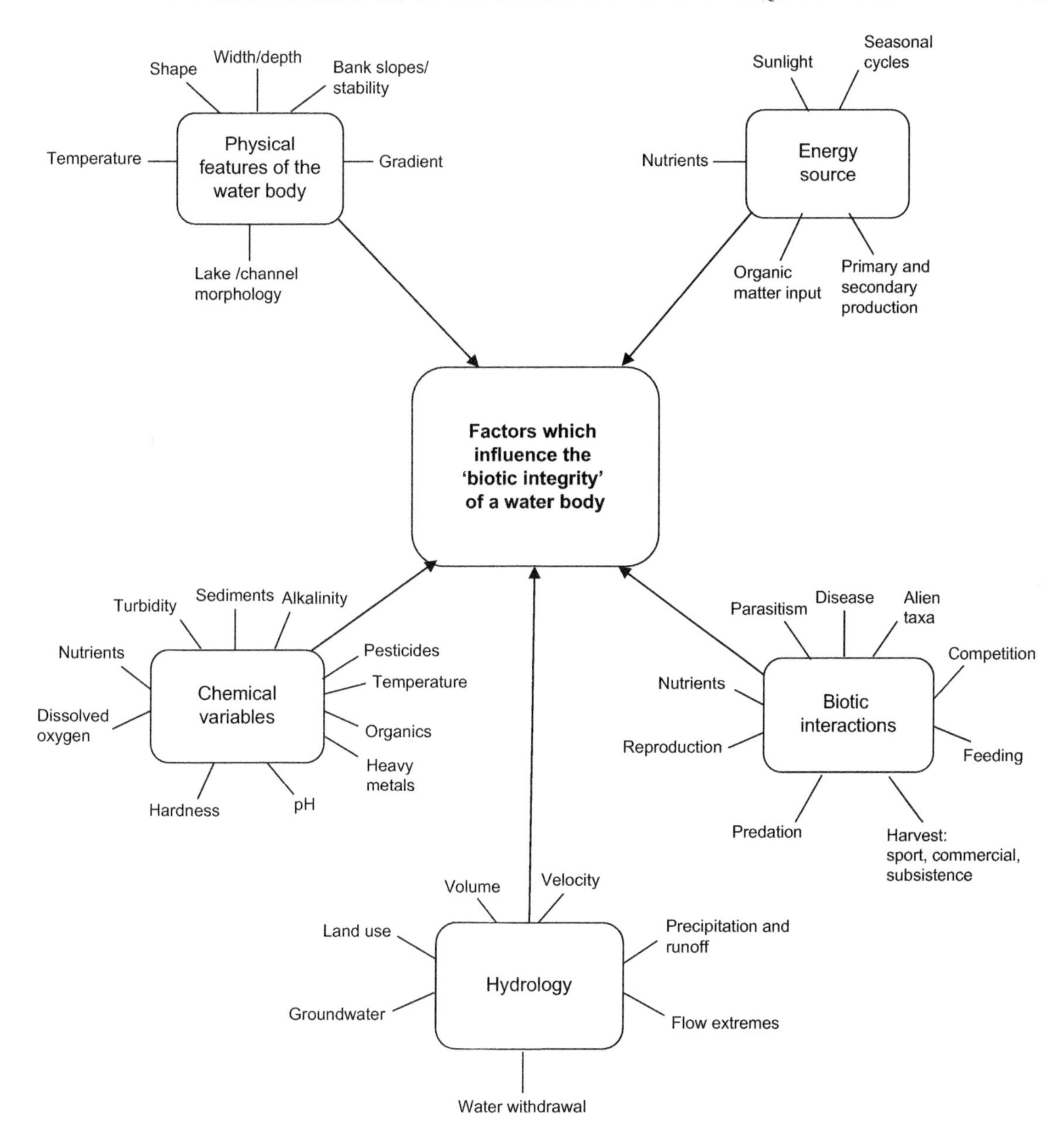

FIGURE 12.1 Factors which influence the biotic integrity of a water body (Abbasi and Abbasi, 2011b).

also in the USA. The subsequent years have seen a slowly increasing reliance on biotic indices as a water-quality management tool, especially in the developed countries. However, the use of biotic indices is yet to catch on in developing countries. India is perhaps the most technologically advanced of the developing countries but there is little advancement here in this field

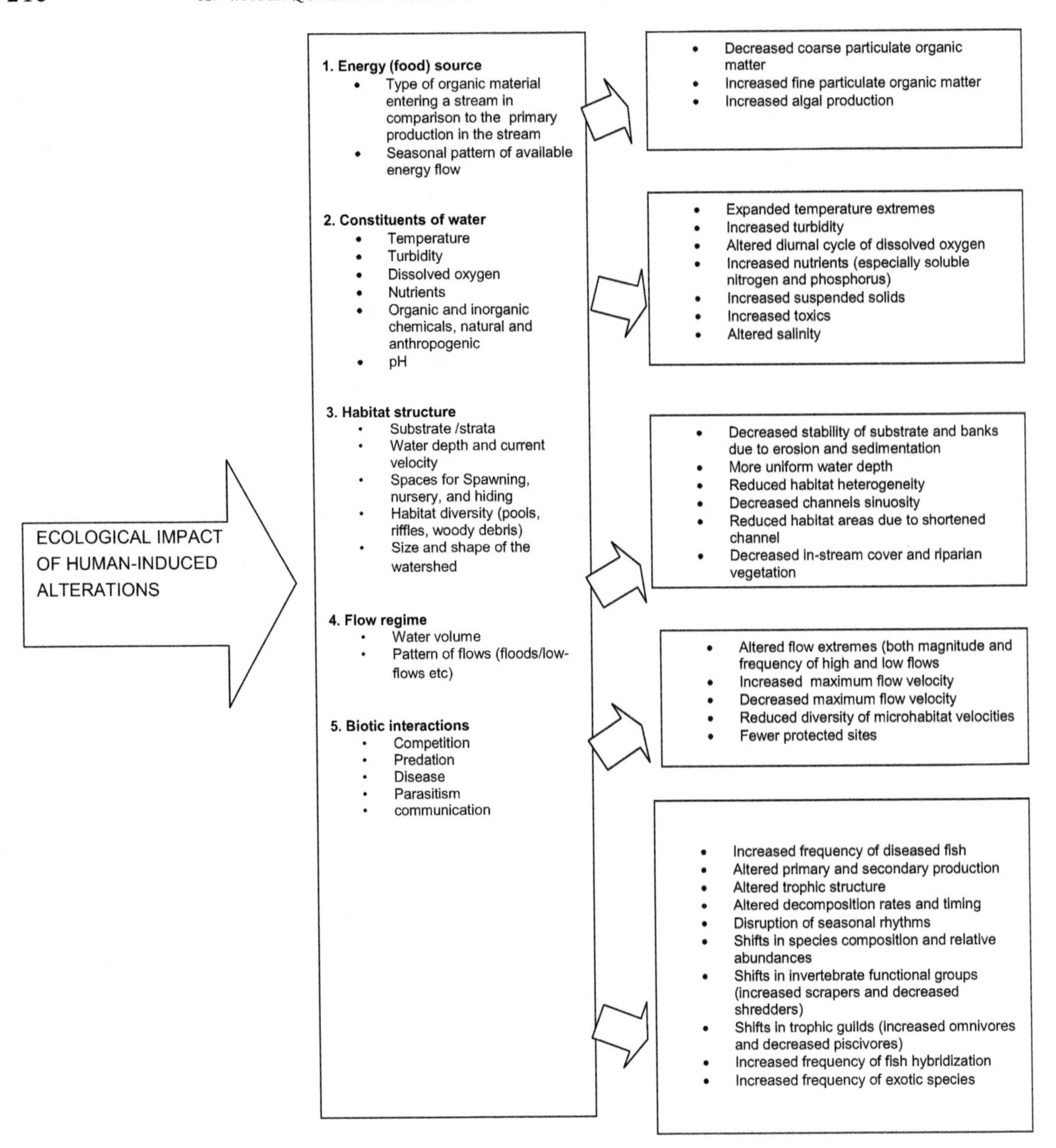

FIGURE 12.2 The myriad factors and cause–effect relationships which influence the biota of a water body and are influenced by it (Abbasi and Abbasi, 2011b).

and there is no accredited biotic index for water-quality assessment.

The importance of even a single species in reflecting the water quality of a water source can be understood from the example of aquatic weeds (Gajalakshmi and Abbasi 2004; Gajalakshmi et al., 2001, 2002). If a pond or a lake is infested with weeds like ipomea or water hyacinth, we can, with just one glance, say with certainty that the pond or the lake does not have clean water. If a lentic habitat is heavily choked with aquatic weeds we can also say, without any further experimentation, that it will be full of the larvae and pupae of mosquitoes and other insects, will have few edible fish, high on BOD and COD, *etc*.

Not all bioindicators are as obvious as aquatic weeds. Also, not all water bodies are so grossly polluted that they are choked with aquatic weeds. But, as explained in the previous section, bio-indicators *are* generally more 'expressive' than physico-chemical parameters. Moreover, as elaborated in the next section, water-quality assessment through physico-chemical parameters represents a stressor-based monitoring approach while the same objective, when addressed through the monitoring of biota, represents the response-based monitoring approach. Both approaches have their distinguishing features and the ideal course is to use both in an integrated fashion. Consequently, biological indicators (or bio-indicators) are increasingly becoming a key element of environmental and water resource management policies in most developed countries (Norris and Norris 1995, Moog and Chovanec 2000). Among the developing countries, only South Africa seems to have used biotic indices extensively, and Serbia to a lesser extent. India does not have any standardised or accredited biotic index and the authors have confined themselves to the use of general indices of species richness, diversity and evenness such as the Shannon Index and the Pieleous' Index (Chari and Abbasi 2003, 2004, 2005; Chari et al., 2003; Ganasan and Hughes 1998; Ingole et al., 2009; Shahnawaz et al., 2010), which are not specific to water quality. Moreover, as detailed later in Section 12.4, these are not 'biotic indices' in the true sense of the term.

12.3. STRESSOR-BASED AND RESPONSE-BASED MONITORING APPROACHES

When the physico-chemical quality of a water-course is measured, it basically represents an attempt to see whether the water is clean or does it carry one or more pollutant. In other words, the monitoring of the water is done for possible 'stressors'. One may also be looking for biological stressors like BOD or pathogens but the known stressors are predominantly physical and chemical. The physico-chemical indices are, therefore, stressor oriented; a stressor being defined as any physical or chemical entity or process that can induce adverse effects on individuals, populations, communities or ecosystems (Thornton et al., 1994).

The stressor-oriented approach attempts, through stressor-specific quality criteria, to link stressors to possible biological responses. This predictive ability is, however, only possible where a known cause—effect relationship exists between a specific stressor and the biological component. But such cause—effect relationships, for a specific suite of conditions, can at best be explored with laboratory bioassays under controlled conditions, and may be far from applicable in real-life situations (Roux et al., 1999). In the stressor-oriented approach, the management focus is on the setting and use of rules for controlling the levels or concentrations of specific stressors, and this approach has a regulatory nature.

The other approach for environmental monitoring is 'response based' wherein the strategy is to assess the environmental health on the basis of the status of the responding organisms. It involves the monitoring of biological or ecological indicators in order to characterise the response of the environment to a disturbance. In turn, 'disturbance' can be defined as any relatively discreet event in time that disrupts ecosystem, community or population structure and that changes the quality of natural resources, availability of substrata or the physical environment. The focus of response monitoring is on the effects resulting from the disturbance. It follows that the response-oriented approach indicates that something has or has not actually gone wrong in response to a stressor.

The collection and use of ecological data in the response-oriented approach are based on an ecosystem management and protection philosophy, in which the focus is on the status and behaviour of the environmental system being monitored and the status of that resource. Environmental response monitoring allows the measurement of how well an ecosystem is functioning, given the degree of perturbation to which it is subject.

From the comparison given in Table 12.1 it would be clear that, for stressor and effects monitoring, the two underlying philosophies and the resulting assessments differ fundamentally. Both approaches have obvious uses and specific benefits in water-quality management. The current thrust is towards operationally integrating the two approaches so that the resulting methodology incorporates the benefits of both.

An offshoot of the stressor-based monitoring approach is the concept of *carrying capacity* or *assimilative capacity* that had gained wide currency during the 1990s. The concept was formulated around the use of the freshwater and marine environments for the disposal of mainly organic wastes and associated effluents. In this context, Cairns (1977) had proposed that the assimilative capacity may be defined as the ability of an ecosystem to cope with certain concentrations of (organic) waste discharges, without suffering any significant deleterious biological effects.

But several assumptions are inherent in the utilisation of the assimilative capacity concept in water-quality management:

1. Each environment has a finite capacity to accommodate some wastes without unacceptable consequences;
2. Such capacity can actually be quantified and subsequently utilised through allocation and management at acceptable impact levels;
3. Unacceptable consequences can be measured and quantified;

TABLE 12.1 Characteristics of Stress-Oriented and Response-Oriented Water-Quality Monitoring Approaches

The Driver	Stress-Oriented Approach	Response-Oriented Approach
Monitoring focus	Stress causing environmental change, i.e., mainly chemical and physical inputs to aquatic systems	Effects (responses) resulting from natural and/or anthropogenic disturbances, e.g., changes in the structure and function of biological communities
Management focus	Water-quality regulation: controlling stressors by regulating their sources (e.g., end-of-pipe focus)	Aquatic ecosystem protection: managing the ecological integrity of aquatic ecosystems (ecosystem or resources focus)
Measurement end points	Concentrations of chemical and physical water-quality variables, e.g., pH, dissolved oxygen, copper	Structural and functional attributes of biological communities, e.g., diversity and abundance of benthic invertebrates
Assessment end points	Compliance or noncompliance with a set criterion or discharge standard	Degree of deviation from a benchmark or desired biological condition

4. The utilisation of the assimilative capacity will not have an injurious effect on those biological processes that contribute significantly to that capacity;
5. Zones of initial mixing or *zones of allowed adverse ecological impact* may be required where significant ecological changes may occur.

Moreover, the assimilative capacity is generally very limited when dealing with toxic substances that are persistent and tend to concentrate in the environment and accumulate in aquatic biota (Abbasi, 1976; Abbasi and Soni 1983; 1984). Also, the utilisation of assimilative capacity must ensure a reference minimum flow condition that will minimise risk. Since cause-and-effect relationships in aquatic ecosystems are not well understood, it is not possible to accurately predict the degree of change that will result from a pollution input, especially when only chemical and physical constituents are being measured. Whereas it is possible to predict the assimilation of conservative constituents like BOD with a fair degree of reliability, that of persistent and toxic substances is fraught with a large degree of uncertainty. The uncertainty is even greater where chemical interaction of multiple stressors, for example, in a complex effluent, occurs. Therefore, stressor-oriented monitoring and management approach should be complemented with the response-oriented approaches (exemplified by the biotic indices) for better water-quality management.

Figure 12.3 illustrates how biomonitoring techniques can be used to quantify biological condition over time. The segment A represents a natural range of variation – this range reflects the condition that will occur, without human interference, within a specific region; it also serves as 'control' or reference region. The segment B signifies a zone of 'acceptable' range of variation or change – the lower end of this zone will determine the lowest quality objective that can be set and managed for. This zone can be subdivided on the basis of more specific biological impact criteria. Political, economic and ethical issues may also be taken into consideration.

The segment C represents an unacceptable range of variation or change – it is essentially a cushion between the lowest allowable management objective and the point where the ecosystem loses its resilience (becomes irreversibly damaged). The cushion provides the safety margin necessary due to the uncertainties

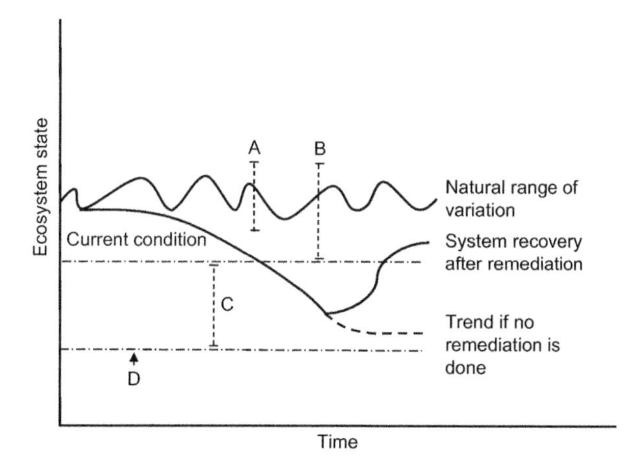

FIGURE 12.3 Naturally functioning ecosystem (A), acceptable range of fluctuations (B), perturbations beyond acceptable range but which can be reversed by remediation (C) and the critical threshold beyond which an ecosystem becomes irretrievable. *(Adapted from Roux et al., 1999).*

associated with ecological factors. The higher the uncertainty and hence unpredictability of what would constitute an irreversible change in a specific region, the bigger the safety margin that should be employed. An example of such ecological uncertainty is the degree to which biota will adapt to selection pressures when a system experiences a certain level of perturbation.

The critical threshold is represented by the segment D. Beyond this threshold the ecosystem will be disturbed irreversibly; it will not be able to recover to its natural equilibrium state (A) or even to an acceptable or desired equilibrium state (B).

From Figure 12.3 it is clear that with only dependable data collected over a long term can a distinction be made between natural and unnatural ranges of variation in an ecological system. The fact that each ecosystem has its own, unique, regime of natural variation adds to the complexity of this task. The manner in which unnatural variations occur also differs according to the type of impact. Furthermore, it may be difficult to identify and characterise variations where there is a collective forcing by anthropogenic and natural impacts.

12.4. BIOTIC INDICES – GENERAL

Biotic indices summarise and present as simple, numeric figures the biological community structure. As with the physico-chemical quality-based indices mentioned earlier, the biotic indices also allow the results to be communicated in a way that is understandable to natural resource managers, decision makers, politicians and the general public (Resh, 1995, Uys et al., 1996, Stark 1998).

Three basic types of indices can be generated (Johnson et al., 1993): diversity indices, comparison (similarity or dissimilarity) indices and biotic indices. The Shannon–Weaver index, the Simpson's diversity and dominance indices, the Pieleous' evenness index, *etc.*, are well-known examples of the first two classes of indices. These indices have been used extensively for aquatic biota and even more extensively for terrestrial and avian biota (Abbasi and Vineethan, 1999; Chari et al., 2003). But these indices overlook many important variables and tend to oversimplify the natural systems which are, in fact, highly complex (Karr, 1981). For example, the Shannon–Weaver index can merely say that two polluted regions are equally diverse but cannot say which of the two contains more beneficial or which one more harmful organisms. Indices have also been developed to assess the risk of pollution due to industrial accidents (Khan and Abbasi 1997a,b; 1998a,b; 1999a,b; 2000a,b; 2001); these indices include impacts on the quality of water resources but none has been specifically applied to aquatic systems. Hence, these indices are not further discussed here. In contrast, biotic indices are more 'expressive' and revealing of ecological health. In biotic indices, each taxon from a particular group of organisms is assigned a sensitivity weighting, or a 'score', based on the tolerance or sensitivity of that taxon to particular pollutants. The scores of all the individual taxa sampled at a site are summed and/or averaged to provide a value by which the ecological health of the biotic community, hence the health of the water body, can be gauged. Some biotic indices include abundance estimates in the scoring system.

Then there are 'Indices of biological integrity' (IBIs), also called 'indices of biotic integrity', which incorporate a suit of indices rather than a single biotic index. These are described in chapter 14. Biological assessment also forms the basis of multivariate approaches like the 'river invertebrate prediction and classification system' (RIVPACS) described further down in chapter 15. An overview of biological methods, which can be used to assess the water quality of an aquatic system, is presented in Table 12.2.

TABLE 12.2 Biological Methods to Assess the Condition of a Water Resource with the Goal of Protecting Human Health, Biotic Integrity or a Specific Resource. An Integrative Index Should Include Metrics Covering all or Most of these Aspects

Technique	Basic Procedure	Variant	Measurables
Bioassay	Exposing test organisms in a laboratory, to various concentrations of suspected toxicants or dilutions of whole effluent	Single Multispecies	• LC_{50} • LC_{25} • Maximum allowable toxic concentration • Safe concentration
Biosurvey	Collecting a representative portion of the organisms living in the water body of interest to determine the characteristics of the aquatic community	Individual/species population (may involve selection of indicator species)	• Tissue analysis for bioaccumulation • Bookmarkers–genetics or physiology • Biomass/yield • Growth rates • Gross morphology (external or internal) • Behaviour • Abundance/density • Variation in population size • population age structure • Disease or parasitism frequency
		Community/ecosystem (may involve indicator taxa or guilds)	*Structure* • Species richness/diversity • Relative abundances among species • Tolerant/intolerants • Abundance of opportunists • Dominant species • Community trophic structure • Extinction *Function* • Production/respiration ratio • Production/biomass ratio • Biogeochemical cycles/nutrients • Decomposition *Landscape* • Habitat fragmentation/patch geometry • Linkages among patches • Cumulative effects across landscapes

References

Abbasi, S.A., Vineethan, S., 1999. Ecological impacts of *Eucalyptus tereticornis-globulus*. (Eucalyptus hybrid) plantation in a mining area. Indian Forester 125 (2), 163–186.

Abbasi, S.A., 1976. Extraction and spectrophotometric determination of vanadium (V) with N-[p-N, N-dimethylanilino-3- methoxy-2-naphtho] hydroxamic acid. Analytical Chemistry 48 (4), 714–717.

Abbasi, S.A., Soni, R., 1983. Stress-induced enhancement of reproduction in earthworm *Octochaetus pattoni* exposed to chromium (VI) and mercury (II) - implications in environmental management. International Journal of Environmental Studies 22 (1), 43–47.

Abbasi, S.A., Soni, R., 1984. Teratogenic effects of chromium (VI) in environment as evidenced by the impact on larvae of amphibian *Rana tigrina*: Implications in the environmental management of chromium. International Journal of Environmental Studies 23 (2), 131–137.

Abbasi, T., Abbasi, S.A., 2011a. Ocean acidification: the newest threat to global environment. Critical Reviews in Environmental Science and Technology 41 (8), 1601–1663.

Abbasi, T., Abbasi, S.A., 2011b. Water quality indices base on bioassessment–the biotic indices. Journal of Water and Health (IWA Publishing) 9 (2), 330–348.

Cairns Jr., J., 1977. Aquatic ecosystem assimilative capacity. Fisheries 2, 5–13.

CCME, 2001. Canadian water quality guidelines for the protection of aquatic life: CCME water quality index 1.0, technical report. In: Canadian Environmental Quality Guidelines. 1999 Canadian Council of Ministers of the Environment, Winnipeg.

Chari, K.B., Abbasi, S.A., 2003. Assessment of Impact of Land use Changes on the Plankton Community of a Shallow Fresh water Lake in South India by GIS and Remote sensing. Chemical and Environmental Research 12, 93–112.

Chari, K.B., Abbasi, S.A., 2004. Implications of environmental threats on the composition and distribution of fishes in a large coastal wetland (Kaliveli). Hydrology Journal 23, 85–93.

Chari, K.B., Abbasi, S.A., 2005. A study on the fish fauna of Oussudu – A rare freshwater lake of South India. International Journal of Environmental Studies 62, 137–145.

Chari, K.B., Abbasi, S.A., Ganapathy, S., 2003. Ecology, habitat and bird community structure at Oussudu lake: Towards a strategy for conservation and management. Aquatic Conservation: Marine and Freshwater Ecosystems 13, 373–386.

Cude, C.G., Smith, D.G., et al., 2002. Reply to the discussion on – Oregon water quality index: A tool for evaluating water quality management effectiveness. Journal of American Water Resources Association 38 (1), 315–318.

Cude, C.G., 2001. 'Oregon water quality index: A tool for evaluating water quality management effectiveness'. Journal of American Water Resources Association 37 (1), 125–137.

Dallas, H.F., 2002. Spatial and Temporal Heterogeneity in Lotic Systems: Implications for Defining Reference Conditions for Macroinvertebrates. PhD thesis, University of Cape Town, South Africa.

Davis, W.S., Simon, T.P. (Eds.), 1995. Biological Assessment and Criteria: Tools for Water Resource Planning and Decision Making. Lewis Publishers, London.

Gajalakshmi, S., Abbasi, S.A., 2004. Neem leaves as a source of fertilizer-cum-pesticide vermicompost. Bioresource Technology 92 (3), 291–296.

Gajalakshmi, S., Ramasamy, E.V., Abbasi, S.A., 2001. Screening of four species of detritivorous (humus – former) earthworms for sustainable vermicomposting of paper waste. Environmental Technology 22 (6), 679–685.

Gajalakshmi, S., Ramasamy, E.V., Abbasi, S.A., 2002. High-rate composting-vermicomposting of water hyacinth (Eichhornia crassipes, Mart. Solms). Bioresource Technology 83 (3), 235–239.

Ganasan, V., Hughes, R.M., 1998. Application of an index of biological integrity (IBI) to fish assemblages of the rivers Khan and Kshipra (Madhya Pradesh). India Freshwater Biology 40, 367–383.

Horton, R.K., 1965. An index number system for rating water quality. Journal of Water Pollution Control Federation 37 (3), 300–306.

Ingole, B., Sivadas, S., Nanajkar, M., Sautya, S., Nag, A., 2009. A comparative study of macrobenthic community from harbours along the central west coast of India. Environmental Monitoring and Assessment 154, 135–146.

Johnson, R.K., Wiederholm, T., Rosenberg, D.M., 1993. Freshwater biomonitoring using individual organisms, populations and species assemblages of Benthic Macroinvertebrates. In: Rosenberg, D.M., Resh, V.H. (Eds.), Freshwater Biomonitoring and Benthic Macroinvertebrates. Chapman and Hall, New York, pp. 40–125.

Kannel, P.R., Lee, S., Lee, Y.S., Kanel, S.R., Kahn, S.P., 2007. Application of water quality indices and dissolved oxygen as indicators for river water classification and urban impact assessment. Environmental Monitoring and Assessment 132, 93–110.

Karr, J.R., 1981. Assessment of biotic integrity using fish communities. Fisheries 6, 21–27.

Karr, J.R., Chu, E.W., 1995. Ecological integrity: reclaiming lost connections. In: Westra, L., Lemons, J. (Eds.), Perspectives on Ecological Integrity. Kluwer Academic Publishers, Dordrecht, pp. 34–48.

Karr, J.R., Chu, E.W., 2000. Sustaining living rivers. Hydrobiologia 422 & 423, 1–14.

Khan, F.I., Abbasi, S.A., 1997a. Accident hazard index: A multi-attribute method for process industry hazard rating. Process Safety and Environmental Protection 75, 217–224.

Khan, F.I., Abbasi, S.A., 1997b. Risk analysis of a chloralkali industry situated in a populated area using the software package MAXCRED-II. Process Safety Progress 16, 172–184.

Khan, F.I., Abbasi, S.A., 1998a. Multivariate Hazard Identification and Ranking System. Process Safety Progress 17, 157–170.

Khan, F.I., Abbasi, S.A., 1998b. DOMIFFECT (DOMIno eFFECT): User-friendly software for domino effect analysis. Environmental Modeling and Software 13, 163–177.

Khan, F.I., Abbasi, S.A., 1999a. Assessment of risks posed by chemical industries – Application of a new computer automated tool MAXCRED-III. Journal of Loss Prevention in the Process Industries 12, 455–469.

Khan, F.I., Abbasi, S.A., 1999b. The world's worst industrial accident of the 1990s: What happened and what might have been – A quantitative study. Process Safety Progress 18, 135–145.

Khan, F.I., Abbasi, S.A., 2000a. Analytical simulation and PROFAT II: A new methodology and a computer automated tool for fault tree analysis in chemical process industries. Journal of Hazardous Materials 75, 1–27.

Khan, F.I., Abbasi, S.A., 2000b. Towards automation of HAZOP with a new tool EXPERTOP. Environmental Modelling and Software 15, 67–77.

Khan, F.I., Abbasi, S.A., 2001. An assessment of the likehood of occurrence, and the damage potential of domino effect (chain of accidents) in a typical cluster of industries. Journal of Loss Prevention in the Process Industries 14, 283–306.

Khan, F.I., Husain, T., Lumb, A., 2003. Water quality evaluation and trend analysis in selected watersheds of the Atlantic region of Canada. Environmental Monitoring and Assessment 88, 221–242.

Moog, O., Chovanec, A., 2000. Assessing the ecological integrity of rivers: walking the line among ecological, political and administrative interests. Hydrobiologia 422 & 423, 99–109.

Norris, R.H., Norris, K.R., 1995. The need for biological assessment of water quality: Australian perspective. Australian Journal of Ecology 20, 1–6.

Parinet, B., Lhote, A., Legube, B., 2004. Principal component analysis: An appropriate tool for water quality evaluation and management – application to a tropical lake system. Ecological Modeling 178, 295–311.

Perry, J., Vanderklein, E., 1996. Water Quality: Management of a Natural Resource. Blackwell Science, Cambridge, MA.

Resh, V.H., 1995. Freshwater Benthic Macro invertebrates and rapid assessment procedures for water quality monitoring in developing and newly industrialized countries. In: Davis, W.S., Simon, T.P. (Eds.), Biological Assessment and Criteria. Tools for Water Resource Planning and Decision-making. Lewis Publishers, Boca Raton, pp. 167–177.

Roux, D.J., Kempster, P.L., Kleynhans, C.J., Vanvliet, H.R., Du preez, H.H., 1999. Integrating stressor and response monitoring into a resource-based water quality assessment framework. Environmental Management 23, 15–30.

SAFE, 1995. Strategic Assessment of Florida's Environment, Florida Stream Water Quality Index, Statewide Summary. available at, http://www.pepps.fsu.edu/safe/pdf/swq3.pdf.

Sarkar, C., Abbasi, S.A., 2006. Qualidex – A new software for generating water quality indice. Environmental Monitoring and Assessment 119, 201–231.

Shahnawaz, A., Venkateshwarlu, M., Somashekar, D.S., Santosh, K., 2010. Fish diversity with relation to water quality of Bhadra River of Western Ghats (INDIA). Environmental Monitoring and Assessment vol. 161 (1–4) February, 83-91.

Stark, J.D., 1998. SQMCI: a biotic index for freshwater macroinvertebrate coded-abundance data. New Zealand Journal of Marine and Freshwater Research 32, 55–66.

Thornton, K.W., Saul, G.E.,Hyatt, D.E., 1994. Environmental monitoring and assessment program assessment framework. Report No. EPA/620/R-94/016. US Environmental Protection Agency, Research Triangle Park, North Carolina.

Uys, M.C., Goetsch, P.A., O'keeffe, J.H., 1996. National Biomonitoring Programme for Riverine Ecosystems: ecological indicators, a review and recommendations. NBP Report Series No. 4. Institute for Water Quality Studies, Department of Water Affairs and Forestry, Pretoria, South Africa.

[illegible] Environmental [illegible] and [illegible]

Khan, F.I., Abbasi, S.A., [illegible]. [illegible] index, [illegible] indicators. [illegible] Application of a new computer automated tool MAXCRED. Journal of Loss Prevention in the Process Industries 12, [illegible].

Khan, F.I., Abbasi, S.A., [illegible]. The world's worst industrial accident of the 1990s. What happened and what might have been: a quantitative study. Process Safety Progress 18, [illegible].

[illegible] simulation and [illegible]. [illegible] Application [illegible]. Journal [illegible] Materials [illegible].

Khan, F.I., Abbasi, S.A., [illegible].

Khan, F.I., Abbasi, S.A., [illegible] of [illegible] and [illegible] of [illegible] industries. Journal of Loss Prevention in the Process Industries [illegible].

[illegible]

[illegible] of [illegible] building [illegible] and administrat[illegible] & [illegible].

[illegible] The need for [illegible] of [illegible]. Journal of [illegible].

[illegible] 2004. [illegible] pollution [illegible] approach to [illegible] quality [illegible] and [illegible].

[illegible] Blackwell Science, Cambridge, [illegible].

[illegible] and rapid [illegible] development of [illegible] in: Davis, W.S., Simon, T.P. (Eds.), Biological Assessment and Criteria: Tools for Water Resource Planning and Decision Making. Lewis Publishers, Boca Raton, pp. 245–272.

Kerans, B.L., Karr, J.R., [illegible], 1999. [illegible] stressor and response [illegible] using [illegible] based [illegible] multimetric [illegible] land [illegible]. Hydrobiologia [illegible].

[illegible] Assessment of [illegible]. [illegible] Vol. [illegible].

[illegible]

Smith, R. [illegible] 2010 [illegible] evaluation of water quality of [illegible] [illegible].

[illegible]

[illegible] index [illegible] study [illegible] Report No. [illegible] [illegible], [illegible].

[illegible] Assessment. [illegible] biological [illegible], and [illegible]. Report No. [illegible] Water Quality [illegible] South Africa.

CHAPTER

13

The Biotic Indices

OUTLINE

Water Quality Indices DOI: 10.1016/B978-0-444-54304-2.00013-0

13.1. INTRODUCTION

The Saprobien or Saprobic system, which stems from the research work of Kolkwitz and Marsson on German rivers in the early 1900s, is generally considered to be the first biological scoring system for the assessment of water quality in river ecosystems (Washington, 1984; Rico et al., 1992; Knoben et al., 1995; Verdonschot, 2000; Sandin et al., 2001). Indices based on the Saprobien System are determined by the presence and absence of specific indicator species from a number of different groups and trophic levels (mainly bacteria, algae, protozoans and rotifers, but including some benthic invertebrates and fish) for which the tolerances to organic pollution have been established (Metcalfe-Smith, 1994). Selected components of the total aquatic community are thus used as an indicator for the degree of organic pollution (Friedrich et al., 1996). Most modern biotic indices, on the other hand, are based on the presence and pollution tolerances of the community of organisms sampled from a particular group, such as the benthic macroinvertebrates (Ollis et al., 2006). The features of the Saprobien system are reflected more in the multimetric indices described in the next chapter.

In recent years, to optimise the use of the time and resources available for ecological assessments, there has been increasing emphasis on the use of biotic indices based on community-level rapid bioassessment techniques (Brown, 2001; Dallas, 2002; Metzeling et al., 2003). The latter, which usually involve qualitative (or semi-quantitative) sampling with few or no replicates and limited taxonomic resolution, have been developed to inexpensively monitor problem areas and thereby provide inputs for decision makers as to where more intensive and quantitative studies for arriving at corrective steps need to be undertaken (Resh et al., 1995; Ollis et al., 2006). Numerical simulations to assess the sensitivity of the values of two biotic indices to the sample size and taxonomic resolution by Bigler et al. (2009) reveal that instead of the stipulated count of 400, a count of just 40 diatom values for 50 streams, and 80 values for 60 streams, were sufficient to obtain the same index classification. Further, excluding rare taxa had negligible effect on the indices. These results indicate that it may be possible to adopt reduced taxonomical resolution for some biotic indices for improving the economics of stream monitoring, without sacrificing precision. But this conclusion is specific to the diatom-based indices studied by the authors and cannot be generalised. It can be said that rapid assessment techniques are not a replacement for more traditional quantitative studies and detailed biological surveys, but rather a precursor to these.

Attempts have also been made to employ machine learning (artificial intelligence) techniques such as artificial neural network and genetic programming for selecting ecologically significant input variables in environmental prediction (Chau et al., 2002; Zhao et al., 2006; Muttil and Chau, 2006, 2007; Wu and Chau, 2006; Faisal et al., 2010). However, all said and done, no short cut has thus far proved reliable or robust over large enough a domain to replace the conventional form of any biotic index.

13.2. THE CHALLENGE OF FINDING 'CONTROL' SITES

One of the most critical issues in any bioassessment is the identification of reference ('control') sites and reference conditions. Such sites should be truly reflective of natural, unpolluted conditions and thus serve as reference or 'control' sites with which the test sites can be compared to know whether a certain impact causes an aquatic assemblage or ecosystem to respond in some way that is outside the natural range of variation (Roux et al., 1999b). In other words, the ultimate objective of any bioassessment programme is to facilitate the detection of disturbance at a site, as reflected by one or more components of the biota. Reference conditions facilitate this by defining what is expected at a site and provide a means of comparing observed conditions with expected conditions so that the degree of impairment or deviation from natural conditions can be determined.

Unfortunately, due to widespread human encroachments everywhere, it is very difficult to find nonimpacted sites in most regions, especially in lowland areas, for use as reference or 'control' sites. Consequently, minimally disturbed or least impacted of the available sites are generally used to determine the *best attainable* reference condition (Roux and Everett, 1994; Reynoldson et al., 1997; Norris and Thoms, 1999; Verdonschot, 2000). Once the best attainable reference conditions have been established for the aquatic ecosystems of a region, these can be used as benchmarks to classify the degree of impairment at monitoring sites (Gerritsen et al., 2000; Dallas, 2002) and can form a scientific basis for setting ecological resource quality objectives (Roux and Everett, 1994; Roux et al., 1999a).

In an attempt to circumvent the problem of the near absence of pristine reference sites in the contemporary world, Lavoie and Campeau (2010) have developed an innovative method for assessing past conditions of streams on the basis of diatom assemblages extracted from the guts of fish stored in museums. By using the Canadian diatom index, they were able to compare stream conditions for the 2003–2007 periods with the conditions prevailing in 1925–48. More work along these lines may be helpful in solving the problem of setting appropriate frames of reference for the present-day monitoring.

13.3. THE COST ASSOCIATED WITH THE USE OF BIOLOGICAL ASSESSMENTS OF WATER

An estimate made by the EPA office of Ohio State, USA (Yoder, 1989) gives an indication (Table 13.1) of the relative cost of physico-chemical assessment of water quality in comparison with bioassays and bioassessments (biotic indices). It is at best an illustrative, region-specific assessment, yet it does reveal that the cost of biotic index-based water-quality assessment is the least of the three options. It must be emphasised, however, that water for physico-chemical parameters can be sampled and analysed more quickly than the conducting of biological sampling and identification. In addition, instrument-based continuous monitoring of physico-chemical quality is possible with real-time transmission of data.

Even continuous computation of water-quality indices is possible (Terrado et al., 2010), as detailed in chapter 8. This attribute has made it possible to design virtual 'water quality meter' (Sarkar and Abbasi, 2006), as detailed in chapter 11. Physico-chemical quality can also be assessed, to some extent, by remote sensing. All these advantages are not available with bioassessment of aquatic

TABLE 13.1 Comparative Costs of Physico-chemical Analysis, Bioassays, and Index-based Bioassessment of the Quality of a Water Resource (Yoder, 1989)

Domain	Per Sample*	Per Evaluation
PHYSICO-CHEMICAL WATER QUALITY		
4 samples/site	$1436	$8616
6 samples/site	$2154	$12924
BIOASSAY		
Screening (acute-48-h exposure)	$1191	$3573
Definitive (LC50* and EC50§-48 and 96 h)	$1848	$5544
7-d (acute and chronic effects-7-d exposure single sample)	$3052	$9156
7-d (as above but with composite sample collected daily)	$6106	$18318
INDEX-BASED BIOASSESSMENT		
Macroinvertebrate community	$824	$4120
Fish community	$740	$3700
Fish and macroinvertebrates (combined)	$1564	$7820

* *At the 1989 values.*

organisms (with the exception of aquatic macrophytes which can be surveyed by remote sensing). Moreover, only physico-chemical analysis can identify specific pollutants that may be stressing the biota of a water body.

13.4. ORGANISMS COMMONLY USED IN BIOASSESSMENT

Various organisms have been used in the bioassessment of the water quality and ecological integrity of aquatic ecosystems, including bacteria, protozoans, diatoms, algae, macrophytes, macroinvertebrates and fish (Dallas and Day, 1993, 2004; Barbour et al., 1999; Milner and Oswood, 2000; Brown, 2001; Meloni et al., 2003; Zgrundo and Bogaczewicz-Adamczak, 2004; Suárez et al., 2005; Moreno et al., 2006; Lavoie et al., 2008; Ayari and Afli, 2008; Maggioni et al., 2009; Girgin, 2010; Guo et al., 2010; Zalack et al., 2010). Of these, benthic macroinvertebrates are the most widely used group (Resh et al., 1995; Dallas, 2002), especially for lotic systems (Moog and Chovanec, 2000; Sandin et al., 2001; Fabela et al., 2001). Besides assessment of biotic conditions in impaired streams, these organisms have also been used to assess the impact of stream restoration measures (Selvakumar et al., 2010) and inter-basin transfer of waters (Zhaia et al., 2010).

Recently, Wu et al. (2010) have explored nematode communities from river water and sediments as bioindicators of water quality. They assessed the nematoda in terms of abundance, feeding types, maturity indices and nematode channel ratio (NCR). The sampling sites studied included different levels of pollution and contamination from agricultural, industrial and sewage sources. In general, greater abundance of nematodes was found in the sediment samples than in the water samples. The lowest nematode abundance in sediment samples and the lowest NCR in water samples were both found at the industrial pollution site. Water samples showed positive correlation between the NCR and river pollution index (RPI). Mean maturity indices in sediment samples were inversely correlated with RPI. The pollutant source determined the relationship between NCR and pollution level, while maturity index always showed negative correlation with pollutant level regardless of the pollutant sources. All in all, the nematode abundance and its community structure both appeared to be reliable bioindicators for monitoring long-term river pollution in both qualitative and quantitative terms.

Links between riparian bird assemblages and stream water quality as reflected by macroinvertebrates have also been found (Larsen et al., 2010).

13.5. BIOTIC INDICES FOR FRESHWATER AND SALINE WATER SYSTEMS BASED ON MACROINVERTEBRATES

There are several advantages in using benthic macroinvertebrates in bioassessment (Ollis et al., 2006; Mugnai et al., 2008; Girgin, 2010; Lavoie and Campeau, 2010; Mazor et al., 2010; Cortelezzi et al., 2011). Benthic macroinvertebrates are largely nonmobile, ubiquitous and relatively abundant inhabitants of both lotic and lentic habitats. There are often many species within a community with varying sensitivities to stresses and relatively quick reaction times, resulting in a spectrum of graded, recognisable responses to environmental perturbations. Also, responses to different types of pollution have been established for many common species. Macroinvertebrates have life cycles that are long enough for temporal changes caused by perturbations to be detected, but short enough to enable the observation of recolonisation patterns following perturbation. They are relatively easy and inexpensive to collect, particularly if qualitative sampling is undertaken, and are well suited to the experiments required for biomonitoring. Studies have shown that the issue of variability in the types of habitats of macroinvertebrates within a water body can be easily resolved by pooling of samples (Chessman et al., 2007).

In addition to the advantages associated with sampling macroinvertebrates, methods of analysing their data are also well established. Consequently, numerous biotic indices have been developed for the assessment of river ecosystems that are based on aquatic macroinvertebrates. Brief descriptions of the more important or widely used indices, listed in chronological order, are provided below.

13.5.1. Beck's Biotic Index (Beck's BI) – 1954

Beck is perhaps the person who coined the term 'biotic index' (Washington, 1984); he surely is the one who popularised it (Davis, 1995). Beck's BI, developed for streams in Florida (USA), is considered to be the first true biotic index; this index is based on the relative tolerances of macroinvertebrates to organic pollution, with field sorting undertaken and identification to species level. Species known to be intolerant to slight organic pollution ('Class I organisms') and those known to be tolerant of moderate organic pollution ('Class II organisms') are distinguished from the rest of a sample. The final index value for a site is calculated by summing the number of species of Class I organisms, multiplied by two, and the number of species of Class II organisms. A single value ranging between 0 and approximately 40 is generated, with values greater than 10 indicating unpolluted sites and values between 1 and 6 indicating moderately polluted sites.

13.5.2. Trent Biotic Index (TBI) – 1964

The TBI (Woodiwiss, 1964), on which several other modern biotic indices are based (Sections 13.5.3–13.5.8, 13.5.13), was developed by the Trent River Authority in England. Qualitative, combined sampling of all available habitats is undertaken for 10 minutes by means of a hand net. A single value is generated by the index, ranging from 0 (grossly polluted) to 10 (unpolluted). The value at a site is determined by the presence or absence of six key types of invertebrates with varying degrees of tolerance to organic pollution, together with the number of specific 'groups' identified to family, genus or species levels.

13.5.3. Indice Biotique (IB) – 1968

The IB (Tuffery and Verneaux, 1968) was derived from the TBI, for use in France. Lotic and lentic habitats are sampled separately using Surber and grab samplers, respectively, and two indices are calculated: a lotic subindex and a lentic subindex. Index values for the IB are determined by the presence of key groups and the number of predefined taxa (or 'systematic units', identified to family, genus and species levels) in each sample, with laboratory-based identification. The IB was modified into the Indice Biologique de Qualité Générale (IBQG), which introduced a greater number of indicator groups and the sampling of eight different habitats at a site, defined on the basis of substrate and velocity conditions. The Indice Biologique Global (IBG), which is based on the IBQG, was adopted as the standard bioassessment method throughout France (Metcalfe-Smith, 1994). The IBG was superseded by an updated version known as the Indice Biologique Global Normalisé (IBGN). With the IBGN, lotic habitats are sampled with a Surber sampler and lentic habitats with a hand net (both 500 μm mesh). The modifications differ from the original IB in that faunal groups are mostly identified to family level.

13.5.4. Chandler's Biotic Score (CBS) – 1970

The CBS (Chandler, 1970), originally developed for upland rivers in the Lothians Region of Scotland, is based on the TBI. However, unlike the TBI, it includes an abundance factor in the final calculation of the index score and only riffle (stones-in-current) areas are sampled with a hand net (1000 μm mesh size) for a total of 5 min. The total score is determined by summing the pollution tolerance scores for each defined 'group' of invertebrates sampled (identified to genus or species), with a sliding scale for individual scores based on the estimated level of abundance. There is no upper limit for the final CBS value, but unpolluted sites generally have scores greater than 3000 (Johnson et al., 1993).

The Average Chandler Biotic Score (Avg. CBS), a modification of the CBS system with the final score for the number of groups present in a sample normalised, was developed because the original system generated low scores for unpolluted, headwater sites (Murphy, 1978; Johnson et al., 1993). This normalised scoring system, which generates values ranging from 0 (severely polluted) to 100 (unpolluted), is more reliable than the original CBS system at discriminating between polluted and unpolluted sites (Washington, 1984) and has been found to be a relatively robust indicator of water quality.

13.5.5. Chutter's Biotic Index (CBI) – 1972

This system, which is loosely based on the TBI (Metcalfe-Smith, 1994), involves sampling the stones-in-current habitat with a hand net or Surber sampler (mesh size 290 μm). A spectrum of 'Quality Values' has been determined for an extensive list of predefined taxa (identified to various taxonomic levels) based on the known occurrence of the defined groups in polluted waters. The final CBI value, which ranges from 0 (unpolluted) to 10 (severely polluted) and represents the average quality value for the organisms sampled, is calculated by dividing the sum of the individual scores for all the taxa sampled by the total number of individuals in the sample. The CBI was never widely used because it requires advanced taxonomic skills, and is time consuming and expensive to apply (Chutter, 1994, 1995, 1998).

13.5.6. Hilsenhoff's Biotic Index (HBI)

The HBI (Hilsenhoff, 1987) is an adaptation of the CBI and was developed for evaluating organic and nutrient pollution in streams in the Wisconsin Region of North America.

The original HBI has been refined in recent years by limiting the number of individuals scored in each taxon to 10, which remedies some problems commonly encountered with the system and reduces seasonal variability in the index value (Hilsenhoff, 1998). The HBI, with tolerance values modified for specific geographic regions, is regularly used for water-quality assessments in many states across North America (Reynoldson and Metcalfe-Smith, 1992).

13.5.7. Biological Monitoring Working Party (BMWP) Score System

In this system, which was introduced in 1978 and modified in 1980 and 1983, all major aquatic habitat types are sampled with a pond net of 90 μm mesh size for a total of 3 min and taxa are identified in the field. The score values for all the predefined invertebrate families present in the sample for a site are summed to give the Total BMWP Score. It is divided by the number of taxa sampled to determine the Average Score Per Taxon (ASPT) for the site. The BMWP–ASPT index has proved to be a relatively robust measure of water quality for rivers in the United Kingdom (Pinder et al., 1987; Metcalfe-Smith, 1994).

13.5.8. Belgian Biotic Index (BBI) – 1983

The BBI (De pauw and Vanhooren, 1983) combines the sampling procedure of the TBI and the scoring system of the IB, but with lotic and lentic habitats scored together. All available habitats are sampled with a 300–500-μm-mesh hand net for a total of 3 min (for rivers less than 2 m wide) or 5 min (for larger rivers). Collected macroinvertebrates are preserved *in situ* and taken back to the laboratory for identification, mainly to family or genus levels. The final index value ranges from 0 (very heavily polluted) to 10 (unpolluted), with values less than 5 indicating that the situation is critical. The BBI has been successfully applied throughout Belgium and in other countries, including Spain, Algeria, Luxemburg, Portugal and Canada (Metcalfe, 1989). It is currently used in Belgium and some surrounding countries (Metcalfe-Smith, 1994; Iversen et al., 2000).

13.5.9. Macroinvertebrate Community Index (MCI) — 1985

The MCI (Stark, 1985), developed for assessing water quality in New Zealand streams, is based on the BMWP method and is similar to the CBI and HBI. Scores are allocated to a list of predefined taxa based on their pollution tolerances, with values from 1 (extremely pollution tolerant) to 10 (extremely pollution sensitive). The final index value for a site is calculated by summing the tolerance values for each taxon present in a sample, dividing by the number of taxa sampled and multiplying by a scaling factor of 20. Although the MCI can theoretically range between 0 and 200, in practice it rarely exceeds 150, with scores greater than 120 indicating pristine conditions and scores less than 50 indicating extreme pollution (Stark, 1993).

13.5.10. Iberian BMWP (IBMWP/BMWP) – 1988

The IBMWP (Bonada, 2003) is also an adaptation of the BMWP System. It is a qualitative or semi-quantitative method that uses a kick net with 250 μm mesh size and field-based macroinvertebrate identification to family level. All available habitats are successively sampled over a 100-m stretch of river until no new taxa are recorded.

The final IBMWP Score, Number of Taxa, and IASPT (IBMWP Score divided by Number of Taxa) are calculated for a site based on all the taxa collected and observed. Separate indices can also be calculated for lotic and lentic habitat groups, if they have been collected and analysed separately. Abundances are estimated according to the following ranks: 1 (1–3); 2 (4–10); 3 (11–100); 4 (>100) (Bonada, 2003). Although these abundance estimates are not used to calculate the final indices, they aid in the interpretation of IBMWP results. The IBMWP has been shown to be effective for the bioassessment of the Spanish rivers and, in 1991, it was adopted by the Spanish Society of Limnology for use throughout the Iberian Peninsula (Zamora-Muñoz and Alba-Tercedor, 1996).

13.5.11. Rivers of Vaud (RIVAUD) Index – 1989, 1995

The RIVAUD Index (Lang et al., 1989) was developed to assess the water quality of rivers in the canton of Vaud in western Switzerland. The method involves the collection of macroinvertebrates from the stones-in-current biotope, using kick-sampling techniques and a hand net with a mesh size of 400 μm. Each sample consists of the macroinvertebrates collected from six areas of 0.1 m^2, with the combined list of taxa from one spring sample and one summer sample used to analyse a sampling site. Macroinvertebrates are identified to family and/or genus level. The final index value, which ranges from zero to 10, is calculated by adding the allocated score for the number of taxa (grouped into six classes of values with allocated scores of 0–5) and that for the number of intolerant taxa (also grouped into six classes of values with allocated scores of 0–5). Intolerant taxa are taken to include Heptageniidae, Plecoptera and case-bearing Trichoptera. Nonhierarchical cluster analysis of the data collected over five years from 162 sampling sites along 51 rivers in western Switzerland was used to delimit classes of values for the total number of taxa and number of intolerant taxa (Lang et al., 1989). The system has been designed so that RIVAUD Index values of 0–3 indicate poor water quality, with values of 4–6 indicating average water quality and values of 7–10 indicating good water quality.

An updated version of the RIVAUD Index was developed by Lang and Reymond (1995), based on additional data collected mostly from the same rivers as those used to initially develop the RIVAUD Index. This updated version was called RIVAUD 95, after the year of its development. The sampling method for RIVAUD 95 is the same as that for the original index system, except that additional late summer samples are collected from rivers in the Alps to ensure that seasonally restricted taxa are captured in this region.

13.5.12. Stream Invertebrate Grade Number – Average Level (SIGNAL) Biotic Index – 1995

The SIGNAL Biotic Index (Chessman, 1995, 2003) was initially developed in 1995 for the assessment of water quality in the Hawkesbury–Nepean River system of New South Wales, eastern Australia (Chessman, 1995) and later modified in 2003 to broaden its applicability to the whole of Australia (Chessman, 2003). Macroinvertebrates are collected from six predefined habitats present at a site. Riffles, pool edges and aquatic macrophytes are sampled with a hand net (250 μm mesh), pool rocks and submerged wood are removed from the stream by hand and soft sediment samples in deep lowland rivers are obtained with a grab sampler and then sieved through 250 mm mesh. The sampling time is not stipulated. Instead, for each habitat type, 100 invertebrates in total are collected with no more than 10 specimens per taxon.

Specimens are preserved and taken back to the laboratory for identification to family level. Sensitivity grades ('SIGNAL 1 grades') ranging from 1 (pollution tolerant) to 10 (pollution sensitive) were initially assigned to widespread families of macroinvertebrates in river systems of south-eastern Australia (Chessman, 1995). Modified 'SIGNAL 2 grades' were subsequently derived for macroinvertebrate families occurring across Australia (Chessman, 2003).

13.5.13. Danish Stream Fauna Index (DSFI) – 1998

The DSFI is based on the TBI, but both positively and negatively scoring diversity groups are used. Also, sampling involves kick sampling of all available habitats along each of three transects, at four equidistant points across the width of the stream, with transects approximately 10 m apart (placed diagonally across the stream if stream width is less than 1 m). The 12 kick samples, which are obtained using a hand net with 500 μm mesh size, are combined for further analysis, and 5 min of handpicking from submerged stones and large wooden debris is carried out. The pooled kick sample and the handpicked sample are preserved separately in the field, with identification (to genus and family level) undertaken in the laboratory, keeping the two groups of samples separate.

The final index value for the DSFI varies from 1 (severely impaired) to 7 (best ecological quality). It is calculated by taking into account the number of diversity groups (i.e., the number of positive groups of taxa minus the number of negative groups of taxa, based on a list of positive and negative taxon groups) and the presence of particular indicator groups of taxa in the total fauna sample (i.e., kick samples plus handpicked sample from each site). The final DSFI index value is obtained from a matrix table that has four categories for the number of diversity groups as columns and six indicator groups (with corresponding lists of indicator taxa) as rows.

13.5.14. Balkan Biotic Index (BNBI) – 1999

The BNBI (Simić and Simić, 1999) was developed on tributaries of the Danube River in Serbia, for river water-quality assessment in the Balkan Peninsula. Loosely based on the CBS, the BNBI requires an estimation of the abundance of sampled macroinvertebrates. It incorporates measures of the dominance and constancy of the taxa sampled, together with a measure of the diversity of the macroinvertebrate community at a sampling site. The BNBI ranges from 0 (for heavily polluted waters) to 5 (for very clean waters).

A rapid reckoner of the attributes and *modus operandi* of these indices is presented in Table 13.2.

13.5.15. Benthic Condition Index (BCI) 1999

The BCI (Engle and Summers, 1999), developed for estuarine environment, includes 1) Shannon–Wiener diversity index adjusted to salinity; 2) mean abundance for Tubificidae; 3) percentages of abundance of the class bivalvia; 4) percentages of abundance of the family Capitellidae; and 5) percentages of abundance of the order amphipoda.

As the step to calculate BCI, the Shannon–Wiener diversity index is calculated, according to the bottom salinity:

$$\begin{aligned} H'_{expected} &= 2.618426 - (0.044795 \times \text{salinity}) \\ &\quad + (0.007278 \times \text{salinity}^2) \\ &\quad + (-0.000119 \times \text{salinity}^3) \end{aligned} \tag{13.1}$$

The final Shannon–Wiener's score is arrived at by dividing the observed by the expected diversity values. After the calculation of the abundance and proportions of the organisms involved, the abundances have to be log transformed and the proportions to be arcsine transformed. The discriminant score is then calculated:

$$\begin{aligned} \text{Discriminant score} = {} & (1.5710 \times \text{proportion of expected diversity}) \\ & + (-1.0335 \times \text{mean abundance of Tubificidae}) \\ & + (-0.5607 \times \text{percent Capitellidae}) + (-0.4470 \times \text{percent Bivalvia}) \\ & + (0.5023 \times \text{percent Amphipoda}) \end{aligned} \tag{13.2}$$

The final BCI index score is obtained by

$$\text{BCI} = \text{discriminant score} - \left(\frac{-3.21}{7.50}\right) \times 10 \tag{13.3}$$

where −3.21 is the minimum, and 7.50 is the range, of the discriminant score.

After the discriminate score transformation, the benthic index can range between 0 and 10 (Table 13.3).

13.5.16. The Marine Biotic Index AMBI of Borja et al., (2000)

This marine biotic index relies on the distribution of individual abundances of the soft-bottom communities into five ecological groups.

Group I: consists of species very sensitive to organic enrichment and present under unpolluted conditions. *Group II:* covers species indifferent to enrichment, always present in low densities with nonsignificant variations with time. *Group III*: is comprised of species tolerant to excess organic matter enrichment. These species may occur under normal conditions; however, their populations are stimulated by organic enrichment. *Group IV:* deals with second-order opportunistic species, adapted to slight-to-pronounced unbalanced conditions. *Group V:* addresses first-order opportunistic species, adapted to pronounced unbalanced situations.

The species are distributed in those groups according to their sensitivity to an increasing stress gradient (enrichment of organic matter). The index is based on the percentages of

TABLE 13.2 An Overview of the Major Biotic Indices, Based on Aquatic Macroinvertebrates

S. No	Biotic Index	Biotopes Sampled	Sampling Equipment	Sampling Protocol	Taxonomic Level	Score Range	Regions in Which Currently Used
1.	Beck's biotic Index	All, combined	Not stipulated	Nonquantitative	Species	0-c.40	–
2.	Trent Biotic Index	All, combined	Hand net	Nonquantitative	Family + genus + species	0–10	–
3.	Indice Biotique	Lotic + lentic, separate	Surber + grab	Semi-quantitative	Family + genus + species	0–10	–
4.	Chandler's Biotic Score	Stones-in-current	Hand net (1000 μm)	Semi-quantitative	Genus + species	0–?	USA
5.	Chutter's Biotic Index	Stones-in-current	Hand net/surber	Quantitative	Family + genus + species	0–10	–
6.	Hilsenhoff's Biotic Index	Stones-in-current	Hand net	Quantitative > 100	Genus + species	0–10	USA
7.	Biological Monitoring working party	All, combined	Hand net	Nonquantitative/ semi-quantitative	Family	0–c.200	UK, Finland, Sweden
8.	Belgian Biotic Index	All, combined	Hand net	Nonquantitative	Family + genus	0–10	Belgium and surrounding countries
9.	Macroinvertebrate Community Index	Stones-in-current	Hand net/surber	Nonquantitative	Genus	0–200	New Zealand
10.	Iberian BMWP	Lotic + lentic, combined/separate	Hand net	Nonquantitative	Family	0–c.200 0–10 (ASPT)	Spain, Italy
11.	Rivers of Vaud Index, 1995 version	Stones-in-current	Hand net	Semi-quantitative	Family + genus	0–20	Western Switzerland
12.	Stream Invertebrate Index	6 per-defined	Hand net	Nonquantitative, 100 organisms	Family	0–10	Australia
13.	Danish Stream Fauna Index	All, combined	Hand net (500 μm)	Semi-quantitative, 12 samples	Family + genus	0–7	Denmark, Sweden
14.	BalkaN Biotic Index	All, combined	Benthos net	Quantitative	Family + sub-family + Genus	0–5	Serbia

TABLE 13.3 Biotic Indices for Saline Waters: A Comparison

Biotic Index	Index Value	Classification	Ecological Quantity Status
AMBI	$0.0 \leq BC \geq 0.2$	Normal	
	$0.2 \leq BC \geq 1.2$	Normal	
	$1.2 \leq BC \geq 3.3$	Slightly polluted	
	$3.3 \leq BC \geq 4.3$	Disturbed	
	$4.3 \leq BC \geq 5.0$	Disturbed	
	$5.0 \leq BC \geq 5.5$	Heavily disturbed	
	$5.5 \leq BC \geq 6.0$	Heavily disturbed	
	Azoic	Extremely disturbed	
BENTIX	$4.5 \leq BENTIX \geq 6.0$	Normal/pristine	High
	$3.5 \leq BENTIX \geq 4.5$	Slightly polluted, transitional	Good
	$2.5 \leq BENTIX \geq 3.5$	Moderately polluted	Moderate
	$2.0 \leq BENTIX \geq 2.5$	Heavily polluted	Poor
	0	Azoic	Bad
BQI	1 to <4		Bad
	4 to <8		Poor
	8 to <12		Moderate
	12 to <16		Good
	16 to <20		High
BCI	<3	Degraded conditions	
	3–5	Transition conditions	
	>5	Non-degraded sites	
BOPA	$0.00000 \leq BOPA \leq 0.06298$	Unpolluted sites	High
	$0.04576 < BOPA \leq 0.19723$	Slightly polluted sites	Good
	$0.13966 < BOPA \leq 0.28400$	Moderately polluted sites	Moderate
	$0.19382 < BOPA \leq 0.30103$	Heavily polluted sites	Poor
	$0.26761 < BOPA \leq 0.30103$	Extremely polluted sites	Bad
BRI	0–33	Marginal deviation	
	34–43	Loss of biodiversity	
	44–72	Loss of community function	
	>72	Defaunation	

TABLE 13.3 Biotic Indices for Saline Waters: A Comparison (*cont'd*)

Biotic Index	Index Value	Classification	Ecological Quantity Status
ISI	>8.75		High
	7.5–8.75		Good
	6.0–7.5		Fair
	4.0–6.0		Poor
	0–4.0		Bad

abundance of each ecological group of one site, or the biotic coefficient (BC), which is given by

$$BC = \left\{\frac{(0 \times \%G_{I}) + (1.5 \times \%G_{II}) + (3 \times \%G_{III}) + (4.5 \times \%G_{IV}) + (6 \times \%G_{V})}{100}\right\} \tag{13.4}$$

The index, also referred to as BC, varies continuously from 0 (unpolluted) to 7 (extremely polluted). It has been widely applied with considerable success. For example, it has been tested in different geographical sites such as the Basque Country coastline, Spain, for which it was originally designed (Borja et al., 2000), the Mondego estuary, Portugal (Salas et al., 2004), three locations on the Brazilian coast and two on the Uruguayan coast (Muniz et al., 2005), and has been tested among different geographical sites (Muxika et al., 2005). It was seen to correctly evaluate the ecosystem conditions at these sites. These and other tests have caused the index to be regarded as a sound tool for management due to its capacity to assess ecosystem health.

A drawback of the AMBI is that mistakes can occur during the grouping of the species according to their response to pollution situations. Once it draws on the response of organisms to organic inputs in the ecosystem it does not detect the effects caused by other types of pollution, as for instance toxic pollution (Marín-Guirao et al., 2005). Moreover, it presents some limitations when applied to semi-enclosed systems (Blanchet et al., 2008).

13.5.17. Benthic Response Index (BRI) – 2001

The benthic response index (BRI) was developed for the Southern California coastal shelf (Smith et al., 2001) and is a marine analogue of the Hilsenhoff index used in freshwater benthic assessments (Hilsenhoff, 1987).

BRI is calculated using a two-step method in which ordination analysis is employed to establish a pollution gradient. Afterwards the pollution tolerance of each species is determined based upon its abundance along the gradient (Smith et al., 1998). The main goal of the index is to establish the abundance-weighted average pollution tolerance of the species in a sample, on the basis of the premise that each species has a tolerance for pollution and if that tolerance is known for a large set of species, then it is possible to infer the degree of degradation from species composition and its tolerances (Gibson et al., 2000). The index is given by

$$I_s = \frac{\sum_{i=1}^{n} p_i \sqrt[3]{a_{si}}}{\sum_{i=1}^{n} \sqrt[3]{a_{si}^f}} \tag{13.5}$$

where I_s is the index value for the sample s, n is the number of species in the sample s, p_i is the tolerance value for species i (position on the gradient of pollution) and a_{si} is the abundance of species i in sample s. The exponent f is for transforming the abundance weights: if $f = 1$, the raw abundance values are used. If $f = 0.5$, the square root of the abundances are used. If $f = 0$, I_s is the arithmetic value of the p_i values greater than zero. The average position for each species (p_i) on the pollution gradient defined in the ordination space is measured as

$$p_i = \frac{\sum_{j=1}^{t_i} g_{ij}}{t_i} \tag{13.6}$$

where t_i is the number of samples to be used in the sum, with only the highest t_i species abundance values included in the sum. The g_{ij} is the position of the species i on the ordination gradient for sample j. The p_i values obtained in Eq. (13.6) are used as pollution tolerance scores in Equation (13.5) to compute the index values.

BRI has a quantitative scale of 0 to 100, where low scores are indicative of healthier benthic communities and the BRI scoring define four levels of response beyond reference condition (Table 13.3). Even as BRI is useful in quantifying disturbances, it is not able to distinguish between natural and anthropogenic disturbance, such as the natural impacts that river flows may have on benthic communities (Bergen et al., 2000). Nevertheless, this index presents the advantage of not underestimating biological effects, as well as possessing low seasonal variability (Smith et al., 2001).

13.5.18. BENTIX

The BENTIX index is based on AMBI and relies on the reduction of macrozoobenthic data from soft-bottom substrata in three wider ecological groups (Simboura and Zenetos, 2002). To accomplish this goal, a list of indicator species has been elaborated, where each species receives a score, from 1 to 3, that represents their ecological group: *Group 1 (GI)*: includes the species that are sensitive or indifferent to disturbances (k-strategies species); *Group 2 (GII)*: includes the species that are tolerant and may increase their densities in case of disturbances, as well as the second-order opportunistic species (r-strategies species); and *Group 3 (GIII)*: includes the first-order opportunistic species.

The index is calculated as

$$\text{BENTIX} = \left\{ \frac{6 \times \%\text{GI} + 2 \times (\%\text{GII} + \%\text{GIII})}{100} \right\} \tag{13.7}$$

This index can range from 2 (poor conditions) to 6 (high ecological quality status, or reference sites). Overall, the BENTIX index considers two major classes of organisms: the sensitive and the tolerant groups. This classification has the advantage of reducing the calculation effort while diminishing the probability of the inclusion of species in inadequate groups (Simboura and Zenetos, 2002). Moreover, for using this index, it is not essential to have expertise for amphipoda identification, since it includes all those organisms (with the exception of individuals from the genus *Jassa*) in the same category of sensitivity to organic matter (Dauvin and Ruellet, 2007).

The BENTIX index has been successfully applied to studies of organic pollution (Simboura and Zenetos, 2002; Simboura et al., 2005), oil spills (Zenetos et al., 2004) and of particulate metalliferous waste (Simboura et al., 2007). The index does not underestimate nor overestimate the role of any of the groups

(Simboura et al., 2005). Nevertheless, some authors consider the BENTIX index to be inadequate for assessing the impact of toxic contaminations (Marín-Guirao et al., 2005). The index also faces limitations when applied to estuaries and lagoons (Simboura and Zenetos, 2002; Blanchet et al., 2008; Pinto et al., 2009).

13.5.19. Indicator Species Index (ISI) – 2002

The indicator species index (Rygg, 2002) is based on the improved version of the Hurlbert index (1971). To calculate ISI, it is necessary to determine the sensitive values for each species as well as the pollution impact factor (ES100min5). The ES100 is the expected number of species among 100 individuals. The average of the five lowest ES100 is defined as the sensitivity value of that taxon, denoted ES100min5. The ISI is then defined as the average of the sensitivity values of the taxa occurring in the sample.

ISI allows an accurate description of environmental quality of the systems but has been applied mostly in the Norway coasts. The reason is that ISI cannot be easily transposed to other geographical regions since the taxonomic list can be significantly different at other regions, and the calculation of the sensitivity factors may require different approaches.

An overview of saltwater BIs is presented in Table 13.3.

13.5.20. Benthic Quality Index (BQI) – 2004

The benthic quality index (BQI) was designed to assess environmental quality according to the WFD (Rosenberg et al., 2004), and includes tolerance scores, abundance and species diversity factors. The main objective of this index is to attribute tolerance scores to the benthic fauna in order to determine their sensitivity to disturbance. The index is expressed as

$$\mathrm{BQI} = \left\{ \sum \left(\frac{A_i}{\mathrm{Tot}\ A} \right) \times ES50_{0.05i} \right\} \times {}^{10}\log(S+1) \tag{13.8}$$

where $A_i/\mathrm{Tot}\ A$ is the mean relative abundance of this species and $ES50_{0.05i}$ the tolerance value of each species, i, found at the station. This metric corresponds to 5% of the total abundance of this species within the studied area. Further, the sum is multiplied by $\log_{10}$ for the mean number of species (S) at the station, as high species diversity is related to high environmental quality. The purpose of using the values calculated from the 5% lowest abundance of a particular species ($ES50_{0.05i}$) is that this value is assumed to be representative for the greatest tolerance level for that species along an increasing gradient of disturbance; implying that if the stress increases further, that species will disappear. $ES50$ is computed as

$$ES50 = 1 - \sum_{i=1}^{S} \frac{(N-N_i)!(N-50)!}{(N-N_i-50)!N!} \tag{13.9}$$

where N is the total abundance of individuals, N_i is the abundance of the i-th species and S is the number of species at the station.

BQI suffers from two methodological lacunae: the sample area is not the same among sampling protocols, and individuals' distribution among species may not be random, particularly when some species appear as strong dominants. To get around these shortcomings, Rosenberg et al. (2004) recommend the use of many stations and replicates for the quality assessment of an area. The index provides strong correlations with environmental variables, such as salinity (Zettler et al., 2007).

13.5.21. The Benthic Opportunistic Polychaeta Amphipoda (BOPA) Index

The benthic opportunistic polychaeta amphipoda (BOPA) index has resulted from the efforts to refine the polychaeta/amphipoda ratio (Gómez-Gesteira and Dauvin, 2000) by Dauvin and Ruellet (2007) in order to make it applicable under the European Water Framework Directive (WFD). The index can be used to assign estuarine and coastal communities into five ecological quality categories (Table 13.3). In accordance with the taxonomic sufficiency principle, the index aims to exploit the polychaeta/amphipoda ratio to determine the ecological quality, using relative frequencies rather than abundances in order to define the limits of the index:

$$\mathrm{BOPA} = \log\left\{\frac{f_P}{f_A+1}+1\right\} \qquad (13.10)$$

where f_P is the opportunistic polychaeta frequency (ratio of the total number of opportunistic polychaeta individuals to the total number of individuals in the sample); f_A is the amphipoda frequency (ratio of the total number of amphipoda individuals, excluding the opportunistic *Jassa* amphipod, to the total number of individuals in the sample), and $f_P + f_A \leq 1$. Its value can range between 0 (when $f_P = 0$) and log 2 (around 0.30103, when $f_A = 0$). The BOPA index will get a null value only when there are no opportunistic polychaetes, indicating an area with a very low amount of organic matter. Hence, when the index provides low scores it is considered that the area has a good environmental quality, with few opportunistic species; the index scores increase when increasing organic matter degrades the environment.

One of the main advantages of this index is its independence of sampling protocols, and specifically of mesh sieve sizes, since it uses frequency data and the proportion of each category of organisms (Pinto et al., 2009). The need for taxonomic knowledge is low, which allows a generalised use and ease of implementation. Moreover, the use of frequencies makes it independent of the surface unit chosen to express abundances. But it takes into account only three categories of organisms – opportunistic polychaetes, amphipods (except Jassa) and other species – and only the first two have a direct effect on the index calculation. It does not consider the oligochaeta influence, which may also include opportunistic species.

13.6. BIOTIC INDICES AS INDICATORS OF WATER SAFETY AND HUMAN HEALTH RISKS

Biotic indices have been very commonly and successfully applied to the monitoring of major stressors such as BOD, COD and plant nutrients *vis a vis* concerns for water safety and human health. But there is strong evidence that water micropollutants (such as pesticides and heavy metals) can also dramatically alter the structure and physiology of benthic communities (Ivorra, 2000; Blanco et al., 2007; Blanco and Becares, 2010; Imoobe and Ohiozebau, 2010), in turn, having their toxic impact manifest in the scores of the concerned indices. Comparative studies have shown that, while macro-invertebrate-based indices are more sensitive to changes affecting structural parameters (i.e., river-bed width and particle-size distribution, flow regime and similar factors), diatom-based indices are more dependent on chemical variables affecting the water, basically nutrients (Soininen and Könönen, 2004; Hering et al., 2006; Blanco et al., 2007; De Jonge et al., 2008; Juttner et al., 2010), although there can also be a close relationship between macro-invertebrates and chemical variables linked with the river basin. Several significant statistical correlations between diatom indices and the water concentrations of heavy metals were noted by Sabater (2000). Zamora-Muñoz et al. (1995) observed significant correlations between BWMP and IBWMP and concentrations of Cu, Zn, pesticides, detergents, fats, and oils. Among toxicological factors in rivers, concentrations of metals and other pollutants such as total PAHs have been seen to influence variance in some macro-invertebrate indices (Pinel-Alloul et al., 1996). Significant correlation of a biotic index with Cu and Pb was also noted by Robson et al. (2006).

In a wide-ranging study, Blanco and Becares (2010) have studied 188 sites in the basin of the River Duero in north-western Spain, in which nineteen diatom and six macro-invertebrate indices were calculated and compared with the concentrations of 37 different toxicants by means of a correlation analysis.

The toxicants included (i) anionic compounds (ACs): chlorides, cyanides, fluorides and linear alkylbenzene sulphonate (LAS); (ii) biocides: atrazine, metolachlor, simazine, terbuthylazine and total pesticides; (iii) fats and oils; (iv) hydrocarbons: dissolved hydrocarbons, petroleum-derived hydrocarbons (PHCs), phenols and total polycyclic aromatic hydrocarbons (PAHs); (v) nitrogen-derived compounds (NDCs): ammonia and ammonium; (vi) potentially toxic elements (PTEs):dissolved As, Ba, Cd, Cu, Cr, Cr (III), Cr (VI), Hg, Ni, Pb, Sb, Se, Zn, and total As, Cd, Cl, Cr, Hg, Zn; and (vii) semi-volatile organic compounds (SVOCs): tetrachloroethene (PCE) and trichloroethene (TCE). The selection of the toxicants was based on their inclusion in the lists of priority substances, dangerous priority substances (European Parliament and European Council, 2001), dangerous substances (List I) or dangerous substances (List II) (European Parliament and European Council, 2006).

The biotic indices that were explored included richness and specific pollution indices (CEMAGREF, 1982); biological diatom index (Lenoir and Coste, 1996). European index (Descy and Coste, 1991); eutrophication pollution index – diatoms (Dell'Uomo, 2004); Sládeček's index (Sládeček, 1986); generic diatom index (Rumeau and Coste, 1988); Swiss diatom index (Hürlimann and Niederhauser, 2006); trophic diatom index (Kelly and Whitton, 1995); Steinberg and Schiefele's index (Steinberg and Schiefele, 1988); Leclercq and Maquet's index (Leclercq and Maquet, 1987) diatom assemblage index for organic pollution (Watanabe et al., 1988); % of pollution tolerant taxa (Schiefele and Kohmann, 1993); artois-picardie diatom index (Prygiel et al. 1996); descy's index (Descy, 1979); Pampean diatom index (Gómez and Licursi, 2001); Lobo's index (Lobo et al., 2002); Rott's Saprobic index (Rott et al., 1997); Rott's trophic index (Rott et al., 1999) biological monitoring working party (Armitage et al., 1983) average score per taxon (Armitage et al., 1983); Shannon's index (Shannon and Weaver, 1949) and equitability, number of families, and total abundance indices.

Several chemical variables analysed correlated significantly with at least one biotic index. Sládeček's diatom index and the number of macro-invertebrate families exhibited particularly high correlation coefficients. Methods based on macro-invertebrates performed better in detecting biocides, while diatom indices showed stronger correlations with potentially toxic elements such as heavy metals. All biotic indices, and particularly diatom indices, were especially sensitive to the concentration of fats and oils and trichloroethene.

In general, shifts from sensitive to tolerant taxa occurring in polluted environments are utilisable as indicators of metallic pollutants and toxic organics by biotic indices as effectively as the monitoring of agents of eutrophication by these indices (Robson et al., 2006). In other words, indices based on invertebrates are well suited to assessing this kind of impact, given a little modification to improve their sensitivity and performance (García-Criado et al., 1999; Blanco and Becares, 2010).

13.7. COMPARISON OF PERFORMANCES OF DIFFERENT BIOTIC INDICES

Several studies have been reported in which performance of different biotic indices (BIs) has been assessed in the interpretation of ecological status of a given water course or a region. The result is a mixed bag; it basically shows that each index has its own special niche or 'zone of influence' within which it performs well and doesn't work so well outside it. Normally, indices developed for one biogeographic region do not work as effectively in other biogeographic regions, barring exceptions.

A few illustrative examples from the reported studies are presented below.

An Assessment of Performance of Three Exotic BIs in Masan Bay, Korea

A comparison of the performance of three benthic biotic indices (BPI, AMBI, BIBI) in indicating the health condition of benthic communities in Masan Bay, Korea, was carried out by Choi and Seo (2007). All the three indices showed that macrozoobenthic communities in the inner bay were in a seriously polluted condition all year round. The macrobenthic fauna in the bay mouth also seemed in an impaired (slightly polluted) condition as per AMBI during summer season. The three indices gave similar assessments and thus each appears potentially useful in assessing ecological status of marine environments in Korea. This case study is one of the few examples when indices developed in USA and Europe have been found to be effective in Southeast Asia.

Response of Different BIs to Eutrophication

Chainho et al. (2007) used BIs based on benthic invertebrate communities to assess the ecological quality of a Portuguese estuary characterised by strong seasonal changes and with eutrophication problems. The studies showed that different indices pointed to different classifications and there was low agreement between indices and index–season interactions. Diversity indices were better correlated to eutrophication related variables than AMBI and ABC. Predictable responses of benthic indices to anthropogenic stress symptoms were stronger during the dry period.

Performance of AMBI and BQI with Regard to Sampling Effort

The comparative performance of ATZI Marine Biotic Index (AMBI) and Benthic Quality Index (BQI) was studied by Fleischer et al. (2007). Both the indices are based on sensitivity/tolerance classification and quantitative information on the composition of soft-bottom macrofauna. Their performance, especially with regard to sampling effort was assessed based on two data sets, one collected in Southern Baltic and one from the Gulf of Lions, Mediterranean.

It was seen that AMBI was not affected by sampling effort but BQI was. Two options were proposed for BQI (1) the removal of the scaling term and (2) the replacement of the scaling term by a different scaling term. When thus modified, both forms of BQIs became largely independent of sampling effort.

An Intra-BI Study

Various BIs were used by Callanan et al. (2008) to test the variability in the ecological quality assessment of headwater streams in Ireland. Various metrics were used including the Irish Q-value and the newly developed Small Streams Risk Score (SSRS). Metrics applied elsewhere in the Atlantic biogeographic region in Europe, including the Biological Monitoring Working Party score (BMWP), the Average Score per Taxon (ASPT), the Ephemeroptera, Plecoptera, and Trichoptera taxa (EPT), the Belgium Biotic Index (BBI) and the Danish Stream Fauna Index (DSFI) were also deployed. The spring and summer data sets were used to test the performance of the metrics with respect to season, and the applicability of their use to assess the ecological quality of wadeable streams.

It was seen that quality status of most sites assigned by the various metrics was high when computed with reference to the spring invertebrate data, but considerable deviation in quality status occurred when the summer data was applied. Seasonal differences were noted using all the biotic indices. The authors have attributed this to the absence of pollution-sensitive groups in summer. Seasonal variability in the water-quality status was particularly evident in acidic streams draining noncalcareous geologies with peaty soils that had relatively lower numbers of taxa. Some indices reflected greater seasonal difference in the quality category assigned; the least variability between seasons was obtained using the ASPT and the SSRS risk assessment system.

The results suggest that reference status was reliably reflected in spring because more pollution-sensitive taxa were present in the season, and that a new ecological quality assessment tool is required for application in summer when impacts may be most severe. The highly heterogeneous freshwater habitat of the kind explored by the authors seems to have too few taxa present in the summer to reliably determine the ecological quality of the stream using the available indices.

A Comparison of Two European BIs using a Large Database

Grémare et al. (2009) used the pan-European MacroBen database to compare the AZTI Marine Biotic Index (AMBI) and the Benthic Quality Index (BQI_{ES}). These two biotic indices rely on two distinct assessments of species sensitivity/tolerance and which up to now have only been compared on restricted data sets.

A total of 12,409 stations were selected from the database. This subset (indicator database) was later divided into 4 marine and 1 estuarine subareas. The authors computed E $(S50)_{0.05}$ in 643 taxa, which accounted for 91.8% of the total abundances in the whole marine indicator database. AMBI EG and $E(S_{50})_{0.05}$ correlated poorly. High values of AMBI were always associated with low values of BQI_{ES}, which underlines the coherence of these two indices in identifying stations with a bad ecological status (ES). Conversely, low values of AMBI were sometimes associated with low values of BQI_{ES} resulting in the attribution of a good ES by AMBI and a bad ES by BQI_{ES}. This was caused by the dominance of species classified as sensitive by AMBI and tolerant by BQI_{ES}. Some of these species are known to be sensitive to natural disturbance, which highlights the tendency of BQI_{ES} to automatically classify dominant species as tolerant.

The studies thus reveal that both indices have weaknesses in their way of assessing sensitivity/tolerance levels (i.e., existence of a single sensitivity/tolerance list for AMBI and the tight relationship between dominance and tolerance for BQI_{ES}). The findings also indicate that future studies should focus on the clarification of the sensitivity/tolerance levels of the species identified as problematic, and assessment of the relationships between AMBI EG and $E(S_{50})_{0.05}$ within and between combinations of geographical areas and habitats.

A Comparison of Four BIs and a Predictive Model

Feio et al. (2009) applied four widely used diatom-based indices: Specific Polluosensitivity Index (SPI), standardised Biological Diatom Index (BDI), European Economic Community Index (CEC) and Generic Diatom Index (GDI), to evaluate stream ecological quality based on diatom communities. Predictive models resulting from the comparison between the communities of the study site and those of a set of reference sites representing undisturbed or the best available conditions of a given region were also explored for evaluating the ecological status of streams. The indices and the predictive model were applied to 54 sites located in central Portugal to assess the sensitivity of the five methods to a range of anthropogenic disturbances cumulatively affecting streams and represented by 27 variables (e.g., organic enrichment, changes in morphology of the channels, integrity of the riparian corridor and land use in the catchment).

The results were analysed comparatively through Spearman correlations, Boxplots and Stepwise Discriminant Analysis.

These findings provide one more evidence of the sensitivity of diatoms to organic and nutrient contamination, as shown by all the four indices MoDi, BDI, CEC and SPI. The studies also reveal the importance of suspended solids (through the MoDi, GDI and SPI). The impact of modifications in land use to diatoms was shown by all the methods applied, except for the GDI. The MoDI also revealed the importance of changes in the structure and morphology of the reach and the channel, like the construction of artificial walls or embankments and connectivity. The BDI could relate its assessments with the riparian zone integrity; but the SPI was not useful in detecting morphological pressures. The GDI produced the most divergent assessments and was less effective in revealing the anthropogenic disturbances.

The use of the predictive model appeared a good method for the assessment of streams in central Portugal because it was able to show up a great diversity of quantitative and qualitative changes in freshwater systems as reflected in the structure (species richness and abundance) of diatom communities.

Individual BI vs. Combination of BIs

Arguing that even as invertebrate biotic indices are used widely to assess river quality, their value in diagnosing reasons of impairment is limited because several potential causes can make an index score unfavourable. Clews and Ormerod (2009) have carried out an investigation to see whether simple combinations of biotic indices can improve diagnostic capability.

The authors have found that in the catchment of the Welsh River Wye, invertebrates varied significantly among groups of 55 streams in taxonomic composition and in the scores of indices representing acidification (AWIC), mild eutrophication/organic pollution (BMWP/ASPT) and flow (LIFE). Although sites impacted by different forms of pollution tended to have reduced BMWP scores, acidified and enriched sites became distinguishable from each other, and from unimpaired streams, when classified with a combination of these indices. Combined indices were also able to differentiate among competing explanations for trends in biological quality through time by revealing how increasing BMWP at some sites reflected local reductions in eutrophication.

These findings illustrate how simple univariate indices, calibrated to respond to specific pressures, have bio-diagnostic capability when used together even in a relatively unpolluted catchment such as the Wye. Moreover, they were useful in identifying specific management needs in different locations – to mitigate acidification in upland base-poor tributaries and to reduce diffuse nutrients in the lower catchment.

The authors advocate (i) the development of more pressure-specific indices, for example, to detect morphological modification, sedimentation and metal impacts; and (ii) further exploration of combined indices from one or more groups of organisms (e.g., diatoms and invertebrates) to increase bio-diagnostic capability in river monitoring.

A Comparison of Three BIs and an IBI

Simboura and Argyrou (2010) have evaluated four indices that have been used for benthic macroinvertebrate ecological quality classification in the Mediterranean Sea. The study was based on the data obtained from the participation of Greece and Cyprus in a Geographical intercalibration exercise. The indices AMBI, M-AMBI, MEDOCC and BENTIX were applied to the available benthic species data, and the succession of the ecological groups along the graded values of each index was plotted. The level of agreement among methods was calculated, and the performance of each method in estimating ecological quality status was evaluated.

It was seen that AMBI, and its derivative IBI, the M-AMBI, overestimated the status, while MEDOCC showed the best level of agreement with BENTIX. BENTIX gave equal weight to tolerant and opportunistic species groups, and correlated them more closely than the other indices. Overall, BENTIX seemed the most sensitive in detecting ecological disturbances in the Eastern Mediterranean basin, where tolerant and opportunistic groups play an equally important role in the response of benthic communities to stressors.

13.8. BIOTIC INDICES AND DEVELOPING COUNTRIES

Biotic indices have been successfully used for the bioassessment of rivers in many parts of the world, mostly the developed countries and some developing ones. But a large number of developing countries have not yet started using biotic indices. For example, biotic indices have not to date been developed or used to any significant extent in Latin America (Pringle et al., 2000), Central and Eastern Asia (Li et al., 2000), or South-east Asia (Dudgeon et al., 2000). Among the exceptions are the use of indices based on benthic macroinvertebrate communities in Nicaragua (Fenoglio et al., 2002), Macedonia, (Lazaridou-Dimitriadou, et al., 2004), Vietnam (Duong et al., 2007) and Brazil (Mugnai et al., 2008). On the Indian subcontinent, no biotic indices are used for assessing the water quality of rivers because none of the currently available biotic indices from other countries have been found to be entirely suitable (Gopal et al., 2000; Sarkar and Abbasi, 2006).

13.9. LIMITATIONS OF BIOTIC INDICES

Despite their proven utility in rapid bioassessments, biotic indices must be carefully interpreted using supplementary data and their significant limitations must always be borne in mind. These include the restricted applicability to a particular geographic area and/or type of stressor (Washington, 1984; Johnson et al., 1993; Norris and Georges, 1993; Metcalfe-Smith, 1994; Friedrich et al., 1996), usually organic pollution but other forms of pollution as well (Gray and Delaney, 2010), and the inability to detect moderate degradation. Moreover, biotic indices should be used in conjunction with, and not as replacements of, conventional indices based on physical and chemical parameters.

A number of alternative approaches to the rapid bioassessment of river ecosystems from a community perspective have been pursued, but few have found as widespread application as biotic indices. This is mainly due to the relative difficulty of using the alternatives compared to biotic indices. The two approaches that have found increasing application are the 'multimetric (or composite) approach', represented by IBI (index of biotic integrity) and the 'multivariate approach' exemplified by RIVPACS (River Invetebrate Prediction And Classification System; Wright, 1995; Wright et al., 2000). These have been discussed in detail in Chapters 14 and 15.

13.10. WQIs AND BIS: AN OVERVIEW

1. Water-quality indices (WQIs) are extensively used all over the world to abstract the numerical values of several water quality characteristics of a sample into a single value. Due to the ease with which WQIs enable an interpretation of the overall quality of water, they play a very important role in the monitoring, comparison and control of water quality. But the WQIs of the modern and post-modern times have been almost exclusively based on physical and chemical characteristics, and have seldom included

'biological' characteristics (other than BOD and faecal coliforms). This approach serves a useful purpose, yet is inadequate in fully reflecting the status of a water body. This is due to the fact that WQIs based on physico-chemical parameters are basically stressor-oriented; they operate on the principle of linking different stressors to possible biological responses. This predictive ability can, however, be precise only where a known cause–effect relationship exists between a specific stressor and the biological component. But such cause–effect relationships, for a specific suite of conditions, can at best be determined with laboratory bioassays under controlled conditions, and are not fully applicable in real life situations.

2. This inherent limitation of stressor-oriented assessment makes it necessary to complement the conventional WQIs with response-based ones; the latter assess environmental health on the basis of the status of the responding organisms. They involve the monitoring of biological or ecological indicators present in a water body to gauge the response of the water body to a disturbance. The 'disturbance' can be any factor that disrupts the water body at ecosystem, community, or population levels. It follows that the response-oriented approach indicates that something has or has not actually gone wrong in response to a stressor.
3. The stressor-oriented as well as the response-oriented approaches have obvious uses and specific benefits in water-quality management. The current thrust is towards operationally integrating the two approaches so that the resulting methodology incorporates the benefits of both.
4. Biotic indices (BIs) have evolved from the response-based approach. In BIs, different taxon from a particular group of organisms are assigned different sensitivity weighting, or 'scores', based on the tolerance or sensitivity of different taxon to particular pollutants. The scores of all the individual taxa sampled at a site are summed and/or averaged to provide a value by which the ecological health of the biotic community, hence the health of the water body, is gauged. Some BIs include abundance estimates in the scoring system. A number of commonly used BIs have been described in the paper, highlighting their distinguishing features as well as limitations.
5. Benthic macroinvertebrates are the most widely used group in biotic indices but various other organisms have also been used including bacteria, protozoa, diatoms, algae, macrophytes, macroinvertebrates, and fish. The advantages with benthic macroinvertebrates are that they are largely nonmobile, ubiquitous, and relatively abundant. There are often many species within a community with varying sensitivities to stresses and relatively quick reaction times. This provides a wide spectrum of graded and recognisable responses to environmental perturbation. The responses of many common macroinvertebrate species to different types of pollution has been extensively documented, which is an added advantage.
6. Even as biotic indices are very effective in complementing conventional WQIs, they are also besieged by several limitations. The most serious one pertains to the difficulty in finding appropriate 'control' sites. Ideally, a 'control' site should be truly reflective of natural, unpolluted conditions and thus serve as reference or 'control' with which the test sites can be compared to know whether a certain impact causes an aquatic assemblage or ecosystem to respond in some way that is outside the natural range of variation. Unfortunately, due to widespread

human encroachments everywhere, it is very difficult to find nonimpacted sites in most regions, especially in lowland areas, for use as 'control' sites. Consequently, minimally disturbed or least impacted of the available sites have to be generally used to determine the *best attainable* reference condition. Moreover, unlike WQIs based on physical and chemical parameters, BIs are unable to detect moderate degradation.

7. Several attempts to modify BIs and make their use simpler and faster have been made, but few have found as widespread application as BIs. The two approaches that have found increasing application are the 'multimetric (or composite) approach', represented by the Index of Biotic Integrity and the 'multivariate approach' exemplified by RIVPACS (River Invertebrate Prediction and Classification System). Neither is simpler or faster than any BI but both are arguably more comprehensive.

References

Armitage, P.D., Moss, D., Wright, J.F., Furse, M.T., 1983. The performance of a new biological water quality score system based on macroinvertebrates over a wide range of unpolluted running water sites. Water Research 17, 333–347.

Ayari, Afli, 2008. Functional groups to establish the ecological quality of soft benthic fauna within Tunis Bay (western Mediterranean). Vie et Milieu 58 (1), 67–75.

Barbour, M.T., Gerritsen, J., Snyder, B.D., Stribling, J.B., 1999. Rapid Bioassessment Protocols for Use in Streams and Wadeable Rivers: Periphyton, Benthic Macro-invertebrates and Fish, second ed. Report EPA 841-B-99–002. US Environmental Protection Agency, Office of Water, Washington DC.

Bergen, M., Cadien, D., Dalkey, A., Montagne, D.E., Smith, R.W., Stull, J.K., Velarde, R.G., Weisberg, S.B., 2000. Assessment of benthic infaunal condition on the mainland shelf of Southern California. Environmental Monitoring and Assessment 64, 421–434.

Bigler, C., Gälman, V., Renberg, I., 2009. Numerical simulations suggest that counting sums and taxonomic resolution of diatom analyses to determine IPS pollution and ACID acidity indices can be reduced. Journal of Applied Phycology 22, 541–548.

Blanchet, H., Lavesque, N., Ruellet, T., Dauvin, J., Sauriau, P., Desroy, N., Desclaux, C., Leconte, M., Bachelet, G., Janson, A., 2008. Use of biotic indices in semi-enclosed coastal ecosystems and transitional waters habitats— implications for the implementation of the European Water Framework Directive. Ecological Indicators 8, 360–372.

Blanco, S., Becares, E., 2010. Are biotic indices sensitive to river toxicants? A comparison of metrics based on diatoms and macro-invertebrates. Chemosphere 79, 18–25.

Blanco, S., Becares, E., Cauchie H.M., Hoffmann, L., Ector, L., 2007. Comparison of biotic indices for water quality diagnosis in the Duero Basin. Spain. Arch. Hydrobiol. Suppl. Large Rivers. 17 267–286.

Bonada, N., 2003. Ecology of Macroinvertebrate Communities in Mediterranean Rivers at Different Scales and Organization Levels. PhD thesis, University of Barcelona, Spain.

Borja, A., Franco, J., Perez, V., 2000. A marine biotic index to establish the ecological quality of soft-bottom benthos within European estuarine and coastal environments. Marine Ecology Progress Series 40 (12), 1100–1114.

Brown, C.A., 2001. A comparison of several methods of assessing river condition using benthic macro-invertebrate assemblages. African Journal of Aquatic Science 26, 135–147.

Callanan, M., Baars, J.-R., Kelly-Quinn, M., 2008. Critical influence of seasonal sampling on the ecological quality assessment of small headwater streams. Hydrobiologia 610, 245–255.

CEMAGREF (1982). Étude des Méthodes Biologiques d'appréciation Quantitative de la Qualité des Eaux. Rapport Q.E. Lyon - A.F. Rhône-Méditerranée-Corse. CEMAGREF, LyonCoring, E., Situation and developments of algal, diatom-based techniques for monitoring rivers in Germany (1999) Use of Algae for Monitoring Rivers III, 122–127, Prygiel J., Whitton B.A., and Bukowska J. (Eds), Agence de l'Eau Artois-Picardie, Douai.

Chainho, P., Costa, J.L., Chaves, M.L., Dauer, D.M., Costa, M.J., 2007. Influence of seasonal variability in benthic invertebrate community structure on the use of biotic indices to assess the ecological status of a Portuguese estuary. Marine Pollution Bulletin 54 (10), 1586–1597. Epub 2007 Aug 2.

Chandler, J.R., 1970. A biological approach to water quality management. Water Pollution Control 69, 415–421.

Chau, K.W., Chuntian, C., Li, C.W., 2002. Knowledge management system on flow and water quality modelling. Expert Systems with Applications 22, 321–330.

Chessman, B.C., 1995. Rapid assessment of rivers using macroinvertebrates: a procedure based on habitat-specific sampling, family level identification and a biotic index. Australian Journal of Ecology 20, 122–129.

Chessman, B.C., 2003. New sensitivity grades for Australian river macroinvertebrates. Marine and Freshwater Research 54, 95–103.

Chessman, B., Williams, S., Besley, C., 2007. Bioassessment of streams with macroinvertebrates: effect of sampled habitat and taxonomic resolution. Journal of the North American Benthological Society 26 (3), 546–565.

Choi, J.W., Seo, J.Y., 2007. Application of biotic indices to assess the health condition of benthic community in Masan Bay, Korea. Ocean and Polar Research 29 (4), P339–348.

Chutter, F.M., 1994. The rapid biological assessment of stream and river water quality by means of the macroinvertebrate community in South Africa. In: Uys, M.C. (Ed.), Classification of Rivers, and Environmental Health Indicators Proceedings of a Joint South African/Australian Workshop, 7–11 February (1994), Cape Town, South Africa. Water Research Commission Report No. TT 63/94. Pretoria. 217–234.

Chutter, F.M., 1995. The role of aquatic organisms in the management of river basins for sustainable utilisation. Water Science and Technology 32, 283–291.

Chutter, F.M., 1998. Research on the rapid biological assessment of water quality impacts in streams and rivers. WRC Report No. 422/1/98. Water Research Commission, Pretoria, South Africa.

Clews, E., Ormerod, S.J., 2009. Improving bio-diagnostic monitoring using simple combinations of standard biotic indices. River Research and Applications. 25 (3), 348–361.

Cortelezzi, A., Paggi, A.C., Rodríguez, M., Capítulo, A.R., 2011. Taxonomic and nontaxonomic responses to ecological changes in an urban lowland stream through the use of Chironomidae (Diptera) larvae. Science of the Total Environment 409 (7), 1344–1350.

Dallas, H.F., 2002. Spatial and Temporal Heterogeneity in Lotic Systems: Implications for Defining Reference Conditions for Macroinvertebrates. PhD thesis, University of Cape Town, South Africa.

Dallas, H.F., Day J.A., 2004. The effect of water quality variables on aquatic ecosystems: a review. WRC Report No. TT 224/04. Water Research Commission, Pretoria, South Africa.

Dallas, H.F., Day, J.A., 1993. The effect of water quality variables on riverine ecosystems: a review. WRC Report No. TT 61/93. Water Research Commission, Pretoria, South Africa.

Dauvin, J.C., Ruellet, T., 2007. Polychaete/amphipod ratio revisited. Marine pollution bulletin 55, 215–224.

Davis, W.S., Simon, T.P. (Eds.), 1995. Biological Assessment and Criteria: Tools for Water Resource Planning and Decision Making. Lewis Publishers, London.

De Jonge, M., Van de Vijver, B., Blust, R., Bervoets, L., 2008. Responses of aquatic organisms to metal pollution in a lowland river in Flanders: a comparison of diatoms and macroinvertebrates. Science of the total environment 407, 615–629.

De pauw, N., Vanhooren, G., 1983. Method for biological quality assessment of watercourses in Belgium. Hydrobiologia 100, 153–168.

Dell'Uomo, A., 2004. L'Indice Diatomico di Eutrofizzazione/Polluzione, EPI-D nel monitoraggio delle acque correnti – Linee guida. APAT, Roma.

Descy, J.P., 1979. A new approach to water quality estimation using diatoms. Nova Hedwigia 64, 305–323.

Descy, J.P., Coste, M., 1991. A test of methods for assessing water quality based on diatoms. Verhandlungen des Internationalen Verein Limnologie 24, 2112–2116.

Dudgeon, D., Choowaew, S., Ho, S.C., 2000. River conservation in south-east Asia. In: Boon, P.J., Davies, B.R., Petts, G.E. (Eds.), Global Perspectives on River Conservation. Science, Policy and Practice. John Whiley and Sons, Chichester, pp. 281–310.

Duong, T.T., Feurtet-Mazel, A., Coste, M., Dang, D.K., Boudou, A., 2007. Dynamics of diatom colonization process in some rivers influenced by urban pollution (Hanoi, Vietnam). Ecological Indicators 7, 839–851.

Engle, V.D., Summers, J.K., 1999. Refinement, validation, and application of a benthic condition index for Gulf of Mexico estuaries. Estuaries 22, 624–635.

European Parliament, European Council, 2001. Decision No 2455/2001/EC of the European Parliament and of the Council of 20 November 2001 establishing the list of priority substances in the field of water policy and amending Directive 2000/60/EC (text with EEA relevance). Official Journal of the European Communities 331, 1–5.

European Parliament, European Council, 2006. Directive 2006/11/EC of the European Parliament and of the Council of 15 February 2006 on pollution caused by certain dangerous substances discharged into the aquatic environment of the Community. Official Journal of the European Communities L64, 52–59.

Fabela, P.S., Sandoval Manrique, J.C., López, R.L., Sánchez, E.S., 2001. Using a diversity index to establish water quality in lotic systems, Utilización de un índice de diversidad para determinar la calidad del agua en sistemas lóticos. Ingenieria Hidraulica En Mexico Impact Factor 16 (2), 57–66 (in Spanish).

Faisal, A., Dondelinger, F., Husmeier, D., Beale, C.M., 2010. Inferring species interaction networks from species abundance data: A comparative evaluation of various statistical and machine learning methods. Ecological Informatics 5, 451–464.

Feio, M.J., Almeida, S.F.P., Craveiro, S.C., Calado, A.J., 2009. A comparison between biotic indices and predictive models in stream water quality assessment based on benthic diatom communities. Ecological Indicator 9 (3), 497–507.

Fenoglio, S., Badino, G., Bona, F., 2002. Benthic macroinvertebrate communities as indicators of river environment quality: an experience in Nicaragua. Revista de Biologia Tropical 50, 1125–1131.

Fleischer, D., Grémare, A., Labrune, C., Rumohr, H., Vanden Berghe, E., Zettler, M.L., 2007. Performance comparison of two biotic indices measuring the ecological status of water bodies in the Southern Baltic and Gulf of Lions. Marine Pollution Bulletin 54, 1598–1606.

Friedrich, G., Chapman, D., Beim, A., 1996. The use of biological material. In: Chapman, D. (Ed.), Water Quality Assessments. A Guide to the Use of Biota, Sediments and Water in Environmental monitoring, second ed. E and FN Spon, London, pp. 175–242.

García-Criado, F., Tomé, A., Vega, F.J., Antolin, C., 1999. Performance of some diversity and biotic indices in rivers affected by coal mining in northwestern. Spain Hydrobiologia 394, 209–217.

Gerritsen, J., Barbour, M.T., King, K., 2000. Apples, oranges and ecoregions: on determining pattern in aquatic assemblages. Journal of the North American Benthological Society 19, 487–496.

Gibson, G.R., Bowman, M.L., Gerritsen, J., Snyder, B.D., 2000. Estuarine and Coastal Marine Waters: Bioassessment and Biocriteria Technical Guidance. EPA 822-B-00–024. U.S. Environmental Protection Agency, Office of Water, Washington, DC.

Girgin, S., 2010. Evaluation of the benthic macroinvertebrate distribution in a stream environment during summer using biotic index. International Journal of Environmental Science and Technology 7 (1), 11–16.

Gómez, N., Licursi, M., 2001. The pampean diatom index (IDP) for assessment of rivers and streams in Argentina. Aquatic Ecology 35, 173–181.

Gomez-Gesteira, J.L., Dauvin, J.C., 2000. Amphipods are good bioindicators of the impact of oil spills on soft-bottom macrobenthic communities. Marine Pollution Bulletin 40 (11), 1017–1027.

Gopal, B., Bose, B., Goswami, A.B., 2000. River conservation in the Indian sub-continent. In: Boon, P.J., Davies, B.R., Petts, G.E. (Eds.), Global Perspectives on River Conservation. Science, Policy and Practice. John Whiley and Sons, Chichester, pp. 233–261.

Gray, N.F., Delaney, E., 2010. Measuring community response of bentic macroinvertebrates in an erosional river impacted by acid mine drainage by use of a simple model. Ecological Indicators 10 (3), 668–675.

Grémare, A., Labrune, C., Vanden Berghe, E., Amouroux, J.M., Bachelet, G., Zettler, M.L., Vanaverbeke, J., Zenetos, A., 2009. Comparison of the Performances of Two Biotic Indices Based on the MacroBen Database. In: Marine Ecology Progress Series, vol. 382, pp. 297–311.

Guo, Q., Ma, K., Yang, L., Cai, Q., He, K., 2010. A comparative study of the impact of species composition on a freshwater phytoplankton community using two contrasting biotic indices. Ecological Indicators 10 (2), 296–302.

Hering, D., Johnson, R.K., Kramm, S., Schmutz, S., Szoszkiewicz, K., Verdonschot, P.F.M., 2006. Assessment of European streams with diatoms, macrophytes, macroinvertebrates and fish: a comparative metric-based analysis of organism response to stress. Freshwater Biology 51, 1757–1785.

Hilsenhoff, W.L., 1987. An improved biotic index of organic stream pollution. The Great Lakes Entomologist 20, 31–39.

Hilsenhoff, W.L., 1998. A modification of the biotic index of organic stream pollution to remedy problems and permit its use throughout the year. Great Lakes Entomologist. 31, 1–12.

Hurlbert, S.H., 1971. The nonconcept of species diversity: a critique and alternative parameters. Ecology 52, 577–586.

Hürlimann, J., Niederhauser, P., 2006. Methoden zur Untersuchung und Beurteilung der Fliessgewässer: Kieselalgen Stufe F (flächendeckend) Bundesamt für Umwelt, BAFU, Bern.

Imoobe, T.O.T., Ohiozebau, E., 2010. Pollution status of a tropical forest river using aquatic insects as indicators. African Journal of Ecology 48 (1), 232–238.

Iversen, T.M., Madsen, B.L., Bogestrand, J., 2000. River conservation in the European Community, including Scandinavia. In: Boon, P.J., Davies, B.R., Petts, G.E. (Eds.), Global Perspectives on River Conservation. Science, Policy and Practice. John Whiley and Sons, Chichester, pp. 79–103.

Ivorra, N., 2000. Metal Induced Succession in Benthic Diatom Consortia, Ph.D. Thesis. University of Amsterdam, Amsterdam.

Johnson, R.K., Wiederholm, T., Rosenberg, D.M., 1993. Freshwater Biomonitoring and Benthic Macroinvertebrates. Freshwater biomonitoring using individual organisms, populations and species assemblages of benthic macroinvertebrates. In: Rosenberg, D.M., Resh, V.H. (Eds.). Chapman and Hall, New York, pp. 40–125.

Jüttner, I., Chimonides, P.J., Ormerod, S.J., 2010. Using diatoms as quality indicators for a newly-formed urban lake and its catchment. Environmental Monitoring and Assessment 162 (1–4), 47–65.

Kelly, M.G., Whitton, B.A., 1995. The trophic diatom index: a new index for monitoring eutrophication in rivers. Journal of Applied Phycology 7, 433–444.

Knoben, R., Roos, C., van Oirschot, M.C.M., 1995. UN/ECE Task Force on Monitoring and Assessment under the Convention on the Protection and Use of Transboundary Watercourses and International Lakes (Helsinki, 1992). vol. 3: biological assessment l'Equipment Rural, Section Pêche et Pisciculture. RIZA Report No. 95.066. RIZA Institute for Inland Water Management and Waste Water Treatment, Lelystad. Available at: http://www.iwac-riza.org/WAC/IWACSite.nsf/.

Lang, C., Reymond, O., 1995. An improved index of environmental quality for Swiss rivers, based on benthic invertebrates. Aquatic Sciences 57, 172–180.

Lang, C.L., Eplattenier, G., Reymond, O., 1989. Water quality in rivers of western Switzerland: application of an adaptable index based on benthic invertebrates. Aquatic Sciences 51, 224–234.

Larsen, S., Sorace, A., Mancini, L., 2010. Riparian Bird Communities as Indicators of Human Impacts Along Mediterranean Streams. Environmental Management, 1–13.

Lavoie, I., Campeau, S., 2010. Fishing for diatoms: Fish gut analysis reveals water quality changes over a 75-year period. Journal of Paleolimnology 43, 121–130.

Lavoie, I., Campeau, S., Darchambeau, F., Cabana, G., Dillon, P.J., 2008. Are diatoms good integrators of temporal variability in stream water quality? Freshwater Biology 53 (4), 827–841.

Lazaridou-Dimitriadou, M., Koukoumides, C., Lekka, E., Gaidagis, G., 2004. Integrative evaluation of the ecological quality of metalliferous streams (Chalkidiki, Macedonia, Hellas). Environmental Monitoring and Assessment 91, 59–86.

Leclercq, L., Maquet, L., 1987. Deux nouveaux indices chimiques et diatomiques de qualité d'eau courante. Application au Samson et à ses affluents, Bassin de la Meuse belge. Comparaison avec d'autres indices chimiques biocénotiques et diatomiques. Inst. Roy. Sci. Nat. Belg. Doc. Trav. 38, 1–113.

Lenoir, A., Coste, M., 1996. Development of a practical diatom index of overall water quality applicable to the French national water board network. In: Whitton, B.A., Rott, E. (Eds.), Use of Algae for Monitoring Rivers II. Institutfur Botanik. University Innsbruck, pp. 29–43.

Li, L., Liu, C., Mou, H., 2000. River conservation in central and eastern Asia. In: Boon, P.J., Davies, B.R., Petts, G.E. (Eds.), Global Perspectives on River Conservation. Science, Policy and Practice. John Whiley and Sons, Chichester, pp. 263–279.

Lobo, E.A., Callegaro, V.L.M., Bender, E.P., 2002. Utilizaçao de Algas Diatomaceas Epiliticas como Indicadores da Qualidade da Agua em Rios e Arroios da Regiao Hidrografica do Guaiba, RS, Brasil. EDUNISC, Santa Cruz do Sul.

Maggioni, L.A., Fontaneto, D., Bocchi, S., Gomarasca, S., 2009. Evaluation of water quality and ecological system conditions through macrophytes. Desalination 246 (1–3), 190–201.

Marín-Guirao, L., Cesar, A., Marın, A., Lloret, J., Vita, R., 2005. Establishing the ecological quality status of soft-bottom mining-impacted coastal water bodies in the scope of the Water Framework Directive. Marine pollution bulletin 50 (4), 374–387.

Mazor, R.D., Schiff, K., et al., 2010. Bioassessment tools in novel habitats: An evaluation of indices and sampling methods in low-gradient streams in California. Environmental Monitoring and Assessment 167 (1–4), 91–104.

Meloni, P., Isola, D., Loi, N., Schintu, M., Contu, A., 2003. Coliphages as indicators of fecal contamination in sea water. (I colifagi come indice di contaminazione fecale nelle acque marine). Ann Ig 15 (2), 111–116 (in Italian).

Metcalfe, J.L., 1989. Biological water quality assessment of running waters based on macroinvertebrate communities: history and present status in Europe. Environmental Pollution 60, 101–139.

Metcalfe-Smith, J.L., 1994. Biological water quality assessment of rivers: use of macroinvertebrate communities. In: Calow, P., Petts, G.E. (Eds.), The Rivers Handbook. Hydrological and Ecological Principles, vol. 2. Blackwell Scientific Publications, Oxford, pp. 144–170.

Metzeling, L., Chessman, B., Hardwick, R., Wong, V., 2003. Rapid assessment of rivers using macroinvertebrates: the role of experience, and comparisons with quantitative methods. Hydrobiologia 510, 39–52.

Milner, A.M., Oswood, M.W., 2000. Urbanization gradients in streams of Anchorage, Alaska: a comparison of multivariate and multimetric approaches to classification. Hydrobiologia 422 & 423, 209–223.

Moog, O., Chovanec, A., 2000. Assessing the ecological integrity of rivers: walking the line among ecological, political and administrative interests. Hydrobiologia 422, 99–109.

Moreno, J.L., Navarro, C., De Las Heras, J., 2006. Proposal of an Aquatic Vegetation Index (IVAM) for assessing the trophic status of the Castilla-La Mancha rivers: A comparison with either indexes. (Propuesta de un índice de vegetation acuática (IVAM) para la evaluación del estado trófico de los ríos de Castilla-La Mancha: Comparación con otros índices bióticos). Limnetica 25 (3), 821–838 (in Spanish).

Mugnai, R., Oliveira, R.B., Do Lago Carvalho, A., Baptista, D.F., 2008. Adaptation of the Indice Biotico Esteso (IBE) for water quality assessment in rivers of Serra do Mar, Rio de Janeiro State, Brazil. Tropical Zoology 21, 57–74.

Muniz, P., Venturini, N., Pires-Vanin., A.M.S., Tommasi, L.R., Borja, A., 2005. Testing the applicability of a marine biotic index (AMBI) to assessing the ecological quality of softbottom benthic communities, South America Atlantic region. Marine Pollution Bulletin 50, 624–637.

Murphy, P.M., 1978. The temporal variability in biotic indices. Environmental Pollution 17, 227–236.

Muttil, N., Chau, K.W., 2006. Neural network and genetic programming for modelling coastal algal blooms. International Journal of Environment and Pollution 28, 223–238.

Muttil, N., Chau, K.W., 2007. Machine-learning paradigms for selecting ecologically significant input variables. Engineering Applications of Artificial Intelligence 20, 735–744.

Muxika, I., Borja, A., Bonne, W., 2005. The suitability of the marine biotic index (AMBI) to new impact sources along European coasts. Ecological Indicators 5, 19–31.

Norris, R.H., Georges, A., 1993. Analysis and interpretation of benthic macroinvertebrate surveys. In: Rosenberg, D.M., Resh, V.H. (Eds.), Freshwater Biomonitoring and Benthic Macroinvertebrates. Chapman and Hall, New York, pp. 234–286.

Norris, R.H., Thoms, M.C., 1999. What is river health? Freshwater Biology 41, 197–209.

Ollis, D.J., Dalls, H.F., Esler, K., Boucher, C., 2006. Bioassessment of the ecological integrity of river ecosystems. African Journal of Aquatic Science 31, 205–227.

Pinder, L.C.V., Ladle, M., Gledhill, T., Bass, J.A.B., Matthews, A.M., 1987. Biological surveillance of water quality, 1. A comparison of macroinvertebrate surveillance methods in relation to assessment of water quality, in a chalk stream. Archiv für Hydrobiologie 109, 207–226.

Pinel-Alloul, B., Methot, G., Lapierre, L., Willsie, A., 1996. Macroinvertebrate community as a biological indicator of ecological and toxicological factors in Lake Saint-Francois, Quebec. Environmental Pollution 91, 65–87.

Pinto, R., Patrıcioa, J., Baeta, A., Fath, B.D., Neto, J.M., Marques, J.C., 2009. Review and evaluation of estuarine biotic indices to assess benthic condition. Ecological indicators 9, 1–25.

Pringle, C.M., Scatena, F.N., Paaby-Hansen, P., Nunez-Ferrera, M., 2000. Conservation in Latin America and the Caribbean. Chapter 2 pages 41–78. In: Boon, P.J., Davies, B., Peets, G.C. (Eds.), Global Perspectives on River Conservation: Science, Policy and Practice. John Wiley and Sons LTD, England.

Prygiel, J., Leveque, L., Iserentant, R., 1996. L'IDP: Un nouvel Indice Diatomique Pratique pour l' evaluation de la qualité des eaux en réseau de surveillance. Revue Des Sciences De L'Eau 9, 97–113.

Resh, V.H., 1995. Freshwater benthic macro invertebrates and rapid assessment procedures for water quality monitoring in developing and newly industrialized countries. In: Davis, W.S., Simon, T.P. (Eds.), Biological Assessment and Criteria. Tools for Water Resource Planning and Decision-making. Lewis Publishers, Boca Raton, pp. 167–177.

Resh, V.H., Norris, R.H., Barbour, M.T., 1995. Design and implementation of rapid assessment approaches for water resource monitoring using benthic macroinvertebrates. Australian Journal of Ecology 20 (1), 108–121.

Reynoldson, T.B., Metcalfe-Smith, J.L., 1992. An overview of the assessment of aquatic ecosystem health, using benthic macroinvertebrates. Journal of Aquatic Ecosystem Health 1, 295–308.

Reynoldson, T.B., Norris, R.H., Resh, V.H., Day, K.E., Rosenberg, D.M., 1997. The reference condition: a comparison of multimetric and multivariate approaches to assess water quality impairment using benthic macroinvertebrates. Journal of the North American Benthological Society 16, 833–852.

Rico, E., Rallo, A., Sevillano, M.A., Arretxe, M.L., 1992. Comparison of several biological indices based on river macroinvertebrate benthic community for assessment of running water quality. Annals Limnology 28, 147–156.

Robson, M., Spence, K., Beech, L., 2006. Stream quality in a small urbanized catchment. Science of the total environment 357, 194–207.

Rosenberg, R., Blomquist, M., Nilsson, H.C., Cederwall, H., Dimming, A., 2004. Marine quality assessment by use of benthic species-abundance distribution: a proposed new Protocol within the European Union Water framework Directive. Marine Pollution Bulletin 49, 728–739.

Rott, E., Pfister, P., Van Dam, H., Pipp, E., Pall, K., Binder, N., Ortler, K., 1999. Indikationlisten für Aufwuchsalgen in österreichischen Fliessgewässern. Teil 2: Trophieindikation sowie geochemische Präferenzen, taxonomische und 24 S. Blanco, E. Bécares / Chemosphere. 79, 18–25.

Rott, E., Hofmann, G. Pall, K. Pfister, P. Pipp, E., 1997. Indikationslisten für Aufwuchsalgen in osterreichischen Fliessgewässern. Teil 1: Saprobielle Indikation, Wasserwirtschaftskataster, Bundesministerium f. Land- u. Forstwirtschaft, Wien.

Roux, D.J., Everett, M.J., 1994. The ecosystem approach for river health assessment: a south African perspective. In: Uys, M.C. (Ed.), Classification of Rivers, and Environmental Health Indicators Proceedings of a Joint South

African/Australian Workshop, 7–11 February 1994, Cape Town, South Africa. Water Research Commission Report No. TT 63/94, Pretoria, South Africa. 343–361.

Roux, D.J., Kempster, P.L., Kleynhans, C.J., Vanvliet, H.R., Du preez, H.H., 1999a. Integrating stressor and response monitoring into a resource-based water quality assessment framework. Environmental Management 23, 15–30.

Roux, D.J., Kleynhans, C.J., Thirion, C., Hill, L., Engelbrecht, J.S., Deacon, A.R., Kemper, N.P., 1999b. Adaptive assessment and management of riverine ecosystems: the Crocodile/Elands River case study. Water SA 25, 501–511.

Rumeau, A., Coste, M., 1988. Initiation à la systématique des diatomées d'eau douce. Bulletin Francais de la Peche et de la Pisciculture 309, 1–69.

Rygg, B., 2002. Indicator species index for assessing benthic ecological quality in marine waters of Norway. Norwegian Institute for Water Research, Report no. 40114, pp. 1–32.

Sabater, S., 2000. Diatom communities as indicators of environmental stress in the Guadiamar River, S-W. Spain, following a major mine tailings spill. Journal of Applied physiology 12, 113–124.

Salas, F., Neto, J.M., Borga, A., Marques, J.C., 2004. Evaluation of the applicability of a marine biotic index to characterize the status of estuarine ecosystems: the case of Mondego estuary (Portugal). Ecological Indicators 4, 215–225.

Sandin, L., Hering, D., BuffagnI, A., Lorenz, A., Moog, O., Rolauffs, P., Stubauer, I., 2001. The development and testing of an Integrated Assessment System for the ecological quality of streams and rivers throughout Europe, using benthic macroinvertebrates, Third Deliverable: experiences with different stream assessment methods and outlines of an integrated method for assessing streams, using benthic macroinvertebrates. AQEM, Contract No. EVK1-CT1999-00027. www.moog.at/downloads/Publikationen.pdf.

Sarkar, C., Abbasi, S.A., 2006. Qualidex – A new software for generating water quality indice. Environmental Monitoring and Assessment 119, 201–231.

Schiefele, S., Kohmann, F., 1993. Bioindikation der Trophie in Fließgewässern. Umweltforschungsplan des Bundesministers für Umwelt, Naturschutz und Reaktorsicherheit, Wasserwirtschaft, Forschungsbericht Nr. 102 01 504. Bayerisches Landesamt für Wasserwirtschaft, München, Berlin.

Selvakumar, A., O'Connor, T.P., Struck, S.D., 2010. Role of stream restoration on improving benthic macroinvertebrates and In-stream water quality in an urban watershed: case study. Journal of Environmental Engineering 136 (1), 127–139.

Shannon, C.E., Weaver, W., 1949. The Mathematical Theory of Communication. University of Illinois Press, Urbana.

Simboura, N., Argyrou, M., 2010. An insight into the performance of benthic classification indices tested in Eastern Mediterranean coastal waters. Marine Pollution Bulletin 60, 701–709.

Simboura, N., Panayotidis, P., Papathanassiou, E., 2005. A synthesis of the biological quality elements for the implementation of the European Water Framework Directive in the Mediterranean ecoregion: the case of Saronikos Gulf. Ecological Indicators 5, 253–266.

Simboura, N., Papathanassiou, E., Sakellariou, D., 2007. The use of a biotic index (Bentix) in assessing long-term effects of dumping coarse metalliferous waste on soft bottom benthic communities. Ecological Indicators 7, 164–180.

Simboura, N., Zenetos, A., 2002. Benthic indicators to use in ecological quality classification of Mediterranean soft bottom marine ecosystems, including a new biotic index. Mediterranean Marine Science 3, 77–111.

Simić, V., Simić, S., 1999. Use of the river macrozoobenthos of Serbia to formulate a biotic index. Hydrobiologia 416, 51–64.

Sládeček, V., 1986. Diatoms as indicators of organic pollution. Acta Hydrochim. Hydrobiol. 14, 555–566.

Smith, R.W., Bergen, M., Weisberg, S.B., Cadien, D., Dalkey, A., Montagne, D., Stull, J.K., Velarde, R.G., 1998. Southern California Bight Pilot Project: Benthic Response Index for Assessing Infaunal Communities on the Mainland Shelf of Southern California. Southern California Coastal Water Research Project. http://www.sccwrp.org.

Smith, R.W., Bergen, M., Weisberg, S.B., Cadien, D., Dankey, A., Montagne, D., Stull, J.K., Velarde, R.G., 2001. Benthic response index for assessing infaunal communities on the Mainland Shelf of Southern California. Journal of Applied Ecology 11, 1073–1087.

Soininen, J., Könönen, K., 2004. Comparative study of monitoring South-Finnish rivers and streams using macroinvertebrate and benthic diatom community structure. Aquatic Ecology 38, 63–75.

Stark, J.D., 1985. A Macroinvertebrate Community Index of Water Quality for Stony Streams, Water and Soil Miscellaneous Publication No. 87. National Water and Soil Conservation Authority, Wellington, New Zealand.

Stark, J.D., 1993. Performance of the Macroinvertebrate Community Index: effects of sampling method, sample replication, water depth, current velocity and substratum on index values. New Zealand Journal of Marine and Freshwater Research 27, 463–478.

Steinberg, C., Schiefele, S., 1988. Biological indication of trophy and pollution of running waters. Zeitschrift für Wasser und Abwasserforschung 21, 227–234.

Suárez, M.L., Mellado, A., Sánchez-Montoya, M.M., Vidal-Abarca, M.R., 2005. Proposal of an index of macrophytes (IM) for evaluation of warm ecology of the rivers of the Segura Basin. (Propuesta de un índice de macrófitos (IM) para evaluar la calidad ecológica de los ríos de la cuenca del Segura). Limnetica 24 (3–4), 305–318.

Terrado, M., Barceló, D., Tauler, R., Borrell, E., Campos, S.D., 2010. Surface-water-quality indices for the analysis of data generated by automated sampling networks. TrAC - Trends in Analytical Chemistry 29 (1), 40–52.

Tuffery, G., Verneaux, J., 1968. Méthode de détermination de la qualité biologique des eaux courantes. Exploitation codifiée des inventaires de la faune du fond. Ministère de l'Agriculture (France), Centre National d'Etudes Techniques et de Recherches Technologiques pour l'Agriculture, les Forets et l'Equipment Rural, Section Pêche et Pisciculture.

Verdonschot, P.F.M., 2000. Integrated ecological assessment methods as a basis for sustainable catchment management. Hydrobiologia. 422 & 423, 389–412.

Washington, H.G., 1984. Diversity, biotic and similarity indices. A review with special relevance to aquatic ecosystems. Water Research 18, 653–694.

Watanabe, T., Asai, K., Houki, A., 1988. Numerical water quality monitoring of organic pollution using diatom assemblage. Proceedings of the Ninth International Diatom Symposium 1986. 123–141. In: Round, F.E. (Ed.). Koeltz Scientific Books, Koenigstein.

Woodiwiss, F.S., 1964. The biological system of stream classification used by the Trent River Board. Chemistry and Industry 83, 443–447.

Wright, J.F., 1995. Development and use of a system for predicting the macroinvertebrate fauna in flowing waters. Australian Journal of Ecology 20, 181–197.

Wright, J.F., Sutcliffe, D.W., Furse, M.T., 2000. Assessing the Biological Quality of Fresh Waters: RIVPACS and Other Techniques. Freshwater Biological Association, UK.

Wu, C.L., Chau, K.W., 2006. Mathematical model of water quality rehabilitation with rainwater utilisation: A case study at Haigang. International Journal of Environment and Pollution 28, 534–545.

Wu, H.C., Chen, P.C., Tsay, T.T., 2010. Assessment of nematode community structure as a bioindicator in river monitoring. Environmental Pollution 158 (5), 1741–1747.

Yoder, C.O., 1989. The cost of biological field monitoring. United States Environmental Protection Agency, Washington, D.C., USA.

Zalack, J.T., Smucker, N.J., Vis, M.L., 2010. Development of a diatom index of biotic integrity for acid mine drainage impacted streams. Ecological Indicators 10 (2), 287–295.

Zamora-Munoz, C., Alba-Tercedor, J., 1996. Bioassessment of organically polluted Spanish rivers, using a biotic index and multivariate methods. Journal of the North American Benthological Society 15, 332–352.

Zamora-Muñoz, C., Sáinz-Cantero, C.E., Sánchez-Ortega, A., Alba-Tercedor, J., 1995. Are biological indices BMPW' and ASPT' and their significance regarding water quality seasonally dependent? Factors explaining their variations. Water Research 29, 285–290.

Zenetos, A., Chadjianestis, I., Lantzouni, M., Simboura, M., Sklivagou, E., Arvanitakis, G., 2004. The Eurobulker oil spill: midterm changes of some ecosystem indicators. Marine Pollution Bulletin 48 (1/2), 121–131.

Zettler, M.L., Schiedek, D., Bobertz, B., 2007. Benthic biodiversity indices versus salinity gradient in the southern Baltic Sea. Marine Pollution Bulletin 55, 258–270.

Zgrundo, A., Bogaczewicz-Adamczak, B., 2004. Applicability of diatom indices for monitoring water quality in coastal streams in the Gulf of Gdańsk region, Northern Poland. Oceanological and Hydrobiological Studies 33 (3), 31–46.

Zhaia, H., Cuia, B., Hua, B., Zhang, K., 2010. Prediction of river ecological integrity after cascade hydropower dam construction on the mainstream of rivers in Longitudinal Range-Gorge Region (LRGR), China. Ecological Engineering 36, 361–372.

Zhao, M.Y., Cheng, C.T., Chau, K.W., Li, G., 2006. Multiple criteria data envelopment analysis for full ranking units associated to environment impact assessment. International Journal of Environment and Pollution 28, 448–464.

CHAPTER

14

Indices of Biological Integrity or the Multi-metric Indices

OUTLINE

Water Quality Indices DOI: 10.1016/B978-0-444-54304-2.00014-2

14.1. INTRODUCTION

As described in Chapter 13, biotic indices are able to provide a better assessment of the overall status of water quality in conjunction with 'physico-chemical' water quality indices, than the latter can achieve in isolation. However, biotic indices are not always adequately eloquent as they are based on the sensitivity to stress of a few indicator species or ecological group of species (Kröncke and Reiss, 2010).

In contrast, indices of biological integrity (IBIs) use a combination of univariate and biotic indices in an endeavour to capture with greater sensitivity the impacts of anthropogenic disturbances on aquatic ecosystems.

Biological integrity has been defined as "... the capability of supporting and maintaining a balanced, integrated, adaptive, community of organisms having a species composition, diversity, and functional organization comparable to that of natural habitat of the region (Karr and Dudley, 1981)." The biological condition attained by a natural system is the consequence of the integration of the chemical, physical and biological parts and processes that maintain the system over time. The organisms that inhabit a natural system, both individually and as communities, are indicators of the actual conditions in that system since their presence as well as their well-being is influenced by the human-induced variations that occur in the system (Mack, 2007).

Whereas biotic indices seek the representation of a natural water body through certain species or group of species, IBIs seek to do it with a suit of metrics that are chosen to reflect the taxonomic composition, tropic relationships, abundance and condition of organisms within an aquatic community (Karr and Yoder, 2004; Dolph et al., 2010). IBIs, thus, aim to convey a more integrated picture of ecosystem health than BIs are capable of. Due to the multi-metric construction of IBIs, the latter are also called 'multi-metric indices'. It can be said that IBIs are perhaps more developed and 'evolved' forms of BIs described in the previous chapter. IBIs can also be called 'indices of indices'.

By virtue of their multi-metric construct, IBIs can reflect important components of ecology: taxonomic richness, habitat and trophic guild composition, besides individual health and abundance. Differences in expected species richness and composition associated with different regions or basins, water body sizes and location in a drainage are factored into metric selection and scoring. Even as the main focus of the IBIs is on the assemblage structure of chosen indicator organisms (or combinations of organisms)—such as fish, plankton, benthos and macrophytes—rather than ecosystem processes, because the latter appear less responsive to stressors (Schindler, 1990), yet both structural and functional metrics are often included in the IBIs.

In a way IBIs translate an aquatic ecologist's assessment of organism assemblage integrity for persons unfamiliar with that organism. However, IBIs cannot, by themselves, convey causal relationships or fundamental ecological processes.

Assessment of biological integrity by the IBIs requires comparison of index scores of test sites with the scores earned by the 'natural' habitat of the region. However, few, if any, undisturbed natural habitats are in existence in any part of the world. Hence, the 'natural' condition has to be estimated from minimally disturbed sites. However, in heavily altered regions even such sites are absent. In those cases, historical data, palaeoecological data, quantitative models and expert judgement have to be used to define the 'natural' condition.

14.2. THE FIRST IBI (KARR, 1981)

The concept of IBI (Index of Biological Integrity) was introduced by Karr (1981). He argued that even though taxa other than fish (macroinvertebrates and diatoms) have been widely used in monitoring because of the availability of a theoretical substructure that allows an integrated ecological approach, the use of diatoms or invertebrates as monitoring targets has major deficiencies. For example, they require specialised taxonomic expertise; they are difficult and time-consuming to sample, sort and identify; background life-history information is often lacking for many species and groups; and the results obtained by using diatoms and invertebrates are difficult to translate into values meaningful to the general public.

On the other hand, fish, argued Karr, have numerous advantages as indicator organisms for biological monitoring programmes:

1. Extensive life-history information is available for most fish species.
2. Fish communities generally include a range of species that represent a variety of trophic levels (omnivores, herbivores, insectivores, planktivores and piscivores) and include foods of both aquatic and terrestrial origin. Their position at the top of the aquatic food web in relation to diatoms and invertebrates also helps to provide an integrative view of the watershed environment.
3. Fish are relatively easy to identify. Technicians require relatively little training. Indeed, most samples can be sorted and identified at the field site, with release of study organisms after processing.
4. The general public can relate to statements about conditions of the fish community.
5. Both acute toxicity (missing taxa) and stress effects (depressed growth and reproductive success) can be evaluated. Careful examination of recruitment and growth dynamics among years can help to pinpoint periods of unusual stress.
6. Fish are typically present, even in the smallest streams and in all but the most polluted waters.
7. Finally, the results of studies using fish can be directly related to the fishable waters mandate by the governmental regulations (for example, The Indian Fisheries Act, 1987/1956).

Karr also cited a number of disadvantages in monitoring fish: for example, the selective nature of sampling, fish mobility on diel and seasonal time scales and manpower needs for field sampling. However, felt Karr, these are disadvantages associated with any major taxa.

Prior efforts to monitor and evaluate biotic communities typically involved use of only one or two criteria, often combined in an index. For example diversity indices, like the one of Shannon and Weaver (1949), discussed in a previous chapter (also Abbasi and Abbasi 2011) consider number of species (richness) and equitability (abundance of species). Others have used production (or biomass) as a single index (Boling et al., 1975) or in combination with diversity (Gammon et al., 1981). Karr argued that these approaches overlook many important variables and thus oversimplify exceedingly complex systems. Hence, he introduced a new index, named 'Index of Biotic integrity' by him.

Karr based his system on the following premises:

1. The fish sample, as noted earlier, is a balanced representation of the fish community at the sample site.
2. The sample site is representative of the larger geographic area of interest.
3. The scientist charged with data analysis and the final classification is a trained, competent biologist with considerable familiarity with the local fish fauna.

In practice, the IBI uses attributes of the fish assemblage in a stream reach to assess the condition of a stream and its catchment, relative to regional and historical standards. It reflects land–water linkages, physical habitat quality, hydrological regime, energy inputs, biological interactions and water quality (Karr et al., 1986; Steedman, 1988; Allan, Erickson and Fay, 1997). The IBI was designed to integrate information from individual, population, assemblage and ecosystem levels into a single numerical indicator and quality rating for water bodies (Karr et al., 1986). Originally proposed by Karr (1981) and later refined by Fausch, et al. (1984) and Karr et al. (1986), the IBI combines twelve fish assemblage attributes, or metrics, classified into three groups: (i) species richness and composition, (ii) trophic composition and (iii) fish abundance and health.

Data are obtained for each of these metrics at a given site and evaluated relative to what would be expected at a set of relatively undisturbed sites located within the same ecoregion (Omernik, 1987) or in a similar ecoregion (Hughes, 1995). When there are no appropriate undisturbed sites on a river, the least-disturbed regional sites can be used as a standard (Hughes, Larsen and Omernik, 1986; Hughes, 1995) or the reference condition can be modelled from knowledge of historical conditions and fish habitat requirements (Hughes, 1995; Hughes et al., 1998). A rating of 5, 3 or 1 is then assigned to each metric, based on whether its value approximates (5), deviates moderately from (3) or strongly deviates from (1) the reference condition (Table 14.1). The IBI is the sum of the twelve ratings, and varies from 12 to 60. A classification of 'no fishes' is assigned when repeated sampling produces no fish. The index score is interpreted as in Table 14.2.

This index was first developed for streams in the midwestern United States, and has been tested and found useful in many other regions

TABLE 14.1 Metrics Used to Assess Biological Integrity of Fish Communities Based on Index of Biotic Integrity, IBI (Karr, 1981). Ratings of 5, 3 and 1 are Assigned to Each Metric According to Whether Its Value Approximates, Deviates Somewhat From or Deviates Strongly From the Value Expected *at a Comparable Site* That is Relatively Undisturbed

Metrics	Rating of Metric*		
	5	**3**	**1**
SPECIES RICHNESS AND COMPOSITION			
1. Total number of fish species* (native fish species)[@]	*Expectations for metrics 1–5 vary with stream size and region*		
2. Number and identity of darter species (benthic species)			
3. Number and identity of sunfish species (water-column species)			
4. Number and identity of sucker species (long-lived species)			
5. Number and identity of intolerant species			
6. Percentage of individuals as green sunfish (tolerant species)	<5	5–20	>20
TROPHIC COMPOSITION			
7. Percentage of individuals as omnivores	<20	20–45	>45
8. Percentage of individuals as insectivorous cyprinids (insectivores)	>45	45–20	<20
9. Percentage of individuals as piscivores (top carnivores)	>5	5–1	<1
FISH ABUNDANCE AND CONDITION			
10. Number of individuals in sample	*Expectations for metric 10 vary with stream size and other factors*		
11. Percentage of individuals as hybrids (exotics or simple lithophiles)	0	>0–1	>1
12. Percentage of individuals with disease, tumors, fin damage and skeletal anomalies	0–2	>2–5	>5

* *Original IBI metrics for midwest United States.*
[@] *Generalized IBI metrics*

TABLE 14.2 Interpretation of the IBI Scores (Karr, 1981)

Total IBI Score (Sum of the 12 Metric Ratings)*	Integrity Class of Site	Attributes
58–60	Excellent	Comparable to the best situations without human disturbance; all regionally expected species for the habitat and stream size, including the most intolerant forms, are present with a full array of age (size) classes; balanced trophic structure.
48–52	Good	Species richness somewhat below expectation, especially due to the loss of the most intolerant forms; some species are present with less than optimal abundances or size distributions; trophic structure shows some signs of stress.
40–44	Fair	Signs of additional deterioration include loss of intolerant forms, fewer species and highly skewed trophic structure (e.g., increasing frequency of omnivores and green sunfish or other tolerant species); older age classes of top predators may be rare.
28–34	Poor	Dominated by omnivores, tolerant forms, and habitat generalists; few top carnivores; growth rates and condition factors commonly depressed; hybrids and diseased fish often present.
12–22	Very poor	Few fish present, mostly introduced or tolerant forms; hybrids common; disease, parasites, fin damage and other anomalies regular.
@	No fish	Repeated sampling finds on fish.

** Sites with values between classes are assigned to appropriate integrity class following careful consideration of individual criteria/metrics by informed biologists.*
[@] No score can be calculated where no fish were found.

of North America (Miller et al., 1988; Simon and Lyons, 1995). It has also been applied to estuaries (Thompson and Fitzhugh, 1986), lakes (Dionne and Karr, 1992; Minns et al., 1994) and rivers outside the United States and Canada (Hughes and Oberdorff, 1998). However, even as an IBI follows an ecoregional approach, it is often required to modify the metrics, or delete some or add some to account for regional differences in fish distribution and assemblage structure. The efforts are not always successful; for example, a modified IBI evaluated in semiarid southeastern Colorado, USA was not able to reflect habitat degradation with adequate sensitivity in other regions. (Bramblett and Fausch, 1991).

Over the years, a number of shortcomings of Karr's IBI have been identified and attempts have been made to overcome those infirmities. In the process, a large number of new IBIs have been developed. Interestingly, despite the variety of adaptations throughout the world, the fundamentals of the original IBI still stand strong (Roset et al., 2007). These aspects are discussed in detail in the subsequent sections.

14.3. THE DRIVER–PRESSURE–STRESS–IMPACT–RESPONSE (DPSIR) PARADIGM AND THE IBI

Borja and Dauer (2008) have described the process of IBI development in the context of what can be called the 'DPSIR paradigm'. DPSIR represents the feedback loop system seen to operate everywhere in which driving forces (D) of social and economic development exert pressure (P) on the environment, thereby stressing it and changing its state (S), potentially resulting in impacts (I) on human health and/or ecosystem function. These, then, elicit an environmental management response (R).

More often than not, the strongest driver indicator is population density, accompanied by different levels of developmental impulses; pressure (stressor) indicators are large-scale anthropogenic pressures which are exemplified by changes in land-use patterns and increase in air–water–soil pollution; state (exposure) indicators include aspects such as extents of organic/inorganic pollution of the environment actually being caused, and impact (ecological response) indicators include changes in biological community structure.

As elucidated by Borja and Dauer (2008), for integrating environmental protection legislation and the DPSIR paradigm, the impact (I) component (Figure 14.1, dashed inset) requires

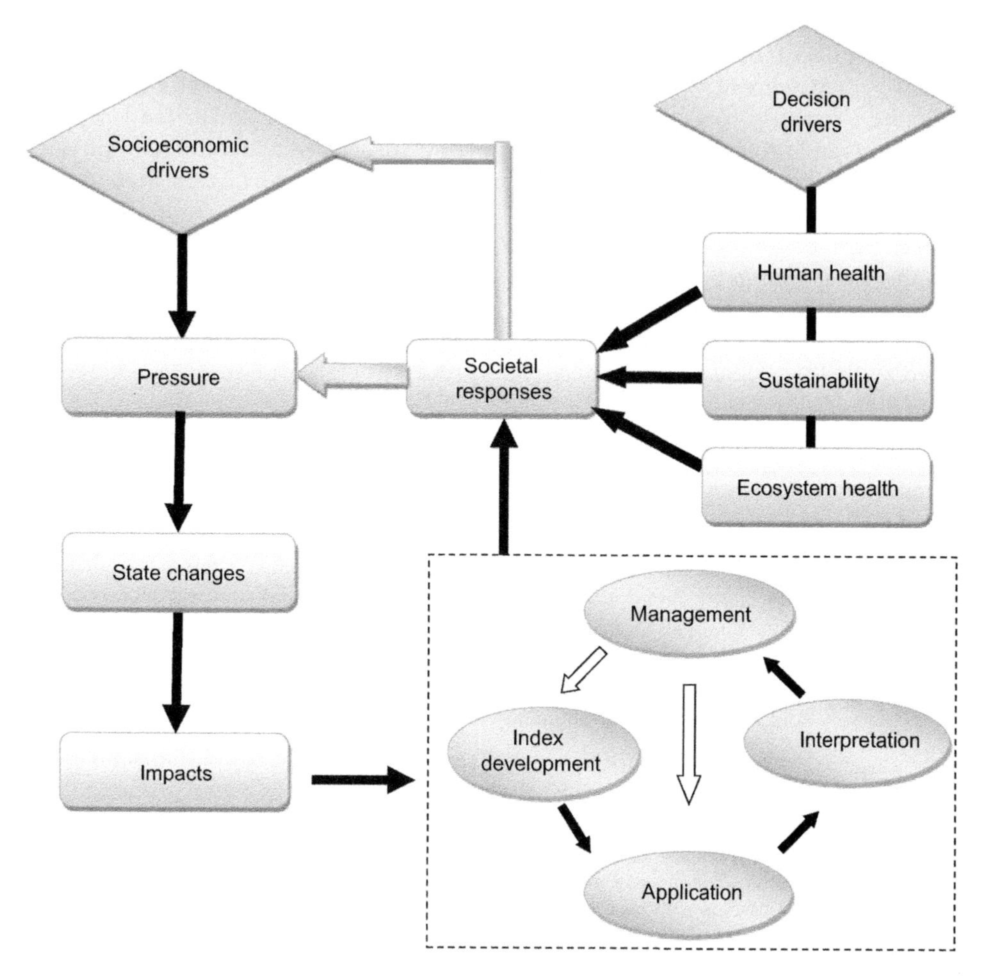

FIGURE 14.1 The Drivers–pressure–State–impact–response (DPSIR) cycle. Societal responses meant to halt, ameliorate, mitigate or reverse unacceptable conditions are shown by hollow arrows. Inset shows the impact assessment components; adaptive monitoring feedback pathways are indicated by hollow arrows (adopted from Borja and Dauer, 2008).

(a) assessing ecological integrity; (b) evaluating if significant ecological degradation has occurred; (c) identifying the spatial extent and location of ecological degradation; and (d) determining causes of unacceptable degradation in order to guide management actions. Feedback loops, which are indicated by hollow arrows within the dashed inset of Figure 14.1, between environmental management and index application–interpretation, represent adaptive monitoring changes that are necessary before developing and finalising societal responses (R). The 'R' component of DPSIR constitutes environmental management strategies to halt, ameliorate, mitigate or reverse unacceptable conditions and protect human health besides maintaining a healthy ecosystem while promoting sustainable development.

An integral part of determining ecological integrity is the measurement of biological integrity, typically emphasising analyses of plankton, benthos, macroalgae and fish. This is accomplished with IBIs. Hence, the impact (I) component of DPSIR requires the use and/or development of indices, their appropriate application in space and time and justifiable and defensible interpretation of results.

14.3.1. Steps to Development of IBI

Building a robust and effective IBI is based on proper selection of measurable attributes that provide reliable and relevant signals about the biological effects of human activities. The biological attributes ultimately incorporated into an IBI — the metrics — are chosen because they reflect specific, predictable responses of organisms to changes in landscape condition. They are sensitive to a range of physical, chemical and biological factors that alter biological systems, and they are relatively easy to measure and interpret. A typical IBI comprises several attributes of the sampled biota, including taxa richness, indicator groups, health of individual organisms and ecological processes (Table 14.3).

TABLE 14.3 Biological Attributes on Which IBI Metrics can be Based

Category	Demonstrated Effective	Need More Testing	Difficult to Measure or Too Theoretical
Taxa richness	Total taxa richness. Richness of major taxa (e.g., mayflies or sunfish)	Dominance (relative abundance of most numerous taxa)	Relative abundance distribution, after Preston (1962)
Tolerance, intolerance	Taxa richness of intolerant organisms; Relative abundance of green sunfish; Relative abundance of tolerant taxa	Number of rare or endangered taxa	Chironomid species (difficult to identify)
Trophic structure	Trophic organisation, e.g., relative abundance of predators or omnivores		Productivity
Individual health	Relative abundance of individual fish with deformities, lesions or tumors Relative abundance of individual chironomids with head-capsule deformities Growth rates by size or age class	Contaminant levels in tissue (biomarkers)	Metabolic rate
Other ecological attributes		Age structure of target species-population	

The most important criterion for choosing a biological attribute as a metric is whether the attribute responds predictably along a gradient of human influence. The vital question is: does the attribute vary systematically with varying degrees of human impact? An effective IBI comprising well-chosen metrics should integrate information from ecosystem, community, population and individual levels (Karr and Chu, 1999) and clearly discriminate the biological 'signal' – including the effects of human activities – from the 'noise' of natural variation. Moreover, because IBI is founded on empirical data, its use does not require resolution of all the higher-order theoretical debates in contemporary ecology (Karr and Chu, 2000) or the formal mathematical models of ecological functions.

Candidate Metrics

Selection of candidate metrics is based on the consideration of ecological relevance as well as feasibility of measurement (Table 14.4). Candidate metrics typically include measures of species diversity, productivity (abundance and biomass), tolerance to and/or indication of association with anthropocentric sources of stress, in other words pollution-indicative or sensitive taxa and taxocene dominance measures because some major taxa are more tolerant of stress than others.

In the early fish-based IBIs (Karr, 1981; Karr and Chu, 2000), metrics were included to characterise taxonomic richness, tolerance/intolerance, trophic structure and individual health (anomalies). IBIs developed later often included other ecological attributes, such as age structure, reproductive guilds, life history/behavioural guilds, habitat guilds, or alien or target-species population size (Hughes et al., 1998). Barbour et al. (1999) recommended grouping macroinvertebrate metrics by categories, such as taxonomic richness, taxonomic composition, tolerance/intolerance, feeding group (e.g., predators, scrapers and filter feeders) and habit type (e.g., clingers and burrowers).

When the ecological relevance is based upon very specific ecological concepts, then candidate metrics are basically *a priori* selected or limited; for example, an index such as AMBI (Borja et al., 2000) that is based upon the relative distribution of sensitive/tolerant species groups or the ABC method (Warwick and Clarke, 1994) based upon k distribution curves consists of predetermined metrics. Indices developed with a utilitarian approach typically begin with a large list of candidate metrics, which is then pruned. Examples of both the approaches to IBI development can be seen in Section 14.4.

TABLE 14.4 Aspects and Attributes of the IBIs that the Candidate Metrics Should Facilitate

Aspect	Attributes
Purpose	• Summarises and simplifies complex data • Conveys information which can be easily understood by the public, media, resource users, and decision-makers
Relevance	• Has ecological relevance as is based upon a conceptual model (theoretically, empirically or heuristically well founded)
Feasibility	• Is feasible to collect the requisite data reliably and cost-effectively to calculate the index
Threshold appreciation	• Users are able to appreciate the significance of the indicator value
Representativeness	• Is able to measure status and trends that are relevant to policy decisions
Sensitivity	• Adequately reflects the responses of the system to management actions

Metric Selection

The list of feasible, ecologically relevant candidates is pruned to retain metrics that are more sensitive (responsive to anthropocentric action – both degradative and restorative) and representative (able to measure status and trends relative to policy decisions and management actions) than others. Theoretically, metric nomination and/or selection is based upon community level characters that represent key community aspects. Such community characteristics are typically part of, or inherent to, the diversity of definitions of biological integrity that include elements of species diversity, abundance, energy-flow–food-web structure, maintenance of complexity and self-organisation, etc. Metric selection protocols include:

1. *a priori* selection based upon a specific ecological foundation and/or best professional judgement;
2. selection based upon univariate statistical tests comparing undegraded and degraded samples from calibration data (Weisberg et al., 1997); and
3. utilitarian selection based upon multivariate tests using a calibration data set (Engle et al., 1994; Paul et al., 2001).

Metric range: The first hurdle which a candidate metric must overcome is the range test. The 'range' is the distribution of metric values across all of the available data and the goal of the range test is to identify metrics that provide a large range of scores with that data. Metrics that have very small ranges (e.g., richness metrics based on only a few taxa) or the ones that have similar values at most sites (e.g., most sites have values = 0) are to be deleted. Assigning scores to metrics with small ranges also is problematic for the same reason.

Reproducibility: A metric providing fairly reproducible values at individual sites is more useful in assessing between-site differences than a metric which is less precise because the latter will be less clear in showing differences in stream condition due to its variation within a site. Sampling variation is estimated from repeat visits to individual sites. Available measures of sampling variation reflect several sources of variability (i.e., short-term index-period temporal variability, spatial variability within the reach and laboratory variability). Low sampling variation is necessary if a metric is to have a high probability of discriminating between sites in good and poor condition, and sampling variation should be small relative to the size if the among-site differences are to be appreciated (Stoddard et al., 2008).

Metric reproducibility is normally quantified by the signal (S): noise (N) ratio. S/N is the ratio of the variance among all sites (signal) to the variance of repeated visits to the same site (noise) (Kaufmann et al., 1999). Metrics with high S/N values are more likely to show consistent responses to stressors than are metrics with low S/N values. A metric that is perfectly correlated with a hypothetical stressor and that has no sampling variability will have an $R^2 = 1.0$ for that stressor. As S/N decreases, the maximum possible R^2 value of the regression decreases because the sampling variability of the metric increases.

There is no fixed threshold below which metrics can be eliminated on the basis of S/N. However, $S/N \leq 1$ values indicate that visiting a single site twice yields as much metric variability as visiting two different sites. In practice, the threshold depends on the inherent level of variability in the assemblages being assessed and might depend on other factors, such as generation times of the organisms in the assemblages (Stoddard et al., 2005). Fish metrics commonly have high S/N values and, therefore, a high threshold for rejection (4 or 5), whereas periphyton metrics have low S/N values and a low threshold for rejection (1 or 1.5). Macroinvertebrate metrics have intermediate S/N values, and their threshold for rejection is close to 2.0.

Adjusting for natural gradients: Metric values can vary with both the stressor gradients being assessed and natural gradients (e.g., elevation, slope and stream size). The stressors themselves might vary along the same natural gradients. Thus, knowledge of how to apportion the variability in metric values between natural and anthropogenic gradients is important. Selection of metrics that appear to respond strongly to some stressor but, in fact, are merely correlated with the same natural gradient with which the stressor is correlated, should be avoided (Stoddard et al., 2008).

One of the techniques for normalising metrics for natural gradients is to remove the stressor gradient from the data by focussing solely on reference-site data and to quantify the remaining correspondence between the metric value and the natural gradient.

Responsiveness: The effectiveness of a metric is directly linked with its ability to distinguish degraded from relatively undisturbed streams. This responsiveness can be tested in a number of ways. For example, metrics can be chosen on the basis of their correlation with specific stressors (e.g., nutrients, organic pollution and sedimentation). Some of the original metrics used by Karr (1981) were chosen on the basis of their hypothesised responses to specific aquatic stressors (e.g., darters and benthic disturbance). However, several difficulties arise when metrics are evaluated in terms of their relationship with specific stressors. First, many stressors are highly correlated with one another, and attributing metric response to any particular stressor may inflate the role of that stressor. Second, not all stressors are well quantified (e.g., short-lived pesticides or herbicides), or even known, at all sites.

In the absence of pristine sites that may serve as frames of reference, the evaluation of the responsiveness of the metric is based on the metric's ability to distinguish least-disturbed (reference) from most-disturbed sites. Metric scoring and threshold selection typically are based on a set of least-disturbed sites.

Final metric selection and check for metric redundancy: All metrics that pass the tests for range and reproducibility are considered for inclusion in the IBIs. Candidate metric that is the most discriminating (i.e., highest t-score) is chosen first. One then proceeds iteratively and adds the most responsive metric from each metric category until all categories are represented. This iterative process is based on the assumption that choosing the most responsive individual metrics will yield the most robust IBI, provided that each metric provides unique information (i.e., that the metrics are not redundant).

What constitutes redundancy? It is not a simple question to answer. Metric redundancy can be defined in at least two very different ways: (1) metrics provide very similar biological information or (2) metrics are highly correlated with other metrics. Philosophically, the first of these definitions may appear as more important, because one would be disinclined to include two metrics that are based on identical (or broadly overlapping) biological or taxonomic information (Stoddard et al., 2008). However, apparently redundant metrics might be, in fact, relatively uncorrelated.

Alternatively, one might avoid including metrics with values that are strongly correlated. If two metrics covary because the same taxa are changing in abundance as levels of disturbance rise and fall, then the metrics are clearly correlated. However, if metrics covary because they respond to similar stressors, one might question whether correlated metrics are necessarily redundant.

Metric Combination

The most difficult challenge in index development is selecting and combining metrics in a manner that is complex enough to capture the dynamics of essential ecological processes but not so complex that its meaning is obscured (NRC, 2000). Without a sound and obvious ecological foundation, an index will not be policy relevant and therefore difficult to use in the DPSIR systems. Once developed, such indices fall into three classes (ICES, 2004), based upon their complexity, information content and method of metric combination:

1. univariate individual-species data, or community structure measures, of the type discussed in Chapter 13;
2. multi-metric indices, or IBIs, combining several measures of community response to stress into a single index; covered in this chapter; and
3. multivariate methods describing the assemblage pattern, including modelling, discussed in Chapter 15.

For IBI development, the first step is to normalise all metrics so that they are scored on a common scale and to combine them in a final IBI. Numerous methods exist for scoring metrics, but the primary decision is whether to score metrics discretely, for example, with values of 1, 3 or 5, based on a subjective assessment of the range of each metric (as was initially done by Karr, 1981 and has since been done extensively), or to score the metrics on a continuous, but consistent scale. Discrete scoring has the disadvantage that it can mask subtle differences by forcing the scores to be in one or the other discrete interval. This could dampen the ability of the index to differentiate among ecological condition classes (Ganasan and Hughes 1998, Blocksom 2003). Continuous scoring (0–10) can avoid the subjective nature of discrete scoring (e.g., the assumption that sites with metric scores in the top of the range of values are all in good condition, whereas sites in the bottom range of values are all in poor condition).

The use of continuous scoring requires consideration of how to set ceiling and floor values for each metric, i.e., how to decide what values of a metric indicate 'ideal' biological condition (eg a score of 10 or 100; usually 10) and what values indicate unacceptably bad condition (eg a score of 0). The 95th percentile of the reference-site distribution of values for each metric is generally used as the scoring ceiling and the 5th percentile of the distribution of values at all sites as the scoring floor. This approach produces an IBI with the highest responsiveness and lowest variability (measured by S/N). Metric values between the ceiling and floor are interpolated linearly to yield intermediate values, and the final MMI for a site is calculated as the sum of its scored metrics. When interpreting final scores, the IBI is generally rescaled to range from 0 to 100.

Index Validation

The robustness of any IBI depends on how widely and repeatedly it has been found to be effective. Index validation should ideally include:

1. Testing of the index using independent data sets, different from the data set used in index development.
2. Setting *a priori* correct classification criteria.
3. Presentation of a strong *a posteriori* justification for use based upon best professional judgement. More commonly after index development, newly collected data are used as validation data. In such cases, strong putative gradients are deliberately selected for validation testing (e.g., Borja et al., 2000, 2003, 2006; Muxika et al., 2005; Quintino et al., 2006). Independent validation, by scientists other than those proposing the index, is highly valued.

Some degree of intercalibration or validation can also be achieved by determining the level of agreement provided by an index with best professional judgement (Weisberg et al., 2008) or by comparing the level of agreement between indices of different geographical origin, such as when comparing results of indices from Europe and USA.

Index Application and Interpretation

Notwithstanding the aim of most index developers to have an index which is applicable across a number of different regions and habitats, it is

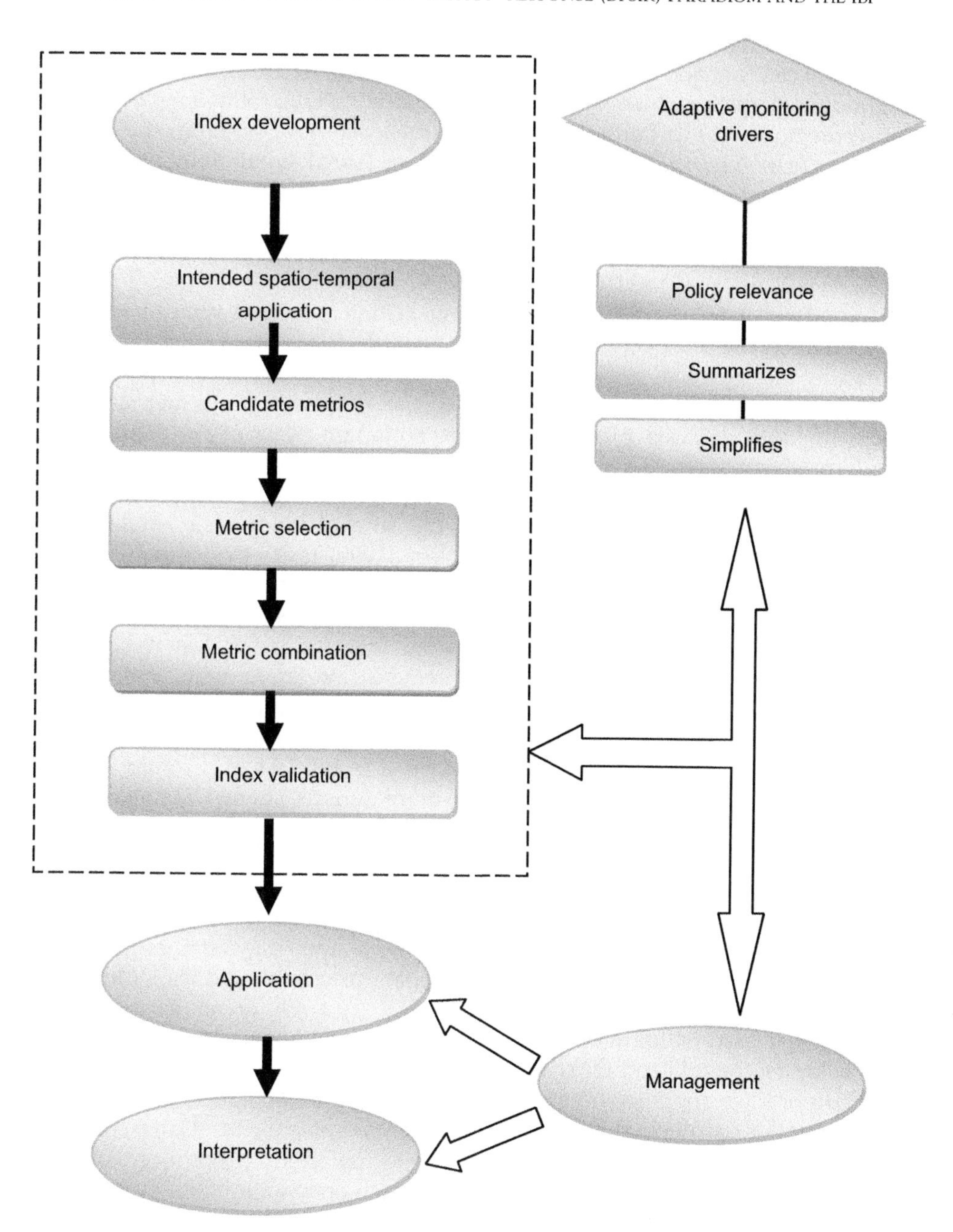

FIGURE 14.2 The index development, application and interpretation process. Primary steps have been enclosed in a dashed rectangle. Adaptive monitoring feedback loops and adaptive change decision drivers are indicated by hollow arrows (adopted from Borja and Dauer, 2008).

generally a major challenge to apply, or even adopt, indices which can function in (a) latitudinally or longitudinally distinct biogeographic provinces or ecoregions; (b) different water depth habitat types (e.g., intertidal *versus* subtidal habitats) and (c) different substratum types.

Adaptive Management of Indices

As public awareness towards the DPSI parts of the DPSIR paradigm increases, and as new scientific knowledge on these aspects emerges, the urgency, nature and extent of response (R) have to undergo changes. With it may change (a) data needs; (b) data analysis procedures; and (c) data interpretation needs. This makes it necessary to make adoptive changes in the indices, and may entail additional effort to demonstrate ecological relevance, representativeness and sensitivity. It may also require further testing of index applicability, and maximising information and understanding by management of the interpretation process.

The key steps associated with IBI development and applications are presented in Figure 14.2.

14.4. ILLUSTRATIVE EXAMPLES OF IBI DEVELOPMENT

In this section, twelve illustrative examples of IBI development are presented which encompass different biogeographic regions, continents, types of water bodies, and intended applications. The summaries are focussed on the methodologies used in developing the IBIs and make only very brief mention of their application because most other sections of this chapter dwell at length on different aspects of applications of IBIs.

14.4.1. An Indian IBI

The first, and to-date the only, attempt to develop an IBI for an Indian situation was made by Ganasan and Hughes (1998). They developed it in the context of a study on rivers Khan and Kshipra, Madhya Pradesh, India (Figure 14.3). They essentially adopted Karr's IBI (Karr, 1981) as follows:

Taxonomic richness. From the original metric of Karr (1981) 'total number of species' was modified by Ganasan and Hughes (1998) to 'number of native species' and 'number of native families' was added (Table 14.5). Number of native species was deemed a measure of biological diversity that typically decreases with increased degradation. Ganasan and Hughes (1998) felt it important to separate native from non-native species in this metric when the latter are abundant (Karr et al., 1986). Although not recommended by Karr et al. (1986) or Miller et al. (1988), 'number of native families' is a measure of biodiversity at the family level (Noss, 1990) that also decreases as anthropogenic disturbance increases. Also, Oberdorff and Hughes (1992) and Witkowski (1992) had found that entire families were eliminated or threatened in areas with long and intensive occupation by humans. Hence, Ganasan and Hughes (1998) argued that loss of a species when it is the last of a family in the basin is a more serious change than loss of a species from a family represented by multiple species.

Habitat composition. One of Karr's (1981) original metrics, 'number of intolerant species' was retained; but four metrics specific to midwestern North America ('number of darter species,' 'number of sunfish species,' 'number of sucker species' and '% green sunfish individuals') were replaced with 'number of benthic species,' 'number of water-column species' and '% individuals as tolerant species.'

Trophic composition. Ganasan and Hughes (1998) replaced 'proportion of individuals as insectivores' with '% of individuals as herbivores.' In their study, insectivore (invertivore) species and individuals were less common than herbivores in less polluted habitats and more common in polluted habitats; therefore, herbivores were deemed more sensitive by them than invertivores.

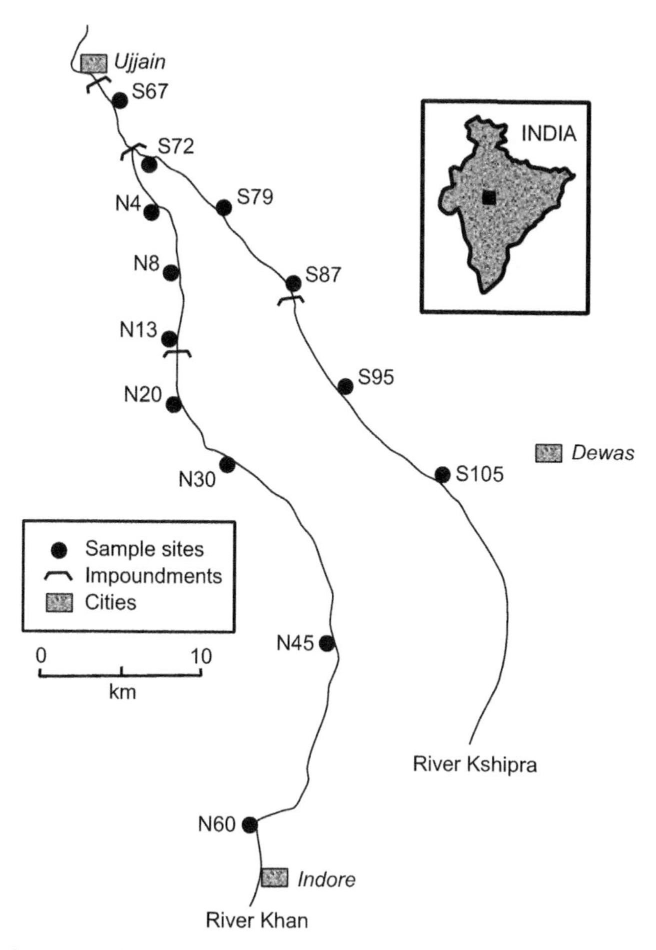

FIGURE 14.3 The sampling network used by Ganasan and Hughes (1998) for developing the First Indian IBI.

Fish health and abundance. Two of the original metrics were unchanged, 'total number of individuals in the sample' and '% of individuals with disease or anomalies.' A new metric was added: '% of individuals as non-native.' It included species introduced for mosquito control in degraded areas and was deemed a measure of the degree by which invasive alien species dominate the assemblage.

Calculation of the IBI Metrics

Because no undisturbed rivers or reaches were sampled in central India, Ganasan and Hughes (1998) used the most 'desirable' metric values, which were obtained at the least-disturbed (furthest downstream) sites in the rivers Khan and Kshipra, as estimates of the reference condition. The authors were aware that such values greatly underestimated the potential biological integrity; and suggested that future research on less disturbed rivers in the same ecoregion may provide more accurate measures of the potential condition.

Salient Findings

The Indian IBI scores increased in the rivers Khan and Kshipra downstream of urban and industrial point sources of pollution (Figure 14.4).

TABLE 14.5 Illustrative Example of Systematisation of a Fisheries Survey as Input to an IBI (Ganasan and Hughes, 1998) *Habitat Guilds*: Subsurface SS, Water Column WC, Benthic B. *Trophic Guilds*: Top Carnivore TC, Herbivore H, Invertivore I, Omnivore O. *Tolerance Guilds*: Intolerant I, Moderately Tolerant M, Tolerant T. *Origin Guilds*: Native N, Non-native NN

Family / Species	Habitat	Trophic	Tolerance	Origin
NOTOPTERIDAE				
Notopterus Notopterus (Pallas)	WC	TC	M	N
CYPRINIDAE				
Chela bacaila (Gunther)	WC	I	M	N
Esomus danricus (Hamilton)	SS	H	M	N
Garra gotyla (Gray)	B	TC	I	N
Labeo calbasu (Hamilton)	WC	O	M	N
Labeo gonius (Hamilton)	WC	H	M	N
Labeo rohita (Hamilton)	WC	H	M	N
Osteobrama cotio (Day)	WC	H	M	N
Puntius conchonius (Hamilton)	WC	O	M	N
Puntius sarana (Hamilton)	WC	O	M	N
Puntius sophore (Hamilton)	WC	O	M	N
Puntius ticto (Hamilton)	WC	O	M	N
Rasbora daniconius (Hamilton)	SS	I	M	N
COBITIDAE				
Botia lohachata (Chaudhuri)	B	O	M	N
Lepidocephalus guntea (Hamilton)	B	O	M	N
Nemacheilus botia (Hamilton)	B	O	M	N
SILURIDAE				
Ompok bimaculatus (Bloch & Schneider)	WC	TC	M	N
Wallago attu (Bloch & Schneider)	WC	TC	M	N
BAGRIDAE				
Aorichthys seenghala (Wu, 1939)	WC	TC	M	N
Mystus bleekeri (Day)	WC	TC	M	N
Mystus tengara (Hamilton)	WC	TC	M	N
Mystus vittatus (Bloch)	WC	TC	M	N
SCHILBEIDAE				
Eutropiichthys vacha (Hamilton)	WC	TC	M	N

TABLE 14.5 Illustrative Example of Systematisation of a Fisheries Survey as Input to an IBI (Ganasan and Hughes, 1998) Habitat Guilds: Subsurface SS, Water Column WC, Benthic B. Trophic Guilds: Top Carnivore TC, Herbivore H, Invertivore I, Omnivore O. Tolerance Guilds: Intolerant I, Moderately Tolerant M, Tolerant T. Origin Guilds: Native N, Non-native NN (*cont'd*)

Family / Species	Habitat	Trophic	Tolerance	Origin
HETEROPNEUSTIDAE				
Heteropneusties fossilis (Bloch)	B	TC	T	N
CLARIIDAE				
Clarias batrachus (Linnaeus, 1758)	B	TC	T	N
BELONIDAE				
Xenentodon cancila (Hamilton)	SS	TC	I	N
CHANNIDAE				
Channa marulius (Hamilton)	B	TC	M	N
Channa orientalis (Schneider)	B	TC	M	N
Channa punctatus (Bloch)	B	TC	T	N
Channa striatus (Bloch)	B	TC	I	N
CHANDIDAE				
Chanda nama (Hamilton)	SS	TC	I	N
Chanda ranga (Hamilton)	SS	TC	M	N
GOBIIDEA				
Glossogobius giurius (Hamilton)	B	TC	M	N
MASTACEMBELIDAE				
Mastacembelus armatus (Lacepede)	WC	TC	M	N
Mastacembelus pancalus (Hamilton)	WC	TC	M	N
POECILIIDAE				
Lebistes reticulatus (Peters)	SS	O	T	NN

The metal and organic pollution from the cities were apparently the major sources of degradation in those rivers, a situation similar to that occurring in many rivers in the world before implementation of regulations like the Clean Water Act (PL 92–500) by the USA in 1972.

The surprising finding of Ganasan and Hughes (1998) was that there were good number of fishes even in stretches with high concentrations of heavy metals and sewage and the poor physical habitat quality (i.e., high silt levels and an absence of riparian trees and large woody debris). Moreover, the variety was also good (Table 14.6). At the less polluted sites, aquatic macrophytes apparently helped the fishes by providing structure, cover and a food base. The macrophytes also possibly reduced the levels of suspended sediments by physical entrapment, and hindered resuspension of deposited sediment from the riverbeds. In this respect,

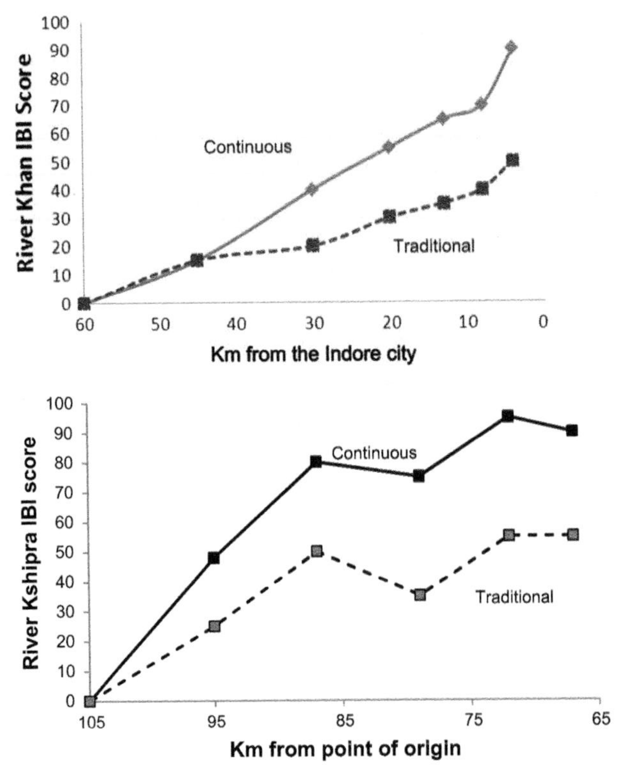

FIGURE 14.4 IBI scores at different points in the two Indian rivers based on a traditional (1, 3, 5 stepwise) and a continuous (0-10) pattern of metric scoring (Ganesan and Hughes, 1998).

macrophytes perhaps acted as *in-situ* treatment systems, a role which was possible solely because the infestation had occurred in flowing waters. Otherwise, in stationary waters, colonisation by macrophytes adds to the pollution load instead of attenuating it (Abbasi et al., 2009).

There was abundance of a non-native species of fish at the most-disturbed site, which indicated a substantial amount of biological disturbance that was perhaps, as much, or more, disruptive of the native fish fauna than the physical and chemical conditions. Ganasan and Hughes (1998) felt that with sufficient socioeconomic support, the chemical habitat could be improved by waste treatment works and the physical habitat structure improved by establishing natural riparian woodlands. However, the elimination or reduction of non-native fishes appeared more difficult because it was not achievable by simple technological or engineering techniques, or by modifications in land use.

The study demonstrated that the logical foundations of the IBI were easily adaptable to fish assemblages in rivers of central India, even though the IBI was first developed for fish assemblages of a totally different agroclimatic region – the Midwestern USA.

14.4.2. A Macrophyte-based IBI for Wetlands

Mack (2007) has described the process of development and testing—refinement—iterations, leading to a robust and versatile IBI. The IBI has since been found to consistently and reliably assess wetland conditions across the whole range

TABLE 14.6 Criteria for Scoring Index of Biological Integrity (IBI) Metrics for Central Indian Rivers by Ganasan and Hughes (1998). Traditional Scoring Criteria Scores of 5, 3 or 1 were Assigned to Each Metric According to Whether Its Value Approximated, Deviated Somewhat from or Deviated Strongly from the Value at the Least-disturbed Site in the Study. Continuous Scoring Criteria Scores were Determined as Proportions (× 10) of the Maximum or 'Best' Value

		Traditional Scoring Criteria			Continuous Scoring Criteria	
Category	**Metric**	**5 (Best)**	**3**	**1 (Worst)**	**10 (Best)**	**0 (Worst)**
Taxonomic richness	1. Number of native species	>24	12–24	<12	32	0
	2. Number of native families	>8	5–8	<5	13	0
Habitat composition	3. Number of benthic species	>7	5–7	<5	11	0
	4. Number of water-column species	>12	6–12	<6	19	0
	5. Number of intolerant species	>2	2	<2	4	0
	6. % Individuals as tolerant species	<10	10–20	>20	5	50
Trophic composition	7. % Individuals as omnivores	<36	36–72	>72	18	50
	8. % Individuals as herbivores	>10	5–10	<5	14	0
	9. % Individuals as top carnivores	>46	23–46	<23	70	0
Fish health and abundance	10. Total number of individuals	>1000	500–1000	<500	1490	0
	11. % Individuals as non-native	<1	1–10	>10	0	50
	12. % Individuals with anomalies or disease	NA	NA	NA	NA	NA

of wetland types encountered throughout the ecological regions of the state of Ohio, the USA.

One of the considerations behind developing this IBI was (Mack, 2007) that prior to this effort, several IBIs were developed around vascular plants as an indicator taxa for the wetlands of the USA, but most were based on data sets from 1 or 2 years and on a single class of wetlands. The wetland types covered by the previous IBIs included depressional emergent marshes in southern Minnesota (Gernes and Helgen, 1999); salt water marshes in Massachusetts (Carlisle et al., 1999); depressional, palustrine wetlands in southeast Wisconsin Till Plains region (Lillie et al., 2002); coastal marshes associated with Southern Lake Michigan in the State of Indiana (Simon et al., 2001); prairie pothole marshes in North Dakota (DeKeyser et al., 2003); some Pennsylvania wetlands (Miller et al., 2004); and small-order stream wetlands in southwest Montana (Jones, 2005). However, the applicability of these IBIs had limited geographic scope and was generally restricted in the types of wetlands that can be assessed.

Site selection: A targeted approach was used in site selection to ensure that wetlands representing the full gradient of disturbance, different plant communities, hydrogeomorphic classes and different ecoregions were adequately represented in the data set.

Classification: The classification system used by Mack (2007) is summarised in Table 14.7. It

TABLE 14.7 Hydrogeomorphic Classes for Wetland Classification System for Ohio Wetlands, Adapted from Cowardin et al. (1979), Anderson (1982), Brinson (1993), Mack (1998, 2001a,b), Cole et al. (1997), by Mack (2007)

	HGM class	(A) Forest	(B) Emergent	(C) Shrub
I	Depression (including areas that could be considered flats, e.g., 'wet woods')	(a) Swamp forest	(a) Marsh	(a) Shrub swamp
II	Impoundment	(b) Bog forest	(b) Wet meadow	(b) Bog shrub swamp
III	Riverine	(c) Forest seep	(c) Open bog	(c) Tall shrub fen
IV	Slope (including hillside fens, mound fens and lacustrine fens)			
V	Fringing (excluding lacustrine fens)			
VI	Coastal			
VII	Bog			

was developed and evaluated by modifying and adopting existing classification systems. To this end an iterative process was employed, which is typical in IBI development: an *a priori* classification approach (Fennessy et al., 2002; Detenbeck, 2002). This approach involves developing a classification scheme, collecting and analysing data from wetlands in the various classes, revising the scheme based on the new information and then repeating this process.

Vegetation sampling methods: At most sites a 'standard' 20 m × 50 m (0.1 ha) plot was established. Where the standard plot would not fit or would not have adequately characterised the plant community being sampled, the size or shape of the plot was modified to obtain a representative sample. In mixed emergent marshes, water depth generally decreases towards the upland boundary and the vegetation is zoned in narrow (to broad) bands. Typically, a narrow shrub zone gives way to a broad emergent zone which grades into a floating-leaved marsh to an open water zone. In such situations, a sampling plot was located in a way that the intensive modules were located within the emergent zone but the 'tails' (ends) of the plot included portions of the shrub and aquatic bed zones. Plots were located in representative areas even if this was well away from the wetland edge.

Attribute selection and scoring: Potential ecological or biological attributes of the taxa group were identified and evaluated, an 'attribute' being defined as a measurable characteristic of the biological community. A useful attribute was expected to have five general characteristics:

1. relevant to the biological community under study and to the specified programme objectives;
2. sensitive to stressors;
3. able to provide a response that can be discriminated from natural variation;
4. environmentally benign to measure in the aquatic environment; and
5. cost-effective to sample.

A 'metric' was defined as an attribute that changes in some predictable way in response to increased human disturbance and that has been included earlier as a component of a multi-metric IBI (Karr and Chu, 1999).

TABLE 14.8 Types of Vascular Plant Attributes for Used in the Wetland IBI Development by Mack (2007)

Type	Possible Attributes
Community structure	Taxa richness, relative cover, density, dominance, abundance of tolerant or sensitive species
Taxonomic composition	Identity, Floristic Quality Index, tolerance or intolerance of key taxa
Individual condition	Disease, anomalies and contaminant levels
Biological processes	Productivity, trophic dynamics, nutrient cycling and forest stand characteristics

Potential vascular plant attributes were categorised by their relationship to community structure, taxonomic composition, individual condition and biological processes (Table 14.7) and classified by type (Table 14.8) using groups previously developed for fish and invertebrate IBIs in streams (Barbour et al., 1995; Karr and Kerans, 1992). Testable hypotheses for these classes of attributes were proposed regarding the direction (increase, decrease and no change) and type of change (linear, curvilinear and threshold) to increasing levels of human disturbance (Table 14.9). Potential attributes were initially proposed as *a priori* hypotheses and included as aspects of the community structure (taxa richness, relative cover, density and dominance), taxonomic composition (species identity, floristic quality and diversity indices) and tolerance or intolerance of particular species to disturbance and ecosystem processes.

Over 400 attributes were extracted from the vegetation data set and evaluated for ecologically meaningful and explainable trends using graphical techniques, descriptive statistics and regression analysis. The short-listed

TABLE 14.9 Types of Metrics Investigated During the Wetland IBI Development by Mack (2007)

Number	Hypothesised Changes Caused by Human Disturbance of Natural Wetlands
1	Total species or generic richness, or richness of individual taxa groups declines
2	Abundance or numbers of sensitive species declines
3	Abundance or numbers of tolerant species increases
4	Proportions or abundance of plants with narrow ecological affinities declines
5	Overall floristic quality of plant community declines
6	Primary productivity increases
7	Proportions or abundances of plants with particular wetland affinities (obligate and facultative) changes based on the type of wetland (forested, emergent, etc.)
8	Proportions or abundances of plants with certain life forms (e.g., forb, graminoid, shrub or tree) changes relative to reference conditions
9	Proportions of individuals (relative density) or relative dominance (basal area) in tree species age classes increases changes relative to reference conditions
10	Proportions or numbers of non-native species or hybrids increases
11	Changes occur in community composition or heterogeneity relative to reference condition

TABLE 14.10 Attribute Classes for Vascular Plants as Used by Mack (2007)

No.	Category	Type
1	Taxa group	Dicots, monocots, certain genera (e.g., Carex), certain families or family groups (e.g., poaceae and cryptogams)
2	Life form	Forb, graminoid, shrub, tree and aquatic
3	Indicator status	Wetland indicator status, e.g., UPL, FACU, FAC, FACW and OBL
4	Age (size) class	The size, and presumably age, class membership of a tree
5	Ecological affinity	Shade tolerance, coefficient of conservatism assigned to plant species by Floristic quality assessment index (Andreas et al., 2004)
6	Indices	Floristic quality index, Simpson's diversity index, Shannon–Weiner, index, etc.

attributes (Table 14.10) had ecologically meaningful linear, curvilinear or threshold relationships to a human disturbance gradient. Attributes and metrics were selected and evaluated in successive refinements of the Vegetation IBI (Fennessy et al., 1998a,b; Lopez and Fennessy, 2002; Mack et al., 2000; Mack, 2001b, 2004b). Metric values were converted to metric scores by quadrisecting the 95th percentile of the metric value (USEPA, 1998, 1999; Hughes et al., 1998).

Human disturbance gradient: Contrary to the development of many stream IBIs, where potential attributes are plotted against stream drainage area, the development of wetland IBIs requires the concomitant development of a meaningful human disturbance gradient with which to evaluate the potential attributes.

The author relied on the Ohio Rapid Assessment Method (ORAM) developed earlier, since refined to ORAM version 5.0 (Mack et al., 2000; Mack, 2001 a, b).

The ORAM score ranges from 0 (very poor condition) to 100 (excellent condition). Its probes are mostly site specific and include size (6 points), buffer width (7 points), dominant land use outside of the buffer (7 points), hydrologic characteristics (up to 18 points), intactness of natural hydrologic regimes (12 points), intactness of natural substrates (4 points), overall wetland development (7 points), habitat characteristics, e.g., heterogeneity, plant community type, microtopography and invasive species abundance (up to 20 points), intactness of natural habitats (9 points) and special wetland communities, e.g., bogs, fens and mature forests (10 points). The point allocations were assigned on the basis of:

a. a consideration of relative importance (e.g., up to 30 points could be awarded for hydrology reflecting its importance in wetland structure and function);
b. the premise that any type of wetland in good (or conversely poor) condition would have a high (or low) score; and
c. the limit that maximum score was 100.

Data analysis: Standard IBI development techniques were used to develop and evaluate the Vegetation IBI (e.g., Karr and Chu, 1999). The VIBI and each metric were individually evaluated by:

1. comparing metric behaviour against the disturbance gradients with linear regression;
2. graphically analysing scatter plots and box and whisker plots;
3. testing for differences in mean metric values with ANOVA and multiple comparison tests for disturbance classes defined by ORAM tertiles (first tertile ORAM scores of 0–33, second tertile scores of 34–66 and third tertile scores of 67–100); and

4. by ordinating metric values using principal components analysis to evaluate differential and overall performance of metrics (Mack, 2004b).

Detrended correspondence analysis (DCA) and cluster analysis were used to evaluate the classification system. Given the large number of species, sites and potential plant communities in the data set, ordination space partitioning was used, which is a polythetic, divisive classification technique recommended by Gauch (1982). After initial ordinations, successive partitions were made to generate a hierarchical classification.

The IBI was subjected to extensive testing and refinement to make it robust and enhance its range of application. Throughout its initial development, and subsequent three testing iterations, the IBI remained significantly correlated with the human disturbance gradient. Eight of the original 10 proposed metrics continued to have significant and interpretable relationships with the disturbance gradient, with 4 metrics remaining completely unchanged, 4 undergoing relatively minor modifications and 2 being replaced.

The IBI and its component metrics were also evaluated against a new disturbance gradient (Landscape Development Index or LDI), derived from land use percentages within a 1 km radius of the wetlands that were not used during the index development.

The IBI score and 9 of 10 metrics were significantly correlated with the LDI disturbance gradient providing separate confirmation of the IBI.

The IBI was seen to consistently and reliably assess wetland condition across the whole range of wetland types throughout Ohio's ecological regions.

14.4.3. Two Fish IBIs for Coldwater and Mixed-water Wadeable Streams

Kanno et al. (2010) report a set of two IBIs for coldwater and mixed-water streams which was applied in the state of Connecticut, USA. The fish assemblage data for the indices were collected from wadeable streams across the state between 1999 and 2007.

A synthetic human disturbance gradient was derived from seven landscape variables: (1) percent of impervious surface, (2) percent of forested land, (3) road density, (4) road crossing density, (5) population density, (6) dam density and (7) density of known water quality issues, such as industrial discharge permits and leachate reports.

Principal component analysis (PCA) was used to reduce the dimensionality of the original variables. Percent of impervious surface and forested land were subjected to arcsine square root transformation, and dam density and density of known water quality issues to log transformation. The remaining variables were not transformed because in their case data transformation did not improve data normality. The resulting human disturbance gradient was then used to classify stream sites into three categories; least, moderately and most-disturbed sites.

For metric selection, forty and fifty-five candidate metrics were chosen for coldwater and mixed-water IBI, respectively. The candidate metrics were then divided into 8 ecological classes in each subset, and a single best metric was selected from each ecological class. For each IBI development, a sequence of screening was applied for each metric: range, signal-to-noise, correlation with stream size, responsiveness to landscape-level human disturbances and redundancy, as explained below.

Range: Given that metrics with small ranges of values are unlikely to discriminate discrepancies among streams (McCormick et al., 2001; Klemm et al., 2003; Bramblett et al., 2005; Whittier et al., 2007), taxonomic richness metrics were eliminated if their range was ≤ 3 species. Any metric yielding zero result for >70% of values in the range test was rejected.

Signal-to-noise: On the premise that good metrics provide repeatable results, within-year variation was tested by comparing the among-reach variance (i.e., signal) to within-year revisitation (i.e., noise). However, this step could be applied only to the mixed-water IBI data set as low revisitation of smaller streams precluded signal-to-noise evaluation in the coldwater IBI development. Seventeen sites (9% of the mixed-water IBI data) were sampled twice within the same summer. Metrics were rejected if their signal-to-noise ratio was <2.

Correlation with stream size: To avoid confusing results arising from anthropogenic effects, only least-disturbed sites were used to derive a linear regression line between drainage area and each metric. Stream-size correction was deemed necessary if 95% prediction intervals of the resulting regression lines had overlapping values at both ends of the stream size gradient and if visual inspection of plotted data confirmed that these intervals were not due to a few influential points. For metrics requiring stream-size correction, residuals from regression lines were calculated, and the original metrics were replaced with the stream-size-corrected metrics.

Responsiveness to human disturbances: One-way analyses of variance (ANOVA) were run to test the ability of each candidate metric to differentiate between the least and the most-disturbed sites. The resulting *F*-statistics were used to select the single best metric from each ecological class.

Redundancy: A redundancy test was done to identify statistically redundant metrics as they add little new information. Two metrics were judged to be redundant if their Spearman correlation coefficients were >0.70. When a metric pair in different ecological classes was redundant, the metric selected for inclusion first (i.e., greater *F*-value) was retained and the other metric was replaced with a non-redundant metric in its class with the next greatest *F*-value.

IBI metric scoring: A continuous scale that ranged 0–100 was used to score each metric. Floor and ceiling values for each metric were bound as the 95th and 5th percentiles, respectively, of all sites. Metric scores were calculated as: (ceiling - floor)/(value - floor) - 100 for positive metrics (i.e., ones of which values are higher in the least-disturbed sites), and: (floor - ceiling)/(floor - value) for negative metrics (i.e., the ones of which values are higher in the most-disturbed sites). Total IBI scores were the averages of the composite metric scores, with a potential range of 0–100.

IBI application to transitional sites: The possibility of miscategorising transitional sites especially near the cutoff, i.e., a true mixed-water site placed in the coldwater IBI data set, and vice versa, was explored. Both IBIs were applied to transitional sites, defined conservatively as streams with drainage areas of 5–40 km^2, and performance was examined with respect to drainage area and assemblage type. Plots of coldwater *versus* mixed-water IBI scores and IBI scores *versus* drainage area were constructed. It was seen that most of the seven landscape variables were correlated to each other and their structure was represented by a single dominant gradient in the PCA. All variables, except dam density, were highly correlated with the first PCA axis.

The coldwater and mixed-water IBIs eventually comprised of 5 and 7 metrics, respectively (Tables 14.11 and 14.12). No metric related to non-native species were selected for either IBI.

Application of IBIs revealed that the mean scores among disturbance categories were significantly different in both IBIs (ANOVA; $p < 0.001$). Mean IBI scores were also different between the most and the least-disturbed sites. The IBIs were also able to distinguish least-disturbed sites from moderately disturbed sites (Tukey's HSD test; $p < 0.05$). Coldwater and mixed-water IBI scores ranged from 5 to 95 (mean 48), and 6 to 79 (mean 47), respectively. However, streams tended to be scored similarly when the two IBIs were applied to transitional sites (i.e., drainage

TABLE 14.11 Metrics Included in the Coldwater IBI Developed for Wadeable Streams (Drainage Area $\leq$ 15 Km^2) in Southern New England by Kanno et al. (2010)

Metric	Ecological Class	F-value	Ceiling	Floor
Brook trout individuals per 100 m^2	Brook trout population	38.9	60.6	0
Fluvial dependent individuals, (%)	Stream flow	19.1	0	71.7
Warmwater species (stream-size-corrected)	Richness	9.1	−2.39	3.06
Warmwater individuals, (%)	Thermal	8.7	0	87.5
Brook trout individuals, (%)	Indicator species and composition	6.3	86.3	0

areas of 5–40 km^2) and many higher-scoring mixed-water streams received intermediate scores for the coldwater IBI.

14.4.4. A Planktonic IBI Applied to Lake Erie

Planktons have been among the lesser favoured organisms with IBI developers, compared to fish and macroinvertebrates, because plankton taxonomic enumeration is time-intensive. However, this disadvantage is balanced by the advantages that plankton is sensitive to environmental changes and inexpensive to collect. Further, plankton samples can be stored for long periods and do not take up large amounts of space, so historical samples can be analysed and compared with current samples.

Relying on these virtues, Kane et al. (2009) have developed one of the plankton-based IBIs (P-IBIs). This IBI has been developed with specific reference to Lake Erie.

The authors began by considering a number of plankton characteristics that can cause 'beneficial use impairments' (BUIs) to the waters of Lake Erie. Based on a United States Environmental Protection Agency (USEPA) (1998) technical document and an extensive literature search they identified planktonic metrics that reflect the BUIs. It led to seven candidate zooplankton metrics (Table 14.13) and six

TABLE 14.12 Metrics Included in the Mixed-water IBI Developed for Wadeable Streams (Drainage Area >15 km^2) in Southern New England by Kanno et al. (2010)

Metric	Ecological Class	F-value	Ceiling	Floor
White sucker individuals, %	Indicator species	19.6	0	43.9
Cyprinidae individuals, %	Composition	15.8	93.7	0.2
Fluvial-specialist individuals, except blacknose dace, %	Stream flow	14.7	64.7	0
Non-tolerant general feeder individuals, %	Trophic	8.7	51.6	0
Native warmwater individuals, %	Thermal	6.3	0	67.9
Intolerant individuals, %	Tolerance	5.1	38.1	0
Fluvial specialist species	Richness	5.1	5	1

TABLE 14.13 Zooplankton Candidate Metrics Considered by Kane et al. (2009) in the Course of Developing a P-IBI. Metrics that were Included in the Final P-IBI are Shown in Bold

Candidate Metrics	Description/Ecological Relevance	Measure of	Hypothesised Response to Degradation	Reference
1. Zooplankton ratio	Abundance of (Calanoida/(Cladocera + Cyclopoida) – low values indicate eutrophication	Trophic status	Decrease	Gannon and Stemberger (1978)
2. Mean zooplankton Size	Larger taxa are preferred food for many fish	Quality of fish food	Decrease or increase	Mills and Schiavone (1982); Mills et al. (1987)
3. Rotifer composition	Taxa are indicative of trophic conditions	Trophic status	Change in taxonomic composition	Gannon and Stemberger (1978)
4. Density of *Limnocalanus macrurus*	Taxon is intolerant to eutrophic/anoxic conditions	Trophic status/oxygen conditions	Decrease	Gannon and Beeton (1971); Kane et al. (2004)
5. % Biomass of predatory invasive zooplankters	Shifts energy away from fish, reduces zooplankton numbers	Fish food quality, assess food web	Increase	Lehman and Caceres (1993); Hoffman et al. (2001); Uitto et al. (1999)
6. Biomass of zooplankton/biomass of phytoplankton	Lower during blooms of nuisance/inedible/toxic algae	Trophic status	Decrease	Havens (1998); Xu et al. (2001)
7. Biomass of crustacean zooplankton	Preferred food for many fish	Quality of fish food	Increase	Gopalan et al. (1998)

candidate phytoplankton metrics (Table 14.14) for further consideration in the P-IBI.

After examining the literature and identifying data availability with reference to the candidate metrics, four metrics were dropped from subsequent analysis. The remaining 9 were included in the discriminant analysis used to form the multi-metric index: (i) zooplankton ratio, (ii) abundance of *L. macrurus*, (iii) % biomass of invasive zooplankters of the total crustacean biomass, (iv) biomass of crustacean zooplankton/biomass of phytoplankton, (v) biomass of crustacean zooplankton, (vi) abundance of centrales/abundance of pennales, (vii) biomass of inedible algae taxa, (viii) % biomass of *Microcystis*, *Anabaena* and *Aphanizomenon* of the total phytoplankton biomass and (ix) biomass of edible algae taxa.

Unfiltered total phosphorus concentrations (μg/L), and chlorophyll-*a* concentrations (μg/L) from 1970 to 1996 (uncorrected for pheophytin), were used to classify sites in Lake Erie with respect to lake trophic status (i.e., oligotrophic–mesotrophic–eutrophic continuum) and thus reflect levels of degradation. Phytoplankton abundance and biomass data from 1996 and for most of 1970 were taken simultaneously with the nutrient, chlorophyll-*a* and zooplankton abundance and biomass data obtained from the same site and date. Of these, the total phosphorus and chlorophyll-*a* concentrations that were classified as eutrophic, mesotrophic and oligotrophic were given metric values of 1, 3 and 5, respectively. The two metrics were then summed to form a trophic

TABLE 14.14 Phytoplankton Candidate Metrics Considered by Kane et al. (2009) in the Course of Developing a P-IBI. Metrics Included in the P-IBI are Shown in Bold

Candidate Metrics	Description/Ecological Relevance	Measure of	Hypothesised Response to Degradation	Reference
8. Generic index of diatoms	% Abundance of (*Achnanthes, Cocconeis* and *Cymbella*)/(*Cyclotella, Melosira* and *Nitszchia*)	Organic pollution	Decrease	Wu (1999)
9. Centrales abundance/pennales abundance	Low values indicate oligotrophy	Trophic status	Increase	Nygaard (1949) and Rawson (1956)
10. Biomass of inedible algae taxa	Large, gelatinous, and colonial algae are of poor food quality	Quality of zooplankton/fish food	Increase	DeMott and Moxter (1991)
11. % Blue-green algae (biomass)	Blue-green algae cause mechanical/chemical interference to zooplankton feeding	Presence of inedible/toxic taxa	Increase	Gliwicz and Siedlar (1980); Gliwicz and Lampert (1990); Carmichael (1986, 1997)
12. % Abundance of *Microcystis, Anabaena* and *Aphanizomenon*	Affect human health and health of aquatic organisms	Presence of toxins	Increase	Carmichael (1986, 1997)
13. Biomass of edible algae taxa	Provide quality nutrition for growth and reproduction of animals	Quality of zooplankton/fish food	Decrease	Kerfoot et al. (1988)

status metric that could have values of 2, 4, 6, 8 or 10. Trophic status metric values of 8 or 10 were classified as oligotrophic, values of 6 were classified as mesotrophic and values of 2 or 4 were classified as eutrophic.

Discriminant function analysis or discriminant analysis (DA) was used to evaluate the ability of plankton metrics to distinguish among levels of degradation.

To calculate individual metric scores, 'box-plots' were constructed of the significant individual plankton metrics frequency distributions. The 95th percentile was used as the upper boundary and zero as the lower boundary, and each of the final individual metrics included in the multi-metric P-IBI (based on significance in the discriminant analyses) trisected into ranges that were assigned a score of 1, 3 or 5.

Two steps were performed to calculate P-IBI:

1. Cutoff scores for each variable were used to calculate individual metric values.
2. Basin or lakewide mean metric scores were estimated.

The calculation of P-IBI was done using the expression:

$$\text{P-IBI} = \frac{1}{B}\sum_{k=1}^{B}\frac{1}{S}\sum_{j=1}^{S}\frac{1}{M}(EA_{jk} + CB_{jk} + RJ_{jk} + LM_{jk} + RA_{jk} + ZB_{jk}) \tag{14.1}$$

where EA_{jk} is the June biomass of edible algae taxa metric score, CB_{jk} is the June % *Microcystis, Anabaena* and *Aphanizomenon* of total phytoplankton biomass metric score, RJ_{jk} is the June zooplankton ratio (Calanoida)/(Cladocera + Cyclopoida) metric score, LM_{jk} is the July *L. macrurus* density metric score, RA_{jk} is the August zooplankton ratio (Calanoida)/(Cladocera + Cyclopoida) metric score, ZB_{jk} is the August crustacean zooplankton biomass metric score, M is the number of metrics, S is the number of sites (within a basin) and B is the number of basins.

Of the nine metrics analysed using discriminant analysis, five achieved significant discrimination between trophic status classes, based on chlorophyll-*a* and total phosphorus, and were included in the final P-IBI.

Considerations that were applied to reject candidate metrics included redundancy, lack of adequate testing, sampling limitation and insignificance in the discriminant analysis. On the other hand, significance in the discriminant analyses was the only criterion for inclusion of a candidate metric in the final index.

The multi-metric P-IBI developed in this manner for Lake Erie was seen to reflect the BIUs in Lake Erie. It showed potential of being a useful, broad-scale way to monitor changes in the offshore water quality of lakes in general. In Lake Erie, the lakewide P-IBI score was seen to increase from 1970 to the mid-1990s and decline subsequently. It reflected the changing trophic status of the lake.

14.4.5. An IBI for Mediterranean Watersheds

Magalhaes et al. (2008) have underscored the special challenges associated with the development of IBI for Mediterranean watersheds in their preamble to describing a new IBI.

Mediterranean streams possess several attributes that make the development and application of IBIs especially challenging. Of these, two prime attributes are low species richness and a high rate of endemism. This limits the range and sensitivity of metrics that can be developed, and results in major natural differences in assemblages among adjacent watersheds. This problem is compounded by the apparent tolerance to the harsh environmental conditions displayed by several species, and by the lack of sound information on their ecological requirements, so that there is a ring of uncertainty around their functional role in assemblages. Moreover, the usual scarcity of pristine sites and shortage of historical records to quantify how assemblages have been affected by environmental degradation complicate the design of management goals specifically directed at these streams.

Earlier efforts to implement an IBI at the large Mediterranean scale have been hampered by weakness in metric responses to human impacts and differences in metric trends among fish assemblages, indicating that specific adaptations may be needed for each watershed and fish type situation (Ferreira et al., 2007a,b). However, previous applications of the IBI approach at smaller regional scales also showed variable degrees of success, despite a general lack of rigorous procedures for selecting appropriate metrics and establishing scoring criteria (Ferreira et al., 1996; Oliveira and Ferreira 2000).

IBI Development and Evaluation

Taking a cue from Hughes et al., (1998), the full set of sites was randomly divided into a development set (n = 95) and a validation set (n = 123), each including sites at least 300 m apart and samples with more than two fish. The development set was used to screen reference sites and candidate metrics and establish scoring criteria for component IBI metrics. The validation set was used for an independent evaluation of metrics and IBI performance. Because of sample size limitations, the development set included samples taken in spring only (March to June), when flow conditions were the most stable and comparable across sites, whereas the validation set included samples spanning all seasons. To check for eventual biases associated with variation in time of sampling, IBI scores were assessed at 22 sites, repeatedly sampled at least once in each season.

Prior to analysis, candidate metrics and environmental variables showing skewed distributions were $\log_{10}(X+1)$ transformed to approach normality and to reduce the influence in the analysis of peak values. Angular transformation was used for proportional data. The significance level was set at $P < 0.01$ throughout, to enable the detection of strong patterns in the data.

Screening of reference sites: As no pristine sites could be located, instead, minimally impacted reference sites were selected from the overall set of sites using a multivariate approach (Reynoldson et al., 1997). Principal component analysis (PCA) was used to identify the main gradients of variation in the environmental data matrix, with principal component (PC) axes being regarded as composite, multivariate, impact stressors. Scores for the first and second PC axes were plotted, and sites located in the quadrant of lower impact created on the PC surface were considered as up to reference standard. Factor loadings for PC axes were used in defining the quadrant of lower impact.

Candidate metrics: A list of candidate metrics was drawn on the basis of five functional attributes: origin, reproduction, trophic, habitat and tolerance. The species known from the study area were assigned for these attributes based on previous grey or published literature, completed by expert judgement when necessary. A simplified system of reproductive guilds, focussing on preferred spawning habitat, was developed based on the classification by Balon (1975). Trophic guilds were defined following Winemiller and Leslie (1992), using dominance of food items in the diet of adults. For habitat, species were classified as benthic, low water and high water column, based on personal observations of their usual position relative to the bottom. For tolerance, species were classified as intolerant, intermediate and tolerant, based on expert judgement of their overall sensitivity to altered flow regime, nutrient regime, water chemistry and habitat structure. Overall, 55 candidate metrics were calculated, following the lead of Karr et al. (1986), Karr (1991), Simon and Lyons (1995), Lyons et al., (2001) and Mccormick et al. (2001), with the modifications required for the Guadiana watershed. Metrics based on both native and exotic species were considered.

Metric screening: Criteria of range, responsiveness, precision and redundancy, were used to screen the candidate metrics. Metrics representing <3 species were considered as having insufficient range to be used in scoring, and rejected. Secondly, the Spearman rank correlation coefficient was used to test the responsiveness of candidate metrics to environmental variables, including the PCA axes derived in the screening of reference sites. Both the significance and sign of the correlations and the plotted distribution of candidate metrics were evaluated, and metrics showing inappropriate responses were excluded. Thirdly, Mann–Whitney tests (Zar, 1996) were used to identify the metrics providing the better discrimination between reference and non-reference sites. Fourthly, associations between metrics were evaluated through the Pearson correlation coefficient (Zar, 1996). Pairs of metrics with strong correlations ($r > 0.70$) were considered redundant and the least responsive metric of the pair was rejected. As exceptions, metrics based on a functional species attribute not yet selected were retained.

Metric and index scoring: The five selected metrics (Table 14.15) were scored on a 0–4 scale, using maximum value lines (MVLs; Karr et al., 1986) plotted against watershed area (as log10). The MVLs were derived considering six classes of watershed area, and adjusting a second-order polynomial curve to the maximum metric values of reference sites in each class. The area under MVL was divided into four sectors to determine metric ratings, with sites scoring the maximum

TABLE 14.15 The Five Component Metrics Chosen for the IBI by Magalhaes et al. (2008), and Their Scoring Ranges

	Scoring Range				
Component Metrics	**0**	**1**	**2**	**3**	**4**
Proportion of native individuals [PN]	≤0.09	0.10–0.35	0.36–0.65	0.66–0.91	>0.91
Number of intolerant and intermediate species [NIInt]	≤25%*	26–50%*	51–75%*	76–100%*	>*
Number of invertivore native individuals [NInN]	≤25%*	26–50%*	51–75%*	76–100%*	>*
Number of phyto-lithophilic and polyphilic species [NPhLiPo]	≤25%*	26–50%*	51–75%*	76–100%*	>*
Catch of exotic individuals per unit sampling effort [CEPUE]	<16.44	16.45–27.02	27.03–44.41	44.42–73.01	>73.02

* *Scoring ranges for these metrics were derived using maximum value lines.*

points (4) when metrics exceed MVL. For metrics showing no relation to watershed area, a similar rating procedure was applied, using the maximum metric values recorded among all reference sites. Negative-scoring metrics were inverted prior to rating. To calculate the final IBI, component metric scores were summed to give the IBI score and adjusted to a 100-point scale with the classification: bad (0–20), poor (20–40), moderate (40–60), good (60–80) and very good (80–100).

Performance of metrics: Internal consistency in metric scoring was assessed using Cronbach's alpha, which is a positive function of the average correlation between items in a combined index (Cronbach, 1951). Restricted alpha values were calculated for each metric by sequentially removing that metric and calculating the alpha coefficient for the remaining metrics, which was then compared with overall alpha. Additionally, Kendall's correlation coefficient was used to analyse the relationships between each metric and the final IBI score, as well as between the final IBI score and the restricted IBI scores derived by sequentially omitting each metric.

Index evaluation: Responsiveness of the final IBI score to environmental variables, including the composite PC degradation axes derived in screening reference sites, was evaluated from the Spearman rank correlation coefficient. Constancy in IBI scoring among seasons was tested using Kendall's coefficient of concordance. The ability of IBI to discriminate between reference and non-reference sites and to produce repeatable scores for development and validation sites was analysed using Mann–Whitney tests.

Application of IBI produced repeatable and consistent results that distinguished differing levels of biotic integrity across the Guadiana watershed. More sites were shown by IBI as poor quality than as excellent, with the distribution of scores being skewed with more low scores than high. Despite the lack of a statistical design precluding quantitative statements about the degree of impairment in the Guadiana watershed, these results suggest that ecological status may be low in many stream reaches. The high scoring sites were the streams that have previously achieved special recognition of their high conservation status, whereas low scoring sites occurred in streams heavily impacted by human interference.

In the view of the authors (Magahaes et al., 2008), their IBI highlights the effectiveness of the IBI approach even with fish assemblages of limited diversity and ecological specialisation, as in Mediterranean streams.

14.4.6. An IBI to Assist Lake Fish Conservation in China

Liu et al., (2008) describe the development of a new fish IBI with specific reference to Lake

Qionghai in China. IBI was applied to historic data of the 1940s and 1980s as well as to recent data experimentally generated by the authors.

The constituent metrics for IBI were selected following analysis of the fish assemblage and examination of previous lake IBIs (Jennings et al., 1999; Thoma, 1999; Lyons et al., 2000; Zhu and Chang 2003; Drake and Valley 2005). Twelve metrics were proposed earlier by Zhu and Chang (2003) for the lakes in the Yangtze River Basin. These were modified and the number of metrics was reduced to 10 (Table 14.16).

The values of the 10 metrics and the IBI were calculated for Lake Qionghai for the data of the 1940s, 1980s and 2003. The total IBI score was seen to decrease from 40 in the 1940s to 26 in the 1980s. It had fallen to 20 in 2003 because of the declining of the metrics M1, M7, M10, M3, M6, M8 and M9, especially M1, M7 and M10. The scores for species richness decreased from 30 in the 1940s to 22 in the 1980s and 16 in 2003, which was seen by the authors as a direct result of extinction of native species and the alternation of the food web. The scores for trophic composition decreased from 13 in the 1940s to 9 in the 1980s to 5 in 2003. The same trend occurred for fish condition, which decreased from 10 in 1940s to 4 in 1980s and 2004. The low IBI scores reflect the decline of the biotic condition in Lake Qionghai.

The authors (Liu et al., 2008) conclude that the IBI results point towards the need for some management measures to restore the biotic integrity and conservation of the remaining native fishes in Lake Qionghai. These measures seem to include the restoration of native species and food-web structure, prevention of alien species invasion and reconstruction of the lake ecosystem. The ability of the lake to provide

TABLE 14.16 IBI Metrics and Scoring Criteria for Lake Qinghai Used for the Lake IBI by Liu et al. (2009)

Constituent Metrics		Scoring Criteria		
		Good (5 Points)	**Fair (3 Points)**	**Poor (1 Point)**
SPECIES RICHNESS AND COMPOSITION (%)				
M1	Native species in total number	>74	50–74	<50
M2	Cyprinidae species in total number	<45	45–60	>60
M3	Bagridae species in total number	2–4	4–6	6–8
M4	Cobitidae species in total number	2–5	6–8	9–12
M5	Crucian carp individuals in total number	<22	23–38	39–54
TROPHIC COMPOSITION (%)				
M6	Omnivorous individuals in total number	<10	10–40	>40
M7	Invertivore individuals in total number	>45	20–45	<20
M8	Top carnivorous individuals in total number	>10	5–10	<5
FISH CONDITION				
M9	Alien invertebrate parasite species in native fish (%)	<25	25–60	>60
M10	Hybrid individuals in total number (%)	0	0–1	>1
	Total score of IBI	>40	20–40	<20

ecosystem services and sustainable fisheries also needs to be considered.

14.4.7. A Macroinvertebrate IBI for Monitoring Rivers in Lake Victoria Basin, Kenya

The water quality of many streams in Lake Victoria Basin, Kenya, is affected by intensive water withdrawal, inflows of agricultural run-off and pollution by municipal and industrial effluents, Livestock and deforestation also cause adverse impacts. However, management efforts in the region have been hampered by lack of standards against which to judge the degree of environmental degradation. To cover this lacuna, a macroinvertebrate-based Index of Biotic Integrity has been developed by Raburu et al. (2009) to monitor ecological integrity of selected rivers occurring in the upper reaches of the Lake Victoria Basin.

Index Development and Scoring Criteria

The authors evaluated the ability of different attributes to distinguish reference sites from the impaired ones using Mann–Whitney *U* tests. Potential metrics were identified when the tests showed significant differences ($p < 0.05$) between site groups. The separation power of potential metrics between impaired and reference sites was evaluated using box plots according to Barbour et al. (1999). Metrics with a separation power <2 were excluded. Redundancy in the remaining metrics was evaluated by Pearson correlation coefficients and visual inspection of scatter plots. Metrics with a correlation coefficient $r \geq 0.7$ were considered redundant. Only one metric with the highest separation power was chosen from a group of redundant metrics and included in the computation of the final index.

The 1, 3, 5 interval scoring system commonly used in developing fish and macroinvertebrate IBIs (Karr, 1981; Kerans and Karr, 1994; Barbour et al., 1999) was used to normalise the ranges (Table 14.17). The scored metrics were then summed to obtain the final IBI score.

Responsiveness of component metrics of the IBI to disturbance was then evaluated by correlating metric values with physico-chemical parameters using visual inspection of scatter

TABLE 14.17 Nine Component Metrics Used in the IBI of Raburu et al. (2009), and Their Scoring Criteria Corresponding with Scores Based on the 1, 3, 5 Scaling System

	Scoring Criteria		
Metric	**1**	**3**	**5**
Number Ephemeroptera genera	<4	4–6	>6
Number Plecoptera genera	<1	2	≥3
Number Trichoptera genera	<3	3–4	>4
Number intolerant genera	<8	8–10	>10
% EPT individuals	<16.6	16.6–51.9	>51.9
% Tolerant individuals	>70.9	31–70.9	<31
% Gatherer individuals	>35.1	4.9–35.1	<4.9
% Predator individuals	<12.4	12.4–15.8	>15.8
% Individuals in dominant genera	>27.3	27.3–28.7	<28.7

TABLE 14.18 Classification Categories of Total Scores of the IBI of Raburu et al. (2009). The Integrity Classes and Narrative Description Based on the Riverine Habitats in the Study Area are Also Given

Total IBI Score and Integrity Classes	**Narrative Description**
Excellent, >41	No human activity within 100 m of the riparian zone, natural vegetation intact, BOD <1 mg/l, % EPT >50%, % Diptera <20%, no intolerant taxa >14. Instream substrate dominated by boulders and vegetal material. Water clear (can see the bottom).
Good, 33–41	No human activity within 50 m of the riparian zone, BOD <2 mg/l, % EPT >50%, % Diptera <20%, no intolerant taxa >10, bottom substrate dominated by stones and vegetal materials. Water clear as the bottom can be seen
Fair, 24–32	Riparian zone >20 m wide with minimal human activity, natural vegetation maintained along the reach with instream cover >50%. BOD <2 mg/l, % EPT >50%, % Diptera <30%, substrate mainly of stones
Poor, 19–23	Riparian zone <20 m, collapsed and eroded river banks, human activity includes agriculture, animal watering points and water abstraction, urbanisation and deforestation, BOD >4 mg/l, % EPT <30%, % tolerant taxa >30% and dominated by chironomids and oligochaetes, bottom dominated by sand and organic materials, water turbid (cannot see the bottom).
Very poor, 9–19	Riparian zone <10 m, collapsed river banks without vegetation, riparian zone under agriculture or urbanisation or industry, human activity include animal watering, water abstraction, bathing, washing and sand harvesting. BOD >5 mg/l, % EPT <20%, % tolerant taxa >60% dominated by chironomids and oligochaetes and number intolerant taxa <8, bottom composed of sand, mud and organic wastes. Water turbid.

plots and Pearson's correlation analysis. The integrity classes for condition categories (Table 14.18) were defined with reference to total IBI scores at both reference and impaired sites in relation to the habitat characteristics at respective sites.

Studies of the sites by the authors revealed that the selected metrics exhibited variability consistent with the level of degradation. In particular, the relative abundance of EPT and pollution-tolerant individuals' metrics (Table 14.17) were able to clearly separate stations according to the level of degradation. Whereas there was a reduction in abundance of intolerant taxa, the reverse was true for tolerant taxa at degraded sites.

The lack of significant difference in taxon richness and diversity indices among the three river systems studied, and similarity of orders in the three river systems, seemed to confirm that the macroinvertebrate community in the region was being regulated by similar ecological conditions.

14.4.8. A Benthic Macroinvertebrate IBI to Assess Biological Condition Below Hydropower Dams

Rehn (2009) has developed an IBI to assess biological condition below hydropower diversion dams on west slope Sierra Nevada streams (California, USA) based on benthic macroinvertebrates.

The author sampled ten streams above the upstream influence of each peak reservoir storage and at five downstream sites sequentially spaced 500 m apart. Reference conditions were defined by screening upstream study sites and 77 other regional streams using quantitative GIS land use analysis, reach-scale physical habitat (PHAB) data and water chemistry data.

Eighty-two metrics were evaluated for inclusion in the IBI based on three criteria:

1. good discrimination between reference and first downstream sites with some indication of recovery over the distance sampled;
2. sufficient range for scoring; and
3. minimal correlation with other discriminating metrics.

Other steps in reaching the IBI were similar to the ones described in preceding case studies.

The IBI showed good discrimination between reference and downstream sites with partial recovery as distance downstream increased, and was validated with an independent data set. Individual metrics, IBI scores and multivariate ordination axes were found to be poorly correlated with PHAB variables across sites. When only reference and first downstream sites were evaluated, decreased IBI scores correlated with lower habitat variability and substrate coarsening below dams.

Lower IBI scores below dams were most strongly associated with altered hydrologic regime, especially non-fluctuating flows as defined by the flow constancy/ predictability index. Hence, the author suggests, flow restoration experiments would be valuable in developing management actions that achieve a sustainable balance between conflicting human and ecological needs for freshwater.

14.4.9. A Diatom-based IBI for Acid Mine Drainage Impacted Streams

Acid mine drainage (AMD) resulting from extensive coal mining since the 1800s has subjected tens of thousands of stream miles across the world, to severe acid and dissolved metal loads. This has critically impacted aquatic life and ecological attributes in countless streams and lakes but there is no index specifically designed to quantify AMD impacts using diatoms. Nor have the response of multiple organism groups been compared for their utility as indices for assessing AMD severity and devising effective management strategy for AMD impacted streams.

In this backdrop, Zalack et al. (2010) have conducted a study towards creating and testing a multi-metric AMD-diatom index of biotic integrity (AMD-DIBI) and compared its response to AMD severity with an already established multi-metric macroinvertebrate community index (ICI). In 2006, 41 sites in southeast Ohio were sampled by the authors that represented an AMD impact gradient and non-AMD impacted reference sites. Metrics comprising the AMD-DIBI were selected based on their responsiveness to AMD and nutrient impacts. In the following year, the AMD-DIBI and its metrics were tested on a validation data set consisting of 18 sites in an AMD impacted watershed.

Results indicate a significant correlation between AMD-DIBI and ICI scores, and both indices and all metrics strongly correlate with water chemistry variables indicative of AMD pollution. Stepwise multiple regression showed up alkalinity and conductivity as most influential to AMD-DIBI and ICI scores. Narrative classes (e.g., Poor, Fair, Good, and Excellent) defined by index scores provided effective classifications of AMD severity.

When tested on the watershed scale, AMD-DIBI and its metrics could successfully quantify AMD gradients and coal mining impacts as indicated by canonical correspondence analysis.

The authors expect that this newly developed AMD-DIBI will be useful for assessing impairment, sensitivity, and recovery of diatom communities in streams damaged or threatened by coal mining activities. In addition, the authors feel, since the AMD-DIBI was responsive to a gradient of AMD pollution, it could be used in future studies measuring the long-term status of streams and effectiveness of various remediation methods. The study also highlights the responsive power of diatom-based metrics.

14.4.10. A Multi-metric Index of Reservoir Habitat Impairment (RHI)

Unlike natural lakes which slowly evolve as aquatic ecosystems and go through from oligotrophic to mesotrophic (and eventually entropic) stages over hundreds of years, man-made reservoirs come up suddenly with large watersheds and large tributaries to capture as much water as possible. The resulting habitat has low diversity and resilience. It gets easily impaired by unnatural water regimes, sedimentation, and nutrient inflows. These happening in many cases bring in rapid changes in the trophic status of the reservoirs and reduce their ability to sustain native fish assemblages and fisheries quality. Rehabilitation of such impaired reservoirs is hindered by the lack of methods suitable for quantifying the impairment status.

To address this issue, an index of reservoir habitat impairment (IRHI) is reported by Miranda and Hunt (2010). In it 14 metrics, which are descriptive of common impairment sources have been merged. The metrics have been given scores of 0 (no impairment) to 5 (high impairment) by fisheries scientists with local knowledge.

The IRHI is scored much like the IBI of Karr et al. (1986), but is different from it in that the suit of IRHI metrics added represent habitat constructs rather than biotic community scores. In conformity with the scale in which the data were collected, average raw scores for each construct ranging from 0 to <1.5 were scored as 1, those ranging from 1.5 to <3.5 scored as 3 and those ranging from ≥3.5 to 5 scored as 5. The scores for the constructs were added to compute an overall IRHI score as:

$$\text{IRHI} = f_i' + f_{i+1}' + \cdots\cdots + f_n' \quad (14.2)$$

$f_i' = 1$ if $0 \le f_i < 1.5$; 3 if $1.5 \le f_i < 3.5$; 5 if $3.5 \le f_i \le 5$

$$f_i = \frac{m_i + m_{i+1} + \cdots\cdots + m_j}{j} \quad (14.3)$$

where f_i' is construct i of the n constructs that make up the IRHI, f_i is the mean score for the j metrics with loadings ≥0.6 in the i factor and m_i is the metric i of the j metrics that make up f_i.

This form of scoring was used instead of an overall total score or average for metrics in the factors because factors are not additive; only internally consistent metrics within factors are additive.

To assess if the IRHI reflected reservoir environments, the authors evaluated how each f_i changed in response to key reservoir descriptors expected to be indicative of reservoir habitat. Data on four commonly use descriptors – reservoir use-type (i.e., primary reason for construction), surface area, maximum depth and age — were taken from existing databases by the authors who applied multivariate analysis of covariance to assess the link between f_i and reservoir use-type (class variable), area (covariate), maximum depth (covariate) and age (covariate).

Application of IRHI to over 482 randomly selected reservoirs dispersed throughout the USA revealed that IRHI reflected five impairment factors including siltation, structural habitat, eutrophication, water regime and aquatic plants. The factors were weakly related to the four key reservoir characteristics used by the authors, suggesting that common reservoir descriptors are poor predictors of fish habitat impairment. The IRHI was found to be rapid and inexpensive to calculate, and seemed to provide an easily understood measure of the overall habitat impairment, allowing comparison of reservoirs and therefore prioritisation of restoration activities. The major limitation of IRHI appears to be its reliance on unstandardised professional judgement rather than standardised empirical measurements.

14.4.11. A Multi-taxa IBI for Assessing the Ecological Status of Wetlands

Multi-taxa IBIs have an advantage over single taxa IBIs in that in the former information from organisms from several trophic levels can be combined to obtain responses that reflect differing sensitivities along an environmental gradient. This may allow the establishment of multiple criteria: on the one hand to protect the most sensitive taxa and on the other hand to protect important, but more tolerant, taxa.

Although a single factor may serve as a primary stressor in a wetland ecosystem, more often it is a combination of factors that result in wetland degradation at the landscape scale (Danielson 2001). Moreover, spatial and temporal variability in stressor concentrations can make it difficult to diagnose the specific stressor or stressor level causing the impairment, especially for sites sampled just once in landscape-scale assessments.

With these considerations, Lougheed et al. (2007) have used a generalised stressor gradient to provide frame of reference for their multi-taxa IBI. It incorporates multiple environmental stressors to ensure that biological responses thoroughly reflect all possible alterations from natural, or relatively undisturbed, conditions.

For their generalised stressor gradient, the authors constructed a multi-metric stressor axis, called wetland disturbance axis (WDA), to integrate and give equal weight to measurements in three primary stressor categories: land use, hydrological modification and water quality. WDA includes three metrics indicative of land use and land cover change (riparian land use, buffer width and distance to the nearest wetland), two metrics indicative of hydrology (hydrological modification and water source) and two water quality metrics (conductivity and contaminants). Distance to the nearest wetland, an indicator of habitat fragmentation due to land use change, was determined using land use maps and GIS. Other land use variables, including riparian land use and buffer width, all hydrological variables and the presence of contaminants were estimated on a semiquantitative scale based on field observations.

A suite of biological metrics or ecological attributes was calculated by the authors for each of the three taxonomic groups used in IBI: plants (Herman et al., 2001, Fennessy et al., 2002), diatoms (van Dam et al., 1994, Stevenson et al., 2002b) and zooplankton (Lougheed and Chow-Fraser 2002). Several new metrics were also calculated for the study. All taxonomic groups had metrics that could be categorised as one or the other of the following:

1. growth form and habitat (e.g., obligate wetland plants, stalked diatoms and vegetation-associated zooplankton);
2. taxonomic-level metrics (e.g., frequency of sedges, percent *Eunotia* + *Pinnularia* and the ratio of cladocerans to rotifers); and
3. trophic state indices or tolerances, similarity to reference communities and metrics representing community composition.

In total, the authors calculated 117 metrics for plants, 84 metrics for zooplankton and 77 metrics for diatoms.

For each taxonomic group, the authors calculated *similarity to undeveloped sites.* This was a comparison between the species that would be expected to occur in the absence of human impacts, relative to the species that actually occurred at a site. For this calculation, undeveloped sites were defined as those with less than 5% developed (agriculture and urban) land in a 1-km buffer around the sampling location and included six of the 35 wetlands. Species from each taxonomic group that were found at more than one-third of the undeveloped wetlands were labelled 'reference species'. The authors assigned each of these species a species-specific value that equated the proportion of undeveloped sites where it occurred. For each of the six undeveloped sites, all species-specific proportional values were summed and then the six sums were averaged to give an expected (E) number of reference species in undeveloped sites.

Metric selection: To shortlist relevant metrics out of the suite of 278 metrics, the authors ran simple linear correlations between the metrics and WDA and retained only those metrics that were significantly correlated to WDA after a Bonferroni correction.

This reduced the total number of metrics to 27 for plants, 13 for zooplankton and 15 for diatoms. The authors then checked for correlations among the metrics that were retained, and removed those metrics that were significantly correlated ($r^2 > 0.70$) with another metric that had a higher correlation with WDA, paying careful attention to retaining what they felt were the most biologically significant metrics.

The authors finally arrived at six plant, two diatom and four zooplankton metrics that showed strong non-linear responses to WDA and were not redundant or substantially correlated with each other. The authors also retained an additional two diatom metrics that had largely linear responses, with only 1%–2% more of the variation explained by the regression tree than linear regression. Van Dam's TSI was retained as it was deemed to be a biologically important metric. The diatom NMDS metric was retained as the regression tree r^2 was substantially higher than the linear regression r^2 when the data was divided into two data sets: data below the upper changepoint and data above the lower changepoint. The final shortlist is presented in Table 14.19.

TABLE 14.19 Final List of Metrics Employed in the Multi-taxa IBI by Lougheed et al. (2007). The Direction of the Relationship Between the Metrics and the WDA is Indicated in the Trend Column

	Trends
PLANTS	
Similarity to undeveloped site	−
Plant NMDS axis	+
Plant Coefficient of Conservatism	−
Dominance of duckweed (*Lemna* spp.)	+
Relative dominance of sensitive plants	−
Relative dominance of tolerant and exotic plants	+
Overall Plant Index	
DIATOMS	
Similarity to undeveloped site	−
Diatom NMDS axis van Dam's Trophic State Index	+
Relative abundance of erect and stalked diatoms	+
Overall Diatom Index	−
ZOOPLANKTON	
Similarity to undeveloped site	−
Zooplankton NMDS axis	−
Zooplankton-inferred conductivity	+
Species richness of chydorids and macrothricids	−
Overall Zooplankton Index	

The metrics were then combined to generate IBI. The authors found that the three taxonomic groups on which IBI was based responded at similar levels of impairment and could be used to classify wetlands into three groups: reference sites representing the highest quality wetlands in the landscape; slightly altered sites where the most sensitive organisms responded (sensitive plants and diatoms); and degraded sites where extensive changes in community structure occurred, which may reflect a shift to an alternate state. For the Muskegon River watershed, in particular, this analysis allowed the authors to identify sites in need of restoration, including approximately one-third of the depressional wetlands in the watershed.

14.4.12. A Multi-taxon IBI for Chesapeake Bay

Three water quality and three biological measures have been combined to formulate a 'Bay Health Index' (BHI) by Williams et al. (2009).

Water quality measures of chlorophyll-*a*, dissolved oxygen (DO) and Secchi depth were averaged by the authors to create Water Quality Index (WQI), and biological measures of the phytoplankton and benthic indices of biotic integrity (P-IBI and B-IBI, respectively) and the area of submerged aquatic vegetation (SAV) were averaged to create Biotic Index (BI). WQI and BI were subsequently averaged to obtain BHI.

BHI is intended for use as a 'spatially explicit management tool' (i.e., a tool suitable for comparing various reporting regions) to evaluate the status of water quality and biotic conditions that are strongly affected by nutrient and sediment loadings. The two multi-metric indices (P-IBI and B-IBI) used in BHI are the ones that have been developed earlier (Weisberg et al., 1997; Lacouture et al., 2006). The metrics and biotic indices were chosen because they have (1) bay-wide coverage that allows discrimination among various reporting regions and (2) defined ecological health-based thresholds that allows discrimination between unimpaired and impaired areas.

Water Quality Index: An annual WQI was generated for each station by averaging the frequencies of passing scores for three water quality metrics: chlorophyll-*a*, DO and Sechhi depth. All the station WQIs within a segment were averaged and the segment WQIs within a reporting region were weighted by the areal proportion of each segment, relative to the reporting region. These were then summed up to obtain a WQI value for the reporting region.

Biotic index: BI combined the frequencies of passing scores for submerged aquatic grasses (SAV), the Benthic Index of Biotic Integrity (B-IBI) and the Phytoplankton Index of Biotic Integrity (P-IBI). Similar to WQI, a BI value was calculated for each segment by averaging the passing frequencies of the three components. BIs for the several segments in a reporting region were weighted by each segment's areal proportion of the reporting region and summed to obtain a BI value for the reporting region.

Estimates of SAV cover for each CBP segment were obtained from the annual aerial surveys of SAV done by VIMS.

The restoration goal for each reporting region was determined by summing the restoration goals of all segments located within the reporting region.

B-IBI was calculated by scoring each of several attributes of benthic community structure and function (abundance, biomass, Shannon diversity, etc.) according to thresholds established from reference data distributions. The scores (on a 1–5 scale) were then averaged across attributes to calculate an index value. Samples with index values of ≥ 3 are deemed to represent good benthic condition and are indicative of good habitat quality.

The individual P-IBI scores were evaluated against a threshold criterion of 3.0 on a scale of 1.0–5.0. Scores ≥ 3.0 were 'pass'; scores < 3.0 'fail'. The annual frequency of passing scores in each CBP segment was weighted by the segment's areal proportion of the reporting region in which it was located. Area-weighted frequencies were then summed to obtain an overall frequency of passing P-IBI score in each reporting region.

Bay Health Index (BHI)

The water quality and biotic indices, both expressed as the average of the percent attainment of their component metrics and biotic indices, were averaged to obtain BHI. The authors used a simple averaging technique for WQI and BI that assumed these indices to be of equal weight in representing ecosystem health. It was based on the rationale that there is no manner in which a weighting scheme can be objectively determined and therefore justified.

BHI of each reporting region was graded according to the following equally divided ranges which are similar to academic grade ranges: 0–20% (grade F), 21–40% (grade D), 41–60% (grade C), 61–80% (grade B) and 81–100% (grade A). Positive and negative qualifiers (i.e., + and −) were used to designate the upper and lower quartiles of each category. Grades similar to those commonly used in academic report cards were chosen because these can be easily understood and appreciated by the public.

Application of BHI indicated that lower chlorophyll-*a* concentrations, higher DO concentrations, deeper Secchi depths, higher phytoplankton and benthic indices relative to ecological health-based thresholds and more extensive SAV area relative to restoration goal areas characterised the least-impaired regions. WQI, P-IBI and BHI were significantly correlated with:

1. regional river flow;
2. nitrogen (N), phosphorus (P) and sediment loads (all positively correlated with flow); and
3. the sum of developed and agricultural land use in most reporting regions.

These findings reveal that BHI is strongly regulated by nutrient and sediment loads from these land uses.

14.4.13. An IBI Based Information–Theoretic Approach

In this section, illustrative examples of IBI development are presented which encompass different biogeographic regions, continents, types of water bodies and intended applications. The summaries are focussed on the methodologies used in developing IBIs and make only very brief mention of their application because most other sections of this chapter dwell at length on different aspects of applications of the IBIs.

Attempts have been made to link community structure of migratory birds with lake ecology (Abbasi and Chari, 2008; Chari et al., 2003) but quantitative studies in this area are few. Recently, Larsen et al. (2010) have made an interesting contribution in the form of their study on riparian bird communities in the backdrop of the assessment of Mediterranean streams using the Italian Extended Biotic Index (IBE). The study was organised on the premise that inclusion of riparian birds in stream bioassessment could add to the information currently provided by existing programmes that monitor aquatic organisms.

To assess if bird community metrics could indicate stream conditions, the authors sampled breeding birds in the riparian zone of 37 reaches in 5 streams draining watersheds representing a gradient of agricultural intensity in central Italy. Simultaneously, macroinvertebrates were sampled for computation of IBE. An anthropogenic index was calculated within 1 km of sampled reaches based on satellite-derived land-use classifications.

Predictive models of macroinvertebrate integrity based on land-use and avian metrics were then compared using an information-theoretic approach (AIC). The authors also determined if stream quality related to the detection of riverine species.

The studies reveal that the apparent bird species diversity and richness peaked at intermediate levels of land-use modification, but increased with IBE values. Water quality did not relate to the detection of riverine species as a guide, but two species, the dipper *Cinclus cinclus* and the grey wagtail *Motacilla cinerea*, were only observed in reaches with the highest IBE values. Small-bodied insectivorous birds and arboreal species were detected more often in reaches with better water quality and in less modified landscapes. In contrast, larger and granivorous species were more common in disturbed reaches. According to the authors' information-theoretic approach, the best model for predicting water quality included the anthropogenic index, bird species diversity and an index summarising the trophic structure of the bird community.

The studies suggest that in combination with landscape-level information, the diversity and trophic structure of riparian bird communities could serve as a rapid indicator of stream-dwelling macroinvertebrates, in turn serving as indicators of the degradation of in-stream biotic integrity.

14.5. OVERVIEW OF IBIs BASED ON DIFFERENT TAXA

14.5.1. Fish-based IBIs

Following the first IBI of Karr (1981), IBI development has been predominantly based on piscifauna (Stevens et al., 2010; Lieffering et al., 2010; Launois et al., 2011; Schmitter-Soto et al., 2011). An illustrative list of fish-based IBIs is presented in Table 14.20.

TABLE 14.20 Metrics Used in Some of the Fish-based IBIs

Region for Which Developed	Author(s)	Constituent Metrics
Midwestern United States	Karr (1981)	**(1)** Number of species **(2)** Presence of intolerant species **(3)** Species richness and composition of darters **(4)** Species richness and composition of suckers **(5)** Species richness and composition of sunfish (except green sunfish) **(6)** Proportion of green sunfish **(7)** Proportion of hybrid individuals **(5)** Number of individuals in sample **(9)** Proportion of omnivorous individuals **(10)** Proportion of insectivorous cyprinids **(11)** Proportion of top carnivores **(12)** Proportion with disease, tumors, fin damage and other anomalies
Great Lakes	Minns et al. (1994)	**(1)** Number of natives **(2)** Number of centrarchids **(3)** Number of intolerants **(4)** Number of non-indigenous **(5)** Number of native cyprinids **(6)** Percent piscivore biomass **(7)** Percent generalist biomass **(8)** Percent specialist biomass **(9)** Number of native individuals **(10)** Biomass of natives **(11)** Percent non-indigenous numbers **(12)** Percent non-indigenous biomass
Northeastern United States	Whittier (1999)	**(1)** Number of non-indigenous fish **(2)** Number of large species **(3)** Percent non-indigenous individuals **(4)** Percent tolerant individuals **(5)** Percent top carnivore individuals **(6)** Percent insectivorous individuals **(7)** Percent omnivorous individuals

TABLE 14.20 Metrics Used in Some of the Fish-based IBIs (*cont'd*)

Region for Which Developed	Author(s)	Constituent Metrics
Wisconsin, United States	Jennings et al. (1999)	**(1)** Number of natives **(2)** Number of centrarchids **(3)** Number of cyprinids **(4)** Number of intolerants **(5)** Number of small benthic fishes **(6)** Percent exotics **(7)** Percent top carnivores **(8)** Percent simple lithophilic lake **(9)** Spawners
Tennessee, United States	McDonough and Hickman (1999)	**(1)** Number of species **(2)** Number of lepomid species **(3)** Number of sucker species **(4)** Number of intolerants **(5)** Percent intolerant individuals **(6)** Percent dominance **(7)** Number of piscivores **(8)** Percent omnivores **(9)** Percent invertivores **(10)** Number of lithophilic spawners **(11)** Total number of individuals **(12)** Percent anomalies
Lake Erie	Thoma (1999)	**(1)** Number of natives **(2)** Number of benthic species **(3)** Number of sunfish species **(4)** Number of cyprinid species **(5)** Number of phytophilic species **(6)** Number of intolerant species **(7)** Percent tolerant individuals **(8)** Percent omnivorous species **(9)** Percent lake individuals **(10)** Percent phytophilic individuals **(11)** Percent top carnivores **(12)** Number of individuals **(13)** Percent non-indigenous species **(14)** Percent diseased individuals
Central Mexico	Lyons et al. (2000)	**(1)** Number of total native species **(2)** Number of common native species **(3)** Number of native Goodeidae species **(4)** Number of native Chirosiomu species **(5)** Number of native sensitive species **(6)** Percent biomass of tolerant species **(7)** Percent biomass of exotic species **(8)** Percent biomass of native carnivores **(9)** Maximum standard length of native species **(10)** Percent of exotic invertebrate parasite species in and (or) on native fish species

(*Continued*)

TABLE 14.20 Metrics Used in Some of the Fish-based IBIs (*cont'd*)

Region for Which Developed	Author(s)	Constituent Metrics
Coldwater streams in Vermont, USA	Langdon (2001)	**(1)** Number of intolerant species **(2)** Proportion of individuals as coldwater stenotherms **(3)** Proportion of individuals as generalist feeders **(4)** Proportion of individuals as top carnivores **(5)** Brook trout density and age class structure
Minnesota, United States	Drake and Pereira (2002); Drake and Valley (2005)	**(1)** Number of native species **(2)** Number of intolerant species **(3)** Number of tolerant species **(4)** Number of insectivorous species **(5)** Number of omnivorous species **(6)** Number of cyprinid species **(7)** Number of small benthic-dwelling species **(8)** Number of vegetation-dwelling species **(9)** Proportion intolerant individuals **(10)** Proportion small benthic-dwelling species **(11)** Proportion vegetation-dwelling species **(12)** Proportion insectivores by biomass **(13)** Proportion omnivores by biomass **(14)** Proportion tolerant individuals by biomass **(15)** Proportion top carnivore species by biomass **(16)** Proportion intolerant individuals by biomass
Austria	Gassner et al. (2003)	**(1)** Native fish species composition **(2)** Total native fish biomass **(3)** Present fish species composition **(4)** Present total fish biomass **(5)** Abundance index **(6)** Reproductive success **(7)** Size frequency of the dominant fish species **(8)** Total length at maturity of the dominant fish species **(9)** Maximum length of the dominant fish species
River Philip, Nova Scotia, Canada	Kanno and MacMillan (2004)	**(1)** Number of fish species **(2)** Percent of individuals that are salmonids **(3)** Percent of individuals that are brook trout (*Salvelinus fontinalis*) **(4)** Percent of individuals that are white sucker (*Catostomus commersoni*) **(5)** Percent of individuals that are catchable salmonids (age 2 years and older)
Northwestern Great Plains Streams of northern USA	Bramblett et al. (2005)	**(1)** Number of native species **(2)** Number of native families **(3)** Number of native catostomid and ictalurid species **(4)** Proportion of tolerant individuals **(5)** Proportion of invertivorous cyprinid individuals **(6)** Number of benthic invertivorous species **(7)** Proportion of litho-obligate reproductive guild individuals **(8)** Proportion of tolerant reproductive guild individuals

TABLE 14.20 Metrics Used in Some of the Fish-based IBIs (*cont'd*)

Region for Which Developed	Author(s)	Constituent Metrics
		(9) Proportion of native individuals **(10)** Number of native species with long-lived individuals
Puget Sound Lowlands, Western Washington, USA	Matzen and Berge (2008)	**(1)** Percent invertivore individuals **(2)** Percent invertivore/piscivore individuals **(3)** Percent coho salmon individuals **(4)** Percent cutthroat trout individuals **(5)** Percent sculpin individuals **(6)** Percent individuals of the most abundant species
French estuaries	Delpech et al. (2010)	**(1)** Total density **(2)** Density of diadromous migrant species **(3)** Density of marine juvenile migrants **(4)** Density of benthic species
Zeeschelde estuary, Belgium	Breine et al. (2010)	For the freshwater zone **(1)** Total number of species **(2)** Total number of individuals **(3)** Percentage of diadromous individuals **(4)** Percentage of specialised spawner individuals **(5)** Percentage of piscivorous individuals **(6)** Percentage of benthic individuals For the oligohaline zone **(1)** Total number of piscivorous species **(2)** Total number of pollution intolerant species **(3)** Total number of diadromous species **(4)** Total number of individuals **(5)** Total number of marine migrating species **(6)** Total number of estuarine species For the mesohaline zone **(1)** Total number of species **(2)** Total number of diadromous species **(3)** Total number of specialised spawners **(4)** Total number of habitat sensitive species **(5)** Percentage of pollution intolerant individuals **(6)** Total number of marine migrating species
Battle river, Alberta, Canada	Stevens et al., 2010	**(1)** Percent older, long-lived individuals **(2)** Catch per 100 s of electrofishing **(3)** Percent deformities, disease, parasites, fin erosion, lesions or tumors **(4)** Percent tolerants **(5)** Percent intolerants
Hondo river basin, Yucatan Peninsula	Schmitter-Soto et al., 2011	**(1)** Relative abundance of *Astyanax aeneus* **(2)** Relative abundance of bentholimnetic species **(3)** Relative abundance of *Cichlasoma urophthalmus* **(4)** Relative abundance of herbivore species **(5)** Numerical evenness

(*Continued*)

TABLE 14.20 Metrics Used in Some of the Fish-based IBIs (*cont'd*)

Region for Which Developed	Author(s)	Constituent Metrics
		(6) Percentage of native species **(7)** Relative abundance of *Poecilia mexicana* **(8)** Relative abundance of *Poecilia* sp. (Calakmul) **(9)** Relative abundance of sensitive species **(10)** Percentage of tolerant species **(11)** Relative abundance of *Xiphophorus hellerii* **(12)** Relative abundance of *Xiphophorus maculatus*

Even as some authors believe that fish assemblages are appropriate ecological indicators of both natural and artificial lakes (Costa and Schulz, 2010; Pei et al., 2010; Miranda and Hunt, 2010; Launois et al., 2011), besides streams (Costa and Schulz, 2010) and saline water environments (Brousseau et al., 2011), the appropriateness of the use of fishes has been questioned from time to time. One of the grounds for the objection is that fish communities exhibit a lag time in manifesting changes in the environment owing to long generation times of many species (Griffith et al., 2005). However, there is evidence to the contrary; for example, Karr et al. (1985) have provided evidence that fish communities were responsive over a 4-year period to changes in the chemical composition (total residual chlorine) of wastewater discharge.

Another criticism is aimed at the mobility of fish communities suggesting that fish may preferentially avoid degraded habitats, thereby affecting the IBI score (Berkman et al., 1986). Fish community responses can also be influenced by stream network position (Hitt and Angermeier, 2011). The problems associated with representativeness of samples, discussed in Chapter 4, may be exacerbated when dealing with highly mobile animals such as fish (Jennings et al., 1999). The advantages and disadvantages of using fish as indicator organisms, in the context of attributes of other organisms, are presented in Table 14.21.

How to weigh information about rare *versus* abundant taxa is another challenge. Wan et al. (2010) have investigated the influence of rare fish taxa (within the lower 5% of rank abundance curves) on IBI metric and total scores for stream sites in two of Minnesota's major river basins, the St. Croix (n = 293 site visits) and Upper Mississippi (n = 210 site visits). The authors artificially removed rare taxa from biological samples by: (1) separately excluding each individual taxon that fell within the lower 5% of rank abundance curves; (2) simultaneously excluding all taxa that had an abundance of one (singletons) or two (doubletons); and (3) simultaneously excluding all taxa that fell within the lower 5% of rank abundance curves. They then compared IBI metric and total scores before and after removal of rare taxa using the normalised root mean square error (nRMSE) and regression analysis.

It was seen that the difference in IBI metric and total scores increased as more taxa were removed. Moreover, when multiple rare taxa were removed, nRMSE was related to sample abundance and to total taxa richness, with greater nRMSE observed in samples with a larger number of taxa or sample abundance. It was also seen that metrics based on relative abundance of taxa were less sensitive to the loss of rare taxa, whereas those based on taxa richness were more sensitive, because taxa richness metrics give more weight to rare taxa compared to the relative abundance metrics.

TABLE 14.21 Advantages and Disadvantages in the Use of Fish, Macrophytes, Macroinvertebrates, Plankton and Periphyton in IBI Development

Bioindicator	Strengths	Weaknesses
Fish	(i) Availability of extensive life-history information (ii) Ease of identification (iii) Information comprehendible by general public (iv) Direct relation to goals of habitat restoration	(i) Slow in reflecting environmental changes (ii) Mobility that may bias index scores (iii) Selective nature of sampling (iv) Possibility of species migrating away from impaired regions of a lake/stream
Macrophytes	(i) Ease of identification (ii) Immobility facilitates sampling (iii) Amenable to rapid sampling methods (e.g. remote sensing and hydroacoustics)	(i) Not representative of a range of trophic levels (ii) Composition can be strongly influenced by hydrologic regime
Macroinvertebrates	(i) Ecologically diverse and widespread, occurring in almost all types of streams (ii) Relatively sedentary (iii) Easy to sample qualitatively and to identify (iv) Sufficiently long life cycles to integrate the environmental stress that have occurred over an extended period (v) A single sample may be adequately representative	(i) Difficulty of quantitative field sampling (ii) Difficulty of identification to species level limits taxonomic resolution (iii) Community highly variable across time
Plankton	(i) Ease of qualitative and quantitative field sampling (ii) Easy to store large number of samples	(i) Community highly variable across time (ii) Community highly variable across space (iii) Difficulty in identification to species level
Periphyton	(i) Ease of qualitative and quantitative field sampling (ii) Immobile (excluding epilimnic species) (iii) Use of palaeolimnological records possible	(i) Community highly variable across time (ii) Heterogeneous even at very small scale (Hollingsworth and Vis, 2010)

14.5.2. IBIs Based on Macroinvertebrates

Next to fish, most IBIs have revolved round macroinvertebrates (Ohio EPA 1987; Plafkin et al., 1989; Kerans and Karr 1994; Barbour et al., 1996; Lewis et al., 2001; Blocksom et al., 2002; Weigel, 2003; Heatherly et al., 2005; Rufer, 2006; Trigal et al., 2006; Rehn, 2009; Benyi et al., 2009; Lang, 2009; Leunda et al., 2009; Delgado et al., 2010; Aura et al., 2010). An illustrative list of macroinvertebrate-based IBIs is presented in Table 14.22.

Macroinvertebrates are generally sedentary and it is relatively easier and simpler to sample them than fish. However, there are several disadvantages in using macroinvertebrates for IBIs (Table 14.21). Firstly individuals are often variably distributed, causing problems in sampling and metric development (Blocksom et al., 2002; Merten et al., 2010). Secondly, a great deal of time and effort is associated with

TABLE 14.22 Metrics Used in Some of the Macroinvertebrate-based IBIs

Region for Which Developed	Author(s)	Constituent Metrics
Rivers of the Tennessee Valley, USA	Kerans and Karr (1994)	**(1)** Total taxa **(2)** Total intolerant snail and mussel species richness **(3)** Total mayfly **(4)** Total caddisfly **(5)** Total stonefly taxa richness **(6)** Relative abundances of *Corbicula* **(7)** Relative abundances of oligochaetes **(8)** Relative abundances of omnivores **(9)** Relative abundances of filterers **(10)** Relative abundances of grazers **(11)** Relative abundances of predators **(12)** Dominance **(13)** Total abundance
Coldwater streams in Wisconsin, USA	Lyons et al. (1996)	**(1)** Number of intolerant species **(2)** Percent of all individuals that are tolerant species **(3)** Percent of all individuals that are top carnivore species **(4)** Percent of all individuals that are native or exotic stenothermal coldwater or coolwater species **(5)** Percent of salmonid individuals that are brook trout (*Salvelinus fontinalis*)
Great Lakes	Burton et al. (1999)	**(1)** Number of Odonota genera **(2)** Relative abundance of (Lake Huron) Odonota genera **(3)** Number of Crustacea and Mollusca genera **(4)** Total number of genera **(5)** Relative abundance of Gastropoda genera **(6)** Relative abundance of Sphaeriidae genera **(7)** Total taxa richness **(8)** Evenness (J) **(9)** Shannon index (H) **(10)** Simpson index (D)
Northeastern United States	Lewis et al. (2001)	**(1)** Hilsenhoff biotic index **(2)** Taxa richness **(3)** Relative abundance **(4)** Percent intolerant taxa **(5)** Percent oligochaetes **(6)** Percent non-insects **(7)** Percent Chironomidae **(8)** Percent dominant taxon **(9)** Community loss index **(10)** Community similarity index **(11)** Trophic condition index **(12)** Dominant-in-common-St
New Jersey, the USA	Blocksom et al. (2002)	**(1)** Number of Diptera taxa **(2)** Percent chironomid individuals **(3)** Percent oligochaetes and (or) leeches **(4)** Percent collector–gatherer taxa **(5)** Hilsenhoff biotic index

TABLE 14.22 Metrics Used in Some of the Macroinvertebrate-based IBIs (*cont'd*)

Region for Which Developed	Author(s)	Constituent Metrics
Mid-Atlantic highland streams, USA	Klemm et al. (2003)	**(1)** Ephemeroptera richness (watershed-adjusted) **(2)** Plecoptera richness (watershed-adjusted) **(3)** Trichoptera richness **(4)** Collector–filterer richness (watershed-adjusted) Negative metrics **(5)** % Non-insect individuals **(6)** MTI (Macroinvertebrate Tolerance Index) **(7)** % Individuals in 5 dominant taxa
Illinois, the USA	Heatherly et al. (2005)	**(1)** Richness **(2)** Percent dominance **(3)** Shannon diversity **(4)** Simpson diversity **(5)** Evenness **(6)** Percent Oligochaela **(7)** Percent Chironomidae **(8)** Percent Insecta taxa
Iberian peninsula	Trigal et al. (2006)	**(1)** Percent Insecta taxa **(2)** Shannon–Weiner diversity index **(3)** Total taxa **(4)** Total taxa of Chironomidae larvae **(5)** Percent Ephemeroptera, Trichoptera and Odonota taxa **(6)** Percent predators **(7)** Percent shredders **(8)** Percent collectors–gatherers
Saline ramblas, SE of the Iberian Peninsula	Cánovas et al. (2008)	**(1)** Family richness **(2)** Coleoptera/Hemiptera coefficient **(3)** Indicator species of naturality **(4)** Indicator species of degradation
Kipkaren and Sosiani Rivers, Kenya	Aura et al. (2010)	**(1)** Abundances of Ephemeroptera, Plecoptera and Trichoptera **(2)** Relative abundances of Diptera **(3)** Ephemeroptera, Plecoptera and Trichoptera:Diptera ratio **(4)** Oligochaeta, Mollusca, Hemiptera and Odonata **(5)** Proportions of tolerance taxa **(6)** Dominant taxa **(7)** The relative proportions of invertebrates that fall into the gatherer and predator feeding groups
Xiangxi river, China	Li et al. (2010)	**(1)** Richness measures **(2)** Composition measures **(3)** Tolerance measures **(4)** Feeding measures **(5)** Habitat measures **(6)** Biodiversity index

(*Continued*)

TABLE 14.22 Metrics Used in Some of the Macroinvertebrate-based IBIs (*cont'd*)

Region for Which Developed	Author(s)	Constituent Metrics
Flanders, Belgium	Gabriels et al. (2010)	(1) Taxa richness (2) Number of Ephemeroptera, Plecoptera and Trichoptera taxa (3) Number of other sensitive taxa (4) Shannon–Wiener diversity index (5) Mean tolerance score

taxonomic identification prior to metric development (Berkman et al., 1986). Thirdly, high temporal variability of macroinvertebrates is a major factor limiting their use in ecological health indices (Tangen et al., 2003).

Ephemoptera, Plecoptera and Tricoptera (EPT) have been commonly used in the metrics of macroinvertebrate-based IBIs. Percent Oligochaetes is another commonly used metric. Whereas the relative abundance of EPT decreases in disturbed water bodies, that of Oligochaetes, which is a pollution-tolerant taxon, increases. Although the aforementioned metrics are the most common, highly specialised IBIs that are limited to specific spatial components or microenvironments of a water body have also been produced (Lewis et al., 2001; Blocksom et al., 2002). Moreover, many studies have been aimed at assessing metric variability rather than developing an integrative index (Pathiratne and Weerasundara 2004; Trigal et al., 2006). This is also reflected in the studies presented in Section 14.7.

14.5.3. Plankton-based IBIs

The functions of lakes, ponds and other lentic water bodies are influenced in major ways by the dynamics of the phytoplankton and zooplankton (Abbasi and Chari; 2008; Chari and Abbasi, 2003, 2005). Phytoplankters are the primary source of energy driving these ecosystems; and the zooplankton is the central trophic link between primary producers and consumers. As a consequence, the two groups of organisms influence the rest of the organisms.

IBIs that rely on phytoplankton or zooplankton communities are much less prevalent than fish-based and macroinvertebrate-based IBIs (Harig and Bain 1998; O'Connor et al., 2000; Lougheed and Chow-Fraser 2002; Kane and Culver 2003; Whitman et al., 2004; Kane et al., 2010; Spartharis and Tsirtsis (2010). The major advantage of a planktonic index is ease of sampling; a single water sample from a pelagic site may be adequate. However, planktonic communities exhibit high temporal and spatial variation (Chari and Abbasi 2003, 2005; Chari et al., 2003). While the phytoplankton metrics lay emphasis on dominant taxa, the zooplankton metrics often focus on Daphnid species; it is believed that more pristine communities are dominated by large-bodied individuals and a markedly increased abundance (Harig and Bain 1998; Beck and Hatch, 2009).

Interestingly, there are more IBIs in existence that focus on diatoms than the ones which cover plankton in general (Zalack et al., 2010; Cejudo-Figueiras et al., 2010). A few examples are given in Table 14.23.

Biomass metrics have been proposed for both phytoplankton and zooplankton, although studies have produced conflicting results of changes in biomass in response to habitat degradation; the top-down impacts of lessened photosynthesis and heightened grazing can be offset

TABLE 14.23 Metrics Used in Two Recent Diatom-based IBIs

Region for Which Developed	Author(s)	Constituent Metrics
Florida, USA	Lane and Brown (2007)	**(1)** Sensitive taxa **(2)** Tolerant taxa **(3)** Elevated pH sensitive **(4)** Elevated pH tolerant **(5)** Salinity sensitive **(6)** Salinity tolerant **(7)** Elevated nitrogen sensitive **(8)** Elevated nitrogen tolerant **(9)** Low D.O. sensitive **(10)** Low D.O. tolerant **(11)** Meso-polysaprobous **(12)** Oligotrophic **(13)** Eutrophic **(14)** Pollution-tolerant
Coastal Galician rivers, northwest Spain	Delgado et al. (2010)	**(1)** Generic Diatom Index **(2)** Specific pollution sensitivity index **(3)** Leclercq & Maquet's pollution index **(4)** Steinber & Schiefele trophic metric **(5)** Sládecek's pollution index **(6)** Trophic Diatom Index **(7)** Percentage abundance of reference taxa **(8)** Percentage of richness of reference taxa

by bottom-up impacts of nutrient release (Siegfried and Sutherland 1992; Harig and Bain 1998).

14.5.4. IBIs Based on Periphyton

IBIs have been developed around periphyton as bioindicators (Fore 2002; Fore and Grafe 2002), but many studies have produced inconsistent results (Hill et al., 2000; Hamsher et al., 2004; Tang et al., 2006; Chessman and Townsend, 2010), thereby querying the use of periphyton as indicators. The advantages and disadvantages of relying on periphyton are similar to those noted for phytoplankton and zooplankton (Table 14.21):

Periphyton require relatively simple sampling protocols, but have high temporal variability which, in view of some authors, makes index development difficult (USEPA 1998).

A case for the use of periphytic diatom communities as biological indicators has been recently made by Cejudo-Figueiras et al. (2010). They studied nineteen shallow permanent lakes from north-west Spain, classifying them into three trophic levels, and studying the epiphytic diatom communities growing on three different macrophytes for each trophic level. They assessed: (1) which of the most common diatom indices provides reliable water quality assessment, (2) how different plant substrata influence the diatom communities growing on them and (3) how these differences affect water quality assessment. Even as similarity tests showed significant differences in the composition of diatom assemblages among nutrient concentrations and host macrophytes, ANOVA results for selected diatom-based metrics showed significant differences among trophic levels but not between different plant substrata. Hence, the findings support the use of epiphytic diatoms as biological indicators for shallow lakes irrespective of the dominant macrophyte.

Hollingsworth and Vis (2010) also found that even as diatoms appeared to be patchily distributed within a reach, and this patchiness often led to varied relative abundance of common species and the introduction or loss of rare species among riffles, yet a multi-metric index could correctly classify a stream based on a one-riffle sample. Hence, variation among riffles of the diatom assemblages does not appear to correspond directly to stream health, but to species richness and diversity.

One special advantage with periphyton is that they provide palaeolimnological evidence, which is an asset in establishing reference conditions.

14.5.5. IBIs Based on Macrophytes

Macrophyte communities present many advantages for biomonitoring of which the most significant ones are ease of identification and immobility (Mack, 2007; Beck et al., 2010). This enables the use of GIS, remote sensing or hydroacoustic sampling (Clayton and Edwards 2006; Valley and Drake 2007). In turn, it makes it possible to survey large areas very quickly. In addition, macrophytes exhibit a more rapid response within an ecologically relevant time frame to environmental changes compared to fish communities (Abbasi and Abbasi 2010; Chari and Abbasi, 2005). However, the major disadvantage of macrophytes is that they represent only one trophic level, the primary level of production. Because of this, it is impossible to use trophic-based metrics in macrophyte-IBIs.

An illustrative list of macrophyte-IBIs is presented in Table 14.24. These include metrics of the coefficient of conservatism as used in the floristic quality index (Nichols 1999), maximum depth of plant growth and the percentage of the littoral zone that is vegetated.

In a study by Hatzenbeler et al. (2004) to compare fish and macrophyte-IBIs, aquatic plant communities were sampled along transects in sixteen lakes in northwestern Wisconsin, USA, whereas fish communities were sampled following procedures of Jennings et al. (1999). Metrics related to structural and compositional components of each community were assessed for their respective correlation with land use within the watershed and amount of lakeshore development. It was seen that macrophyte metrics correlated with shoreline development, but none of the fish metrics were correlated with levels of disturbance. Metrics from both bioindicators were not correlated with land use within the watershed, possibly due to the lack of urban development within the watersheds of the study lakes. The authors suggest that a macrophyte IBI would be more appropriate for the northwestern Wisconsin area, given the naturally low diversity of fish communities in its small lakes.

Macrophyte-IBIs can be particularly useful when employed in conjunction with fish IBIs because the importance of aquatic macrophytes to the well-being of fish communities has been well established. Besides offering habitat and prey for many species of fish, macrophytes facilitate their reproductive success by providing protection from predators (Brazner and Beals 1997). When eutrophic conditions cause runaway growth of macrophytes, it seriously harms the fish community. Given these strongly inter-dependent relationships between macrophytes and fish, a fish-based IBI is likely to have many correlations to a macrophyte-based IBI. Ecological forces governing each community can be determined from an analysis of metrics associated with each index. As such, a macrophyte-based IBI would offer an appropriate complement to a fish-based IBI.

TABLE 14.24 Constituent Metrics of Some Macrophyte-based IBIs

Region for Which Developed	Author(s)	Constituent Metrics
Wisconsin, the USA	Nichols et al. (2000)	**(1)** Maximum depth of plant growth **(2)** Percent of littoral area vegetated **(3)** Relative frequency of submersed species **(4)** Relative frequency of exotic species **(5)** Relative frequency of sensitive species **(6)** Simspon's diversity index **(7)** Taxa number
Great Lakes	Wilcox et al. (2002)	**(1)** Percent wetland in sedge vegetation type **(2)** Percent wetland in invasive vegetation types **(3)** Percent wetland obligate species **(4)** Floristic quality index **(5)** Number of native taxa **(6)** Sum of mean percent cover turbidity-tolerant taxa in SAV vegetation type **(7)** Sum of mean percent cover invasive taxa in sedge vegetation type
North Dakota, the USA	DeKeyser et al. (2003)	**(1)** Number of native perennials **(2)** Number of genera of native perennials **(3)** Number of native grass and grass-like species **(4)** Percentage of annual, biennial and introduced species **(5)** Number of native perennials (in wet meadow zone) **(6)** Number of species with C value >5 **(7)** Number of species with C value >4 (wet meadow zone) **(8)** Average C value **(9)** Floristic quality index
New Zealand	Clayton and Edwards (2006)	**(1)** Native condition index **(2)** Invasive condition index **(3)** Total lake SP1 index
Pennsylvania, the USA	Miller et al. (2006)	**(1)** Adjusted floristic quality assessment index **(2)** Percent cover of tolerant species **(3)** Percent annual species **(4)** Percent non-native species **(5)** Percent invasive species **(6)** Percent trees **(7)** Percent vascular cryptograms **(8)** Percent cover of Phalarisantadinacea
Indiana, the USA	Rothrock et al. (2008)	**(1)** Total number of species **(2)** Number of submergent species **(3)** Number of floating-leaved species **(4)** Number of emergent species **(5)** Number of sensitive species **(6)** Percent tolerant and exotic species **(7)** Relative abundance of obligate wetland species **(8)** Relative abundance of pioneer species **(9)** Relative abundance of woody species

(*Continued*)

TABLE 14.24 Constituent Metrics of Some Macrophyte-based IBIs (*cont'd*)

Region for Which Developed	Author(s)	Constituent Metrics
		(10) Average cover **(11)** Relative abundance of exotics
Minnesota lakes	Beck et al., 2010	**(1)** Maximum depth of plant growth, 95% occurrence (m) **(2)** Percentage of littoral vegetated **(3)** Number of species with relative frequency over 10% **(4)** Relative frequency of submersed species **(5)** Relative frequency of sensitive species **(6)** Relative frequency of tolerant species **(7)** Number of native taxa

14.5.6. Multi-taxa IBIs

Considering that strengths and weaknesses are associated with each community used in bioassessment (Table 14.21), several studies have examined the possibility of the integration of several bioindicators into multi-taxon indices (Harig and Bain 1998; O'Connor et al., 2000; Wilcox et al., 2002; Laugheed et al., 2007; Williams et al., 2009).

The logic behind selection of combination of taxa in multi-taxa IBIs can be illustrated by the example of the IBI developed by Harig and Bain (1998), who examined the efficacy of six zooplankton metrics, two phytoplankton and fish metrics each and one benthic invertebrate metric in a study of small lakes in Adirondack Park, USA. One of the phytoplankton metrics chosen by the authors was a dominant taxa type and the other was a dominant biomass type. Disturbed systems were expected to be dominated by green and blue-green filamentous algae and dino-flagellates, while undisturbed lakes were expected to be dominated by golden-brown algae and cryptomonads. Phytoplankton biomass was expected to be high in disturbed systems because of diminished grazing pressure. The dominant phytoplankton taxa metric responded as hypothesised, whereas the dominant biomass metric did not, possibly due to bottom-up effects.

The four zooplankton metrics chosen by the authors were expected to show predictable responses to changes in recreational use, in terms of abundance of *Daphnia* sp., community richness, morphology of body size and biomass. Abundance of *Daphnia* was expected to decrease in disturbed systems along with other large-body taxa, and the overall diversity of zooplankton. As a corollary, zooplankton biomass was expected to decrease with disturbance. All four proposed metrics responded as expected to these impacts and were included in the final multi-taxon index. Among the remaining three metrics explored by Harig and Bain (1998), relying on fish and benthic invertebrates, one fish metric (abundance of native species) was included in the nine-metric multi-taxa IBI.

Two other examples of multi-taxa IBI development have been given in Sections 14.4.11 And 14.4.12.

Besides enhancing the sensitivity and coverage of an IBI, integrating highly responsive organisms into a multi-taxon as illustrated in these examples, IBI could facilitate early detection and swifter response to disturbances within a watershed. For example, periphyton that colonise nearshore areas of lakes have been shown to be highly responsive to disturbance (Lambert et al., 2008). Both epipelic and epilithic periphyton have been suggested to provide a means of early detection within a lake system given the rapid response to disturbance, particularly in relation to disturbances originating from the immediate shoreline. Macroinvertebrates may also be sensitive indicators of nearshore disturbance (DeSousa et al., 2008). On the other hand, organisms with long generational times, such as fish, could be more useful in picking up long-term changes within a watershed. Hence, multi-taxon indices that use organisms representative of multi-organisational levels could provide much more information useful in detection of disturbances as well as reflect longer-term trends, than mono-taxa IBIs. In this context, it is very important to see the trade-off between extra costs of implementing multi-taxa IBIs and the benefits; the latter must be substantial enough to justify the former.

Studies on relative sensitivities of different organisms to different stressor situations provide key inputs for metric selection in multi-taxa IBIs. For example, Growns (2009) explored bioregional classifications which are used extensively for conservation management and monitoring programmes. He used generalised dissimilarity modelling (GDM) to test the ability of different regional classifications of four groups of aquatic biota to be used as surrogates for each other: aquatic macrophytes, macroinvertebrates, freshwater fish and frogs. It was seen that regional classifications differed markedly between different biotic groups because the environmental drivers that were related to species turnover throughout the region differed among groups. Altitude and rainfall were the strongest drivers of species turnover among the groups. The author suggests that physiographic variables should be incorporated in reserve design and monitoring programmes to explicitly address differences in classifications between similar biotic groups. In another study on the impact of pharmaceutical wastewaters on Ebro river, Spain, Gros et al. (2010) found the susceptibility of the organism to follow the order algae > daphnia > fish.

An illustrative list of metrics used in a few multi-taxa IBIs is given in Table 14.25.

14.6. IBIS FOR DIFFERENT AQUATIC SYSTEMS

The first-ever IBI (Karr, 1989) was intended for lotic water bodies and since then IBI development has predominantly been for rivers and streams. However, lotic-water IBIs cannot be used effectively for lentic systems or saline environments because of difference in the structure and functioning of the corresponding ecosystems. Hence, attention has been given to IBIs for freshwater lakes, seasonal wetlands and saline environments, though to a much lesser extent than given to lotic systems.

The necessity that different IBIs must be developed for each set of reasonably comparable aquatic ecosystems arises from the fact that index performance is based on the expected community assemblage as influenced by a specific range of ecosystem parameters (Plafkin et al., 1989). Streams are distinguished primarily by ecoregion, temperature and size, while lakes are defined by additional attributes that include, but are not limited to, greater variation in depth, surface area, chemical composition and effect of seasons. At times, even two largely unimpacted, adjacent rivers can have different autocorrelation patterns as observed for assemblages of benthic invertebrates in two upland rivers in Australia by Lloyd et al. (2006). Hence, assuming dependence or independence of any two sites *a priori* is fraught with risks.

As species diversity in space and time is generally higher and more varied in poorly mixed multidimensional habitats like lakes (Jackson and Harvey 1997), compared to streams, the challenge associated with obtaining representative samples (Chapter 4) is that much bigger for the lakes. Because standing water systems exhibit a wide range of

TABLE 14.25 Metrics Used in Some of the Multi-taxa-based IBIs

Region for Which Developed	Author(s)	Constituent Metrics
Coldwater streams of western Oregon and Washington, USA	Hughes et al. (2004)	**(1)** Adjusted number of coldwater species **(2)** Percent coldwater species **(3)** Adjusted number of coldwater individuals **(4)** Percent anadromous individuals **(5)** Adjusted number of age classes **(6)** Percent coolwater individuals **(7)** Adjusted number of tolerant individuals **(8)** Percent alien species
Mountain ecoregion of the western USA	Whittier et al. (2007)	**(1)** Sensitive rheophilic species **(2)** Assemblage tolerance index **(3)** Sensitive invertivores–piscivores **(4)** Lithophilic spawners **(5)** Salmonidae **(6)** Native sensitive long-lived species **(7)** Alien vertebrate species
Xeric ecoregion of the western USA	Whittier et al. (2007)	**(1)** Assemblage tolerance index **(2)** Lithophilic spawners **(3)** Omnivores **(4)** Native sensitive lotic species **(5)** Alien vertebrate species
Plains ecoregion of the western USA	Whittier et al. (2007)	**(1)** Assemblage tolerance index **(2)** Non-tolerant vertebrate species richness **(3)** Non-tolerant native benthic species **(4)** Non-tolerant invertivores–piscivores **(5)** Ictaluridae **(6)** Lithophilic spawners **(7)** Alien vertebrate species
Muskegon river watershed, Michigan, USA	Lougheed et al. (2007)	**Plants** **(1)** Similarity to undeveloped site **(2)** Plant NMDS axis **(3)** Plant Coefficient of **(4)** Conservatism **(5)** Dominance of duckweed **(6)** (*Lemna* spp.) Relative dominance of **(7)** sensitive plants Relative dominance of **(8)** tolerant and exotic plants **(9)** Overall plant Index **Diatoms** **(10)** Similarity to undeveloped site **(11)** Diatom NMDS axis van Dam's Trophic State Index **(12)** Relative abundance of erect **(13)** and stalked diatoms **(14)** Overall Diatom Index

TABLE 14.25 Metrics Used in Some of the Multi-taxa-based IBIs (*cont'd*)

Region for Which Developed	Author(s)	Constituent Metrics
		Zooplankton (15) Similarity to undeveloped site (16) Zooplankton NMDS axis Zooplankton-inferred conductivity species richness of chydorids and macrothricids (17) Overall Zooplankton Index (18) Overall Index of Wetland (19) Quality

environmental heterogeneity, development of lake IBIs have had to contend with issues related to determining how much sampling effort would be adequate to accurately characterise biotic communities. Stream IBI development for fish has not encountered such a degree of difficulty attributed to sampling effort as the scope of assessment is usually only a reach that is accurately characterised by electrofishing, whereas lakes require subsamples across the entire basin (Whittier, 1999 Pei et al., 2010; Launois et al., 2011). Furthermore, the chosen sampling method can have a profound effect on estimates of species richness (Minns 1989: Jackson and Harvey 1997). In a study by Jennings et al. (1999) seine sampling was seen to produce the most accurate and precise richness-based metrics, while electrofishing was shown to sample top carnivores most effectively.

Because different sampling equipment differ in their ability to characterise biotic communities, lake IBI development has also encountered problems related to the combining of data obtained from different methods (Weaver et al., 1993; Jennings et al., 1999; Whittier, 1999; Lyons et al., 2000). Hence, the use of equipment-specific metrics that retain the advantages of using different sampling equipment without introducing imponderables when combining data has been suggested. Suggestions to use standardised sampling protocols have also been advanced (Lyons et al., 2000).

A large number of man-made lakes or reservoirs exist which are generally more prone to ecodegradation than natural lakes (Abbasi and Abbasi, 2011; Abbasi 2001; Chari et al., 2005 a, b). However, much lesser number of IBIs have been developed for reservoirs than natural lakes (Jennings et al., 1995; Launois et al., 2011). One of the few reservoir-specific IBIs has been described in Section 14.4.10.

Adaptation of IBIs to seasonal water bodies (referred here as 'seasonal wetlands') entails problems similar to the ones encountered with lakes. Successful efforts for wetland IBIs have largely relied on macrophyte communities as bioindicators (DeKeyser et al. 2003; Miller et al. 2006). High levels of diversity of macrophyte communities typically encountered in seasonal wetlands appear to contribute to the success of these indices by providing sensitive metrics. However, variations in hydrologic regime in seasonal wetlands due to natural fluctuations of water levels are often much greater than variations arising from anthropogenic stress. This may cause ambiguity in the index responses (Wilcox et al. 2002).

Efforts to develop IBIs suited to estuarine environment have dealt with fish, benthic invertebrates and phytoplankton (Thompson and Fitzhugh 1986; Deegan et al. 1997; Weisberg et al. 1997; Lacouture et al., 2006; Williams et al., 2009). For example, Thompson and Fitzhugh (1986) have developed a prototype IBI designed to assess Louisiana estuaries using fish communities. The IBI includes metrics related to species richness and composition, trophic function, abundance and condition. This metric construction is similar to that of the IBIs developed for freshwater systems, with the exception that effects of the salinity regime on metric function have been incorporated. Weisberg et al. (1997) have developed a benthic invertebrate IBI for Chesapeake Bay, Maryland, by sampling seven habitats defined by salinity and substrate. Similarly, Lacouture et al. (2006) have developed a phytoplankton IBI for Chesapeake Bay using metrics that varied by two seasons and four salinity regimes. Both IBIs were able to produce predictable results, but considerable time and effort had to be spent in either case to limit the effects of environmental noise on index behaviour. Three Scandinavian marine IBIs were also found to produce converging results by Josefson et al. (2009), as detailed in Section 14.7.

It can be said that benthic invertebrate and phytoplankton IBIs seem to be able to successfully discriminate between impaired and reference sites in estuaries, but the studies have been too few and far between to be conclusive.

14.7. INTER-IBI COMPARISON

Studies have been done from time to time wherein different IBIs have been used to process similar data to see how different IBI constructs respond to a given situation. There are also studies comparing the performances of biotic indices (BIs) with IBI and 'bootstrapping' studies aimed to assess the response of IBIs to random sampling variation. Studies have been done to see how the interpretations of a given data set by an IBI differ from the multivariate (MV) approach. The IBI–MV comparison is discussed in the next chapter. A few examples of intra-IBI comparison, BI–IBI comparison and bootstrapping studies are given in this section.

14.7.1. A Comparison Between a Fish-based IBI and a Benthos-based IBI

Freund and Petty (2007) quantified the response of a fish-based index (Mid-Atlantic Highlands Index of Biotic Integrity, MAH-IBI) and a benthic invertebrate-based index (West Virginia Stream Condition Index, WV-SCI) to acid mine drainage (AMD)-related stressors in 46 stream sites within the Cheat River Watershed, West Virginia. They also identified specific stressor concentrations at which biological impairment was always or never observed.

WV-SCI was seen to be highly responsive across a range of AMD stressor levels. Furthermore, impairment to macroinvertebrate communities was observed at relatively low stressor concentrations, especially when compared to state water quality standards. In comparison, MAH-IBI was significantly less responsive to local water quality conditions. Low fish diversity was observed in several streams that possessed relatively good water quality. This pattern was especially pronounced in highly degraded subwatersheds, suggesting that regional conditions may have a strong influence on fish assemblages in this system.

The authors (Freund and Petty, 2007) feel that biomonitoring programmes in mined watersheds should include both benthic invertebrates, which are consistent indicators of local conditions, and fishes, which may better reflect the regional conditions.

14.7.2. A Comparison of IBIs Developed in the USA and in Europe

Borja et al., (2008) compared two widely used measures of ecological integrity, the Benthic Index of Biotic Integrity (B-IBI) developed in the USA, and the European Marine Biotic Index (AMBI) and its multivariate extension, the M-AMBI. Specific objectives of the study were to identify the frequency, magnitude and nature of differences in assessment of Chesapeake Bay sites as 'degraded' or 'undegraded' by these indices. A data set of 275 subtidal samples taken in 2003 from Chesapeake Bay was used in this comparison.

Linear regression of B-IBI and AMBI accounted for 24% of the variability; however, when evaluated by salinity regimes, the explained variability increased in polyhaline (38%), high mesohaline (38%) and low mesohaline (35%) habitats, remained similar in the tidal freshwater (25%) and decreased in oligohaline areas (17%). Using M-AMBI, the explained variability increased to 43% for linear regression and 54% for logarithmic regression. By salinity regime, the highest explained variability was found in high mesohaline and low polyhaline areas (53–63%), while the lowest explained variability was in the oligohaline and tidal freshwater areas (6–17%). The total disagreement between methods, in terms of degraded–undegraded classifications, was 28%, with high spatial levels of agreement. Overall, the study suggests that different methodologies in assessing benthic quality can provide similar results even though these methods have been developed within different geographical areas.

14.7.3. Comparison of Three Scandinavian IBIs

Josefson et al. (2009) report a comparative study of the performance of three Scandinavian IBIs, which share the same underlying rationale that the quality of the faunal communities is reflected by both the contribution of sensitive and tolerant species, and the species diversity. However, the relative weighting of the diversity and sensitivity components in the three indices are different, and the indices use two different classification systems of the sensitivity of the species, and three different measures of species diversity. In all the three indices, very low numbers (i.e., <10) of individuals in the samples are in all cases considered as a low quality feature and this is accounted for by using different factors to modify the indices against the number of individuals in the samples. All three indices are calculated on data from a sample area of the bottom of 0.1 m^2 and species larger than 1 mm (macrofauna).

The Danish Index (DKI): The index uses the AMBI index (described in Chapter 12) where species are classified according to sensitivity or tolerance to disturbance (Borja et al., 2000), the Shannon diversity (H'; Shannon and Weaver, 1963), number of species (S) and the number of individuals (N). The value of *AMBI* is calculated from the proportions of individuals of sensitive or tolerant species within a sample. The diversity component (H') and the sensitivity component (AMBI) are both normalised to attain a value between 0 and 1, where the diversity is normalised against the highest diversity observed in the sea area. The two components are weighted equally in the calculation of DKI, and the index value is modified by the factor including S and N, which compensates for low species numbers and low numbers of individuals:

$$\mathrm{DKI} = \left(\frac{\left(1-\frac{\mathrm{AMBI}}{7}\right)+\frac{H'}{H'_{\max}}}{2}\right) * \left(\frac{\left(1-\frac{1}{N}\right)+\left(1-\frac{1}{S}\right)}{2}\right) \quad (14.4)$$

where H' is the Shannon index with log base 2 and H'_{max} is the reference value which H is normalised against and is the highest value that H reaches in undisturbed conditions. H'_{max} was set to 5.6 which was the highest value observed in this material, except for Maarmorilik where H'_{max} was set to 4.0, N the number of individuals and S the number of species.

DKI can attain value between 0 and 1. For very high number of species, the value of DKI approaches the mean of the normalised sensitivity and diversity components. If $S=1$ then $0<\text{DKI}<0.25$ and if $S=1$ and $N=1$ then $\text{DKI}=0$. DKI varies linearly with AMBI.

The Norwegian index (NQI): This index also uses the AMBI index (Borja et al., 2000) as a measure of sensitivity and is normalised to attain values between 0 and 1 in the same way as in the DKI index. The diversity component is described by a factor, *SN* (Rygg, 2006), which is normalised to attain values between 0 and 1 and the diversity component is also modified by a factor to compensate for low densities. The normalised sensitivity and diversity components are weighted equally in NQI:

$$\text{NQI} = 0.5 * \left(1 - \frac{AMBI}{7}\right) + \left[0.5 * \frac{SN}{2.7} * \frac{N_{tot}}{N_{tot}+5}\right] \tag{14.5}$$

where AMBI is the sensitivity component and $SN=\ln(S)/\ln(\ln(N))$ is a diversity index. N is the number of individuals in the sample and S the number of species in the sample. The diversity component is normalised with a factor of 2.7, which is the maximum value of SN observed in the samples. NQI can be calculated for values of $N>1$ and $S<N$ and can attain values between 0 and 1. However, for $S=N$ and $S<4$ SN is not a continuous function and calculation of NQI results in outlying index values. Calculation of the diversity component is independent of the relative dominance (unevenness) of species in the sample.

The Swedish Index (IBQI): This index (Rosenberg et al., 2004; Blomqvist et al., 2006; Anonymous, 2008) has a sensitivity component based on a classification of sensitive or tolerant species which is used to calculate a weighted average of the sensitivity of a species assemblage. It accounts for species richness by a factor that scales with logarithm to the number of species and the index is modified for low densities:

$$\text{BQI} = \left[\sum_{i=1}^{S_{classified}} \left(\frac{N_i}{N_{totalclassified}} * Sensitivityvalue_i\right)\right] * \log_{10}(S+1) * \left(\frac{N_{total}}{N_{total}+5}\right) \tag{14.6}$$

where $S_{classified}$ represents the number of species classified; $Sensitivityvalue_i$ denotes the sensitivity of the ith species which ranges between 1 and 15. Low values indicate a high proportion of tolerant species, and high values indicate a high proportion of sensitive species. S is the number of species (including not classified) per sample. N_i is the number of individuals of the ith species; $N_{totalclassified}$ is the total number of classified individuals and N_{total} is the total number of individuals per 0.1 m^2.

Benthic macrofaunal data from seven marine coastal areas with gradients of different pollution sources were analysed by the authors (Josefson et al., 2009) with these IBIs to assess ecological benthic quality. The indices were found to respond in a similar way to the stressors, irrespective of whether the stressor was organic pollution, oxygen deficiency or heavy metal contamination. Correlations between the three indices were generally high.

14.7.4. Relative Efficacy of Five Fish-based Indices in Assessing the Ecological Status of Marine Environment

Henriques et al. (2008) adapted five estuarine multi-metric indices to the marine environment and applied them in three types of substrates, analysing the metrics responsible for the obtained patterns of ecological status. The indices were:

1. The community depradation index, CDI (Ramm, 1988);

2. The biological health index, BHI (Cooper et al., 1994);
3. The estuarine biotic integrity index, EBI (Deegan et al., 1997);
4. The estuarine fish community index, EFCI (Harrison and Whitefield, 2004); and
5. The transitional fish classification index, TFCI (Coates et al., 2007).

The performance of the indices was tested in the evaluation of ecological status at 15 sites: five from sandy subtidal zones, five from rocky subtidal zones and the remaining five from rocky intertidal zones. The choice was based on habitat characteristics (type of substrate), sampling method used, geographical localisation and depth of each area. It was expected to facilitate the comparison of the ecological status possible, resulting in the application of the adapted indices at each substrate. All the sampling methods of the chosen studies were the standard ones.

The metrics of the EBI, EFCI and TFCI were adapted to a marine environment by replacing estuarine metrics with equivalent marine functional groups (marine metrics). However, for some of the metrics it was not possible to obtain a direct equivalence between estuarine and marine metrics because of the peculiarities of the marine environment and their fish communities. For these cases, the marine metrics chosen as equivalent were based on the initial concept that each author used to measure ecological status.

The results obtained in the application of the adapted estuarine indices (Table 14.26) show that different indices gave different grading to identical sites; in other words, the assessments were inconsistent, sometimes even contradictory. Additionally, the indices seemed to be swayed by the substrate type. Generally, the index scores gave lower ecological status to sandy substrate and the highest to the rocky intertidal areas.

Among the sandy substrate zones evaluated (Table 14.26, A), the highest values were obtained for Algarve, where none of the indices gave scores below moderate. The lowest ecological status corresponded to the deep sandy zones 1, 2 and 3 (20–100 m). For this substrate,

TABLE 14.26 Ecological Status Obtained with Different Indices Tested by Henriques et al. (2008). A Represents Sandy Zones; B Rocky Zones; and C Intertidal Rocky Zones. Index Scores are in Parenthesis

Index		Sand-Zone 1	Sand-Zone 2	Sand-Zone 3	Sand-Tejo	Sand-Algarve
A	CDI	Poor (7.56)	Poor (6.80)	Poor (7.10)	Poor (7.08)	Moderate (5.28)
	BHI	Poor (2.00)	Poor (2.77)	Poor (2.56)	Poor (2.49)	Moderate (4.29)
	EBI	Poor (20)	Good (45)	Good (50)	Good (55)	Excellent (70)
	EFCI	Poor (38)	Good (46)	Moderate (44)	Good (48)	Good (56)
	TFCI	Poor (0.48)	Moderate (0.63)	Moderate (0.56)	Moderate (0.58)	Good (0.71)
B	CDI	Good (3.87)	Poor (7.13)	Moderate (5.35)	Poor (7.86)	Excellent (1.90)
	BHI	Moderate (5.64)	Poor (2.40)	Moderate (4.18)	Bad (1.67)	Good (7.63)
	EBI	Excellent (65)	Moderate (30)	Good (55)	Bad (10)	Excellent (65)
	EFCI	Good (54)	Moderate (44)	Good (56)	Poor (38)	Excellent (64)
	TFCI	Good (0.71)	Poor (0.48)	Moderate (0.67)	Poor (0.46)	Excellent (0.83)
C	CDI	Good (3.77)	Excellent (1.95)	Good (3.77)	Excellent (1.34)	Excellent (0.73)
	BHI	Moderate (5.47)	Good (7.30)	Moderate (5.47)	Good (7.90)	Excellent (8.51)
	EBI	Good (55)	Excellent (65)	Good (55)	Excellent (75)	Excellent (65)
	EFCI	Good (50)	Good (62)	Good (52)	Excellent (64)	Excellent (66)
	TFCI	Good (0.73)	Excellent (0.88)	Good (0.69)	Excellent (0.88)	Excellent (0.90)

CDI and BHI were in agreement, although these indices gave scores indicative of the lowest ecological status. Among the remaining indices, the most demanding one (for lower ecological status) was TFCI, followed by EFCI and EBI.

CDI and BHI were unable to assess the ecological status, compared to the remaining indices. This seemed to be due not merely to the simplicity of these indices, as they are only based on presence/absence data, but also to the strong influence of sample size in the determination of species richness estimates, which affects the computation of the Jaccard coefficient.

14.7.5. A Comparison of Biotic Indices and an IBI

Lavesque et al. (2009) compared the ability of three univariate biotic indices (BIs) and an IBI (Table 14.27) to assess the ecological impact of the destruction of a *Zostera noltii* seagrass bed in Arcachon Bay (France) following sediment deposits.

In the author's study, none of the four indices passed muster. The classification of the sites by AMBI was not consistent with *in situ* observations. There were no real differences in AMBI values between control and impacted sites. At certain times, the control sites were even shown as more perturbed than impacted sites. Moreover, the ecological status of one of the sites was always shown to be 'acceptable' ('good' or higher ecological quality status), whereas that site appeared to be the most perturbed according to visual observations such as disappearance of vegetation and change in sediment type. The index BENTIX classified both control sites and impacted sites as 'not acceptable' ('moderate' or worse ecological quality status) and the impacted sites did not even receive lower scores of BENTIX (i.e., more degraded) than control sites (except at $t+8$ months). Hence, the index was unable to detect any perturbation in impacted sites. On the other hand, it scored the majority of the sites as 'acceptable', even the obviously perturbed ones.

The authors have then gone on to describe a new IBI acronymed MISS (Macrobenthic Index of Sheltered Systems). It uses a set of 16 metrics gathered into three categories: community descriptors, trophic composition and pollution indicators.

In another study, Prato et al., (2009), who examined two of the indices which were also

TABLE 14.27 Indices Used in a Comparative Study by Lavesque et al. (2009)

The Indices	Number of Ecological Groups	Author(s)	Computation of the Indices
AMBI	5	Borja et al. (2000) Glémarec and Hily (1981)	$0\ EG_I + 1.5\ EG_{II} + 3\ EG_{III} + 4.5\ EG_{IV} + 6\ EG_V$ based on percentage of ecological groups
BENTIX	2	Borja et al. (2000) Glémarec and Hily (1981)	$6\ EG_{I\&II} + 2\ EG_{III\text{-}V}$ based on percentage of ecological groups
BOPA	2	Borja et al. (2000) Dauvin and Ruellet (2007) Glémarec and Hily (1981)	$\log_{10}[(fp/fa + 1) + 1]$ based on ratio of ecological groups
M-AMBI	5	Borja et al. (2000) Glémarec and Hily (1981)	Multi-metric analysis using AMBI, H′ and S based on percentage of ecological groups

EG: ecological groups; fp: opportunistic polychaetes frequency; fa: amphipods frequency.

used by Lavesque et al. (2009) in the study described above – AMBI and BENTIX – along with the same multivariate IBI that Lavesque et al. (2009) used, viz., M-AMB, also recorded a lack of sensitivity on the part of the indices in distinguishing the extent of pollution of two different lagoons.

Interestingly, in these studies, the univariate AMBI seemed to provide a more suitable evaluation of the ecological quality status of 'slightly polluted lagoons' than the multi-metric M-AMBI.

14.7.6. A Comparison of the Performance of BIs, IBI, RIVPACS and Expert Judgement

Ranasinghe et al., (2009) have evaluated the performance of five benthic indices that rely on different sets of community or species composition measures, and have compared their site assessments to the professional judgement of nine benthic experts.

The five indices used by the authors are:

1. The Relative Benthic Index (RBI; Hunt et al., 2001),
2. The Index of Biotic Integrity (IBI; Thompson and Lowe, 2004),
3. The Benthic Response Index (BRI; Smith et al., 2001, 2003),
4. The River Invertebrate Prediction and Classification System (RIVPACS; Wright et al., 1993; Van Sickle et al., 2006) and
5. The Benthic Quality Index (BQI; Rosenberg et al., 2004).

Of these, RBI and IBI are based on community measures, BRI and RIVPACS on species composition and BQI on both.

The comparisons were conducted in two ecologically and geographically distinct habitats: (a) the marine bays of southern California and (b) the polyhaline San Francisco Bay. The objective was to evaluate the relative performance of these indices alone and in combination in each habitat, in relation to assessments by nine benthic experts.

The performance of the indices was evaluated in four steps:

1. Data for sampling sites in each of the two habitats were identified, acquired and adjusted to create consistency across sampling programmes.
2. The five benthic indices were calibrated using a common set of data for all indices.
3. Threshold values were selected for each index to assess benthic condition on a four-category scale.
4. Performance of the indices, and all possible index combinations, was evaluated by applying them to independent data and comparing the condition assessments to that of nine benthic experts. The experts were given species abundances, together with habitat, depth, salinity and sediment grain-size information for 35 sites that were not used in index development or calibration. The experts were asked to (a) rank the sites in each habitat from best to worst condition and (b) classify each site on the four-category scale of benthic condition to which the benthic indices were calibrated.

Index condition rank order was evaluated against the average expert rank order using Spearman rank correlation coefficients based on index values for all 35 evaluation samples. Associations among the five indices were also evaluated, and compared to associations among the experts, using Spearman rank correlation coefficients. All the index values used for the index condition rank order evaluation were used in this analysis.

The studies revealed that indices that include measures of species composition generally outperformed indices that include only community measures. This appeared consistent with Weisberg et al. (1997), who had found that relative dominance of pollution-tolerant and pollution-sensitive species are the metrics that have the best relationship to pollution gradients. Pearson and Rosenberg (1978) suggest that the initial benthic response to low levels of stress is a shift in species composition, with shifts in community metrics, such as loss of species richness and biomass, manifesting at later stages of stress. Thus, indices based on community metrics should be more effective at differentiating sites subject to high levels of stress, but less effective at differentiating sites with low to intermediate levels of stress that are more typical of the estuarine sites encountered in California.

Combinations of indices consistently outperformed individual indices. It may be due to the fact that use of multiple indices incorporates a larger number of metrics and presumably balances the occasional erratic behaviour of individual metrics. Some of the individual indices had showed biases, with RBI assessing samples as more disturbed than the experts and IBI behaving the opposite, but the use of multiple indices apparently evened out these kinds of biases.

Interestingly, none of the individual indices performed as well as the average expert in ranking sample condition or evaluating whether benthic assemblages exhibited evidence of disturbance. However, several index combinations outperformed the average expert. When results from both habitats were combined, two four-index combinations and one three-index combination performed best. It must be emphasised that this performance difference among several combinations was perhaps not high enough to justify substantial extra input of effort and costs that might entail when multi-index combinations are used.

14.7.7. Comparison of Two Richness Metrics, Three Biotic Indices and Two IBIs

Sanchez-Montoya et al. (2010) compared two richness metrics (total number of families and number of the Ephemeroptera, Plecoptera and Trichoptera families), three biotic indices (IBMWP, IASPT and t-BMWQ) and two recently proposed IBIs for Mediterranean streams (ICM-9 and ICM-11a or IMMi-L). The sensitivity of these indices and metrics to a multiple stressor gradient, which reflected the main pressures present in the study area was studied. For this purpose, data from 193 sites belonging to five different Mediterranean stream types present in 35 basins were studied.

The results showed that the adjusted regression coefficients for all seven metrics in the exponential regression models were higher than linear ones, thus indicating an exponential relationship between metrics and the environmental alteration. The two IBIs presented higher regression coefficients ($r^2 = 0.590-0.669$) than the three biotic indices ($r^2 = 0.524-0.574$) and the two metrics ($r^2 = 0.471-0.525$), thereby showing a better response to a stressor gradient in Mediterranean streams than simpler indices. Between the two IBIs, ICM-11a provided higher regression coefficients.

In another study (Tataranni and Lardicci, 2010) which compared the performance of biotic indices and an IBI in Mediterranean coastal areas, it was found that the response of the indices had 'complex trends'.

14.7.8. How Sensitive IBIs and Their Interpretations are to the Variability that Occurs in Repeat Biological Samples?

Even as new IBIs continue to be developed and the existing ones continue to be applied to ever larger geographic contexts and scales, much fewer studies have directed attention to the inherent statistical properties of the IBIs —; their precision, accuracy and sensitivity to natural variation in biological samples (Fore et al., 1994; Carlisle and Clements, 1999; Blocksom, 2003). To cover this knowledge gap, Dolph et al. (2010) have used a 'bootstrapping' approach to quantify the response of IBIs to random sampling variation.

Bootstrapping, originally described by Efron (1979, 2003), is a computer-intensive statistical technique used to estimate variability of a statistic when the actual distribution is unknown, such as when that statistic is determined from a single random sample.

For this study, the authors used IBIs developed earlier to assess the health of fish communities in the river basins of Minnesota by the Minnesota Pollution Control Agency (MPCA, 2007) as part of the state's biological monitoring programme (Niemela and Feist, 2000, 2002). Each IBI comprised of a slightly different set of metrics and metric scoring criteria (Table 14.28).

Gist of the Studies Conducted

Bootstrapping: Bootstrapping, as explained by Dolph et al. (2010), creates replicate samples from a single sample by randomly resampling from the original sample with replacement (Efron, 2003; Manly, 2007). For example, in the case of a fish sample from a single stream site, the bootstrapping algorithm randomly selects one individual specimen at a time, adds it to a new replicate sample and replaces it in the original sample, where it again has the same probability of being selected as any other specimen. Resampling is repeated in this manner until the number of individuals in the replicate sample is equivalent to the number of

TABLE 14.28 Metrics of Which the IBIs Used in the Comparative Study of Dolph et al., (2010) Were Constructed

Metric	Anticipated Response to Disturbance	Stream Classes Using Metric in IBI Score
Total no. of species	Decrease	All classes
No. of benthic invertivore species	Decrease	2–5
No. of darter species	Decrease	3–5
No. of darter, sculpin and madtom species	Decrease	8–9
No. of invertivore species[a]	Decrease	1; 6–9
No. of minnow species[a]	Decrease	1–2; 7
No. of omnivore species	Increase	3–5
No. of sensitive species	Decrease	2–5; 7–9
No. of wetland species[a]	Decrease	6–8
No. headwater species[a]	Decrease	1
% of total abundance comprised of individuals of the two most abundant taxa	Increase	1–2; 6–7
% of individuals with deformities, lesion or tumors	Increase	All classes
% of individuals classified as omnivore species	Increase	9
% of individuals classified as lithophilic spawners	Decrease	1–5; 7–9
% of individuals classified as piscivores	Decrease	3–5; 8–9
% of individuals classified as tolerant	Increase	All classes
No. of individuals per metre[a]	Decrease	All classes

[a] *These metrics exclude species that are considered tolerant to disturbance.*

individuals in the original sample. The result is a series of samples that contain 'collections of fish that could have been caught at the same site and time by electrofishing' differing only by random variation. An IBI score can be determined for each individual bootstrap replicate sample, and the mean and variance can be calculated across all bootstrap replicates for a given site visit. The end result is the range of possible IBI scores that a stream site could receive, given variability in the fish collection that may arise from random sampling effects.

Dolph et al. (2010) created 1000 bootstrap samples for each original fish sample using statistical software as this number of replicates is generally considered sufficient for confidence interval generation (Carpenter and Bithell, 2000).

Of the many methods available for determining confidence intervals from bootstrap data, Dolph et al. (2010) used the percentile method based on simplicity of use and to enable comparisons with an earlier study by Fore et al. (1994). To estimate a 95% percentile confidence interval for an IBI score at a given stream site, the authors first sorted IBI scores from the replicate bootstrap samples into ascending order. Statistically, for 1000 replicate scores, the 25th ordered value represents the lower bound of the confidence interval, and the 975th ordered value represents the upper bound (Carpenter and Bithell, 2000). Confidence interval length was determined by subtracting the lower bound from the upper bound value.

Implications of variability for impairment status: As the objective was to evaluate whether stream impairment decisions based on IBI scores are likely to change as a result of random sampling error, the authors determined how many of the 1000 IBI scores generated for each site visit indicated a different impairment status ('impaired', 'potentially impaired' or 'unimpaired') than the original IBI score. This number was divided by 1000 to calculate the proportion of bootstrap scores that diverged from the original score for a given site visit. Finally, these proportions were averaged across all 513 site visits.

Identifying covariates of IBI sensitivity: In addition to quantifying the IBIs' sensitivity to random sampling errors, the authors sought to identify aspects of fish samples that were related to this sensitivity. In particular, they sought to investigate whether IBI sensitivity was affected by aspects of community abundance or richness. For this, they used simple linear regression to assess the relationship between confidence interval length and several possible covariates, including the total number of fish in a sample (i.e., sample size), the total number of taxa in a sample (i.e., species richness), Pielou's evenness and the number of species that occurred as single individuals in a sample (i.e., the number of singletons).

Comparative performance of IBIs: The authors evaluated whether IBI confidence interval length varied among the nine different stream classes using ANOVA followed by Dunnett's modified Tukey–Kramer (DTK) multiple comparison test (Dunnett, 1980) to determine whether some combinations of metrics were more sensitive to random sampling error than others.

Continuous scoring of metrics and bias among IBI scores: The component metrics of the fish IBIs used by the authors have discretised scales which are discontinuous; the metric values typically have intervals of 0, 2, 5, 7 or 10. The authors converted these methods to continuous scoring methods and evaluated whether this led to reduced bias among replicate IBI scores compared to the discontinuous method.

To achieve this conversion, a linear piecewise polynomial was defined in which continuous scores were anchored to the discontinuous discrete scores at the midpoint of the metric values for a given discrete score (i.e., continuous and discrete scores were the same at this midpoint value) and were everywhere continuous (de Boor, 1978). The authors then recalculated IBI scores for each bootstrap replicate sample using the new scoring system.

Random sampling error versus variability over time: In one of the basins studied by the authors (the St. Croix River basin), fish samples were available over four consecutive years at 12 unique stream sites. The authors compared confidence intervals derived from bootstrapping for these sites with those determined by calculating variance over time. This was done to understand the relative contributions of random sampling error and temporal variation to overall variability in IBI scores.

Salient Findings

It was seen that the IBI sensitivity as measured by the IBI confidence interval length in response to random sampling error ranged from 0 to 40 points (mean = 11) across all 513 stream site visits included in study. In 510 (99.4%) of these site visits, this random sampling variability was not sufficient to change the degree of impairment from 'unimpaired' to 'impaired', or vice versa, relative to the site's original IBI score. Only 0.3%, 0.4% and 1.0% of replicate samples at three sites were classified as 'impaired' when the original sample was classified as 'unimpaired'.

Within the impairment outcomes, also, only 11.3% of bootstrap replicate samples yielded IBI scores that indicated a different degree of impairment than the original IBI score. For sites with original scores that fell within 20 points of the impairment threshold, 16.0% of bootstrap replicate samples yielded a different impairment outcome than the original sample. Confidence interval length was not significantly related to species richness or Pielou's evenness for sites in either river basin.

Across all sites, mean IBI scores of bootstrap replicates were significantly lower (i.e., exhibited negative bias) than the original IBI score but replicate samples for sites with low original scores were more likely to exhibit zero bias (i.e., replicate samples gave approximately the same score relative to the original sample) than sites with high original scores in which almost uniformly negative bias was displayed (i.e., replicate samples underestimated the score).

When IBIs were recalculated using the continuous scoring method, a significant negative correlation between IBI score and bias was still evident but the mean bias was substantially reduced and mean IBI scores of bootstrap replicates were no longer significantly different from the original IBI scores.

Nearly a quarter of IBI scores varied by more than 15 points as the result of random sampling error, with the most variable score having a range of 40 points.

Only one in 10 replicate IBI scores for a given site, on average, indicated an impairment outcome different from the original score. Moreover, random sampling variability was not sufficient to change a site's status from 'unimpaired' to 'impaired' or vice versa in over 99% of stream sites even though, in some cases, the IBI score changed from 'unimpaired' or 'impaired' to 'potentially impaired', or vice versa.

The number of replicate samples that produced different outcomes relative to the original sample increased only for sites with IBI scores close to the impairment threshold. Overall, it was seen that the effects of random sampling error on IBI score was not likely to change the impairment status of a stream site in most cases.

When random sampling variability did change the impairment outcome, type I error (underestimating stream health) appeared to be more common than type II error (underestimating stream impairment). Only the quality of less-impacted sites tended to be underestimated, whereas the quality of highly degraded sites tends to be more accurately conveyed by bootstrap replicate samples.

The findings imply that impairment decisions based on IBI scores were conservative in terms of protecting stream health. In other words, by using IBI scores to determine site impairment, management agencies are more likely to list unimpaired sites as impaired or potentially impaired than they are to fail to list impaired sites. If the goal among resource managers were to protect water resources before they become severely degraded, this type of conservatism would be appropriate.

A significant finding was that when metrics were scored using linear piecewise continuous curves, mean bootstrap IBI scores matched more closely with the original IBI scores. Hence, this study makes another case to justify the adoption by management agencies of continuous scoring methods instead of the more conventional discrete scoring method for new IBIs.

Few significant differences in confidence interval length were found among the nine different stream classes, suggesting that the use of different combinations of metrics for IBIs in each stream class did not result in dramatic differences in IBI sensitivity to sampling error.

14.7.9. In Summary

In summary, the eight examples of BI–IBI and inter-IBI comparison invoke the following pointers:

1. On balance, IBIs do reveal greater capability to assess the impact of different stressors on the ecological health of water bodies than BIs, but exceptions do exist. Much depends on whether at least one constituent metric of an IBI is as, or more, responsive to a given stressor as the competing BI.
2. More often than not, different IBIs are able to 'read' the ecological status with similar judgement; different IBIs can differ on degrees of impairment but it is not often that a site classified as 'bad' by an IBI is classified as 'good' by another IBI.

A few studies, for example of Herbst and Silldorff (2006), bring out that different IBIs are likely to generate similar bioassessments even if the data used by different IBIs were based on different sampling methods and other experimental protocols. An IBI and observed: expected (OA) estimates of macroinvertebrates in Montana streams showed that both methods had similar consistency and repeatability (Stribling et al., 2008).

14.8. THE PRESENT AND THE FUTURE OF IBI

In its basic approach, an IBI aims to integrate multiple biological indicators to measure and communicate the biological condition of a water body, and thereby the ecological health of that water body. Just as a physician relies on several diagnostic tests, not just one, to diagnose an illness, so do IBIs to diagnose the condition of different types of watercourses. Connecting many previously disconnected dimensions of water science and policy, the indices have been tested and refined over nearly two decades and now provide the foundation of biological monitoring programmes – in USA and, increasingly, in other countries (Atazadeh et al., 2007; Liu et al., 2008; Acosta et al., 2009; Raburu et al. 2009; Costa and Schulz, 2010; Schmitter-Soto et al., 2011). IBIs have been applied on every continent except Antarctica; in developing as well as developed countries; and in basic science, resource management, engineering, public policy, legal and community volunteer arenas (Hughes and Oberdorff, 1999; Simon, 1999; Lussier et al., 2008; Southerland et al; 2009; Schwartz et al., 2009; Pollack et al., 2009; Pei et al., 2010; Launois et al., 2011).

Some of the IBIs such as the one of Karr and Chu (2000) based on benthic invertebrates (B-IBI) have been applied in areas from the United States to Japan (Rehn, 2009).

Based on empirically defined 'dose–response' relationships between particular human influences and the biological responses they provoke, IBIs are generally regarded as among the most commonly used – and among the most effective (Simon, 1999 Wu et al., 2009; Smucker and Vis, 2009; Zalack et al., 2010; Miranda and Hunt, 2011) – biological monitoring approaches.

The range of application of IBIs has been very broad as reflected from the following illustrative examples.

Bhagat et al., (2007) evaluated two IBIs developed for fish in Great Lakes coastal wetlands dominated (>50% cover) by *Typha* (cattail) and *Schoenoplectus* (bulrush) vegetation. Sites subject to low levels of anthropogenic influence had favourable IBI scores. The *Typha*-specific IBI showed a marginally significant negative correlation with population density and residential development The *Schoenoplectus*-specific IBI negatively correlated most strongly with nutrient and chemical inputs associated with agricultural activity and point-source pollution. All in all, the IBIs appeared to indicate effects of some, but not all, classes of anthropogenic disturbance on fish communities. The authors feel that calibrating these measures against specific stress gradients is likely to allow one to interpret the sources of impairment, and thereby use the measures beyond a simple identification of impaired sites.

Application of a diatom-based IBI on the Gharasou River in western Iran, by Atazadeh et al., (2007) showed that a Triophi Diatom Index (TDI) significantly correlated with measures of human disturbance at the sites (e.g., PO_4–P, NO_3–N and dissolved oxygen) as well as to biomass measures (chlorophyll-*a*, ash-free dry mass and biovolume). The sensitivity of TDI and its component metrics to environmental stressors reflected the potential of this index for monitoring ecological conditions in streams in Iran and to aid diagnosis of the cause of their impairment.

Two studies, by Catalano et al., (2007) and Maloney et al., (2008), employed IBIs to see whether dam removal improves the ecological status of the concerned streams. Catalano et al., (2007) found that after dam removal, biotic integrity scores of the fish-based IBI (possible range = 0–100) increased by 35–50 points at three of the four former impoundments as a result of decreases in percent tolerant species, increases in the number of intolerant species and in some cases, increases in species richness. Fish assemblage shifts were muted at a fourth, lower-gradient impoundment site, indicating that responses differed among dam sites within a river system. In tailwater areas, post-removal assemblage shifts were transient; biotic integrity and species richness declined initially but then recovered at two of the three sites within 2 years after dam removal.

While reviewing the existing fish assemblage indicators and methodologies, Roset et al. (2007) have made an interesting comparison of the types of dominant metrics used in European IBIs with North-American IBIs. As may be seen from Figure 14.6, the mean relative number of metrics belonging to the trophic group is roughly the same (about 20%), but the relative importance of abundance and reproduction metrics is lower in North American than in European IBIs (7.6% against 16.3%, respectively), while it is the opposite for species composition metrics (55% in North American samples against 41% in Europe).

The authors (Roset et al., 2007) have attributed the differences to the difference in species richness in the continents. Assessing biotic integrity in the generally species-poor European rivers requires focussing more frequently on abundance metrics and reproduction. For example, in the European IBIs, more attention has been paid to metrics dealing with age or size class distributions for the dominant and/or most intolerant species, than in North-American IBIs.

A study of Maloney et al., (2008) reveals that after the breach of a low-head dam, relative abundance of Ephemeroptera, Plecoptera and Trichoptera (EPT) increased, whereas relative abundance of Ostracoda decreased, in the former impoundment (IMP) to levels comparable to free following (FF) sites. High variation in other metrics (e.g., total taxa and diversity) precluded determination of an effect of the breach on these aspects of the assemblage. However, non-metric multidimensional scaling (NMDS) ordinations indicated that overall macroinvertebrate assemblage structure at the former IMP shifted to a characteristically FF assemblage 2 years following the breach. Total fish taxa and a regional fish index of biotic integrity became more similar in the former IMP to FF sites following the breach. Effects of the breach to the site immediately below the former dam included minor alterations in habitat (decreased flow rate and increased particle size) and short-term changes in several macroinvertebrate metrics (e.g., decreased assemblage diversity and EPT richness for first post-year), but longer-term alterations in several fish metrics (e.g., decreased assemblage richness for all three post-years; decreased density for first two post-years).

Both the studies indicate that IBIs reflect the impact of dam removal and that dam removal is feasible as a restoration practice for impaired streams and rivers.

Castela et al. (2008) combined benthic macroinvertebrate-derived metrics and a biotic index as measures of structural integrity with oak litter decomposition and associated fungal sporulation rates as measures of functional integrity to study habitat degradation gradient in a small stream. The biotic index, invertebrate metrics, invertebrate and fungal communities' structure and sporulation rates all were able to discriminate upstream sites from the downstream ones. Although both functional and structural approaches gave the same results for the most impacted site, they were complementary for moderately impacted sites thereby indicating the benefit of incorporating functional measures with biotic indices in evaluations of stream ecological integrity.

The possibility of developing biologically based criteria for linking sediment amounts with aquatic vertebrate response was explored by Bryee et al. (2008). They related an aquatic vertebrate index of biotic integrity with a measure of the areal percentage of streambed surficial fines ($\leq$0.06 mm). The association suggested that fine sediment limits the biological potential of mountain streams. The use of quantile regression to model the upper limit of IBI response predicted a 4.7% decline in IBI for each 10% increase in areal surficial fines. Further studies indicated that streambed areal surficial fine sediment (particle size $\leq$ 0.06 mm) levels of 5% or less retain habitat potential for sediment-sensitive aquatic vertebrates in mountain streams. Hence, it is possible to link sediment amounts with IBI scores.

Genet and Olsen (2008) assessed depressional wetland quantity and quality of 40 randomly selected sample sites using plant and macroinvertebrate indices of biological integrity (IBI). The wetlands chosen for the study have experienced an estimated 56% reduction in number, equivalent to a 21% decrease in area in the watershed over the last 20 years. Of the remaining wetland area, an estimated 91% has been officially designated as impaired. The findings bring it out that management practices with the goal of increasing suitable habitat for native wetland plant and animal communities should focus on restoration of drained wetlands as well as improvement of existing wetlands to maximise outcomes.

Setting out to determine the relative strength of relationships between reach-scale habitat characteristics and important biological metrics, Fayram and Mitro (2008) used artificial neural network models to examine the relationships between 11 reach-scale habitat variables and (1) catch per effort (CPE) for brook trout *Salvelinus fontinalis*, (2) CPE for brown trout *Salmo trutta* and (3) coldwater fish index of biotic integrity (IBI) scores. The authors also performed quantile regression to evaluate whether any one of the three response metrics of interest was limiting to any of the other variables. The results show that IBI scores and trout densities have strong positive correlation. The studies build upon a previous report of Griffin and Fayram (2007) in which mean length of brown trout had negatively correlated with the IBI score while mean length of brook trout had shown no correlation.

Gomez et al. (2008) explored the effects of a textile industry effluent on water quality, habitat quality and structural and functional responses of benthic communities in a lowland stream. The effluent was seen to modify the structure of the microbenthic assemblages downstream, increase the density of organisms and the biomass of primary producers but diminish the species richness. The oxygen consumption of the microbenthic community was 3 times higher downstream of the effluent and abnormal frustules of diatoms were noticed. The richness and abundance of invertebrate taxa were lower at the impacted site. The invertebrate modes of existence and the functional feeding groups were also significantly affected. This study provides a useful frame of reference for assessment of lowland streams with high water residence time and a notable development of hydrophytes.

An IBI along with hydrologic indices has been used to develop a rapid protocol for the evaluation of the ecological status of Andean rivers (CERA) from the Northern Andes, Venezuela, through the Altiplano in the Central Andes, Bolivia (Acosta et al., 2009). This protocol had already been used in 45 sampling sites in the Guayllabamba River Basin in Ecuador and in 42 sampling sites in the Cañete River Basin in Peru. Besides evaluating the benthic macroinvertebrates, the river habitat and riparian vegetation were also evaluated. The CERA protocol was able to identify the anthropogenic perturbation gradients as well as the natural variability of the reference sites in both countries. In another effort cutting across national boundaries, a combination of benthos-based biotic indices and IBIs was employed in an intercalibration exercise of the surface water ecological quality assessment involving six countries: Cyprus, France, Greece, Italy, Slovenia and Spain (Ambrogi et al., 2009).

Cookson and Schorr (2009) examined relationships of watershed housing density with instream environmental conditions and fish assemblage attributes in Tennessee, the USA, using an IBI. Housing density was directly correlated with stream temperature, variation in discharge, fine sediment depth and abundances of introduced and tolerant species. It was inversely correlated with dissolved oxygen, pH, variation in depth, substrate diversity and native species richness. Most reaches exhibited signs of degradation. The results underscore the negative effects of residential development on water quality, hydrology, habitat complexity and fish assemblages in a suburban stream.

In another study originating from the USA, DeGasperi et al. (2009) used the Pacific Northwest Benthic Index of Biological Integrity (B-IBI) to represent the gradient of response of stream macroinvertebrates to urbanisation. The authors found that eight hydrologic metrics were significantly correlated with B-IBI scores.

Doustshenas, et al. (2009) compared macrobenthos and organic content to assess the condition of the ecological health of Khowr-e-Musa bay, Pakistan, as impacted by two developmental periods, and found evidence of severe impairment.

Helms et al. (2009), when studying seasonal variability in land-use impacts in streams of Georgia, USA, found that biotic integrity declined with decreasing % forest cover throughout the seasons. Multiple regression models and partial correlation analysis revealed that physico-chemical and benthic habitat variables explained more variation in macroinvertebrate metrics throughout the seasons than did hydrological variables at most sites. The influence of anthropogenic land-use impacts on macroinvertebrate assemblages appeared to be consistent throughout the year and had the effect of dampening seasonal changes in the benthic assemblages.

Lenhardt et al. (2009) demonstrated that IBIs could be effective in the assessment of reservoir ageing by applying a fish-based IBI to the Medjuvrsje Reservoir, which is one of the oldest Serbian reservoirs, formed in 1953. It was seen that the relative abundance of omnivorous, phytophilic and tolerant species increased, while that of lithophilic, intolerant and rheophilic species decreased during the 45 years of the reservoir life. The total IBI decreased from 44 in 1955 to 24 in 2000, while the sediment deposition rate increased from 26.8% in 1963 to 70.4% in 2005. There was a significant negative correlation between IBI and sediment deposition rate (Figure 14.5).

Llansó et al. (2009) used the Chesapeake Bay benthic index of biotic integrity (B-IBI) to develop a method for impairment decisions based on B-IBI. The method was applied to 1430 probability-based benthic samples in 85 Chesapeake Bay segments. It indicated that twenty-two segments were impaired.

Pollack et al. (2009) developed the Freshwater Inflow Biotic Index (FIBI) to determine how changes in freshwater inflow affect benthic populations, which in turn reflect the ecological condition of an estuary. Based on benthic succession theory and long-term data, 12 biotic metrics were chosen that characterised benthic community structure in response to inflow regimes. FIBI and hydrological variables were significantly correlated, indicating that benthic communities respond to changes in salinity and do so in a relatively predictable manner. When inflow is reduced (i.e., salinity increased), upstream communities seem to take on characteristics of downstream communities. FIBI appeared to successfully characterise effects of a salinity gradient in the Lavaca-Colorado estuary.

Chin et al. (2010) have examined whether geomorphological responses are linked to ecological responses in restored urban streams. The study covered three stream channels in Austin, Texas, which have been restored since 1998

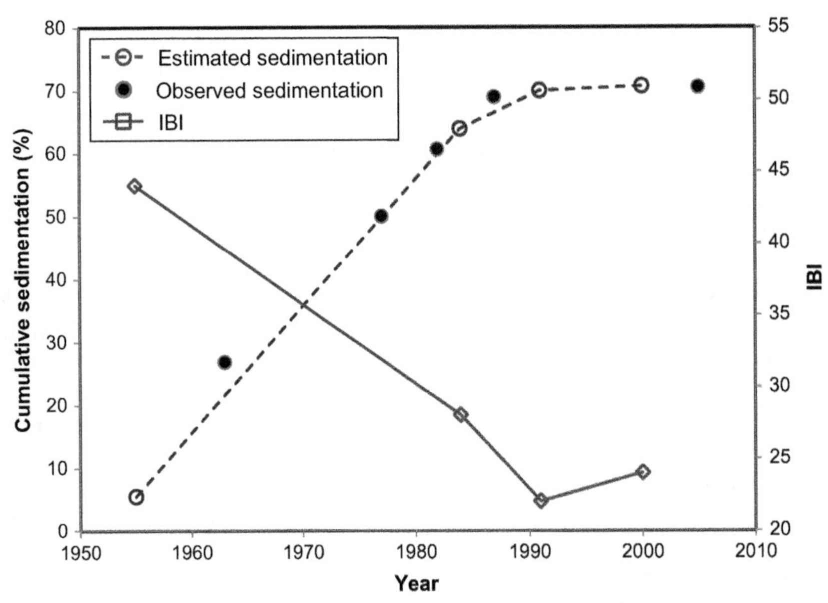

FIGURE 14.5 Cumulative sedimentation in the Medjuvrsje reservoir and the concurrent fall in the IBI as recorded by Lenhardt et al. (2009).

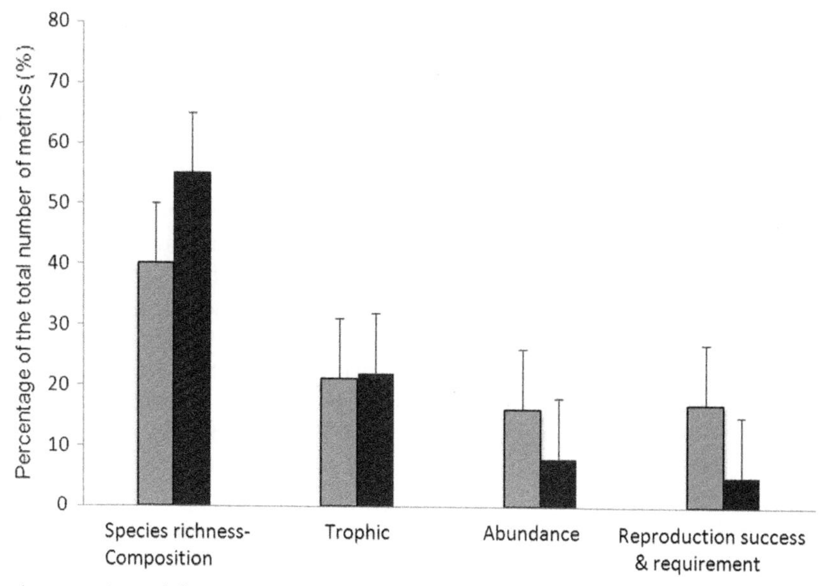

FIGURE 14.6 A comparison of the mean (and SE) relative importance (%) of metric categories in European (grey bars) and North-American (black bars) IBIs, as reported by Roset et al. (2007).

with riffles and steps and riparian planting along graded banks. These restoration measures have led to increased channel widths and depths, larger cross-sectional areas and inferred lowered velocities and unit stream power. It was seen that the geomorphological changes have induced changes for the better in functional feeding groups of macroinvertebrate communities as reflected in metrics of Benthic Index of Biotic Integrity. The changes include greater percentages of grazers, filterers and collector–gatherers in the restored streams. A link between improvement in key ecological response variables (taxa richness, % EPT, % grazers and % chironomids) and better conditions in habitats (lower embeddedness, greater epifaunal cover, greater riparian vegetative width and more velocity–depth regimes) was seen when a multivariate analysis was performanced.

Davis et al. (2010) report an IBI-based study on the impact of release of water from coalbed natural gas (CBNG) development plants in two rivers. Several of their results suggest that CBNG development did not affect fish assemblages as the species richness and index of biotic integrity (IBI) scores were similar in streams with and streams without CBNG development. Nor was the overall biotic integrity related to the number or density of CBNG wells. Fish occurred in one stream that was composed largely or entirely of CBNG product water, and sentinel fish survived in cages at treatment sites where no or few fish were captured, suggesting that factors such as lack of stream connectivity rather than water quality limited fish abundance at these sites. Biotic integrity had declined from 1994 to 2006; however, declines occurred at both impact and reference sites, possibly because of long-term drought. Notwithstanding these general findings, the authors did find evidence that in some instances CBNG development seems to negatively affect fish assemblages, or may do so over time.

In a study exploring the response of indices based on invertebrates and diatoms to wetland condition in the West Coast of New Zealand, Suren et al. (2011) found a lack of strong correlation between measured wetland condition indices and either diatom or invertebrate community composition. It was possibly because neither index was dominated by variables directly influencing the aquatic component of wetland biota. The study underscores the need to identify the critical variables and to develop complementary wetland scoring systems that better reflect the status of small aquatic organisms.

Seasonal floods are part of the natural disturbance regime of many streams, but urbanisation increases their frequency and magnitude. In order to assess the extent to which such events can reduce biological integrity in urban streams, Coleman et al. (2011) evaluated the impact of hydrologic disturbance on fish and aquatic macroinvertebrates in 81 (56 urban/25 reference) Ohio streams. Hydrologic variables included annual and monthly 24-h rainfall maxima and computed annual peak discharge, with computation supported by GIS-based drainage area delineation and land cover characterisation. It was seen that the Ohio biological criteria for fish and macroinvertebrates measured during the late spring and summer were negatively impacted by annual peak discharge in urban streams as compared to reference streams. Marce and Fornells (2010) studying the Llebregat river basin in Spain also found that floods had some impact on the values of the biological indices but the magnitude of the impact was marginal and reversed quickly. This has prompted the authors (Marce and Fornells, 2010) to suggest that biotic integrity is influenced more by pollution than by hydrology.

Doll (2011) has employed an IBI and a Qualitative Habitat Evaluation Index to develop a methodology that can predict the probability of biological impairment based on routinely collected habitat assessments. Two models were constructed from a validation data set. The first predicted a binary outcome of impaired (IBI < 35) or non-impaired (IBI ≥ 35) sites and the second predicted a categorical gradient of impairment. The models were then validated with an independently collected data set, and successfully predicted biological integrity of the validation data set with an accuracy of 0.84 (binary) and 0.75 (categorical). The author feels that his models can be easily applied to other data sets from the Eastern Corn Belt Plain to aid in stressor identification by predicting the probability of observing an impaired fish community based on habitat. Predicted probabilities from the models can also be used to check the robustness of the conclusions that have already been reached.

Among most reports of falling IBIs, an exception is seen in the study of Brousseau et al. (2011). Their application of an IBI on fish survey data collected from 1988 to 2009 in the Bay of Quinte and elsewhere has indicated that (1) the Bay of Quinte has a relatively healthy fish habitat; (2) the Bay's IBI scores increased significantly between 1990 and 1999 due to changes in relative species richness; and (3) differences in fish communities were correlated with physical habitat attributes at survey locations. Data from both nearshore electrofishing and trap net surveys confirmed that the Bay of Quinte is able to support a highly productive and diverse fish community.

14.8.1. Thrust Areas

Some of the aspects that need to be addressed on a priority basis in the ongoing and future efforts associated with IBI development, include:

1. estimating and integrating temporal variability of the assemblages of the chosen taxa when describing or modelling reference conditions;
2. standardising the organism sampling methods at the scale of the assessment, for both reference and study sites;
3. electing reference sites based on standardised criteria, including wherever appropriate reliable historical information is available;
4. integrating statistical methods and techniques of artificial intelligence in reducing sampling efforts while enhancing

representativeness and reproducibility of the assemblage surveys;

5. developing methods to define objectively the optimum number of metrics to be included in the final index for different types of water bodies and stressor assessments;
6. preferring continuous scoring methods over discrete scoring methods, to lower the risk of type I error: continuous scoring methods also allow for a more sensitive composite index; and
7. providing an assessment of the uncertainty inherent in the new IBIs: assessing variability and placing a bound on uncertainty may help identify where methods are lacking. Bootstrap statistical analyses may help bound uncertainty by assessing the variability of index results. A qualitative and quantitative analysis of an extensive list of potential metrics with levels of disturbance may also reduce uncertainty.

14.9. THE NOW WELL-RECOGNISED ATTRIBUTES OF IBI

It is now well recognised that IBI, along with its component metrics, has the following multiple attributes:

- It tells us whether we are maintaining water bodies, water supply and flow through the water cycle, along with the vital resources the water cycle supports.
- It provides us with a system that is applicable to streams, rivers, lakes and wetlands, as well as estuaries and coastal marine systems.
- It integrates elements that conventionally have been fragmented in water policy and decision making, such as water quality and water quantity and surface water and groundwater.
- It can stand up to prior legal scrutiny and is actually improving the legal and regulatory approaches.
- It is used by governments to protect the public's interest in water resources.
- It is simple to develop and use, demanding no advanced technologies beyond the reach of those with limited financial means, such as citizen groups seeking to understand the condition of their local and regional watersheds.
- It defines the health of a water resource system and aids in diagnosing and identifying causes of any detected degradation.
- It enables us to prioritise for protection the places most deserving of conservation; and defines places where restoration is possible and practical.
- It allows a comparison of the effects of single acts with the cumulative effects of many activities; and provides results that can be used to assess the effectiveness of particular resource management decisions and to set funding priorities.
- It can be applied with a broad range of taxa, from plankton and vascular plants to invertebrates, fishes and waterbirds.

14.9.1. Range of Applications

The strength of IBI is its adaptability – to diverse contexts and geographies and diverse environmental conditions, such as point-source and nonpoint pollution or the effects of urbanisation (Karr and Chu 2000). As IBIs are tested worldwide, they prove more sensitive to human impacts than many conventional measures of these impacts. IBI has detected the effects of effluents from a bauxite plant in Guinea (Hugueny et al., 1996) and from salmonid aquaculture on small streams in France (Oberdorff and Porcher, 1994); the effects of channelisation and chemical effluents in small Venezuelan rivers (Gutierrez, 1994); metal and organic pollution in central Indian rivers (Ganasan and Hughes, 1998); the cumulative effects of channelisation, agricultural run-off and urbanisation in France's Seine River basin and Kenya's Lake Victoria basin (Raburu et al., 2009); effect of river training in China (Pei et al., 2010); and damming in Serbia (Lenhardt et al., 2009); and the impact of diverse land uses on streams in dry west–central Mexico (Lyons et al., 1995). IBIs have been used to reflect the biological effects of chemical contamination and the impacts of human activities on watershed or landscape scales and have served as a framework for reinventing biological assessments for water resource investigations. They have provided foundations for citizen water-monitoring programmes and to support the goals of the Clean Water Act. An IBI has helped to evaluate watersheds in the entire Sierra Nevada for setting conservation priorities; it was applied in a long-term evaluation of dammed streams and to evaluate a relatively undisturbed stream heavily invaded by alien species (Moyle and Randall, 1998). IBIs have formed the basis of international benchmarking and protocol development (Acosta et al., 2009; Amprogi et al., 2009).

Attempts to enhance the applicability of IBIs continue to be made. For example, Manolakos et al. (2008) have attempted to build up an efficient data analysis and visualisation tool to assess the simultaneous effects of anthropogenic stressors on the fish population through the fish metrics and the associated IBI. It employs self-organising feature maps (SOMs) and unsupervised neural networks to pattern the sampling sites based on similar metric characteristics. Canonical correspondence analysis (CCA) then allows conclusions to be drawn about the role of the environmental variables in maintaining the perfect abode for fishes. Different visualisations superimposed with SOM clustering are realised to explore the complex interrelationships in the aquatic system. It is hoped that this tool may aid watershed managers to better comprehend the effects of the environment on fish. Indeed, even greater use of statistical and machine-learning methods in facilitating IBI development and enhancing the IBI utility is a major thrust area at present (Faisal et al., 2009).

14.10. THE SHORTCOMINGS OF IBI

Notwithstanding their virtues, IBIs are besieged with a number of problems as well, of which some are the broad ones endemic to all forms of sampling associated with water (Chapter 4). Some other problems in IBI use are similar to the ones encountered with biotic indices. Then, there are problems unique to the IBI concept.

14.10.1. The Problem of Finding a Reference Site

As has been detailed in the preceding chapter, it is necessary to study undisturbed 'reference' sites in order to establish how anthropogenic disturbances alter the biological community in disturbed sites of similar physical–chemical–hydrological characteristics occurring in geographically similar locations.

All sites that are assessed are compared to the conditions prevailing in the reference site that serves as a benchmark or 'frame of reference' for assessing the extent of departure from it. However, undisturbed or even 'minimally' disturbed sites are hard to find (Herlihy et al., 2008. More often, all sites within a study region are significantly degraded to the point where no 'natural' conditions exist. Hence, a reference condition has to be predicated for the sake of comparison. However, the reference condition is often only the best observable condition, which may not always be indicative of historical water quality. Hence, a major uncertainty gets introduced at the outset. To get around this problem, historical frames of reference such as the use of past records of water quality or palaeoecological data have been suggested (Lavoie and Campeau, 2010). However, a great deal of work remains to be done before this option becomes widely utilisable.

In the absence of any other option, may IBIs have been developed without defining an accurate reference condition. In such cases, IBI scores are at best relative indications of the water quality that provide no genuine context of evaluation against natural integrity.

In an attempt to get around this problem, Kosnicki and Sites (2007) have introduced a 'least-desired index (LDI) multi-metric approach'. It is based on the logic that when reference conditions are not available for the use of an IBI, anti-reference sites, representing least-desired conditions, can be used in constructing an LDI. However, it is far from a well-tried and well-tested option.

14.10.2. The Problem of Distinguishing Between Impacts of Natural Variation Across Time and Space with Variations Caused by Anthropogenic Stress

Any index seeking to detect changes in biological assemblage owing to anthropogenic disturbances may provide misleading information if the effects cannot be distinguished from those caused by natural variation in community structure. Due to this aspect, problems are caused because inter-annual and seasonal variations in IBI scores occur as biotic communities exhibit natural responses to environmental changes. Indeed, IBIs reliant on biotic communities that are highly variable in abundance and composition (such as macroinvertebrates and plankton) are particularly susceptible to this problem (Hill et al., 2000; Kane and Culver 2003). One of the ways to address this issue is to examine the effects of inter-annual and seasonal changes on IBI and metric scores using multiple sample dates during index development (Kerans and Karr 1994; Hamsher et al., 2004). Limiting the sampling to a single time period to create a seasonal range and applying the index to that context is another approach (Fore et al., 1996). However, developing an index range in this manner effectively limits the influence of environmental heterogeneity on index score, even when annual applicability is limited (Beck and Hatch, 2009).

Attempts to enhance the applicability of the IBIs have led to the introduction by some authors of scoring thresholds that differ by season (Lacouture et al., 2006). However, developing a season-specific IBI is comparable to developing a unique IBI for each season and may not be the most efficient approach towards index development.

The capability of the IBIs to accurately reflect community thresholds across a broad range of anthropogenic gradients has also been challenged by some authors (King and Baker, 2010).

14.10.3. Changes in Species Richness with Size of Watercourse

Natural variations in species richness occur with changes in the size of the watercourse (MacArther and Wilson 1967). This problem is covered by normalising the species richness metrics for stream size, or by correlating IBI scores with the size of the system: most IBIs adopt richness-based metrics for different scales to account for stream size (Fausch et al., 1984), or plots of expected species richness against increasing stream size are drawn and used (Beck and Hatch, 2009).

14.10.4. Problems Associated with Metric Adaptation

IBIs developed for one region are not always appropriate for other regions, necessitating adaptation, despite efforts to develop robust metrics with carefully framed data collection and analytical contexts to accommodate differences among regions (Back and Hatch, 2009). A key component of regional adaptation is establishing which species are appropriate for specific metrics as the response of a species to environmental degradation can vary from region to region. A species that is tolerant in one region may not necessarily be tolerant in another region, owing to the complex interaction of factors that govern community composition (Lacoul and Freedman 2006). Indeed, tolerance-based metrics often have different sensitivity measures in different regions for the same species because of localised adaptations of an organism in response to environmental stressors.

14.10.5. Problems of Metric Scaling

The 'raw' metrics obtained from experimental data are scaled using a standardised scoring method to allow for inclusion into the composite index. Karr (1981) proposed the use of a discrete scaling system whereby the cumulative distribution of raw metric values is divided into three equal categories of 1, 3 and 5 (for 12 metrics with a minimum IBI score of 12 and a maximum of 60). Most of the IBIs utilise this scaling technique despite the obvious problems associated with this kind of discrete scaling which causes a decrease in the sensitivity of the composite IBI as a result of sequence gaps within metric values (Minns et al., 1994; Dolf et al., 2010). Often a significant degree of metric variability is found within a single scaled metric score. To overcome this drawback of the discrete method, a continuous method of scaling has been proposed. Here, a score from 0 to 10 is assigned to the raw metric scores and a linear relationship is assumed between the raw metric and the scaled metric. Continuous scaling is seen to allow for a greater range of scores, avoid sequence gaps and minimise bias in metric scores as compared to discrete scaling (Dolf et al., 2010).

14.10.6. Ambiguity of the Composite Score

The single composite score of an IBI can obscure information needed by ecosystem managers. For an IBI with 12 metrics, there are 8074 ways to obtain a score of 48 (Stewart-Oaten, 2008).

However, factors related to ecological health may be entirely different between two watercourses with the same index score. Fortunately, this ambiguity is serious only at moderate IBI scores; the extremes of ecological health are more clearly reflected by IBI. Furthermore, close examination of the configuration of values across the component metrics can reveal what is masked in the total IBI scores (Beck and Hatch, 2009).

The concept of IBI has not been developed to identify cause and effect relationships; hence, individuals seeking this information may find IBI scores to be confusing in certain circumstances.

14.10.7. Some Yet-to-be-answered Basic Questions About the Significance of IBI

The first-ever IBI, of Karr (1981), pioneered subsequent work on IBIs but was received with scepticism as well. Suter (1993) is commonly cited as the primary critic not only of heterogeneous indices, such as IBI, but also of the very concept of ecosystem health that is the goal of the environmental indices. The most serious criticism is that IBIs give unrealistic portrayals of ecological information. Suter (1993) has argued that the units of the indices are meaningless because they lack a relevant scale and hence lack of ability to moderate management actions. Moreover, how much of a decrease in overall scale would be considered a detriment to biological integrity? IBI scores are often relative to other aquatic systems with no accurately defined frame of reference because the best observable condition is often used as the benchmark for restoration and is not necessarily representative of ecological health.

References

Abbasi, T., Chari, K.B., Abbasi, S.A., 2009. A geospatial modelling-based assessment of water quality in and around Kaliveli watershed. Research Journal of Chemistry and Environment 13 (4), 48–55.

Abbasi, T., Abbasi, S.A., 2010. Factors which facilitate waste water treatment in presence of aquatic weeds-the machanism of the weed's purifying action. International Journal of environmental studies (Taylor and Francies) 67 (3), 349–371.

Abbasi, T., Abbasi, S.A., 2011. Water quality indices based on bioasessment: the biotic indices. Journal of Water and Health (IWA Publishing) 9 (2), 330–348.

Abbasi, S.A., Chari, K.B., 2008. Environmental Management of Urban Lakes. Discovery publishing House, New Delhi. Viii + 269 pages.

Acosta, R., Ríos, B., Rieradevall, M., Prat, N., 2009. Proposal for an evaluation protocol of the ecological quality of Andean rivers (CERA) and its use in two basins in Ecuador and Peru. Limnetica 28 (1), 35–64.

Allan, J.D., Erickson, D.L., Fay, J., 1997. The influence of catchment land use on stream integrity across multiple spatial scales. Freshwater Biology 37, 149–161.

Ambrogi, A.O., Forni, G., Silvestri, C., 2009. The Mediterranean intercalibration exercise on soft-bottom benthic invertebrates with special emphasis on the Italian situation. Marine Ecology 30 (4), 495–504.

Andreas, B.K., Mack, J.J., McCormac, J.S., 2004. Floristic Quality Assessment Index for Vascular Plants and Mosses for the State of Ohio. Ohio Environmental Protection Agency, Division of Surface Water, Wetland Ecology Group, Columbus, OH.

Anonymous, 2008. Naturvårdsverkets föreskrifter (2008:1) och allmänna råd om klassificering och miljökvalitetsnormer avseende ytvatten. http:// www. naturvardsverket.se/Documents/foreskrifter/nfs2008/ nfs_2008_01.pdf (in Swedish).

Atazadeh, I., Sharifi, M., Kelly, M.G., 2007. Evaluation of the trophic diatom index for assessing water quality in River Gharasou, Western Iran Hydrobiologia 589, 165–173.

Aura, C.M., Raburu, P.O., et al., 2010. A preliminary macroinvertebrate Index of Biotic Integrity for bioassessment of the Kipkaren and Sosiani Rivers, Nzoia River basin, Kenya. Lakes and Reservoirs: Research and Management 15 (2), 119–128.

Balon, E.K., 1975. Reproductive guilds of fishes: a proposal and definition. Journal of Fisheries Research Board of Canada 32, 821–864.

Barbour, M.T., Gerrilsen, J., Griffith, G.E., Frydenborg, R., McCarron, E., White, J.S., Baslian, M.L., 1996. A framework for biological criteria for Florida streams using benthic macroinvertebrates. Journal of the North American Benthological Society 15, 185–211. doi:10.2307/ 1467948.

Barbour, M.T. Stribling, J.B. Karr, J.R., 1995. Multimetric approach for establishing and measuring biological condition.

Beck, M.W., Hatch, L.K., 2009. A review of research on the development of lake indices of biotic integrity.

Beck, M.W., Hatch, L.K., et al., 2010. Development of a macrophyte-based index of biotic integrity for Minnesota lakes. Ecological Indicators 10 (5), 968–979.

Benyi, S.J., Hollister, J.W., Kiddon, J.A., Walker, H.A., 2009. A process for comparing and interpreting differences in two benthic indices in New York Harbor. Marine Pollution Bulletin 59 (1–3), 65–71.

Berkman, H.E., Rabeni, C.F., Boyle, T.P., 1986. Biomonitors of stream quality in agricultural areas: fish versus invertebrates. Environ. Manage. 10, 413–419.

Blocksom, K.A., 2003. A performance comparison of metric scoring methods for a multimetric index for mid-Atlantic highlands streams. Environmental Management 31 (5), 670–682.

Blocksom, K.A., Kurtenbach, J.P., Klemm, D.J., Fulk, F.A., Cormier, S.M., 2002. Development and evaluation of the Lake Macroinvertebrate Integrity Inde (LMII) for New Jersey lakes and reservoirs. Environment Monitoring, and Assessment 77, 311–333. doi: 10. 1023/A: 1016096925401. PM1D: 12194418.

Blomqvist, M., Cederwall, H., Leonardsson, K., Rosenberg, R., 2006. Bedömningsgrunder för kust och hav. Bentiska evertebrater 2006. Rapport till Naturvårdsverket 2006-03-21, p. 70. (In Swedish with English summary).

Boling, R.H., Petersen, R.C., Cummins, K.W., 1975. Ecosystem modeling for small woodland streams. In: Patten, B.C. (Ed.), Systems analysis and simulation in ecology, 3. Academic Press, New York, pp. 183–204.

Borja, A., Dauer, D.M., 2008. Assessing the environmental quality status in estuarine and coastal systems: comparing methodologies and indices. Ecological Indicators 8, 331–337.

Borja, A., Franco, J., Pérez, V., 2000. A marine biotic index to establish the ecological quality of soft-bottom benthos within European estuarine and coastal environments. Marine Pollution Bulletin 40, 1100–1114.

Borja, A., Muxika, I., Franco, J., 2003. The application of a Marine Biotic Index to different impact sources affecting softbottom benthic communities along European coasts. Marine Pollution Bulletin 46, 835–845.

Bramblett, R.G., Fausch, K.D., 1991. Variable fish communities and the index of biotic integrity in a western Great Plains river. Transactions of the American Fisheries Society 120, 752–769.

Bramblett, R.G., Johnson, T.R., Zale, A.V., Heggem, D.G., 2005. Development and evaluation of a fish assemblage index of biotic integrity for northwestern Great Plains streams. Transactions of the American Fisheries Society 134, 624–640.

Brazner, J.C., Beals, E.W., 1997. Patterns in fish assemblages from coastal wetland and beach habitats in Green Bay, Lake Michigan: a multivariate analysis of abiotic and biotic forcing factors. Canadian Journal of Fisheries and Aquatic Sciences 54, 1743–1761. doi: 10.1 139/cjfas54-8-1743.

Breine, J., Quataert, P., Stevens, M., Ollevier, F., Volckaert, F.a.M., Van Den Bergh, E., Maes, J., 2010. A zone-specific fish-based biotic index as a management tool for the Zeeschelde estuary (Belgium). Marine Pollution Bulletin 60 (7), 1099–1112.

Brousseau, C.M., Randall, R.G., et al., 2011. Fish community indices of ecosystem health: how does the Bay of Quinte compare to other coastal sites in Lake Ontario? Aquatic Ecosystem Health and Management 14 (1), 75–84.

Carlisle, B.K., Hicks, A.L., Smith, J.P., Garcia, S.R., Largay, B.G., 1999. Plants and aquatic invertebrates as indicators of wetland biological integrity in Waquoit Bay watershed, Cape Code. Environment Cape Code 2, 30–60.

Carlisle, D.M., Clements, W.H., 1999. Sensitivity and variability of metrics used in biological assessments of running waters. Environmental Toxicology and Chemistry 18, 285–291.

Carmichael, W.W., 1986. Algal toxins. Advances in Botanical. Research 12, 47–101.

Carmichael, W.W., 1997. The cyanotoxins. Advances in Botanical. Research 27, 211–256.

Carpenter, J., Bithell, J., 2000. Bootstrap confidence intervals: when, which, what? A practical guide for medical statisticians. Statistics in Medicine 19, 1141–1164.

Castela, J., Ferreira, V., Graça, M.A.S., 2008. Evaluation of stream ecological integrity using litter decomposition and benthic invertebrates. Environmental Pollution 153 (2), 440–449.

Cejudo-Figueiras, C., Warey-Blaco, I., Bécares, E., Blanco, S., 2010. Epiphytic diatoms and water quality in shallow lakes: the natural substrate hypothesis renisited? Marine and Freshwater Research 16 (12), 1457–1467.

Chari, K.B., Abbasi, S.A., Ganapathy, S., 2003. Ecology, habitat and bird community structure at Oussudu lake: towards a strategy for conservation and management. Aquatic Conservation: Marine and Freshwater Ecosystems 13, 373–386.

Chari, K.B., Abbasi, S.A., 2003. Assessment of Impact of Land use Changes on the Plankton Community of a Shallow Fresh water Lake in South India by GIS and Remote sensing. Chemical and Environmental Research 12, 93–112.

Chari, K.B., Abbasi, S.A., 2005. A study on the fish fauna of Oussudu - A rare freshwater lake of South India. International Journal of Environmental Studies 62, 137–145.

Chari, K.B., Sharma, R., Abbasi, S.A., 2005a. Comprehensive Environmental Impact Assessment of Water Resources Projects, vol. 1. Discovery Publishing House, New Delhi.

Chari, K.B., Sharma, R., Abbasi, S.A., 2005b. Comprehensive Environmental Impact Assessment of Water Resources Projects, vol. 2. Discovery Publishing House, New Delhi.

Chessman, B.C., Townsend, S.A., 2010. Differing effects of catchment land use on water chemistry explain contrasting behaviour of a diatom index in tropical northern and temperate southern Australia. Ecological Indicators 10 (3), 620–626.

Chin, A.F., Gelwick, F., Laurencio, D., Laurencio, L.R., Byars, M.S., Scoggins, M., 2010. Linking geomorphological and ecological responses in restored urban pool-riffle streams. Ecological Restoration 28 (4), 460–474.

Clayton, J., Edwards, T., 2006. Aquatic plants as environmental indicators of ecological condition in New Zealand lakes. Hydrobiologia 570, 147–151. doi:10.1007/s10750-006-OI74-4.

Coates, S., Waugh, A., Anwar, A., Robson, M., 2007. Efficacy of a multi-metric fish index as an analysis tool for the transitional fish component of the water framework directive. Marine Pollution Bulletin 55, 225–240.

Coleman II, J.C., Miller, M.C., Mink, F.L., 2011. Hydrologic disturbance reduces biological integrity in urban streams. Environmental Monitoring and Assessment 172 (1-4), 663–687.

Cookson, N., Schorr, M.S., 2009. Correlations of watershed housing density with environmental conditions and fish assemblages in a tennessee ridge and Valley stream. Journal of Freshwater Ecology 24 (4), 553–561.

Cooper, J.A.G., Ramm, A.E.L., Harrison, T.D., 1994. The estuarine health index: a new approach to scientific information transfer. Ocean and Coastal Management 25, 103–141.

Costa, P.F., Schulz, U.H., 2010. The fish community as an indicator of biotic integrity of the streams in the Sinos River basin, Brazil | [A ictiofauna como indicadora da integridade biótica dos arroios da bacia do Rio dos Sinos, Brasil]. Brazilian Journal of Biology 70 (Suppl. 4), 1195–1205.

Danielson, T.J., 2001. Methods for Evaluating Wetland Condition: Introduction to Wetland Biological Assessment, U.S. Environmental Protection Agency, Office of Water, Washington, DC, USA. EPA 822-R-01–007a.

Dauvin, J.C., Ruellet, T., 2007. Polychaete/amphipod ratio revisited. Marine Pollution Bulletin 55, 215–224.

Davis, P.A., Brown, J.C., Saunders, M., Lanigan, G., Wright, E., Fortune, T., Burke, J., Connolly, J., Jones, M.B., Osborne, B., 2010. Assessing the effects of agricultural management practices on carbon fluxes: spatial variation and the need for replicated estimates of net ecosystem exchange. Agricultural and Forest Meteorology 150 (4), 564–574.

Deegan, L.A., Finn, J.T., Ayvazian, S.G., Ryder-Kieffer, C.A., Buonaccorsi, J., 1997. Development and validation of an estuarine biotic integrity index. Estuaries 20, 601–617.

DeGasperi, C.L., Berge, H.B., Whiting, K.R., Burkey, J.J., Cassin, J.L., Fuerstenberg, R.R., 2009. Linking hydrologic alteration to biological impairment in urbanizing streams of the Puget Lowland, Washington, USA. Journal of the American Water Resources Association 45 (2), 512–533.

DeKeyser, E., Kirby, D., Ell, M., 2003. An index of plant community integrity: development of the methodology for assessing prairie wetland plant communities. Ecological Indicators 3, 119–133. doi:10.1016/S1470-160X(03)00015-3.

DeKeyser, E.S., Kirby, D.R., Ell, M.J., 2003. An index of plant community integrity: development of the methodology for assessing prairie wetland plant communities. Ecological Indicators 3, 119–133.

Delgado, C., Pardo, I., Garcia, L., 2010. A multimetric diatom index to assess the ecological status of coastal Galician rivers (NW Spain). Hydrobiologia 644, 371–384.

Delpech, C., Courrat, A., Pasquaud, S., Lobry, J., Le Pape, O., Nicolas, D., Boët, P., Girardin, M., Lepage, M., 2010. Development of a fish-based index to assess the ecological quality of transitional waters: The case of French estuaries. Marine Pollution Bulletin 60 (6), 908–918.

DeMott, W.R., Moxter, F., 1991. Foraging on cyanobacteria by copepods: responses to chemical defenses and resource abundance. Ecology 72, 1820–1834.

DeSousa, S., Pinel-Alloul, B., Cattaneo, A., 2008. Response of littoral macroinvertebrate communities on rocks and sediments to lake residential development. Canadian Journal of Fisheries and Aquatic Sciences 65, 1206–1216. doi: 10.1139/F08-031.

Detenbeck, N.E., 2002. Methods for evaluating wetland condition: wetland classification. EPA-822-R-02–017, Office of Water, U.S. Environmental Protection Agency, Washington, DC.

Dionne, M., Karr, J.R., 1992. In: McKenzie, D.H., Hyatt, D.E., McDonald, V.J. (Eds.), Ecological monitoring of fish assemblages in Tennessee River reservoirs. Ecological Indicators, 1. Elsevier, New York, pp. 259–281.

Doll, J.C., 2011. Predicting biological impairment from habitat assessments. Environmental Monitoring and Assessment, pp. 1–19.

Dolph, L., Aleksey, Y., Sheshukov, Christopher, J., Chizinski, Vondracek, B., Wilson, B., 2010. The Index of Biological Integrity and the bootstrap: Can random sampling error affect stream impairment decisions? Ecological Indicators 10, 527–537.

Doustshenas, B., Savari, A., Nabavi, S.M.B., Kochanian, T., Sadrinasab, M., 2009. Applying benthic index of biotic integrity in a soft bottom ecosystem in north of the persian gulf. Pakistan Journal of Biological Sciences 12 (12), 902–907.

Drake, M.T., Pereira, D.L., 2002. Development of a fish-based index of biotic integrity for small inland lakes in central Minnesota. North American Journal of Fisheries Management 22, 1105–1123.

Drake, M.T., Valley, R.D., 2005. Validation and application of a fish-based index of biotic integrity for small central Minnesota lakes. North American Journal of Fisheries Management 25, 1095–1111. doi:10.1577/ M04-128.1.

Efron, B., 1979. Bootstrap methods: another look at the jackknife. Annals of Statistics 7, 1–26.

Efron, B., 2003. Second thoughts on the bootstrap. Statistical Science 18, 135–140.

Engle, V.D., Summers, J.K., Gaston, G.R., 1994. A benthic index of environmental condition of Gulf of Mexico estuaries. Estuaries 17, 372–384.

Fausch, K.D., Karr, J.R., Yant, P.R., 1984. Regional application of an index of biotic integrity based on stream fish communities. Transactions of the American Fisheries Society 113, 39–55.

Fayram, A.H., Mitro, M.G., 2008. Relationships between reach-scale habitat variables and biotic integrity score, brook trout density, and brown trout density in Wisconsin streams. North American Journal of Fisheries Management 28 (5), 1601–1608.

Fennessy, M.S., Geho, R., Elfritz, B., Lopez, R., 1998a. Testing the Floristic Quality Assessment Index as an indicator of riparian wetland disturbance. Final Report to U.S. Environmental Protection Agency for Grant CD995927. Ohio Environmental Protection Agency, Division of Surface Water, Columbus, Ohio. http://www.epa.state.oh.us/dsw/wetlands/WetlandEcologySection_ reports.html.

Fennessy, M.S., Gray, M.A., Lopez, R.D., 1998b. An ecological Assessment of wetlands using reference sites, Final Report to U.S. Environmental Protection Agency for Grant CD995761. Ohio Environmental Protection Agency, Division of Surface Water, vol. 1. Columbus, Ohio.

Fennessy, S., Gernes, M., Mack, J., Wardrop, D.H., 2002. Methods for evaluating wetland condition: using vegetation to assess environmental conditions in wetlands. EPA-822-R-02–020. U.S. Environmental Protection Agency; Office of Water, Washington, DC.

Ferreira M.T., Caiola N., Casals F., Cortes R., Economou A., Garcia-Jalon D., Ilheu M., Martinez-Capel F., Oliveira J., Pont, D., Prenda J., Rogers C., Sostoa A., Zogaris, S., 2007a. Ecological traits of fish assemblages from Mediterranean Europe and their responses to human disturbances. Fisheries Management and Ecology. 14, 473–481.

Ferreira, M.T., Caiola, N., Casals, F., Oliveira, J., Sostoa, A., 2007b. Assessing perturbation of river fish communities in the Iberian ecoregion. Fisheries Management and Ecology 14, 519–530.

Ferreira, M.T., Cortes, R.M., Godinho, F.N., Oliveira, J.M., 1996. Indicators of the biological quality of aquatic ecosystems for the Guadiana basin. Recursos Hidricos 17, 9–20.

Fore, L.S., 2002. Response of diatom assemblages to human disturbance: development and testing of a multimetric index for the Mid-Atlantic Region (U.S.A.). In: Simon, T.P. (Ed.), Biological Response Signatures: Patterns in Biological Integrity for Assessment of Freshwater Aquatic Assemblages. CRC Press, Boca Raton, Fla, pp. 445–480.

Fore, L.S., Grafe, C., 2002. Using diatoms to assess the biological condition of large rivers in Idaho (U.S.A.). Freshwater Biology 47, 2015–2037. doi:10.1046/j.1365-2427.2002.00948.x.

Freund, J.G., Petty, J.T., 2007. Response of fish and macroinvertebrate bioassessment indices to water chemistry in a mined Appalachian watershed. Environmental Management 39 (5), 707–720.

Gabriels, W., Lock, K., De Pauw, N., Goethals, P.L.M., 2010. Multimetric Macroinvertebrate Index Flanders (MMIF) for biological assessment of rivers and lakes in Flanders (Belgium). Limnologica 40 (3), 199–207.

Ganasan, V., Hughes, R.M., 1998. Application of an index of biological integrity (IBI) to fish assemblages of the rivers Khan and Kshipra (Madhya Pradesh), India. Freshwater Biology 40, 367–383.

Gannon, J.E., Beeton, A.M., 1971. The decline of the large zooplankter, Limnocalanus macrurus Sars (Copepoda:-Calanoida), in Lake Erie In: Proceedings of the 14th Conference on Great Lakes Research, pp. 27–38.

Gannon, J.E., Stemberger, R.S., 1978. Zooplankton (especially crustaceans and rotifers) as indicators of water quality. Transactions of the American Microscopical Society 97 (1), 16–35.

Gassner, H., Tischler, G., Wanzenbock, J., 2003. Ecological integrity assessment of lakes using fish communities-Suggestions of new metrics developed in two Austrian prealpine lakes. International Review of Hydrobiology 88, 635–652.

Gauch Jr., H.G., 1982. Multivariate Analysis in Community Ecology. Cambridge University Press 298.

Gammon, J.R., Spacie, A., Hamelink, J.L., Kaesler, R.L., 1981. Role of electrofishing in assessing environmental quality of the Wabash River. ETATS-UNIS, American Society for Testing and Materials, Philadelphia, PA.

Genet, J.A., Olsen, A.R., 2008. Assessing depressional wetland quantity and quality using a probabilistic sampling design in the Redwood River watershed, Minnesota, USA. Wetlands 28 (2), 324–335.

Gernes, M.C., Helgen, J.C., 1999. Indexes of biotic integrity (IBI) for wetlands: vegetation and invertebrate IBI's. Final Report to U.S. EPA, Assistance Number CD995525–01. Minnesota Pollution Control Agency, Environmental Outcomes Division, St. Paul, Minnesota.

Glémarec, M., Hily, H., 1981. Perturbations apporteés á la macrofaune benthique de la baie de Concarneau. Acta Ecologica 2, 139–150.

Gliwicz, Z.M., Lampert, W., 1990. Food thresholds in Daphnia species in the absence and presence of blue–green filaments. Ecology 7, 691–702.

Gliwicz, Z.M., Siedlar, E., 1980. Food size limitation and algae interfering with food collection in Daphnia. Archiv fur Hydrobiologie 88, 155–177.

Gomez, L.D., Steele-King, C.G., McQueen-Mason, S.J., 2008. Sustainable liquid biofuels from biomass: the writing's on the walls. New Phytologist 178 (3), 473–485.

Gopalan, G., Culver, D.A., Wu, L., Trauben, B.K., 1998. The effect of recent ecosystem changes on the recruitment of young-of-year fish in western Lake Erie. Canadian Journal of Fisheries and Aquatic Sciences 55, 2572–2579.

Griffin, J.D.T., Fayram, A.H., 2007. Relationships between a fish index of biotic integrity and mean length and density of brook trout and brown trout in Wisconsin streams. Transactions of the American Fisheries Society 136 (6), 1728–1735.

Griffith, M.B., Hill, B.H., McCormick, F.H., Kaufman, P.R., Herlihy, A.T., Selle, A.R., 2005. Comparative application of indices of biotic integrity based on periphyton, macroinvertebrates, and fish to Rocky Mountain streams. Ecological Indicators 5, 117–136. doi:10.1016/j.ecolind.2004.11.001.

Gros, M., Petrović, M., Ginebreda, A., Barceló, D., 2010. Removal of pharmaceuticals during wastewater treatment and environmental risk assessment using hazard indexes. Environment International 36 (1), 15–26.

Growns, I., 2009. Differences in bioregional classifications among four aquatic biotic groups: implications for conservation reserve design and monitoring programs. Journal of Environmental Management 90 (8), 2652–2658.

Gutiérrez-Cánovas, C., Velasco, J., Millán, A., 2008. SAL-INDEX: a macroinvertebrate index for assessing the ecological status of saline "ramblas" from SE of the Iberian Peninsula. Limnetica 27, 299–316.

Hamsher, S.E., Verb, R.G., Vis, M.L., 2004. Analysis of acid mine drainage impacted streams using a periphyton index. Journal of Freshwater Ecology 19, 313–324.

Harig, A.L., Bain, M.B., 1998. Defining and restoring biological integrity in wilderness lakes. Ecological Applications 8, 71–87.

Harrison, T.D., Whitfield, A.K., 2004. A multi-metric fish index to assess the environmental condition of estuaries. Journal of Fish Biology 65, 683–710.

Hatzenbeler, G.R., Kampa, J.M., Jennings, M.J., Emmons, E.E., 2004. A comparison of fish and aquatic plant assemblages to assess ecological health of small Wisconsin lakes. Lake Reservior Manage 20, 211–218.

Havens, K.E., 1998. Size structure and energetics in a plankton food web. Oikos 81, 346–358.

Heatherly, T., Whiles, M.R., Knuth, D., Garvey, J.E., 2005. Diversity and community structure of littoral /one macroinvertebrates in southern Illinois reclaimed surface mine lakes. The American Midland Naturalist 154, 67–77. doi: 10.1674/0003-0031(2005) 154[0067:DACSOL]2.0.CO;2.

Helms, B.S., Schoonover, J.E., Feminella, J.W., 2009. Assessing influences of hydrology, physicochemistry, and habitat on stream fish assemblages across a changing landscape. Journal of the American Water Resources Association 45 (1), 157–169.

Henriques, S., Pais, M.P., Costa, M.J., Cabral, H., 2008. Efficacy of adapted estuarine fish-based multimetric indices as tools for evaluating ecological status of the marine environment. Marine Pollution Bulletin 56, 1696–1713.

Herbst, D.B., Silldorf, E.L., 2006. Comparison of the performance of different bioassessment methods: similar evaluations of biotic integrity from separate programs and procedures. Journal of the North American Benthological Society 25, 513–530.

Herlihy, A.T., Paulsen, S.G., van sickle, J., Stoddard, J.L., Hawkins, C.P., Yuan, L.L., 2008. Striving for consistency in a national assessment: the challenges of applying a reference condition approach at a continental scale. Journal of the North American Benthological Society 27, 860–877.

Herman, K.D., Masters, L.A., Penskar, M.R., Reznicek, A.A., Wilhelm, G.S., Brodovich, W.W., Gardiner, K.P., 2001. Floristic quality assessment with wetland categories and examples of computer applications for the state of Michigan – Revised 2nd edition. Michigan Department of Natural Resources, Wildlife, Natural Heritage Program, Lansing, MI, USA.

Hill, B.H., Herlihy, A.T., Kaufmann, P.R., Stevenson, R.J., McCormick, F.H., Johnson, C.B., 2000. Use of periphyton assemblage data as an index of biotic integrity. Journal of the North American Benthological Society 19, 50–67. doi: 10.2307/1468281.

Hoffman, J.C., Smith, M.E., Lehman, J.T., 2001. Perch or plankton: top-down control of Daphnia by yellow perch (Perca flavescens) or Bythotrephes cederstroemi in an inland lake? Freshwater Biology 46 (6), 759–775.

Hughes, R.M., 1995. In: Davis, W.S., Simon, T.P. (Eds.), Defining acceptable biological status by comparing with reference conditions Biological Assessment and Criteria: Tools for Water Resource Planning and Decision Making. Lewis Publishers, Boca Raton, FL, pp. 31–47.

Hughes, R.M., Oberdorff, T., 1998. Applications of IBI concepts and metrics to waters outside the United States and Canada. In: Simon, T.P. (Ed.), Assessment Approaches for Estimating Biological Integrity using Fish Assemblages. Lewis Press, Boca Raton, FL, pp. 79–83.

Hughes, R.M., Oberdorff, T., 1999. Applications of IBI concepts and metrics to waters outside the United States. In: Simon, T.P. (Ed.), Assessing the Sustainability and Biological Integrity of Water Resources Using Fish Communities. CRC Press, Boca Raton, FL, pp. 79–93.

Hughes, R.M., Howlin, S., Kaufmann, P.R., 2004. A biointegrity index for coldwater streams of western Oregon and Washington. Transactions of the American Fisheries Society 133, 1497–1515.

Hughes, R.M., Larsen, D.P., Omernik, J.M., 1986. Regional reference sites: a method for assessing stream potentials. Environmental Management 10, 629–635.

Hughes, R.M., Kaufman, P.R., Herlihy, A.T., Kincaid, T.M., Reynolds, L., Larsen, D.P., 1998. A process for developing and evaluating indices of fish assemblage integrity. Canadian Journal of Fisheries and Aquatic Sciences 55, 1618–1631.

Hunt, J.W., Anderson, B.S., Phillips, B.M., Tjeerdema, R.S., Taberski, K.M., Wilson, C.J., Puckett, H.M., Stephenson, M., Fairey, R., Oakden, J.M., 2001. A large-scale categorization of sites in San Francisco Bay, USA, based on the sediment quality triad, toxicity identification evaluations, and gradient studies. Environmental Toxicology and Chemistry 20, 1252–1265.

Jackson, D.A., Harvey, H.H., 1997. Qualitative and quantitative sampling of lake fish communities. Canadian Journal of Fisheries and Aquatic Sciences 54, 2807–2813. doi:10.1139/cjfas-54-12-2807.

Jennings, M.J., Lyons, J., Emmons, E.E., 1999. Toward the development of an index of biotic integrity for inland lakes in Wisconsin. In: Simon, T.P. (Ed.), Assessing the Sustainability and Biological Integrity of Water Resources Using Fish Communities. CRC Press, Boca Raton, pp. 541–562.

Jones, W.M., 2005. A vegetation index of biotic integrity For small-order streams In Southwest Montana and a Floristic Quality Assessment For Western Montana Wetlands. Montana Natural Heritage Program, Natural Resource Information System. Montana State Library, Helena, MT.

Josefson, A.B., Blomqvist, M., Hansen, J.L.S., Rosenberg, R., Rygg, B., 2009. Assessment of marine benthic quality change in gradients of disturbance: comparison of different Scandinavian multi-metric indices. Marine Pollution Bulletin 58, 1263–1277.

Kane, D.D., Culver, D.A., 2003. The Development of a Planktonic Index of Biotic Integrity for the Offshore Waters of Lake Erie. Final Report to the Lake Erie Protection Fund. The Ohio State University.

Kane, D.D., Gordon, S.I., Munawar, M., Charlton, M.N., Culver, D.A., 2009. The Planktonic Index of Biotic

Integrity (P-IBI): an approach for assessing lake ecosystem health. Ecological Indicators 9, 1234–1247.

Kane, D.D., Gannon, J.E., Culver, D.A., 2004. The Status of Limnocalanus macrurus (Copepoda:Calanoida:Centropagidae) in Lake Erie. Journal of Great Lakes Research 30, 22–30.

Kanno, Y., Macmillan, J.L., 2004. Developing an index of sustainable coldwater streams using fish community atiributes in river philip, nova scotia. Proceedings of the Nova Scotian Institute of Science 42 (2), 319–338.

Kanno, Y., Vokoun, J.C., Beauchene, M., 2010. Development of dual fish multi-metric indices of biological condition for streams with characteristic thermal gradients and low species richness. Ecological Indicators 10, 565–571.

Karr, J.R., 1981. Assessment of biotic integrity using fish communities. Fisheries 6, 21–27.

Karr, J.R., 1991. Biological integrity: a long neglected aspect of water resource management. Ecological Applications 1, 66–84.

Karr, J.R., Chu, E.W., 2000. Sustaining living rivers. Hydrobiologia 422 & 423, 1–14.

Karr, J.R., Dudley, D.R., 1981. Ecological perspective on water-quality goals. Environmental Management 5, 55–68. doi: 10.1007/ BFO1866609.

Karr, J.R., Fausch, K.D., Angermeier, P.L., Yant, P.R., Schlosser, I.J., 1986. Assessing Biological Integrity in Running Waters. A Method and its Rationale. Illinois Natural History Survey Campaigne. Special Publications, Illinois. 28.

Karr, J.R., Kerans, B.L., 1992. Components of biological integrity: their definition and use in development of an invertebrate IBI. In: Simon, T.P., Davis, W.S. (Eds.), Proceedings of the 1991 Midwest Pollution Control Biologists Meeting: Environmental Indicators Measurement Endpoints. EPA 905/R-92/003. U.S. Environmental Protection Agency, Region V. Environmental Sciences Division, Chicago, IL.

Karr, J.R., Yoder, C.O., 2004. Biological assessment and criteria improve total maximum daily load decision making. Journal of Environmental Engineering 130, 594–604.

Kaufmann, P.R., Levine, P., Robison, E.G., Seeliger, C., Peck, D., 1999. Quantifying physical habitat in wadeable streams. EPA/620/R-99/003. Office of Research and Development. US Environmental Protection Agency, Washington, DC.

Kerans, B.L., Karr, J.R., 1994. A benthic index of biotic integrity (B-IBI) for rivers of the Tennessee Valley. Ecological Applications 4, 768–785.

Kerfoot, W.C., Levitan, C., DeMott, W.R., 1988. Daphnia–phytoplankton interactions: density-dependent shifts in resource quality. Ecology 69, 1806–1825.

King, R.S., Baker, M.E., 2010. Considerations for analyzing ecological community thresholds in response to anthropogenic environmental gradients. Journal of the North American Benthological Society 29 (3), 998–1008.

Klemm, D.J., Blocksom, K.A., Fulk, F.A., Herlihy, A.T., Hughes, R.M., Kaufmann, P.R., Peck, D.V., Stoddard, J.L., Thoeny, W.T., Griffith, M.B., 2003. Development and evaluation of a macroinvertebrate biotic integrity index (MBII) for regionally assessing Mid-Atlantic Highlands streams. Journal of Environmental Management 31, 656–669.

Kosnicki, E., Sites, R.W., 2007. Least-desired index for assessing the effectiveness of grass riparian filter strips in improving water quality in an agricultural region. Environmental Entomology 36 (4), 713–724.

Kröncke, I., Reiss, H., 2010. Influence of macrofauna long-term natural variability on benthic indices used in ecological quality assessment. Marine Pollution Bulletin 60 (1), 58–68.

Lacoul, P., Freedman, B., 2006. Environmental influences on aquatic plants in freshwater ecosystems. Environmental Reviews 14, 89136. doi:10.1139/A06-001.

Lacouture, R.V., Johnson, J.M., Buchanan, C., Marshall, H.G., 2006. Phytoplankton index of biotic integrity for Chesapeake Bay and its tidal tributaries. Estuaries and Coasts 29, 598–616.

Lane, Ch.R., Brown, M.T., 2007. Diatoms as indicators of isolated herbaceous wetland condition in Florida, USA. Ecological Indicators 7, 521–540.

Langdon, R.W., 2001. A preliminary index of biological integrity for fish assemblages of small coldwater streams in Vermont. Northeastern Naturalist 8, 219–232.

Lambert, D., Cattaneo, A., Carignan, R., 2008. Periphyton as an early indicator of perturbation in recreational lakes. Canadian Journal of Fisheries and Aquatic Sciences 65, 258–265. doi:10.1139/F07-168.

Larsen, S., Sorace, A., Mancini, L., 2010. Riparian Bird Communities as Indicators of Human Impacts Along Mediterranean Streams. Environmental Management, 1–13.

Launois, L., Veslot, J., et al., 2011. Selecting fish-based metrics responding to human pressures in French natural lakes and reservoirs: towards the development of a fish-based index (FBI) for French lakes. Ecology of Freshwater Fish 20 (1), 120–132.

Lavesque, N., Blanchet, H., de, X., 2009. Montaudouin, Development of a multimetric approach to assess perturbation of benthic macrofauna in Zostera noltii beds. Journal of Experimental Marine Biology and Ecology 368 (2), 101–112.

Lavoie, I., Campeau, S., 2010. Fishing for diatoms: fish gut analysis reveals water quality changes over a 75-year period. Journal of Paleolimnology 43 (1), 121–130.

Lehman, J.T., Caceres, C.E., 1993. Food-web responses to species invasion by a predatory invertebrate: Bythotrephes in Lake Michigan. Journal of Great Lakes Research 38, 879–891.

Lenhardt, M., Markovic, G., Gacic, Z., 2009. Decline in the Index of Biotic Integrity of the Fish Assemblage as a Response to Reservoir Aging. Water Resour Manage 23, 1713–1723.

Leunda, P.M., et al., 2009. Longitudinal and seasonal variation of the benthic macroinvertebrate community and biotic indices in an undisturbed Pyrenean river. Ecological Indicators 9 (1), 52–63.

Lewis, P.A., Klemm, D.J., Thoeny, W.T., 2001. Perspectives on use of a multimetric lake bioassessment integrity index using benthic macroinvertebrates. Northeastern Naturalist 8, 233–246.

Li, F., Cai, Q., Ye, L., 2010. Developing a benthic index of biological integrity and some relationships to environmental factors in the subtropical xiangxi river, China. International Review of Hydrobiology 95 (2), 171–189.

Lillie, R.A., Garrison, P., Dodson, S.I., Bautz, R.A., Laliberte, G., 2002. Refinement and expansion of wetland biological indices for Wisconsin. Final Report to the U.S. Environmental Protection Agency Region V Grant No. CD975115.Wisconsin Department of Natural Resources, Madison, WI.

Liu, Y., Zhou, F., Guo, H., Yu Y., Zou Y., 2008. Biotic condition assessment and the implication for lake fish conservation: a case study of Lake Qionghai, China. Water and Environment Journal.

Llansó, R.J., Dauer, D.M., Vølstad, J.H., 2009. Assessing ecological integrity for impaired waters decisions in Chesapeake Bay. USA. Marine Pollution Bulletin 59, 48–53.

Lopez, R.D., Fennessy, M.S., 2002. Testing the Floristic Quality Assessment Index as an indicator of wetland condition. Ecological Applications 12, 487–497.

Lougheed, V.L., Chow-Fraser, P., 2002. Development and use of a zooplankton index of wetland quality in the Laurentian Great Lakes basin. Ecological Applications 12, 474–486.

Lougheed, V.L., Parker, C.A., Stevenson, R.J., 2007. Using non-linear responses of multiple taxonomic groups to establish criteria indicative of wetland biological condition. Wetlands 27 (1), 96–109.

Lussier, S.M., da Silva, S.N., Charpentier, M., Heltshe, J.F., Cormier, S.M., Klemm, D.J., Chintala, M., Jayaraman, S., 2008. The influence of suburban land use on habitat and biotic integrity of coastal Rhode Island streams. Environmental Monitoring and Assessment 139 (1-3), 119–136.

Lyons, J., Gutierrez.-Hernandez, A., Diaz-Pardo, E., Soto-Galera, E., Medina-Nava, M., Pineda-Lopez, R., 2000. Development of a preliminary index of biotic integrity (IBI) based on fish assemblages to assess ecosystem condition in the lakes of central Mexico. Hydrobiologia 418, 57–72. doi: 10.1023/A: 1003888032756.

Lyons, J., Piette, R.R., Niermeyer, K.W., 2001. Development, validation and application of a fish-based index of biotic integrity for Wisconsin_s large warmwater rivers. Transactions of the American Fisheries Society 130, 1077–1094.

MacArther, R.H., Wilson, E.O., 1967. The Theory of Island Biogeography. Princeton University Press, NJ.

Mack, J., 2007. Developing a wetland IBI with statewide application after multiple testing iterations. Ecological Indicators 7, 864–881.

Mack, J.J., 2001b. Vegetation Indices of Biotic Integrity (VIBI) for Wetlands: Ecoregional, Hydrogeomorphic, and Plant Community Comparison with Preliminary Wetland Aquatic Life Use designations. Final Report U.S. EPA Grant No. CD985875, Ohio EPA, Division of Surface Water. Wetland Ecology Group, vol. 1. Columbus, Ohio. http://www.epa.state.oh.us/dsw/wetlands/ etlandEcologySection_reports.html.

Mack, J.J., 2004b. Integrated Wetland Assessment Program. Part 4: Vegetation Index of Biotic Integrity (VIBI) for Ohio Wetlands. Ohio EPA Technical Report WET/2004-4. Ohio EPA, Division of Surface Water. Wetland Ecology Group, Columbus, Ohio. http://www.epa.state.oh.us/ dsw/wetlands/WetlandEcology Section_reports.html.

Mack, J.J., Micacchion, M., Augusta, L.D., Sablak, G.R., 2000. Vegetation indices of biotic integrity (VIBI) for wetlands and calibration of the Ohio rapid assessment method for wetlands v. 5.0. Final Report to U.S. EPA Grant No. CD985276. Ohio Environmental Protection Agency, Division of Surface Water. Wetland Ecology Group, Columbus, Ohio. http://www.epa.state.oh.us/ dsw/ wetlands/WetlandEcologySection_reports.html.

Magalhaes, M.F., Ramalho, C.E., Collares-pereira, M.J., 2008. Assessing biotic integrity in a Mediterranean watershed: development and evaluation of a fish-based index. Fisheries Management and Ecology 15, 273–289.

Maloney, K.O., Feminella, J.W., Mitchell, R.M., Miller, S.A., Mulholland, P.J., Houser, J.N., 2008. Landuse legacies and small streams: identifying relationships between historical land use and contemporary stream conditions. Journal of the North American Benthological Society 27, 280–294.

Manly, B.F.J., 2007. Randomization, Bootstrap and Monte Carlo Methods in Biology, third ed. Chapman & Hall/ CRC, Boca Raton, FL.

Manolakos, D., Mohamed, E.Sh., Karagiannis, I., Papadakis, G., 2008. Technical and economic comparison between PV-RO system and RO-Solar Rankine system. Case study: Thirasia island. Desalination 221(1–3) 37–46.

Marce, M.N., Fornells, N.P., 2010. Effects of droughts and floods on the biotic index in the Llobregat river. Efectos de la sequía y las crecidas en los índices biológicos en el río Llobregat 30 (320), 46–55.

Matzen, D.A., Berge, H.B., 2008. Assessing small-stream biotic integrity using fish assemblages across an urban landscape in the Puget Sound Lowlands of western Washington. Transactions of the American Fisheries Society 137, 677–689.

McCormick, F.H., Hughes, R.M., Kaufmann, P.R., Peck, D.V., Stoddard, J.L., Herlihy, A.T., 2001. Development of an index of biotic integrity for the Mid-Atlantic Highlands region. Transactions of the American Fisheries Society 130, 857–877.

Merten, E.C., Hemstad, N.A., et al., 2010. Relations between fish abundances, summer temperatures, and forest harvest in a northern Minnesota stream system from 1997 to 2007. Ecology of Freshwater Fish 19 (1), 63–73.

Miller, D.L., Leonard, P.M., Hughes, R.M., Karr, J.R., Moyle, P.B., Schrader, L.H., Thompson, B.A., Daniels, R.A., Fausch, K.D., Fitshugh, G.A., Gammon, J.R., Haliwell, D.B., Angermeier, P.L., Orth, D.J., 1988. Regional applications of an index of biotic integrity for use in water resource management. Fisheries 13, 12–20.

Mills, E.L., Schiavone Jr., A., 1982. Evaluation of fish communities through assessment of zooplankton populations and measures of lake productivity. North American Journal of Fisheries Management 2, 14–27.

Mills, E.L., Green, D.M., Schiavone Jr., A., 1987. Use of zooplankton size to assess the community structure of fish populations in freshwater lakes. North American Journal of Fisheries Management 7, 369–378.

Miller, S.J., Wardrop, D.H., Mahaney, W.M., Brooks, R.P., 2004. Plant-based indices of Biological Integrity (IBIs) for Wetlands in Pennsylvania Monitoring and assessing Pennsylvania wetlands. In: Brooks, R.P. (Ed.), Final report for Cooperative Agreement No. X-827157 submitted to U.S. Environmental Protection Agency, Office of Wetlands, Oceans and Watersheds. Pennsylvania State University, Cooperative Wetlands Center, Stateline, PA.

Miller, S.J., Wardrop, D.H., Mahaney, W.M., Brooks, R.R., 2006. A plant-based index of biological integrity (IBI) for headwater wetlands in central Pennsylvania. Ecological indicators 6, 290–312. doi:10.1016/j.eeolind.2005.03.011.

Minns, C.K., 1989. Factors affecting fish species richness in Ontario lakes. Transaction of the American fisheries society 118, 533–545. doi:10.1577/15488659(1989)118<0533:FAFSRI>2.3.CO;2.

Minns, C.K., Cairns, V.W., Randall, R.G., Moore, J.E., 1994. An index of biotic integrity (IBI) for fish assemblages in the littoral zone of Great Lakes areas of concern. Canadian Journal of Fisheries and Aquatic Sciences 51, 1804–1822.

Miranda, L.E., Hunt, K.M., 2010. An index of reservoir habitat impairment. Environmental Monitoring and Assessment. DOI 10.1007/s10661-010-1329-3.

Miranda, L.E., Hunt, K.M., 2011. An index of reservoir habitat impairment. Environmental Monitoring and Assessment 172 (1–4), 225–234.

MPCA, 2007. Guidance Manual for Assessing the Quality of Minnesota Surface Waters for the Determination of Impairment: 305(b) Report and 303(d) List. Minnesota Pollution Control Agency, Environmental Outcomes Division, St. Paul, MN, St. Paul, Minnesota.

Muxika, I., Borja, A., Bonne, W., 2005. The suitability of the marine biotic index (AMBI) to new impact sources along European coasts. Ecology. Indicators 5, 19–31.

Nichols, S., 1999. Floristic quality assessment of Wisconsin lake plant communities with example applications. Lake Reservior Manage 15, 133–141.

Nichols, S., Weber, S., Shaw, B., 2000. A proposed aquatic plant community biotic index for Wisconsin lakes. Environ. Manage 26, 491–502.

Niemela, S., Feist, M., 2002. Index of Biotic Integrity (IBI) Guidance for Coolwater Rivers and Streams of the Upper Mississippi River Basin in Minnesota. Minnesota Pollution Control Agency, Biological Monitoring Program, St. Paul, MN.

Niemela, S., Feist, M., 2000. Index of Biotic Integrity (IBI) Guidance for Coolwater Rivers and Streams of the St. Croix River Basin in Minnesota. Minnesota Pollution Control Agency, Biological Monitoring Program, St. Paul, MN.

NRC (National Research Council), 2000. Ecological Indicators for the Nation. National Academy Press, Washington D.C. 180.

Nygaard, G., 1949. Hydrobiological studies in some lakes and ponds. Part II. The quotient hypothesis and some new or little known phytoplankton organisms. Kongelige Danske videnskabernes selskab, Biologiske skrifter 7, 1–293.

Oberdorff, T., Porcher, J.-P., 1994. An index of biotic integrity to assess biological impacts of salmonid farm effluents on receiving waters. Aquaculture 119, 219–235.

O'Connor, R.J., Walls, T.E., Hughes, R.M., 2000. Using multiple taxonomic groups to index the ecological condition of lakes. Environmental. Monitoring and Assessment 61, 207–228. doi:10.1023/A: I006119205583.

Ohio, E.P.A., 1987. Biological criteria for the protection of aquatic life, Surface Water section. Ohio EPA, Columbus, OH.

Oliveira, J.M., Ferreira, M.T., 2000. Desenvolvimento de um índice de integridade biotica para a avaliacao da qualidade ambiental de rios ciprinícolas. Revista de Ciȇncias Agrárias 25, 198–210.

Omernik, J.M., 1987. Ecoregions of the conterminous United States. Annals of the Association of American Geographers 77, 118–125.

Pathiratne, A., Weerasundara, A., 2004. Bioassessment of selected inland water bodies in Sri Lanka using benthic oligochaetes with consideration of temporal variations. International Review of Hydrobiology 89, 305–316. doi:10.1002/iroh.200310676.

Paul, J.F., Scott, K.J., Campbell, D.E., Gentile, J.H., Strobel, C.S., Valente, R.M., Weisberg, S.B., Holland, A.F., Ranasinghe, J.A., 2001. Developing and applying a benthic index of estuarine condition for the Virginian biogeographic province. Ecological Indicators 1, 83–99.

Pearson, T.H., Rosenberg, R., 1978. Macrobenthic succession in relation to organic enrichment and pollution of the marine environment. Oceanography and Marine Biology: An Annual Review 16, 229–311.

Pei, X., Niu, C., et al., 2010. The ecological health assessment of Liao River Basin, China, based on biotic integrity index of fish. Shengtai Xuebao/Acta Ecologica Sinica 30 (21), 5736–5746.

Plafkin, J.L., Barbour, M.T., Porter, K.D., Gross, S.K., Hughes, R.M., 1989. Rapid bioassessment protocols for use in streams and rivers: Benthic macroinvertebrates and fish. U.S. Environmental Protection Agency, Wash DC, EPA/444/4-89-001.

Pollack, J.B., Kinsey, J.W., Montagna, P.A., 2009. Freshwater inflow biotic index (FIBI) for the Lavaca-Colorado Estuary, Texas. Environmental Bioindicators 4 (2), 153–169.

Prato, S., Morgana, J.G., Valle, P.L., Finoia, M.G., Lattanzi, L., Nicoletti, L., Ardizzone, G.D., Izzo, G., 2009. Application of biotic and taxonomic distinctness indices in assessing the Ecological Quality Status of two coastal lakes: Caprolace and Fogliano lakes (Central Italy). Ecological indicators 9, 568–583.

Quintino, V., Elliott, M., Rodrigues, A.M., 2006. The derivation, performance and role of univariate and multivariate indicators of benthic change: case studies at differing spatial scales. Journal of Experimental Marine Biology and Ecology 330, 368–382.

Raburu, P.O., Masese, F.O., Mulanda, C.A., 2009. Macroinvertebrate Index of Biotic Integrity (M-IBI) for monitoring rivers in the upper catchment of Lake Victoria Basin. Kenya Aquatic Ecosystem Health and Management 12 (2), 197–205.

Ramm, A.E.L., 1988. The community degradation index: a new method for assessing the deterioration of aquatic habitats. Water Research 22, 293–301.

Ranasinghe, J.A., et al., 2009. Calibration and evaluation of five indicators of benthic community condition in two California bay and estuary habitats. Marine Pollution Bulletin 59 (1-3), 5–13.

Rawson, D.S., 1956. Algal indicators of trophic lake types. Limnology and Oceanography 1, 18–25.

Rehn, A.C., 2009. Benthic macroinvertebrates as indicators of biological condition below hydropower dams on west slope Sierra Nevada streams, California. USA. River Research and Applications 25 (2), 208–228.

Reynoldson, T.B., Norris, R.H., Resh, V.H., Day, K.E., Rosenberg, D.M., 1997. The reference condition: a comparison of multimetric and multivariate approaches to assess water quality impairment using benthic macroinvertebrates. Journal of the North American Benthological Society 16, 833–852.

Rosenberg, R., Blomqvist, M., Nilsson, H.C., Cederwall, H., Dimming, A., 2004. Marine quality assessment by use of benthic species-abundance distributions: a proposed new protocol within the European Union Water Framework Directive. Marine Pollution Bulletin 49, 728–739.

Roset, N., Grenouillet, G., Goffaux, D., Pont, D., Kestemont, P., 2007. A review of existing fish assemblage indicators and methodologies. Fisheries Management and Ecology 14, 393–405.

Rothrock, P.E., Simon, T.P., Stewart, P.M., 2008. Development, calibration, and validation of a littoral zone plant index of biotic integrity (PIBI) for lacustrine wetlands. Ecol. Ind 8, 79–88.

Rufer, M.M., 2006. Chironomidae emergence as an indicator of trophic state in urban Minnesota lakes. University of Minnesota, St. Paul, Minn.

Rygg, B., 2006. Developing indices for quality-status classification of marine softbottom fauna in Norway. NIVA report 5208–2006. p. 33.

Schindler, D.W., 1990. Experimental perturbations of whole lakes as tests of hypotheses concerning ecosystem structure and function. Oikos 57, 25–41.

Schmitter-Soto, J.J., Ruiz-Cauich, L.E., Herrera, R.L., González-Solís, D., 2011. An Index of Biotic Integrity for shallow streams of the Hondo River basin, Yucatan Peninsula. Science of the Total Environment 409 (4), 844–852.

Shannon, C.E., Weaver, W., 1949. The Mathematical Theory of Communication. The University of Illinois Press, Urbana IL.

Shannon, C.E., Weaver, W., 1963. The Mathematical Theory of Communication. University of Illinois Press, Urbana, p. 117.

Siegfried, C.A., Sutherland, J.W., 1992. Zooplankton communities of Adirondack Lakes- Changes in community structure associated with acidification. Journal of Fresh Water Ecology 7, 97–112.

Simon, T.P. (Ed.), 1999. Assessing the Sustainability and Biological Integrity of Water Resources Using Fish Communities. CRC Press, Boca Raton, FL.

Simon, T.P., Lyons, J., 1995. Application of the index of biotic integrity to evaluate water resource integrity in freshwater ecosystems. In: Davis, W.S., Simon, T.P. (Eds.), Biological Assessment and Criteria: Tools for Water Resource Planning and Decision. Lewis Publishers, Boca Raton, pp. 245–262.

Simon, T.P., Stewart, P.M., Rothrock, P.E., 2001. Development of multimetric indices of biotic integrity for riverine and palustrine wetland plant communities along Southern Lake Michigan. Aquatic Ecosystem Health & Management 4, 293–309.

Smith, R.W., Bergen, M., Weisberg, S.B., Cadien, D.B., Dalkey, A., Montagne, D.E., Stull, J.K., Velarde, R.G., 2001. Benthic response index for assessing infaunal communities on the southern California mainland shelf. Ecological Applications 11, 1073–1087.

Smith, R.W., Ranasinghe, J.A., Weisberg, S.B., Montagne, D.E., Cadien, D.B., Mikel, T.K., Velarde, R.G., Dalkey, A., 2003. Extending the southern California Benthic Response Index to assess benthic condition in bays Southern California Coastal Water Research Project. Technical Report 410, Westminster, CA.

Smucker, N.J., Vis, M.L., 2009. Use of diatoms to assess agricultural and coal mining impacts on streams and a multiassemblage case study. Journal of the North American Benthological Society 28 (3), 659–675.

Southerland, M.T., Vølstad, J.H., Weber, E.D., Klauda, R.J., Poukish, C.A., Rowe, M.C., 2009. Application of the probability-based Maryland Biological Stream Survey to the state's assessment of water quality standards. Environmental Monitoring and Assessment 150 (1-4), 65–73.

Steedman, R.J., 1988. Modification and assessment of an index of biotic integrity to quantify stream quality in Southern Ontario. Canadian Journal of Fisheries and Aquatic Sciences 45, 492–500.

Stevens, C.E., Counci, T., et al., 2010. Influences of Human Stressors on Fish-Based Metrics for Assessing River Condition in Central Alberta. Water Quality Research Journal of Canada 45 (1), 35–46.

Stevenson, R.J., McCormick, P.V., Frydenborg, R., 2002b. Methods for Evaluating Wetland Condition: Using Algae To Assess Environmental Conditions in Wetlands. U.S. Environmental Protection Agency, Office of Water, Washington, DC, USA. EPA-822-R-02–021.

Stewart-Oaten, A., 2008. Chance and randomness in design versus model-based approaches to impact assessment: comments on Bulleri et al. (2007). Environmental Conservation 35 (1), 8–10.

Stoddard, J.L., Peck, D.V., Paulsen, S.G., Van Sickle, J., Hawkins, C.P., Herlihy, A.T., Hughes, R.M., Kaufmann, P.R., Larsen, D.P., Lomnicky, G., Olsen, A.R., Peterson, S.A., Ringold, P.L., Whittier, T.R., 2005. An ecological assessment of western streams and rivers. EPA 620/R- 05/005. Office of Research and Development. US Environmental Protection Agency, Washington, DC.

Stoddard, J.L., Herlihy, A.T., Peck, D.V., Hughes, R.M., Whittier, T.R., Tarquinio, E., 2008. A process for creating multimetric indices for large-scale aquatic surveys. Journal of the North American Benthological Society 27 (4), 878–891.

Stribling, J.B., Jessup, B.K., Feldman, D.L., 2008. Precision of benthic macroinvertebrate indicators of stream condition in Montana. Journal of the North American Benthological Society 27, 58–67. doi:10.1899/07-037R.1.

Suren, A., Kilroy, C., Lambert, P., Wech, J., Sorrell, B., 2011. Are landscape-based wetland condition indices reflected by invertebrate and diatom communities? Wetlands Ecology and Management 19 (1), 73–88.

Suter II, G. W, 1993. A critique of ecosystem health concepts and indexes. Environmental Toxicology and Chemistry 12, 1533–1539. doi:10.1897/1552-8618 (1993)12 [1533:ACOEHC]2.0.CO;2.

Tang, T., Cai., Q.H., Liu, J.K., 2006. Using epilithic diatom communities to assess ecological condition of Xiangxi River system. Environmental Monitoring and Assessment 112, 347–361. doi: 10.1007/ s10661-006-7666-6. PMID:16404550.

Tangen, B.A., Butler, M.G., Michael, J.E., 2003. Weak correspondence between macroinvertebrate assemblages and land use in Prairie Pothole Region wetlands, USA. Wetlands 23, 104–115. doi: 10.1672/0277-5212(2003)023 [0104:WCBMAA]2.0.CO;2.

Tataranni, M., Lardicci, C., 2010. Performance of some biotic indices in the real variable world: a case study at different spatial scales in North-Western Mediterranean Sea. Environmental Pollution 158 (1), 26–34.

Thoma, R.F., 1999. Biological monitoring and an index of biotic integrity for Lake Erie's nearshore waters. In: T.P, Simon (Ed.), Assessing the Sustainability and Biological Integrity of Water Resources Using Fish Communities. CRC Press, Boca Raton, Fla, pp. 417–461.

Thompson, B., Lowe, S., 2004. Assessment of macrobenthos response to sediment contamination in the San Francisco Estuary, California. USA. Environmental Toxicology and Chemistry 23, 2178–2187.

Thompson, B.A., Fitzhugh, G.R., 1986. A Use Attainability Study: An Evaluation of Fish and Macroinvertebrate Assemblages of the Lower Calcasieu River, Louisiana. LSU-CFI-29. Center for Wetland Resources. Coastal Fisheries Institute, Louisiana State University, Baton Rouge, La.

Trigal, C., Gareia-Criado, F., Fernandez-Alaez, C., 2006. Among-habitat and temporal variability of selected macroinvertebrate based metrics in a Mediterranean shallow lake (NW spain). Hydrobiologia 563, 371–384. doi:10.1007/s10750-0060031-5.

Uitto, A., Gorokhova, E., Valipakka, P., 1999. Distribution of the non-indigenous Cercopagis pengoi in the coastal waters of the eastern Gulf of Finland. ICES Journal of Marine Science 56 (Suppl.), 49–57.

USEPA, 1998. Lake and Reservoir Bioassessment and Biocriteria. Technical Guidance Document. EPA 841-B-98–007. Office of Water, U.S. Environmental Protection Agency, Washington, DC.

USEPA, 1999. Rapid Bioassessment Protocols for Use in Wadeable Streams and Rivers. Periphyton, Benthic Macroinvertebrates and Fish. EPA 841-B-99–002., second ed. Office of Water, U.S. Environmental Protection Agency, Washington, DC.

van Dam, H., Mertenes, A., Sinkeldam, J., 1994. A coded checklist and ecological indicator values of freshwater diatoms from the Netherlands. Netherlands Journal of Aquatic Ecology 28, 117–133.

Van Sickle, J., Huff, D.D., Hawkins, C.P., 2006. Selecting discriminant function models for predicting the expected richness of aquatic macroinvertebrates. Freshwater Biology 51, 359–372.

Wan, H., Chizinski, C.J., Dolph, C.L., Vondracek, B., Wilson, B.N., 2010. The impact of rare taxa on a fish index of biotic integrity. Ecological Indicators 10 (4), 781–788.

Warwick, R.M., Clarke, K.R., 1994. Relating the ABC: taxonomic changes and abundance/biomass relationship in disturbed benthic communities. Marine Biology 118, 739–744.

Weaver, M.J., Magnuson, J.J., Clayton, M.K., 1993. Analyses for differentiating littoral fish assemblages with catch data from multiple sampling gears. Transaction of the American fisheries society 122, 1111–1119. doi:10.1577/1548-8659(1993)122<1111:AFDLFA>2.3.CO;2.

Weigel, B.M., 2003. Development of stream macroinvertebrate models that predict watershed and local stressors in Wisconsin. Journal of the North American Benthological Society 22, 123–142. doi:10.2307/1467982.

Weisberg, S.B., Ranasinghe, J.A., Dauer, D.M., Schaffner, L.C., Diaz, R.J., Frithsen, J.B., 1997. An estuarine benthic index of biotic integrity (B-IBI) for Chesapeake Bay. Estuaries 20 (1), 149–158.

Weisberg, S.B., Thompson, B., Ranasinghe, J.A., Montagne, D.E., Cadien, D.B., Dauer, D.M., Diener, D., Oliver, J., Reish, D.J., Velarde, R.G., Word, J.Q., 2008. The level of agreement among experts applying best professional judgment to assess the condition of benthic infaunal communities. Ecology Indicators 8, 389–394.

Whitman, R.L., Nevers, M.B., Goodrich, M.L., Murphy, P.C., Davis, B.M., 2004. Characterization of Lake Michigan coastal lakes using zooplankton assemblages. Ecology Indicators 4, 277–286. doi:10.1016/j.ecolind.2004.08.001.

Whittier, T.R., 1999. Development of IBI metrics for lakes in southern New England. In: Simon, T.P. (Ed.), Assessing the Sustainability and Biological Integrity of Water Resources Using Fish Communities. CRC Press, Boca Raton, Florida, pp. 563–582.

Whittier, T.R., Hughes, R.M., Stoddard, J.L., Lomnicky, G.A., Peck, D.V., Herlihy, A.T., 2007. A structured approach for developing indices of biotic integrity: three examples from streams and rivers in the western USA. Transactions of the American Fisheries Society 136, 718–735.

Wilcox, D.A., Meeker, J.E., Hudson, P.L., Armiiage, B.J., Black, M.G., Uzarski, D.G., 2002. Hydrologic variability and the application of index of biotic integrity metrics to wetlands: a Great Lakes evaluation. Wetlands 22, 588–615. doi:10.1672/ 0277-5212(2002)022[0588:HVATAO]2.0.CO;2.

Williams, M., Longstaff, B., Buchanan, C., Llansó, R., Dennison, W., 2009. Development and evaluation of a spatially-explicit index of Chesapeake Bay health. Marine Pollution Bulletin 59, 14–25.

Winemiller, K.O., Leslie, M.A., 1992. Fish assemblages across a complex tropical freshwater/marine ecotone. Environmental Biology of Fishes 34, 29–50.

Wright, J.F., Furse, M.T., Armitage, P.D., 1993. RIVPACS: a technique for evaluating the biological water quality of rivers in the UK. European Water Pollution Control 3, 15–25.

Wu, J.-T., 1999. A generic index of diatom assemblages as bioindicator of pollution in the Keelung River of Taiwan. Hydrobiologia 397, 79–87.

Wu, N., Tang, T., Zhou, S., Jia, X., Li, D., Liu, R., Cai, Q., 2009. Changes in benthic algal communities following construction of a run-of-river dam. Journal of the North American Benthological Society 28, 69–79.

Xu, F.-L., Dawson, R.W., Tao, S., Cao, J., Li, B.-G., 2001. A method for lake ecosystem health assessment: an Ecological Modeling Method (EMM) and its application. Hydrobiologia 443, 159–175.

Zalack, J.T., Smucker, N.J., Vis, M.L., 2010. Development of a diatom index of biotic integrity for acid mine drainage impacted streams. Ecological Indicators 10 (2), 287–295.

CHAPTER

15

Multivariate Approaches for Bioassessment of Water Quality

OUTLINE

15.1. INTRODUCTION

Multivariate approaches for bioassessment are basically methods which use statistical tools to develop relationships between fauna and environmental characteristics for an 'ideal' or high-quality reference site. The relationships are used to predict the fauna at any test site in the absence of pollution (or other type of stress). The observed fauna at the test site is then compared with the fauna expected at that site to derive assessment of ecological quality.

Common ingredients of multivariate analyses include but are not limited to cluster analyses, ordination techniques and discriminant analyses. These methods develop statistical relationships among samples for which two or more variables have been measured. Typically, samples are grouped based on their similarity to each other, often through the use of complex algorithms.

Multivariate approaches utilise deductive testing of hypotheses on the condition of the community as impacted by disturbance gradients

Water Quality Indices DOI: 10.1016/B978-0-444-54304-2.00015-4

derived from analysis results. In contrast, IBIs described in the preceding chapter follow a more inductive approach to hypothesis testing that relies on *a posteriori* assumptions.

A number of multivariate approaches have been standardised and consequently adopted for widespread use on an international level. Interestingly, these methods have been largely limited to benthic macroinvertebrate communities, and mostly applied to lotic habitats so far. The first, and, to date, the most widely used multivariate approach is the River Invertebrate Prediction And Classification System (RIVPACS) developed in the UK using discriminant function models (Wright et al. 1984). In recent years, the RIVPACS model has been applied to other taxa, for example, diatoms (Cao et al., 2007) and fish (Joy and Death, 2002; Dolph et al., 2011); it has also been applied to lentic habitats (Davis et al., 2006). But such reports are few and far between.

15.2. RIVPACS

The River Invertebrate Prediction and Classification System (RIVPACS) was originally developed for rivers in Great Britain. Since 1990, the RIVPACS approach and its associated software system has been adopted by the UK government's agencies in England, Wales, Scotland and Northern Ireland as their principal tool for assessing the ecological quality of their rivers (Clarke et al., 2003). Several general aspects of the RIVPACS approach have since been incorporated into the prescribed methods of the Water Framework Directive (WFD) for assessing the ecological quality and ecological status of the surface waters of other European countries (Davy-Bowker et al., 2006; Feio et al., 2009).

15.2.1. The Basic Approach

The reference sites for RIVPACS are carefully chosen to cover short river stretches, which are considered to be of high ecological and chemical quality, and representative of the best examples of their particular river type. They are selected to encompass a wide range of physical types of running-water sites across a geographical region. The reference sites are initially classified into groups based solely on their macroinvertebrate composition. Discriminant analysis is then performed to obtain equations that represent the best fit between the biological classification and measured values of standard environmental variables at each reference site. The same environmental variables are then measured for the test site and the values are used in the discriminant equations from the reference sites database to predict the macroinvertebrate fauna to be expected at the test site, assuming that the test site is also unstressed. A macroinvertebrate sample is then collected at the test site using the same standardised sampling protocols as were used to survey the reference sites. The observed fauna is compared with the expected fauna to derive an index of ecological quality for that stretch of the river.

The main steps involved in this approach are detailed below.

15.2.2. Step 1: Selection of Reference Sites

The initial choice of sites for RIVPACS took place early in 1978 and was made as described by Clarke et al. (2003). The sites ranged from mid-sized to large streams and rivers, but excluded small streams. A 100 river systems were identified as being of high quality with respect to their chemical, physical and biological attributes. After collating selected chemical and physical data for each river system, a subset of 41 was chosen to reflect the range of conditions found in the full set. Subsequent site selections for RIVPACS I, RIVPACS II, RIVPACS III and RIVPACS III+ were conducted during 1981–84, 1984–88, 1991–95 and 1995–2003, respectively. Some reference sites chosen in earlier phases were rejected in subsequent phases because with further knowledge they were found to be inadequate to qualify as reference sites. Thus, site selection for each phase of the project has been an iterative process.

As is only to be expected, the RIVPACS reference sites are 'clean'/'unpolluted'/'unstressed'/'top quality', but are not quite perfect. They are best regarded as examples of the highest class of a quality banding system. In other words, they represent the best sites available in the population for each physical type of site and provide a realistic 'target' fauna for each type of site (Clarke et al., 2003).

15.2.3. Step 2: Macroinvertebrate Sampling

The macroinvertebrate samples collected at the RIVPACS reference sites and at sites being assessed are all based on a standardised 3-minute active pond-net sampling regime in which all habitats, including substrata and macrophytes, are sampled in proportion to their occurrence at the site (Murray-Bligh et al., 1997; Clark et al., 2003). For the reference sites, a single sample is taken in each of three seasons: spring (March–May), summer (June–August) and autumn (September–November).

15.2.4. Step 3: Classification of Reference Sites into Biological Groups

In early versions of RIVPACS, the clustering of reference sites was based on the hierarchical divisive classification method TWINSPAN (Two-way Indicator Species Analysis; Hill, 1979). For subsequent versions of RIVPACS, several alternative multivariate clustering methods were explored but the best classification method, in terms of the one which led to the most accurate prediction of biotic indices for the reference sites, was found to be TWINSPAN (Moss, 2000; Clarke et al., 2003).

A novel development for RIVPACS III was the use of data on both the presence–absence of species and the log abundance classes of families in developing the TWINSPAN classification. This helped to improve the classification in terms of the consistency of species presence–absence within groups, whilst also taking account of the abundance of families used in some biotic indices (Wright, 2000).

A separate classification and prediction system was developed for Northern Ireland because it had a less species-rich fauna than Great Britain (Wright et al., 2000). When applying elsewhere in Europe, it is required to gauge the extent of the individual biogeographic regions over which single prediction models can work effectively and accurately.

15.2.5. Step 4: Predicting the Expected Fauna

The classification groups of reference sites, previously defined by their biological attributes, are linked to their environmental characteristics using the multivariate statistical technique of multiple discriminant analysis (MDA), also known as canonical variates analysis (Krzanowski, 1988).

In developing RIVPACS, the aim has been to select a set of appropriate environmental predictor variables, which can be measured in a standardised way at any type of river site (Clarke et al., 2003). The predictor variables involved in the current version of RIVPACS are given in Table 15.1. They were selected from a much larger set (Moss et al., 1987) to give the best discrimination of the biological groups. The ease and cost of obtaining or measuring values for each variable at new sites was also taken into account in the selection procedure.

Whereas in most applications of MDA (Krzanowski, 1988), a new sample (or site) is simply allocated to the group for which the probability of belonging to that group is highest, in RIVPACS a different approach is followed which is based on the viewpoint that

TABLE 15.1 Environmental Variables Used in RIVPACS Predictions (Clarke et al., 2003)

TIME INVARIANT	
Map Location (National Grid Reference)	→ latitude, longitude → mean air temperature, air temperature range
Altitude at site (m)	–
Distance from source (km)	–
Slope (m km^{-1})	–
Discharge category (1–10)	
(long-term historical average)	
(1 = ≤ 0.31, 2 = 0.31–0.62, 3 = 0.62–1.25, ..., 9 = 40–80, 10 = 80–160 $m^3 s^{-1}$)	
ESTIMATED AT SITE AT TIME OF SAMPLING (AVERAGED ACROSS THREE SEASONS)	
Stream width (m).	–
Stream depth (cm).	–
Substratum composition:	→ mean particle size (in phi units)
% cover of clay/silt, sand, gravel/pebbles, cobbles/boulders	
Water geo-chemistry: alkalinity (mg l^{-1} $CaCO_3$)	–

→ Denotes derived variable.

the changes in community structure both downstream and across the full range of UK rivers are best regarded as a continuum, and that sites do not naturally fall into completely distinct ecological types. Therefore, when predicting the fauna to be expected at a site, the biological classification of the reference sites into groups is treated as an intermediate stage. New test sites are not simply assigned to their most likely group based on their environmental attributes but, instead, they are assigned probabilistically to each of the 35 classification groups using the predicted probabilities. Typically, a test site in RIVPACS will have a predicted probability of >1% of belonging to between one and five of the 35 classification groups. This probabilistic assignment of sites to biological groups makes the predictions of the expected community resilient to the precise choice of method of clustering the reference sites. Importantly, this use of MDA provides an automatic means of assessing whether or not a new site is within the environmental scope of the reference database (Clarke et al., 2003).

Estimating the expected fauna: If taxon i occurs in r_{ij} of the n_{ij} reference sites in group j, then the expected probability, P_i, of finding a particular taxon i (species or family) at a test site, if the site is unstressed, is estimated from the proportions $\{q_{ij}=/r_{ij}/n_{ij}\}$ of reference sites in each group i with taxon j present, weighted by the test site's probabilities $\{G_j\}$ of belonging to each group, namely,

$$P_i = \sum_{j=1}^{c} G_j q_{ij} \tag{15.1}$$

Because samples are taken from each reference site in each of spring, summer and autumn, predictions of the expected probabilities of individual taxon occurrence can be made for any single season sample or for any two or three seasons combined samples. Thus, the predictions can be both site and season specific.

15.2.6. Step 5: Indices Comparing the Observed and Expected Fauna

With presence–absence data, the statistical log likelihood, log L, of the observed sample $\{Z_i, Z_i = 1$ if taxon i is present, 0 if absent$\}$, conditional on the model expectations $\{P_i\}$ in Eq. (15.1), is (Clarke et al., 2003):

$$\log L = \sum_i [Z_i \log(P_i) + (1 - Z_i)\log(1 - P_i)] \quad (15.2)$$

The conditional distribution of log L for any particular sample, given the expectations $\{P_i\}$, can be obtained by Monte Carlo simulation. A simulated sample is generated by treating each taxon i as present if random uniform (0–1) deviate R_i is less than P_i. If Q_1 is the proportion of, say, 10,000 simulated values of log L which are less than the observed log L, small values of Q_1 (less than say 0.05 or 0.01) indicate lack of agreement between the observed and expected fauna.

This index, Q_1, was proposed by Clarke et al. (1996) but a problem in using Eq. (15.2) for test sites is that if any one taxon occurs for which the expected probability $P_i = 0$, or any one taxon is absent for which $P_i = 1$, then the likelihood L is zero and $Q_1 = 0$ (the lowest value possible), regardless of the degree of agreement between observed and expected values for the remainder of the fauna. This problem can be overcome by using the empirical logistic transform for proportions to estimate q_{ij} in Eq. (15.1) by

$$q_{ij} = \exp\frac{(z)}{(1 + \exp(z))} \quad (15.3)$$

where

$$z = \log\frac{\left(r_{ij} + 0.5\right)}{\left(n_{ij} - r_{ij} + 0.5\right)}$$

This allows for the finite number of references sites involved in estimates of the $\{P_{ij}\}$ in Eq. (15.1).

15.3. VARIANTS OF RIVPACS

15.3.1. General

The RIVPACS approach has been adopted in Australia where the Australian River Assessment System (AUSRIVAS) has been developed under their National River Health Program during the 1990s (Norris and Norris, 1995; Davies, 2000; Turak and Koop, 2003; Growns et al., 2006; Halse et al., 2007). Each state and territory has been treated as a bio-geographic region and separate reference site classifications and prediction systems have been developed for each region (Smith et al., 1999). All the Australian prediction systems are based on unweighted pair-group mean arithmetic averaging (UPGMA) agglomerative clustering of reference sites using Bray–Curtis pairwise similarity (Simpson and Norris, 2000). This is followed by MDA and the probabilistic assignment of test sites to site groups, as used in RIVPACS.

A variation of RIVPACS, referred to as the BEAST (Benthic Assessment of Sediment), has been used in North America to assess sediment quality in the Great Lakes (Reynoldson et al., 1995) and in the Fraser River catchment in British Columbia (Rosenberg et al., 2000). The important difference between RIVPACS and BEAST is that in the latter, test sites are merely assigned to the most probable site group (i.e., the group with the highest probability, G_j). The test site and the reference sites in the most probable group are then compared in ordinations derived from pairwise Bray–Curtis similarities using the ordination technique semi-strong hybrid multi-dimensional scaling (Belbin, 1991). The impairment level of the test site is estimated by its position in ordination space in relation to probability confidence ellipses for the selected reference sites.

RIVPACS and its variants are essentially empirical and descriptive. Their primary use is as tools for estimating and monitoring the ecological quality of river sites using standardised protocols, which enable comparisons to be made across a wide range of sites. But these systems cannot be used with any confidence as dynamic models to predict the impact of environmental changes (Clarke et al., 2003). Further efforts ought to be invested in developing complimentary models based on data from both reference sites and impaired sites, including chemical, land use and habitat modification variables considered to represent the range of impacts which can operate on river sites (Clarke et al., 2003).

Other predictive multivariate models based on, and similar to, RIVPACS, have been developed in Sweden ('SWEDEPACS'), the Czech Republic ('PERLA'), and for the Nordic region ('NORDPACS'). RIVPACS-type models have also been used in North America (Van Sickle et al., 2005; Hargett et al., 2007; Dolph et al., 2011; Kosnicki and Sites, 2011) and New Zealand (Joy and Death, 2002; 2003) but there are no reports of their use in developing countries even though exploratory efforts, like the ones made by Hart et al. (2001) and Sudaryanti et al. (2001) with reference to Indonesian rivers, have indicated that the models are 'highly applicable' in those regions.

15.3.2. ANNA

Linke et al. (2005) have argued that the classification step of RIVPACS/AUSRIVAS, in which reference sites are grouped and these groups are associated with environmental features of the sites, appears artificial and can be problematic. By classifying or grouping sites based on similarities in their biota, these models assume that macroinvertebrate communities occur in discrete groups. Yet, species assemblages have been commonly reported as changing continuously. Furthermore, seven different clustering strategies for the classification step for RIVPACS have shown marked differences in group sizes and predictions between the methods (Moss et al., 1999). It has also been found that classifications can be sensitive to the input order of the data (Podani, 1997). Also, the final groups are determined by 'visual inspection' of the dendrogram (Simpson and Norris, 2000). Linke et al. (2005) have felt that a good way of removing these theoretical concerns about describing the stream communities as groups of sites, as well as the sources of associated error, would be to remove the classification altogether. To this end, they have proposed a new prediction method called ANNA (Assessment of Nearest Neighbour Analysis).

The philosophy behind ANNA is similar to the philosophy of RIVPACS or AUSRIVAS. The probabilities that particular taxa will occur at a test site are calculated and summed, to compute an expected number of taxa, which is then compared with the numbers of taxa observed. The difference (Figure 15.1) arises from the fact that the ANNA model avoids the classification and discriminant function analyses when matching test sites with reference sites. ANNA finds the reference sites that most resemble the test sites in their values for environmental predictor variables. It then predicts community composition of these test sites based on the community composition of those 'nearest neighbours' (Figure 15.2), thus treating the macroinvertebrate assemblage as a continuum instead of discrete groups.

Once the probability of a taxon occurring has been estimated, observed *vs.* expected (O/E) scores are computed in the same way as in

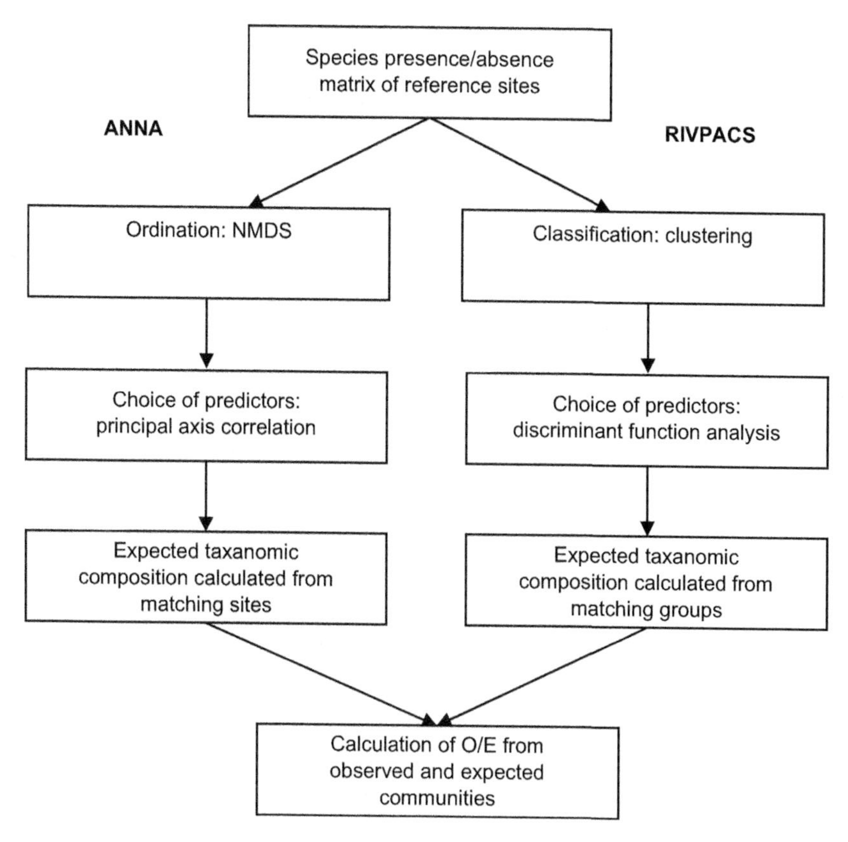

FIGURE 15.1 Differences and similarities between ANNA and AUSRIVAS models to predict taxon occurrence (Linke et al., 2005).

AUSRIVAS or RIVPACS. To reduce noise created by random occurrence of taxa in reference sites and to be consistent with AUSRIVAS, only taxa with $P > 0.5$ are considered.

The authors (Linke et al., 2005) compared AUSRIVAS and ANNA models on 17 data sets representing a variety of habitats and seasons. They found that the two approaches were generally equivalent in performance, but ANNA appeared potentially more robust for the O/E regressions and also appeared potentially more accurate on the trace metal gradient sites.

Hermoso et al. (2010) used ANNA to assess the deviation of the observed and expected community composition at reference condition when studying piscifauna of Mediterranean

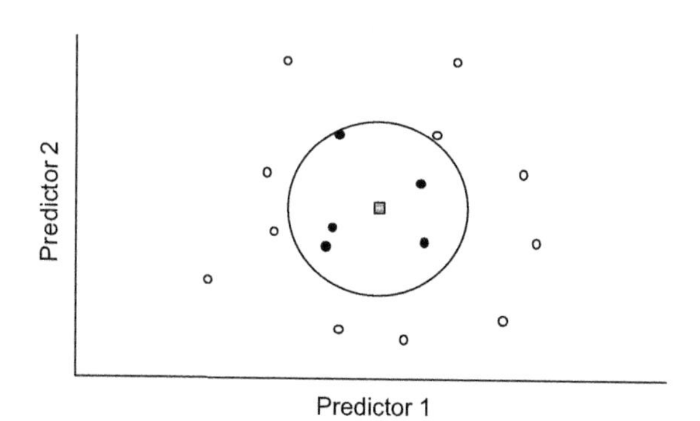

FIGURE 15.2 Conceptual diagram of Assessment by Nearest Neighbour Analysis (ANNA) approach of Linke et al. (2005). Reference sites (dots) that are within a fixed distance (circle) of a test site (square) based on environmental predictor variables are considered in the prediction.

rivers. It was seen that ANNA allowed incorporating in the index some rare species, though not all of them were present in the basin. Deviations were transformed into probabilities of belonging to a reference site and species-by-species measures were then integrated in a final score.

15.3.3. Use of Neural Networks

Biological monitoring of river water quality in the United Kingdom and several other European and Commonwealth countries is based on the Biological Monitoring Working Party (BMWP) system. Central to the application of this system is the prediction of 'unpolluted' average score per taxon (ASPT) and number of families present (NFAM). To facilitate such predictions, Walley and Fontama (1998) have developed predictors of ASPT and NFAM using neural networks. The authors found that neural networks performed marginally better than RIVPACS III; ASPT and NFAM can be predicted directly, without reference to site type or biological community, from a few key environmental variables. It appeared that there is scope for improved predictions in the new procedure if additional relevant environmental data are collected.

In another study, Hoang et al. (2001) report that the use of predictive Artificial Neural Network (ANN) models for 37 macroinvertebrate taxa based on 896 stream data sets from the Queensland stream system led to 73% as lowest rate and 82% as average rate of correct ANN predictions of stream site habitats. The increase of correct predictions was 30%, if ANNs and the statistical stream model AUSRIVAS were compared based on the same data sets.

Olden et al. (2006a) used data for 195 sites in the western United States to develop single-scale, multi-scale and hierarchical multi-scale neural networks relating EPT (Ephermeroptera, Plecoptera, Trichoptera) richness to environmental variables quantified at 3 spatial scales: entire watershed, valley bottom (100s–1000s m) and local stream reach (10s–100s m). Results showed that models based on multiple spatial scales greatly outperformed single-scale analyses and that a hierarchical ANN, which accounts for the fact that valley- and watershed-scale drivers influence local characteristics of the stream reach, provided greater insight into how environmental factors interact across nested spatial scales than did the nonhierarchical multi-scale model.

The same authors (Olden et al., 2006b) have compared multi-response artificial neural network (MANN) to two traditional approaches used to predict community composition: (1) a species-by-species approach using logistic regression analysis (LOG) and (2) a 'classification-then-modelling' approach in which sites are classified into assemblage types using two-way indicator species analysis and multiple discriminant analysis (MDA). For freshwater fish assemblages of the North Island, New Zealand, MANN was shown to outperform all other methods for predicting community composition based on multi-scaled descriptors of the environment. The simple-matching coefficient comparing predicted and actual species composition was, on average, greatest for the MANN (91%), followed by MDA (85%) and LOG (83%). Mean Jaccard's similarity (emphasising model performance for predicting species' presence) for the MANN (66%) exceeded both LOG (47%) and MDA (46%). The MANN also correctly predicted community composition (i.e., a significant proportion of the species membership based on a randomisation procedure) for 82% of the study sites compared to 54% (MDA) and 49% (LOG). MANN also seemed to provide valuable explanatory power by simultaneously quantifying the nature of the relationships between environment and both individual species and the entire community (composition and richness), which is not readily available from traditional approaches.

15.3.4. Use of Self-organising Maps and Evolutionary Algorithms

Walley and O'Connor (2001) have described a novel pattern recognition system (MIR-max) that was developed by them to facilitate the construction of a river pollution diagnostic system for the British Environment Agency. It is a non-neural self-organising map based on information theory, which, unlike Kohonen's self-organised map (SOM), separates the processes of clustering and ordering. It first clusters the input samples into a predefined number of classes by maximising the mutual information between the samples and the classes. The classes are then ordered in a two-dimensional output space by maximising the correlation coefficient between the Euclidean distances separating the classes in data space and their corresponding distances in output space. This produces a map of the classes which when labelled can be used for the classification/diagnosis of new samples.

According to the authors (Walley and O'Connor, 2001) MIR-max permits the disaggregation of the classes in the output map, thus enabling exceptional classes to separate from their neighbours. MIR-max was used to classify biological samples into 100 river quality classes for five different site types. The classifiers were then tested against two corresponding neural network classifiers, and were shown to provide better performance. The advantages of the system over RIVPACS and related methods have been outlined by the authors concluding that MIR-max has considerable potential for use in the visualisation and interpretation of multivariate ecological data.

In contrast, Maier et al. (2006) feel that even as MIR-max method overcomes some of the limitations of AUSRIVAS by using mutual information as the performance measure, it uses a hill-climbing approach for optimising the mutual information, which could become trapped in local optima of the search space. The authors have explored the potential of using evolutionary algorithms (EAs), such as genetic algorithms and ant colony optimisation algorithms, for maximising the mutual information of ecological data clusters. The MIR-max and EA-based approaches were applied by them to the South Australian combined season riffle AUSRIVAS data, and the results obtained were compared with those obtained using the unweighted pair-group arithmetic averaging (UPGMA) method. The results indicate that the overall mutual information values of the clusters obtained using MIR-max and the EA-based approaches are significantly higher than those obtained using the UPGMA method, and that the use of genetic and ant colony optimisation algorithms is successful in determining clusters with higher overall mutual information values compared with those obtained using MIR-max for the case study considered.

15.4. THE MULTIVARIATE APPROACHES AND THE IBI

The underlying principle of the indices of biotic integrity (IBI) or the multimetric indices, and the multivariate approaches typified by RIVPACS is that the biota is able to integrate and reflect anthropogenic stressors. The two approaches overlap in many important ways. For example:

1. Both focus on the composition of biological assemblages to define the health of the concerned water body.
2. Both use the concept of reference condition as a benchmark.
3. Both organise sites into classes with a select set of environmental characteristics.
4. Both assess change in biota caused by human effects as distinct from natural variations.
5. Both require standardised sampling, laboratory and analytical methods.
6. Both score sites numerically to reflect site condition.

7. Both define "bands," or condition classes, representing degrees of degradation.
8. Both furnish needed analyses for selecting high-quality areas as acquisition and conservation priorities.

Despite these common attributes, RIVPACS and IBI also differ in several important ways enumerated in Table 15.2. Of these, perhaps the most important is the biological information used in the structuring of the assessment process. The early goal of RIVPACS was to improve site selection for conservation (Wright et al., 1984). As a result, recognising patterns of species composition was and continues to be the core of RIVPACS analyses. In contrast, IBI was developed primarily to measure river condition, or river health, and included development of measurements for diagnosing causes of degradation. The biological signals that feed IBI analyses are broader, including taxa richness and composition, trophic or other aspects of ecological organisation, presence and relative abundance of tolerant and intolerant taxa and presence of diseased individuals or individuals with other anomalies (Karr, 1991; Karr and Chu, 1999).

A primary goal of IBI is to define the attributes of living systems that change systematically in diverse situations when exposed to the activity of humans. RIVPACS emphasises species identity and abundances. IBI searches for a broader array of biological signals from species to higher taxa, including systematic shifts in the natural history of assemblages of species. Because RIVPACS analyses emphasise probability statements about species presence, rare taxa are excluded from analyses; this is not done in IBI. Finally, RIVPACS depends on large data sets for its multivariate models, whereas smaller data sets can produce an effective IBI.

Given the special capabilities of the two approaches, it has been suggested that multivariate methods (MM) should be used in conjunction with IBIs to obtain greater information about, and deeper insights into, the ecological status of a test site than is possible by either of the method used in isolation.

Reynoldson et al. (1997) assessed the strengths and deficiencies of the multimetric and the multivariate methods for establishing water-quality status. A data set of environmental measurements and macroinvertebrate collections from the Eraser River, British Columbia, was used in the comparison. It was seen that the precision and accuracy of the two multivariate methods tested (AUSRIVAS and BEAST) were consistently higher than for the multimetric assessment. Classification by ecoregion, stream order and biotic group yielded precisions of 100% for the AUSRIVAS, 80–100% for the BEAST and 40–80% for multimetrics; and accuracies of 100%, 100% and 38–88%, respectively.

Comparing the relative merit of the MM and the IBI approaches, the authors (Reynoldson et al., 1997) noted that IBIs are attractive because they produce a single score that is comparable to a target value and they include ecological information. However, not all information collected is used, metrics are often redundant in a combination index, errors can be compounded and it is difficult to acquire current procedures. In contrast, MM appear attractive because they require no prior assumptions either in creating groups out of reference sites or in comparing test sites with reference groups. However, potential users of MM may be discouraged by the complexity of initial model construction. The complementary emphases in the two multivariate methods examined (presence/absence in AUSRIVAS, abundance in BEAST) led the authors to recommend that they be used together, and in conjunction with IBIs.

Studies have also shown that MM can be used to enhance the accuracy and precision of the IBIs. For example, Cao et al. (2007) used modelling to develop, evaluate and compare two types of diatom-based indicators for Idaho streams: an observed/expected (O/E) ratio of

TABLE 15.2 A Comparison of the Conceptual, Sampling, and Analytical Characteristics of Multivariate (RIVPACS) and Multimetric (IBI) Approaches

Attribute	RIVPACS (Multivariate)	IBI (Multimetric)
Site classification or characterization	Stream size, geography, substrate particle size, water chemistry	Geology, stream size and temperature, altitude
Reference standard	Sites without serious pollution (RP); other human influences not considered. Broader in recent applications (AR)	Sites with little or no human influence
Model foundations	Multivariate associations between environmental variables an species present	Empirically defined measures; dose–response graphs plotting human influence against biological response
Decision criteria	Species presence or absence (observed vs. expected rations) from probabilistic models; sometimes tolerance measures	Biological attributes such as taxa richness, relative abundance, taxa composition, tolerance and intolerance
Microhabitat sampled	Multiple (e.g., riffle, edge, pool-rock: RP). Riffles (AR)	Riffles
Subsampling	Varies regionally and with application	Full samples counted
Stream application available	Benthic invertebrates	Benthic invertebrates, fish, algae
Data set required	Extensive, hundreds of sites Regional foundation; some smaller-scale applications being developed	Fifteen to 20 sites representing a broad gradient of human disturbance; can be developed locally or over larger areas
Transferability	Extensive, data sets and new species-specific models needed for each region	Consistency in selected measures and biological responses across regions
Treatment of rare species	Often excluded from analyses	Included in analyses
Sampling period	RP: combined 3-season model AR: model for each season	Defined period for sampling
Analytical basis	Species presence–absence	Diverse dimensions of biological and natural history patterns
Human influence	Largely chemical contamination	Full spectrum of human influences
Diagnostic capability	Not explored	Moderately well developed
Communication	Statistical foundations difficult Breadth of signal narrow (O/E ratio); pollution tolerance	Simple dose–response curves similar to toxicology; broad range of biological signal

taxon loss derived from a model similar to the RIVPACS and a multimetric index (MMI). Finding that diatom assemblage structure varied substantially among reference-site samples, but neither ecoregion nor bioregion accounted for a significant portion of that variation, the authors used Classification and Regression Trees (CART) to model the variation of individual metrics with natural gradients. On average, 46% of the total variance in 32 metrics was explained by CART models, but the predictor variables differed among the metrics and often showed evidence of interacting with one another. The use of CART residuals (i.e., metric values adjusted for the effect of natural environmental gradients) affected whether or how strongly many metrics discriminated between reference and test sites. The authors used cluster analysis to examine redundancies among candidate metrics and then selected the metric with the highest discrimination efficiency from each cluster. This step was applied to both unadjusted and adjusted metrics and led to inclusion of 7 metrics in MMIs. It was found that the adjusted MMIs were more precise than unadjusted ones. Use of unadjusted MMIs probably resulted in higher rates of both type I and type II errors than use of adjusted metrics; this appeared to be a logical consequence of the inability of unadjusted metrics to distinguish the confounding effects of natural environmental factors from those associated with human-caused stress. The RIVPACS-type model for diatom assemblages performed similarly to models developed for invertebrate assemblages. The O/E ratio was as precise as the adjusted MMI, but rated a lower proportion of test sites as being in nonreference condition, implying that taxon loss was less severe than changes in overall diatom assemblage structure. In summary, the authors feel that MM-based modelling appears to be an effective means of developing more accurate and precise MMIs. Furthermore, this type of modelling enabled them to develop a single MMI for use throughout an environmentally heterogeneous region. Several reports have since appeared in which IBIs and MMs have been used in conjunction (Furnish and Gibson, 2007; Beck and Hatch, 2009) or MMs have been used to enhance the applicability of the IBIs (Clarke et al., 2003; Novotny et al., 2009).

References

Beck, M.W., Hatch, L.K., 2009. A review of research on the development of lake indices of biotic integrity. Environmental Reviews 17, 21–44.

Belbin, L., 1991. Semi-strong Hybrid Scaling, a new ordination algorithm. Journal of Vegetation Science 2, 491–496.

Cao, Y., Hawkins, C.P., Olson, J., Kosterman, M.A., 2007. Modeling natural environmental gradients improves the accuracy and precision of diatom-based indicators. Journal of the North American Benthological Society 26, 566–585.

Clarke, R.T., Furse, M.T., Wright, J.F., Moss, D., 1996. Derivation of a biological quality index for river sites: comparison of the observed with the expected fauna. Journal of Applied Statistics 23, 311–332.

Clarke, R.T., Wright, J.F., Furse, M.T., 2003. RIVPACS models for predicting the expected macroinvertebrate fauna and assessing the ecological quality of rivers. Ecological Modelling 160 (3), 219–233.

Davies, P.E., 2000. Development of a national river bioassessment system (AUSRIVAS). In: Wright, J.F., Sutcliffe, D.W., Furse, M.T. (Eds.), Assessing the Biological Quality of Freshwaters: RIVPACS and Other Techniques. Freshwater Biological Association, Ambleside, pp. 113–124.

Davis, J., Horwitz, P., Norris, R., 2006. Are river bioassessment methods using macroinvertebrates applicable to wetlands? Hydrobiologia 572, 115–128.

Davy-Bowker, J., Clarke, R.T., Johnson, R.K., Kokes, J., Murphy, J.F., Zahradkova, S., 2006. A comparison of the European Water Framework Directive physical typology and RIVPACS-type models as alternative methods of establishing reference conditions for benthic macroinvertebrates. Hydrobiologia 566, 91–105.

Dolph, C.L., Huff, D.D., Christopher, J., Chizinski, Vondracek, 2011. Implications of community concordance for assessing stream integrity at three nested spatial scales in Minnesota, U.S.A. Freshwater Biology 56, 1652–1669.

Feio, M.J., Norris, R.H., Graça, M.A.S., Nichols, S., 2009. Water quality assessment of Portuguese's streams: regional or national predictive models? Ecological Indicators 9, 791–806.

Furnish, J., Gibson, D., 2007. Identifying watersheds at risk using bioassessment techniquest based onmultimetric Index of Biotic Integrity (IBI) and multivariate RIVPACS methods, Report – University of California Water Resources Center (109). p. 104.

Growns, I., Schiller, C., O'Connor, N., Cameron, A., Gray, B., 2006. Evaluation of four live-sorting methods for use in rapid biological assessments using macroinvertebrates. Environmental Monitoring and Assessment 117 (1–3), 173–192.

Halse, S.A., Scanlon, M.D., Cocking, J.S., Smith, M.J., Kay, W.R., 2007. Factors affecting river health and its assessment over broad geographic ranges: the Western Australian experience. Environmental Monitoring and Assessment 134, 161–175.

Hargett, E.G., ZumBerge, J.R., Hawkins, C.P., Olson, J.R., 2007. Development of a RIVPACS-type predictive model for bioassessment of wadeable streams in Wyoming. Ecological Indicators 7, 807–826.

Hart, B.T., Davies, P.E., Humphrey, C.L., Norris, R.N., Sudaryanti, S., Trihadiningrum, Y., 2001. Application of the Australian river bioassessment system (AUSRIVAS) in the Brantas River, East Java, Indonesia. Journal of Environmental Management 62, 93–100.

Hermoso, V., Clavero, M., et al., 2010. Assessing the ecological status in species-poor systems: A fish-based index for Mediterranean Rivers (Guadiana River, SW Spain). Ecological Indicators 10 (6), 1152–1161.

Hill, M.O., 1979. TWINSPAN—A FORTRAN program for arranging multivariate data in an ordered two-way table by classification of the individuals and the attributes. In: Ecology and Systematics. Cornell University, Ithaca, NY.

Hoang, H., Recknagel, F., Marshall, J., Choy, S., 2001. Predictive modelling of macroinvertebrate assemblages for stream habitat assessments in Queensland (Australia). Ecological Modelling 146 (1–3), 195–206.

Joy, M.K., Death, R.G., 2002. Predictive modelling of freshwater fish as a biomonitoring tool in New Zealand. Freshwater Biology 47 (11), 2261–2275. DOI 10.1046/j.1365-2427.2002.00954.x.

Joy, M.K., Death, R.G., 2003. Biological assessment of rivers in the Manawatu-Wanganui region of New Zealand using a predictive macroinvertebrate model. New Zealand Journal of Marine and Freshwater Research 37 (2), 367–379.

Karr, J.R., 1991. Biological integrity: a long neglected aspect of water resource management. Ecological Applications 1, 66–84.

Karr, J.R., Chu, E.W., 1999. Restoring Life in Running Waters. Island Press, Washington, DC, USA, p. 206.

Kosnicki, E., Sites, R.W., 2011. Seasonal predictability of benthic macroinvertebrate metrics and community structure with maturity-weighted abundances in a Missouri Ozark stream, USA. Ecological Indicators 11 (2), 704–714.

Krzanowski, W.J., 1988. Principles of Multivariate Analysis: A User's Perspective. Clarendon Press, Oxford, p. 563.

Linke, S., Norris, R.H., Faith, D.P., Stockwell, D., 2005. ANNA: a new prediction method for bioassessment programs. Freshwater Biology 50, 147–158.

Maier, H.R., Zecchin, A.C., Radbone, L., Goonan, P., 2006. Optimising the mutual information of ecological data clusters using evolutionary algorithms. Mathematical and Computer Modelling 44 (5–6), 439–450.

Moss, D., 2000. Evolution of statistical methods in RIVPACS. In: Wright, J.F., Sutcliffe, D.W., Furse, M.T. (Eds.), Assessing the Biological Quality of Freshwaters: RIVPACS and Other Techniques. Freshwater Biological Association, Ambleside, pp. 25–37.

Moss, D., Furse, M.T., Wright, J.F., Armitage, P.D., 1987. The prediction of the macro-invertebrate fauna of unpolluted running-water sites in Great Britain using environmental data. Freshwater Biology 17, 41–52.

Moss, D., Wright, J.F., Furse, M.T., Clarke, R.T., 1999. A comparison of alternative techniques for prediction of the fauna of running water sites in Great Britain. Freshwater Biology 41, 167–181.

Murray-Bligh, J.A.D., Furse, M.T., Jones, F.H., Gunn, R.J.M., Dines, R.A., Wright, J.F., 1997. Procedure for Collecting and Analysing Macroinvertebrate Samples for RIVPACS. The Institute of Freshwater Ecology and the Environment Agency, p. 162.

Norris, R.H., Norris, K.R., 1995. The need for biological assessment of water quality: Australian perspective. Australian Journal of Ecology 20, 1–6.

Novotny, V., Bedoya, D., Virani, H., Manolakos, E., 2009. Linking indices of biotic integrity to environmental and land use variables: multimetric clustering and predictive models. Water Science and Technology 59 (1), 1–8.

Olden, J.D., Joy, M.K., Death, R.G., 2006b. Rediscovering the species in community-wide predictive modeling. Ecological Applications 16 (4), 1449–1460.

Olden, J.D., Poff, N.L., Bledsoe, B.P., 2006a. Incorporating ecological knowledge into ecoinformatics: an example of modeling hierarchically structured aquatic communities with neural networks. Ecological Informatics 1 (1), 33–42.

Podani, J., 1997. On the sensitivity of ordination and classification methods to variation in the input order of data. Journal of Vegetation Science 8, 153–156.

Reynoldson, T.B., Bailey, R.C., Day, K.E., Norris, R.H., 1995. Biological guidelines for freshwater sediment based on benthic assessment of sediment (the BEAST) using a multivariate approach for predicting biological state. Australian Journal of Ecology 20, 198–219.

Reynoldson, T.B., Norris, R.H., Resh, V.H., Day, K.E., Rosenberg, D.M., 1997. The reference condition: a comparison of multimetric and multivariate approaches to assess water-quality impairment using benthic macroinvertebrates. Journal of the North American Benthological Society 16, 833–852.

Rosenberg, D.M., Reynoldson, T.B., Resh, V.H., 2000. Establishing reference conditions in the Fraser River catchment, British Columbia, Canada, using the BEAST (Benthic Assessment of Sediment). In: Wright, J.F., Sutcliffe, D.W., Clarke, R.T., et al (Eds.), Ecological

Modelling, vol. 160 (2003) 219/233. Furse, M.T. (Ed.), Assessing the Biological Quality of Freshwaters: RIVPACS and Other Techniques, Freshwater Biological Association, Ambleside, pp. 181–194.

Simpson, J., Norris, R.H., 2000. Biological assessment of water quality: development of AUSRIVAS models and outputs. In: Wright, J.F., Sutcliffe, D.W., Furse, M.T. (Eds.), RIVPACS and Similar Techniques for Assessing the Biological Quality of Freshwaters. Freshwater Biological Association and Environment Agency, Ableside, Cumbria, U.K., pp. 125–142.

Smith, M.J., Kay, W.R., Edward, D.H.D., Papas, P.J., Richardson, S.t.J., Simpson, J.C., Pinder, A.M., Cale, D.J., Horwitz, P.H.J., Davis, J.A., Yung, F.H., Norris, R.H., Halse, S.A., 1999. AUSRIVAS: using macroinverteb rates to assess ecological condition of rivers in Western Australia. Freshwater Biol. 41, 269–282.

Sudaryanti, S., Trihadiningrum, Y., Hart, B.T., Davies, P.E., Humphrey, C., Norris, R., Simpson, J., Thurtell, L., 2001. Assessment of the biological health of the Brantas River, East Java, Indonesia using the Australian River Assessment System (AUSRIVAS) methodology. Aquatic Ecology 35, 135–146.

Turak, E., Koop, K., 2003. Use of rare macroinvertebrate taxa and multiple-year data to detect low-level impacts in rivers. Aquatic Ecosystem Health and Management 6 (2 Spec. Iss.), 167–175.

Van Sickle, J., Hawkins, C.P., Larsen, D.P., Herlihy, A.T., 2005. A null model for the expected macroinvertebrate assemblage in streams. Journal of the North American Benthological Society 24, 178–191.

Walley, W.J., Fontama, V.N., 1998. Neural network predictors of average score per taxon and number of families at unpolluted river sites in Great Britain. Water Research 32 (3), 613–622.

Walley, W.J., O'Connor, M.A., 2001. Unsupervised pattern recognition for the interpretation of ecological data. Ecological Modelling 146 (1–3), 219–230.

Wright, J.F., Gunn, R.J.M., Blackburn, J.H., Grieve, N.J., Winder, J.M., Davy-Bowker, J., 2000. Macroinvertebrate frequency data for the RIVPACS III sites in Northern Ireland and some comparisons with equivalent data for Great Britain. Aquatic Conservation: Marine and Freshwater Ecosystem 10, 371–389.

Wright, J.F., Moss, B., Armitage, P.D., Furse, M.T., 1984. A preliminary classification of running-water sites in Great Britain based on macro-invertebrate species and the prediction of community type using environmental data. Freshwater Biology 14, 221–256.

Wright, J.F., 2000. An introduction to RIVPACS. In: Wright, J.F., Sutcliffe, D.W., Furse, M.T. (Eds.), Assessing the Biological Quality of Freshwaters: RIVPACS and Other Techniques. Freshwater Biological Association, Ambleside, pp. 1–24.

CHAPTER

16

Water-Quality Indices: Looking Back, Looking Ahead

OUTLINE

16.1. INTRODUCTION

It is about 160 years since the concept of water quality was advanced to categorise water of different streams and lakes according to the degree of purity/impurity of the water course (Lumb et al., 2011, citing Sladecek, 1973 and others). The first modern water-quality index (WQI) – that of Horton (1965) – came about 46 years ago. It heralded a new era because it presented a simple mathematical procedure which integrated physical, chemical and (some) biological parameters into a single score. It was an approach which has influenced the development of all subsequent WQIs, predominantly based on physico-chemical parameters. The first modern bioassessment-based WQI – the 'Trent Biotic Index' (TBI) – was introduced just a little before the Horton's index (in 1964). TBI (Abbasi and Abbasi 2011) was intended for the streams of Florida, USA. Sixteen years later, Karr (1981) was to present the first-ever 'index of biotic integrity' (IBI) which was to stimulate, and continues to do so, enormous useful work in bioassessment-based categorisation of water quality. By a coincidence all the three classes of WQIs were introduced by scientists working in the USA.

After the introduction of the TBE and the Horton's index, the remaining 3½ decades of the 20th century witnessed a rapid growth in the popularity of WQIs, especially in developed countries. These WQIs were based on 'crisp' and 'deterministic' mathematical treatment of water quality data or biological assemblage information and the advancements were aimed at enhancing the objectivity (in the choice of representative parameters and assignment of

Water Quality Indices DOI: 10.1016/B978-0-444-54304-2.00016-6

weightage), sensitivity (to changes in water quality), clarity (in showing a water source up as bad/fair/good/very good, etc), and reach (appropriateness for larger number of regions and types of water use) of the indices. In the course of achieving these objectives, some of the shortcomings of the different indices came to the fore especially the ones relating to the aggregation methods used for indices based predominantly on physico-chemical characteristics, and the sampling/scaling methods used in the BIs and the IBIs.

A great deal of work was done in solving the problems of ambiguity, eclipsing, rigidity, etc, faced by the WQIs. More and more techniques of statistics were pressed into service to achieve this objective. Its most spectacular use was in the development of the RIVPACS (River Pollution Assessment and Classification System) in Great Britain (Clarke et al., 2003).

As the world moved into the last decade of the 20th century, there was increasing realization about the fuzziness and stochasticity associated with the steps of sampling, analysis and demarcation of water quality. It was increasingly realised that 'crisp' and deterministic approaches were not able to capture the ground reality; the resulting index scores tended to give unrealistically sharp cut-offs due to their inability to 'sense' grey areas.

These concerns and the lead-up studies (Kung et al., 1992; Lu et al., 1999; Silvert, 1997; 2000; Chang et al., 2001; Lu and Lo, 2002; Haiyan, 2002, and others) resulted in the first fuzzy WQI (Ocampo-Duque et al., 2006) and the first stochastic WQI (Beamonte et al., 2005). The increasing application of the concepts of artificial intelligence in water resource systems also saw the first Al-based WQI – the one in which genetic algorithm was applied (Peng, 2004). During the last 5 years, the application of fuzzy rules in WQI development has intensified. Several other techniques and tools of Al and statistics are also being increasingly applied to WQIs based on physico-chemical parameters. In the bioassessment-based methods, also, ever greater sophistication is being introduced.

16.2. THE BEST WQI?

Despite a plethora of indices which have been developed, and used, across the world it is not possible to say which index is the best or even list 'ten best' or 'twenty best' indices. One does find that some indices are more popular than some others. For example, the US National Sanitation Foundation's WQI, which is commonly referred as NSF-WQI (Brown et al., 1970), is used not only in the country of its origin but also in several other countries spanning several continents (Brazil, Mexico, Guinea-Bissau, Poland, Egypt, Portugal, Italy and India, among others). The WQI of the Canadian Council of Ministers of the Environment, called CCME-WQI (CCME, 2001), which has come 31 years after NSF-WQI, is another index which is used in many countries besides the country of its origin (Lumb et al., 2011). CCME-WQI has also been shown to be an index particularly suited for use with continuous water-quality monitoring networks (Torredo et al., 2010). An index for groundwater-quality assessment proposed by Tiwari and Mishra (1985) from India is another WQI used extensively in many countries. There are takers for the WQI of Bharagava (1985) outside India (Lumb et al., 2011).

While each index has its special virtues and shortcomings, no attempt has been made so far to quantitatively 'weigh' different indices and suggest which pulls how much weight. Hence, it is not possible to say exactly why some indices are more popular than some other. Fernandez et al. (2004) compared 36 WQIs to observe that appreciable differences exist between classifications given by different indices on the same water sample. These differences arise primarily because of differing parameter types and numbers, weightage assignments and aggregation formulae on which different indices are based. As WQIs have been developed in different geographic, regional and management contexts, and there is no procedure yet in place to compare their performance, all one can do is to look at complementarities of the information, credibility of measurements, transparency in indice formulation, relevancy of key parameters selected and comparability of results, to make qualitative judgement on the suitability or otherwise of a WQI.

16.3. THE PATH AHEAD

The situation described in the preceding section points towards a need to devise a procedure with which the performance of various WQIs can be compared in terms of efficiency, adequacy, inexpensiveness, reach and flexibility. Enormous historical water-quality data are lying archived in many countries; such data can be used in testing the efficacy of different WQIs and in developing a universal WQI (Lumb et al., 2011). The concept of 'virtual water-quality metre (VIRWQIM)' introduced by Sarkar and Abbasi (2006) can come handy in this initiative because, with it, virtual 'dash boards' can be created wherein different VIRWQIMs can show classification of a water source as per different indices at a given time, besides providing an integrated, overall, assessment.

Besides increasing use of AI, advanced statistics and probability theory, the field of WQI has also witnessed increasing integration of the virtues of these approaches. For example, multivariate methods of the kind used in RIVPACS have been integrated with IBI development and there is increasing use of tools such as artificial neural networks and self-organizing maps in interpreting water-quality scores arrived by BIs and IBIs.

Major avenues of future R&D thrust include developing the ability of the indices to account for nonmeasured parameters *via* correlation modelling. This will, on the one hand, reduce the costs of water-quality monitoring and, on the other hand, will make it increasingly possible to use the indices in conjunction with automated water quality monitoring networks.

Just as some 'global' water-quality standards are available in the form of standards set by the World Health Organization, global WQIs for some of the water use would be very helpful as they would enable the water quality of a region being seen in a globally acceptable context.

There is a case for the development of multivariate analysis methods with which 'weights' can be assigned to different physico-chemical parameters instead of doing it, as is common, with Delphi or with some kind of ad hoc formula using water-quality standards. The former is very cumbersome and inevitably carries an element of subjectivity while the latter approach is totally ad hoc.

16.4. THE LAST WORD

There is one future trend about which it is not possible to have any doubt: That is of rapidly increasing importance of WQIs in people's lives. With water demand running ahead of water supply, and the quality of water that is available steadily declining, the importance of water in everyone's life across the world is set to increase by the day. Water of good quality will be at increasing premium and there will be ever greater need all over the world to segregate water usable for drinking from the water usable for other contact applications and from the water usable only for irrigation, industry etc. These compulsions, in turn, would enhance the necessity that water quality is quantified in a manner that it is intelligible to everyone. For example, a national water-quality index may say that when any water has index score 80 or higher, the water is fit for drinking and that closer the score is to the upper limit of 100, the better the water is. It may also say, for instance, that a water with index score between 70 and 80 may also be used for drinking provided that it has been boiled for 5 minutes. Companies supplying water will be providing its score for the customer's knowledge and choice. All these happenings are expected to make WQI a household word in the near future.

References

Abbasi, T., Abbasi, S.A., 2011. Water quality indices based on bioassessment: the biotic indices. Journal of Water and Health 9 (2), 330–348.

Beamonte, E., Bermúdez, J.D., Casino, A., Veres, E., 2005. A global stochastic index for water quality: the case of the river Turia in Spain. Journal of Agricultural, Biological, and Environmental Statistics 10 (4), 424–439.

Bhargava, D.S., 1985. Water quality variations and control technology of Yamuna river. Environmental Pollution Series A: Ecological and Biological 37 (4), 355–376.

Brown, R.M., McClelland, N.I., Deininger, R.A., Tozer, R.G., 1970. A water quality index – do we dare? Water Sewage Works 117, 339–343.

Canadian Council of Ministers of the Environment (CCME), 2001. Canadian Water Quality Index 1.0 Technical Report and User's Manual, Canadian Environmental Quality Guidelines Water Quality Index Technical Subcommittee. Gatineau, QC, Canada.

Chang, N.-B., Chen, H.W., Ning, S.K., 2001. Identification of river water quality using the fuzzy synthetic evaluation approach. Journal of Environmental Management 63 (3), 293–305.

Clarke, R.T., Wright, J.F., Furse, M.T., 2003. RIVPACS models for predicting the expected macroinvertebrate fauna and assessing the ecological quality of rivers. Ecological Modeling 160, 219–233.

Fernandez, N., Ramirez, A., Sonalo, F., 2004. Physicochemical water quality indices—a comparative review. Bistua, Rev Fac Cienc Basic 2 (1), 19–30. ISSN 0120–4211.

Haiyan, W., 2002. Assessment and prediction of overall environmental quality of Zhuzhou City, Hunan Province, China. J. Environ. Manage 66, 329–340.

Horton, R.K., 1965. An index number system for rating water quality. Journal of Water Pollution Control Federation 37 (3), 300–306.

Karr, J.R., 1981. Assessment of biotic integrity using fish communities. Fisheries 6 (6), 21–27.

Kung, H., Ying, L., Liu, Y.C., 1992. A complementary tool to water quality index: fuzzy clustering analysis. Water Resources Bull. 28 (3), 525–533.

Lu, R.S., Lo, S.L., 2002. Diagnosing reservoir water quality using self-organizing maps and fuzzy theory. Water Res. 36 (9), 2265–2274.

Lu, R.S., Lo, S.L., Hu, J.Y., 1999. Analysis of reservoir water quality using fuzzy synthetic evaluation. Stochastic Environmental Research and Risk Assessment 13 (5), 327–336.

Lumb, A., Sharma, T.C., Bibeault, J.-F., 2011. A Review of genesis and evolution of water quality index (WQI) and some future directions. Water Quality, Exposure and Health, 1–14.

Ocampo-Duque, W., Ferré-Huguet, N., Domingo, J.L., Schuhmacher, M., 2006. Assessing water quality in rivers with fuzzy inference systems: a case study. Environment International 32 (6), 733–742.

Peng, L., 2004. A universal index formula suitable to multiparameter water quality evaluation. Numerical Methods for Partial Differential Equations 20 (3), 368–373.

Sarkar, C., Abbasi, S.A., 2006. Qualidex-A new software for generating water quality indice. Environmental Monitoring and Assessment 119 (1–3), 201–231.

Silvert, W., 1997. Ecological impact classification with fuzzy sets. Ecological Modelling 96, 1–10.

Silvert, W., 2000. Fuzzy indices of environmental conditions. Ecological Modelling 130 (1–3), 111–119.

Sladecek, V., 1973. System of water quality from biological point of view. In: Mai, V. (Ed.), Archiv fur Hydrobiologie. Advances in Limnology, vol. 7. E. Schweizerbart'sch Verlagsbuch-handling, Stuttgart, p. 218. ISBN: 978-3-510-47005-1 paperback.

Terrado, M., Borrell, E., Campos, S.D., Barcelo', D., Tauler, R., 2010. Surface-water-quality indices for the analysis of data generated by automated sampling networks. Trends in Analytical Chemistry 29 (1), 40–52. doi:10.1016/j.trac.2009.10.001.

Tiwari, T.N., Mishra, M., 1985. A preliminary assignment of water quality index to major Indian rivers. Indian Journal of Environmental Protection 5 (4), 276–279.

Index

Note: Page numbers followed by "f" indicate figures and "t" indicate tables.

Printed and bound by CPI Group (UK) Ltd, Croydon, CR0 4YY
08/06/2025

01896872-0011